Crystalline Electric Field Effects in *f*-Electron Magnetism

Crystalline Electric Field Effects in *f*-Electron Magnetism

Edited by

Robert P. Guertin
Tufts University
Medford, Massachusetts

and

Wojciech Suski and
Zygmunt Zołnierek
Polish Academy of Sciences
Wrocław, Poland

PLENUM PRESS • NEW YORK AND LONDON

Library of Congress Cataloging in Publication Data

International Conference on Crystalline Electric Field Effects in *f*-Electron Systems (4th: 1981: Wrocław, Poland)
Crystalline electric field effects in f-electron magnetism.

"Proceedings of the Fourth International Conference on Crystalline Electric Field and Structural Effects in *f*-Electron Systems, held September 22-25, 1981, in Wrocław, Poland"—Verso t.p.
Includes bibliographical references and indexes.
1. Crystalline electric field—Congresses. 2. Lattice dynamics—Congresses. 3. Intermetallic compounds—Magnetic properties—Congresses. 4. Nuclear magnetism—Congresses. 5. Trzebiatowski, Włodzimierz. I. Guertin, Robert P. II. Suski, Wojciech. III. Zołnierek, Zygmunt. IV. Title.
QC176.8.E35I58 1981 530.4′1 82-5267
ISBN 978-1-4684-8648-3 ISBN 978-1-4684-8646-9 (eBook) AACR2
DOI 10.1007/978-1-4684-8646-9

Proceedings of the Fourth International Conference on Crystalline Electric Field and Structural Effects in *f*-Electron Systems, held September 22-25, 1981, in Wrocław, Poland

Softcover reprint of the hardcover 1st edition 1982

A Division of Plenum Publishing Corporation
233 Spring Street, New York, N.Y. 10013

FOREWORD

The present conference, the fourth successive on this subject, was organized to commemorate the 75th birthday of Professor Włodzimierz Trzebiatowski, one of the pioneers in the field of f-electron materials structure, particularly in the magnetism of actinides.

This volume contains 64 papers presented at the conference held in Wrocław, Poland, September 22-25, 1981. Twenty-one were invited talks. About 100 participants from 13 countries attended the meeting during four days of lecture presentation (note these two numbers have been constant for the last two conferences). The conference consisted of sessions devoted to the investigation of crystalline electric fields and structural effects by spectroscopic techniques, neutron diffraction, magnetic, thermodynamic and electrical measurements all over broad temperature, magnetic field and pressure ranges. Materials investigated included rare earth intermetallics, hydrides, diluted systems and actinides, and among them some exhibited singlet ground state behavior. The experimental results were supplemented by theory.

It is our pleasure to mention those persons who helped us make the conference successful. The International Advisory Committee included W.J.L. Buyers, B.R. Cooper, J.E. Crow, P. Fulde, A. Furrer, T. Kasuya, L. Kowalewski, G.H. Lander, R. Lemaire and D. Wohlleben. We thank them for valuable suggestions concerning invited speakers. We also wish to thank the co-workers of the Institute for Low Temperature and Structure Research of the Polish Academy of Sciences in Wrocław, especially A. Baran, M. Grzebyk, K. Hoffman, B. Janus, P.J. Markowski, B. Preś, K. Tarchała, A. Zygmunt, and Z. Zołnierek for their handling of day to day operations before and during the conference. We express our thanks to Robert P. Guertin, our American coeditor, for his initiative and cooperation, whose assistance was indispensable for this volume to be published at all. The conference was supported by the Polish Academy of Sciences.

Bohdan Stalinski
Chairman, Int. Advisory Committee

Wojciech Suski
Chairman, Local Organizing Committee

Wrocław, Poland, September 1981

PREFACE

The sixty-four selections in this volume represent much of the work currently going on in f-shell magnetism, an active field of condensed matter physics and chemistry. An impressive arsenal of experimental and theoretical techniques have now been brought to bear on problems associated with dilute and concentrated f-shell systems, and we feel the papers herein reflect accurately the latest work in the field. Many of the papers tend to be review-like, but they frequently show up areas for future research. Many represent the degree of sophistication attained in f-shell magnetism work. For example, in common usage are various resonance techniques such as ESR, Mössbauer and μSR. Also we note the incorporation of crystal field effects into problems associated with unstable, or nearly unstable, moment systems, an area of great complexity and one which requires much future theoretical and experimental effort. The book contains a chapter devoted entirely to cerium compounds and one to uranium compounds, thus illustrating the activity in systems where the f-shell and the conduction electrons are strongly coupled.

The major portion of the papers in this book were written by scientists for whom English is not their native language. While recasting the writing of many manuscripts into more idiomatic English, we made every effort to retain accurately all the physical content. It was our desire that the work read evenly but with the same fine scientific input originally imposed by the authors. The papers were assigned to chapters by either the type of material discussed or by the experimental technique employed. The discussions at the end of the papers were taken from those question and answer sheets submitted to the editors by conference participants. The discussions at the conference were wide-ranging and intensive, and they often hinted at future research trends. The materials index as well as the usual subject index should be useful to workers interested in reviewing what work has been done on particular systems.

It is a great pleasure to thank those who made this conference (and volume) possible. I am grateful to Prof. B. Stalinsky and to the students and staff of the Institute for Low Temperature and Structure Research for their fine hospitality in Wrocław. The conference was well-organized and a most memorable experience for all who attended; I thank the session chairmen: Y. Berthier, B.R. Cooper, B. Coqblin, J.J.M. Franse, P. Fulde, T. Kasuya, J. Keller, L. Kowalewski, J. Leciejewicz, B. Luthi, I.A. Smirnov, B. Stalinski, W. Trzebiatowski and D. Wohlleben. To my co-editor, Wojciech Suski, I am especially grateful. Not only did he serve as central organizer for the conference, which took place at a difficult time, but he also contributed enormously to the publication of this volume. His trip to the US following the conference was most useful in that regard. Finally, to Celia Mees, who retyped most of the manuscripts, I owe much thanks. Without Celia's participation the publication of these proceedings would have been greatly delayed. She not only did her work with great enthusiasm and dedication, but she also contributed to correcting the authors' (and my own) English in no small way.

Robert P. Guertin

Medford, Massachusetts, January 1982

OPENING ADDRESS

I am greatly honored to open the present conference on crystal field and structural effects in f-electron systems which has been organized to celebrate the 75th birthday of Professor W. Trzebiatowski, one of the pioneers in the field of actinide magnetism, and the organizer and first director of our Institute. The activities of Professor Trzebiatowski have been associated with Wrocław for many years. Here he spent his childhood, and he came here again directly after World War II. Prof. Trzebiatowski graduated from the Technical University in Lwow where he received the Ph.D. degree in 1930 and the DSc. degree in 1935. In 1938 he was appointed extraordinary professor at the Ioannes Casimirus University in Lwów as a head of the inorganic chemistry department.

Before the war, Prof. Trzebiatowski was engaged in the research of the structure of metals and intermetallic compounds, and his research was supplemented by visits to Zürich, Berlin, Freiburg and Stockholm. During this period, he published several pioneering papers on the mechanism of metal powder sintering, which were of basic importance in the development of powder metallurgy. This work was highly appreciated in the world and resulted in his being awarded an Honorary Membership in the International Team for Sintering Studies. After the war, Prof. Trzebiatowski created in Wrocław the most important research center in the country devoted to the physico-chemistry of solids, particularly in the fields of x-ray crystallography, magneto-chemistry and the structure and thermodynamics of metal systems. He began his magnetism research in 1949 during his stay as a visiting professor in P.W. Selwood's laboratory in the U.S.A.. There he studied the magnetic properties of urania-thoria solid solutions, and he continued the research in this field in Wrocław. Soon afterwards, he discovered the ferromagnetic properties of uranium hydride. This was near the end of 1951, i.e. thirty years ago, thus making the occasion of this conference an additional jubilee. In 1951, uranium hydride was the first known example of a magnetically ordered substance among the actinide compounds. In subsequent years the investigation of the other simple metallic or semimetallic uranium compounds was developed by Prof. Trzebiatowski and his students, and as a result, the ferro- or antiferromagnetic properties of over 50 compounds of this element have been established. The results obtained became known world-wide and stimulated closely related studies in other research centers abroad.

Finally, we mention that Prof. W. Trzebiatowski highly contributed to the organization of science in Poland, being Vice President (1969-71) and President (1972-77) of the Polish Academy of Sciences. He is now the director of the International Laboratory of High Magnetic Fields and Low Temperatures, and has been since its creation in Wrocław in 1968.

On behalf of the organizing committee I would like to congratulate Prof. W. Trzebiatowski on the occasion of his 75th birthday and wish him good health and long years of further activity as a scientist and teacher. I would like to also wish all the participants of this conference a pleasant stay in Wrocław and full success in their scientific discussions.

Prof. B. Stalinski

CONTENTS

SECTION 3: THEORY

SECTION 4: LATTICE EFFECTS

SINGLET GROUND STATE SYSTEM IN AMORPHOUS RARE EARTH ALLOYS

A.K. Bhattacharjee and B. Coqblin

Laboratoire de Physique des Solides, Université
Paris-Sud, 91405 Orsay, France

INTRODUCTION

Amorphous alloys containing rare earths have been extensively studied in the last years [1]. To explain their magnetic properties, Harris, Plischke and Zuckermann (HPZ) [2] have introduced the random anisotropy model in which each rare earth moment is subjected to a local uniaxial anisotropy, the anisotropy axes being randomly oriented. The HPZ model can account for the magnetic properties of amorphous alloys such as $DyNi_{3.4}$, by assuming a ferromagnetic nearest neighbor interaction. On the other hand, a model describing random anisotropy antiferromagnetism has been then developed to account for the properties of amorphous alloys such as TbAg [3].

The random anisotropy model can account fairly well for the properties of amorphous alloys containing most of the rare earths, in particular the heavy ones. On the other hand, it cannot explain recent experiments [4-8] on amorphous alloys containing Praseodymium, because of the importance of the singlet character of the ground state for the Praseodymium ion at low temperatures.

Magnetic measurements in amorphous Pr_xAg_{100-x} alloys (with x = 50 and 21.5) suggest that about 30% of Pr ions are in a singlet ground state at 4.2 K [4]. The magnetic susceptibility χ of dilute $Pr_xLa_{80-x}Au_{20}$ alloys follows a Curie-Weiss law above 100 K, while the low temperature slope of the $1/\chi$ versus T curve is twice that of the high temperature value and the saturation magnetic moment is roughly 1.4 μ_B [5]. An analysis [6] of these data suggests that roughly 50% of Pr ions are in a singlet ground state. Specific heat measurements of $Pr_{21}Ag_{79}$ [7] and dilute $Pr_xLa_{80-x}Au_{20}$ alloys [6,8] down to 0.3 K indicate a predominance of singlet ground state and the existence of very low "crystal field" excitation energies.

On the other hand, the superconducting temperature of $Pr_xLa_{80-x}Au_{20}$ remains roughly constant up to x = 1 and then decreases for larger x values. This suggests an induced moment type behaviour characteristic of singlet ground state systems, implying a threshold exchange strength for the onset of magnetism. Such a conclusion is also drawn from the nuclear hyperfine specific heat contribution which is important for x = 10 and disappears for x = 1 [8].

On the theoretical side, the existence of a singlet ground state for Pr is easily understood in amorphous systems as in crystalline ones. The most important point seems to be that the energy Δ of excitation to the first excited state varies from site to site. Cochrane et al. [9], using the point charge model on computer-generated clusters of dense random packing of hard spheres, have shown that the distribution of Δ and of the overall splitting has almost a gaussian form. Concerning induced magnetism in amorphous rare earth alloys, the uniaxial random anisotropy model [2] assumes a fixed Δ value, which disagrees completely with the low temperature specific heat data and the calculations of Cochrane et al. [9] . A calculation of the magnetic susceptibility using a distribution of the crystal field coefficients has been also performed to explain the magnetic properties of amorphous alloys with Praseodymium [10].

In the present work, we propose a singlet-singlet model with the energy splitting Δ varying randomly over the rare earth sites according to a Gaussian distribution, with either a ferromagnetic exchange interaction or an antiferromagnetic one. As susceptibility data indicate that in amorphous alloys like Pr_xAg_{100-x} and $Pr_xLa_{80-x}Au_{20}$ the exchange interaction is predominantly antiferromagnetic, it is certainly more appropriate to apply to these alloys the antiferromagnetic calculation where we assume that the induced order is spin-glass-like. Thus, we will study here the paramagnetic susceptibility in the dilute regime, the magnetic ordering induced by the exchange field in more concentrated alloys, and the 4f shell contribution to the specific heat in both regimes.

THEORY

The Hamiltonian of the system is given by :

$$H = \sum_i H^i_{CF} - \sum_{i,j} J_{ij} \vec{J}_i \cdot \vec{J}_j - g\mu_B H \sum_i J_{iz} \qquad (1)$$

where the first term is the crystalline field (CF) part, the second the Heisenberg Hamiltonian and the last term corresponds to an applied field H.

In the singlet-singlet model, both the ground state $|0_i>$ and the first excited state $|1_i>$ of H^i_{CF} are assumed to be singlet and the contribution of higher states is assumed to be negligible at low temperatures. In an amorphous system, we assume that the energy splitting Δ_i between the states $|0_i>$ and $|1_i>$ varies randomly over the rare earth (RE) sites according to a Gaussian distribution

$$P(\Delta) = \frac{1}{\sqrt{2\pi}\,\Delta_1} \exp\left[-\frac{(\Delta-\Delta_o)^2}{2\Delta_1^2}\right]. \qquad (2)$$

Our results will be presented for different values of the relative width $\gamma = \Delta_1/\Delta_0$ between 0 and 1. In the following we will replace the polarization matrix elements $\alpha_i = <0_i|J_{iz}|1_i>$ by their average value α. Physically, α gives the saturation moment at 0 K.

Paramagnetic Susceptibility

The "CF susceptibility" (zero exchange) is given by the average of the Van Vleck susceptibility

$$\chi_{CF} = \frac{2g^2\mu_B^2\alpha^2}{\Delta_0} \overline{\left[\frac{\Delta_0}{\Delta} \tanh\left(\frac{\Delta}{2kT}\right)\right]} \qquad (3)$$

where $\overline{F(\Delta)}$ means an average of $F(\Delta)$ over the gaussian distribution $P(\Delta)$.

In the presence of an exchange interaction, the paramagnetic susceptibility χ is given in the molecular field approximation [11] by

$$\frac{2g^2\mu_B^2\alpha^2}{\Delta_0}\frac{1}{\chi} = \frac{2g^2\mu_B^2\alpha^2}{\Delta_0}\frac{1}{\chi_{CF}} - A_0 \qquad (4)$$

where

$$A_0 = \frac{4\,\mathcal{J}(0)\ \alpha^2}{\Delta_0} \tag{5}$$

characterizes the relative strength of the exchange field with respect to the CF splitting and here $\mathcal{J}(0)$ is equal to $\sum_i \mathcal{J}_{ij}$.

Induced Magnetic Ordering

In the effective field approximation, for a given configuration of the RE ions, the exchange Hamiltonian is replaced by

$$\mathcal{H}_{ex} = - \sum_i h_i\ J_{iz} \tag{6}$$

where

$$h_i = 2 \sum_{j\neq i} \mathcal{J}_{ij}\ \ \langle J_{jz} \rangle, \tag{7}$$

the angular bracket indicating thermal and quantum-mechanical averaging. It should be noted that we retain only the z-component of the effective field. In the absence of external field (H=0), the effective Hamiltonian of the i-th ion is then $- h_i\ H_{iz} + \mathcal{H}^i_{CF}$, which is a 2×2 matrix that can be diagonalized exactly to obtain the induced moment :

$$J(\Delta_i, h_i) \equiv \langle J_{iz}\rangle = \frac{2\alpha^2\ h_i}{D_i}\ \tanh \frac{D_i}{2kT}, \tag{8}$$

where

$$D_i = \sqrt{\Delta_i^2 + (2\alpha\ h_i)^2} \tag{9}$$

is the singlet-singlet separation in the presence of ordering.

Let us first approximate Eq. (7) by the "nearest-neighbour exchange" form :

$$h_i = 2\,\mathcal{J} \sum_{j=i}^{v} \langle J_{jz} \rangle \tag{10}$$

where v is the mean magnetic coordination number and $\mathcal{J}$ the average exchange parameter. This is consistent with the spirit of configuration averaging that is implicit in the following. In the random effective field approximation, h_i is assumed to be a random variable with a probability distribution P(h). As Δ_i is also treated as a random variable, $\langle J_{iz}\rangle$ becomes a random variable. Thus, Eq. (10) presents the effective field variable h as a sum of v independent random variables, each with the same probability distribution (by neglecting correlations). Elementary probability theory then tells us that the mean value of h is

$$\bar{h} = 2 v J \bar{J}, \tag{11}$$

where

$$\bar{J} = \int d\Delta\ P(\Delta) \int dh\ P(h)\ J(\Delta, h), \tag{12}$$

and the standard deviation σ is given by

$$\sigma^2 = 4v J^2 \int d\Delta\ P(\Delta) \int dh P(h)\ \{J(\Delta,h) - \bar{J}\}^2. \tag{13}$$

If v is not too small, we can assume for P(h) a Gaussian form centered at $\bar{h}$ and with a width equal to σ. Thus equations (11)-(13) then allow us to calculate h and σ self-consistently.

In the ferromagnetic case ($J > 0$) all the $\langle J_{iz} \rangle$ are positive and we can futher simplify our calculations by neglecting fluctuations, i.e., by making the simple mean field approximation $P(h) = \delta(h - \bar{h})$. The mean induced moment $\bar{J}$ is then given by:

$$\frac{\bar{J}}{\alpha} = \overline{\frac{2\alpha\bar{h}}{D} \tanh \frac{D}{2kT}} = \overline{\frac{2\alpha\ \bar{h}}{D} \tanh \left(\frac{\sqrt{\Delta^2 + (2\alpha\bar{h})^2}}{2kT}\right)} \tag{14}$$

and the Curie temperature T_c is given by

$$\frac{1}{A_o} = \overline{\frac{\Delta_o}{\Delta} \tanh \frac{\Delta}{2kT_c}} \tag{15}$$

In the antiferromagnetic case ($J < 0$) the $\langle J_{iz} \rangle$ are randomly positive or negative and $\bar{h} = 0$. The standard deviation σ of the effective field distribution then represents the "order parameter" of the spin glass like order. The detailed calculation of σ can be found elsewhere [12] and the ordering temperature T_N is given by

$$1 = \frac{A_o^2}{v} \overline{\left(\frac{\Delta_o}{\Delta} \tanh \frac{\Delta}{2kT_N}\right)^2} \tag{16}$$

Specific Heat

In the paramagnetic regime, the specific heat per mole of RE is given by the average of the Schottky formula

$$\frac{C_v}{R} = \overline{\left(\frac{\Delta}{2kT} \operatorname{sech} \frac{\Delta}{2kT}\right)^2} \tag{17}$$

In the ferromagnetic case, we finally obtain in the mean field approximation [12]

$$\frac{C_v}{R} = F_1 + A_o\left(F_2 + \frac{1}{2}\,\frac{\bar{J}}{\alpha}\right)\frac{\partial}{\partial\left(\frac{kT}{\Delta_o}\right)}\left(\frac{\bar{J}}{\alpha}\right) \tag{18}$$

where

$$F_1 = \overline{\left(\frac{D}{2kT}\ \mathrm{sech}\ \frac{D}{2kT}\right)^2} \tag{19}$$

$$F_2 = -\overline{\frac{\alpha\bar{h}}{D}\left(\tanh\frac{D}{2kT} + \frac{D}{2kT}\ \mathrm{sech}^2\frac{D}{2kT}\right)}\ . \tag{20}$$

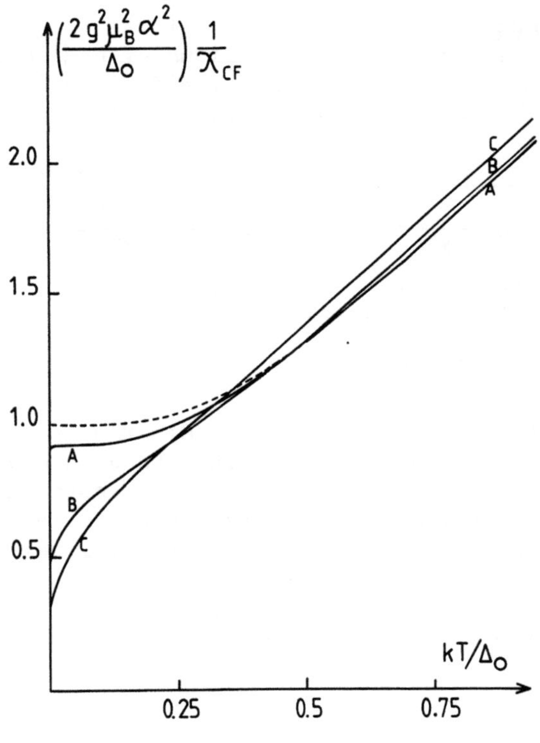

Fig. 1 : Plot of $\frac{2g^2\mu_B^2\alpha^2}{\Delta_0}\ \frac{1}{\chi_{CF}}$ versus $\frac{kT}{\Delta_0}$ for the crystalline $\gamma = 0$ case (dotted curve), $\gamma = 1/4$ (curve A), $\gamma = 1/2$ (B) and $\gamma = 1$(C).

RESULTS AND DISCUSSION

We present some results of the theoretical model in figures 1 to 4. Figure 1 shows the reduced inverse susceptibility $1/\chi_{CF}$ versus kT/Δ_0 for several values of $\gamma = \Delta_1/\Delta_0$. Curve A (for $\gamma = 1/4$) deviates only slightly from the Van Vleck behaviour of a crystalline system given by the dotted curve ($\gamma = 0$). The curves B ($\gamma = 1/2$) and C ($\gamma = 1$) have a behaviour intermediate between Van Vleck and Curie. We find that curve C is very similar to the experimental $1/\chi$ versus T curves for dilute $Pr_xLa_{80-x}Au_{20}$ alloys [5] or $Pr_x(CuLu)_{1-x}$ and $Pr_x(CuZr)_{1-x}$ alloys [10]. In particular, the low temperature slope of curve C (between $kT/\Delta_0 = 0.025$ and 0.05) is about 2 times the high temperature slope.

Figure 2 corresponds to induced ordering in the ferromagnetic case. It shows the reduced zero-field induced magnetic moment

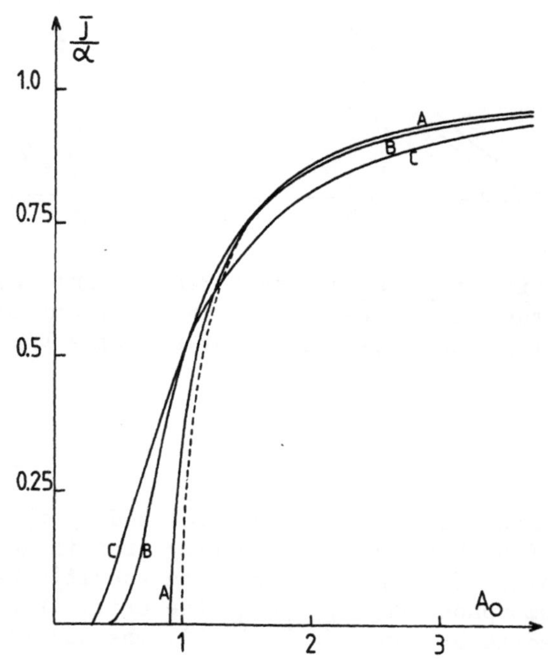

Fig. 2 : Ferromagnetic case : Plot of the reduced zero-field induced moment $\bar{J}/\alpha$ at 0 K versus A_0 for the crystalline $\gamma = 0$ case (dotted curve), $\gamma = 1/4$ (curve A), $\gamma = 1/2$ (B) and $\gamma = 1$ (C).

$\bar{J}/\alpha$ at 0 K as a function of the relative exchange strength A_0 for different values of γ. Curve A (γ = 1/4) is similar to the dotted curve (γ = 0) representing the crystalline case. It has a threshold value of A_0 close to 1 and a sharp rise above that value. The curves B and C have a different behavior with both a much smaller threshold value and a much slower rise above it, in qualitative agreement with experimental data. Experimentally, the

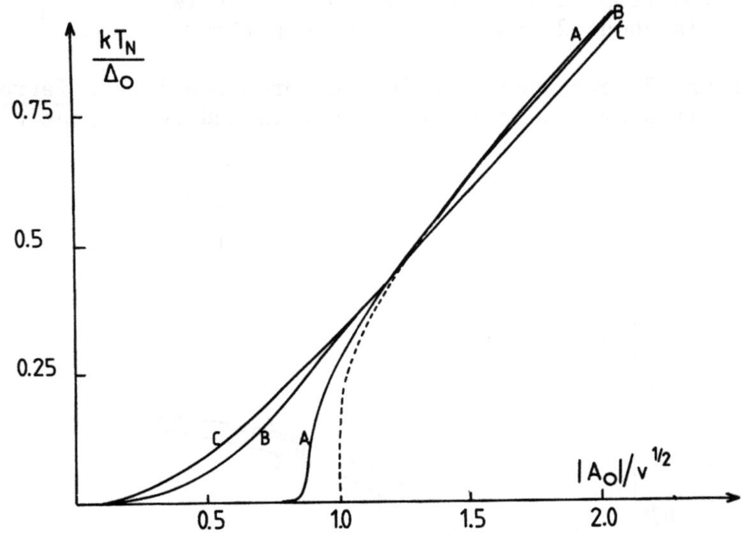

Fig. 3 : Antiferromagnetic case : reduced ordering temperature kT_N/Δ_0 as a function of relative exchange strength $|A_0|/v^{1/2}$ for γ= 0 (dotted curve), γ = 1/4 (curve A), γ = 1/2 (curve B) and γ = 1 (curve C).

existence of an exchange threshold is evidenced by the concentration dependence of both the superconducting transition temperature [5], and the onset of a nuclear hyperfine specific heat contribution [8] . In amorphous $Pr_xLa_{80-x}Au_{20}$ the critical concentration of Pr is estimated to be about 2%, whereas in crystalline (dhcp) LaPr alloys it is about 16% [13]. It should be noted that figure 2 also allows us to read off the reduced Curie temperature kT_c/Δ_0 as a function of A_0.

Figure 3 shows the reduced spin-glass ordering temperature kT_N/Δ_0 as a function of $|A_0|/V^{1/2}$ for different values of γ, in the antiferromagnetic case. The curves giving the reduced "order parameter" $2\alpha\sigma/\Delta_0$ at 0 K as a function of $|A_0|/V^{1/2}$ are similar to those presented in figure 3 and are presented elsewhere [12]. The case $\gamma = 0$ (dotted curve) represents a model for a crystalline induced moment spin glass. This problem has been recently discussed by Sherrington [14]. Once again, the curve for $\gamma = 1/4$ is relatively similar to the $\gamma = 0$ curve. For $\gamma = 1/2$ or 1, on the other hand, the appearance of induced moment is so much more progressive that both T_N and σ acquire substantial values at rather low values of the relative exchange strength. Note, however, that now the abscissa is proportional to $v^{1/2}$ rather than v, which corresponds to the ferromagnetic case. The concentration (x) dependence of the root-mean-square internal field in $Pr_xLa_{80-x}Au_{20}$, estimated from the nuclear hyperfine specific heat [8] is in better agreement with curves B or C or figure 3 (antiferromagnetic case) than those of figure 2 (ferromagnetic case). Moreover, the decrease of the critical concentration in passing from the crystalline system (16%) to the amorphous one (2%) is in better agreement with Fig. 3. However, such differences between the ferromagnetic and antiferromagnetic cases may be related to the neglect of fluctuations in the former.

Figure 4 shows the reduced specific heat C_V/R per mole of the RE ion versus kT/Δ_0 in the presence of induced magnetic ordering for the ferromagnetic case. The dotted curves correspond to the case in the absence of any induced magnetism. When γ increases from 0 to 1, the low temperature decrease of C_v changes from exponential to almost linear and the experimental data [8] on dilute $Pr_xLa_{80-x}Au_{20}$ seem to be closer to the $\gamma = 1/2$ curve which has still upward curvature at low temperatures than to the γ=1 one. The full curves correspond to the case in the presence of a ferromagnetic induced magnetism. The most prominent feature is a sharp peak corresponding to the finite jump at T_c that arises from the sharp fall of the induced moment as a function of temperature. Experimentally no sharp peak has been observed. Anyway, near T_c it is not easy to separate this 4f-shell part from the nuclear hyperfine contribution. Nevertheless, according to Garoche et al. [8], as the temperature decreases below 1 K, the C_v(T) curve for $Pr_{10}La_{70}Au_{20}$ rises above that for $Pr_1La_{79}Au_{20}$ and goes through an inflection point around 0.5 K, which corresponds to a broad maximum of the difference between the two curves. This is perhaps a reminescence of the peak at the ordering temperature. The case $\gamma = 1/2$ and $A_0 = 0.75$ in figure 4 appears to be appropriate for a comparison. Another effect of ordering is the reduction of the low temperature C_v owing to the increase of the two singlet separation in a molecular field. Such a depression of the low temperature tail of $C_v(T)$ can be also induced by an applied field, as observed by Garoche et al [8].

We do not present specific heat curves for the antiferromagnetic case, because in the random effective field approximation, they are qualitatively similar to those for the ferromagnetic case. In particular, there is a jump at T_N. In fact, this specific heat jump at the ordering temperature is a general feature of all effective-field models. The rounded maximum often observed in disordered systems can be obtained only in more sophisticated theories. Let us mention, in particular, that in the paramagnetic regime our singlet-singlet model is equivalent to an Ising model in a transverse field of randomly distributed magnitude. The phase transition of the latter Hamiltonian in 1D has been recently studied by Uzelac et al [15] in the real-space renormalization group method ; they indeed obtain a rounded maximum of the specific heat at the transition point.

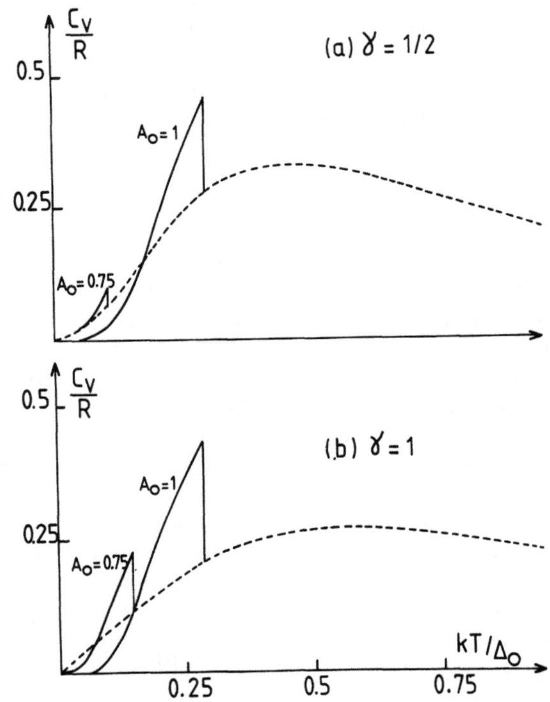

Fig. 4 : Plot of the specific heat C_v/R versus kT/Δ_0 for a) $\gamma = 0.5$ and b) $\gamma = 1$ and for $A_0 = 0.75$ and $A_0 = 1$. The dotted curves correspond to a zero magnetic moment $\bar{J} = 0$.

CONCLUDING REMARKS

We have developed a singlet-singlet model for an amorphous system by assuming a Gaussian distribution of the energy splitting. A comparison between our theoretical curves for the dilute regime and the experimental susceptibility and specific heat data on dilute $Pr_xLa_{80-x}Au_{20}$ suggests a value of the relative width γ of the order 1/2.

As for induced magnetic ordering in more concentrated alloys, we have separately treated the cases of ferromagnetic and antiferromagnetic exchange interaction. In the ferromagnetic case the order parameter is the induced magnetization and calculations are carried out in the simple mean field approximation. In the antiferromagnetic case, the induced order is assumed spin-glass-like; the order parameter is the standard deviation of the random effective field distribution, which is approximated by a Gaussian. Both cases predict a large decrease of the exchange threshold for $\gamma = 1/2$ or 1 with respect to the crystalline case ($\gamma = 0$). Moreover, the rise of the order parameter as a function of the relative exchange strength beyond the "threshold" is much slower in the amorphous case. Both effects are more accentuated in the antiferromagnetic case. The observed concentration (x) dependence of the superconducting transition temperature and the nuclear hyperfine specific heat in $Pr_xLa_{80-x}Au_{20}$ appear to be in better agreement with the antiferromagnetic case than the ferromagnetic one. This is consistent with the concentration dependence of the dilute regime susceptibility that indicates that the exchange interaction in such alloys is predominantly antiferromagnetic. The rounded maximum of the 4f shell specific heat at the ordering temperature also fits in with a spin-glass-like order. A more direct evidence is the observation of a relatively sharp peak in the a.c. susceptibility of concentrated alloys ($x > 30$) [16]. It would be very interesting to extend their curve of the temperature T_M of the susceptibility maximum versus x to lower values of x. If we assume $T_M = 0.5$ K for $x = 10$, as suggested by the specific heat data, we find that the T_M versus x curve shows an initial upward curvature, in good agreement with the curves for $\gamma = 1/2$ or 1 in figure 3. Thus, the experimental results on amorphous $Pr_xLa_{80-x}Au_{20}$ alloys indicate that these alloys can be well described by the present theoretical model with an antiferromagnetic exchange interaction.

REFERENCES

1. R.W. Cochrane, R. Harris and M.J. Zuckermann, Physics Reports <u>48</u>, 1 (1978).

2. R. Harris, M. Plischke and M.J. Zuckermann, Phys. Rev. Lett. 31, 160 (1973).
3. A. K. Bhattacharjee and B. Coqblin, Solid State Comm., 27, 599 (1978).
4. C. Pappa, Thesis 3ème Cycle, Université Paris VI (1979).
5. N. Hassanain, A. Berrada, J. Durand and B. Loegel, J.M.M.M., 15-18, 1377 (1980).
N. Hassanain, Thesis 3ème Cycle, Université Louis Pasteur, Strasbourg, (1980).
6. A. Fert, P. Garoche, B. Boucher and J. Durand, in:"Crystalline Electric Field and Structural Effects in f-Electron Systems," Plenum, New York, 1980. p. 491.
7. P. Garoche, A. Fert, J.J. Veyssié and B. Boucher, J.M.M.M. 15-18, 1397 (1980).
8. P. Garoche, J.J. Veyssié and J. Durand, J. Physique Lettres 41, L 357 (1980).
9. R.W. Cochrane, R. Harris, M. Plischke, D. Zobin and M.J. Zuckermann, J. Phys. F. : Metal Phys. 5, 763 (1975).
10. J.B. Bieri,Thesis 3ème Cycle, Université Paris Sud, Orsay (1981).
J.B. Bieri and A. Fert. To be published.
11. B.R. Cooper in "Magnetic Properties of Rare-Earth Metals", ed. R.J. Elliott, Plenum Press London, and New York (1972) pp. 44.45.
12. A.K. Bhattacharjee and B. Coqblin. To be published.
13. S. Legvold, R.W. Green, B.J. Beaudry and J.E. Ostenson, Solid State Comm. 18, 725 (1976).
14. D. Sherrington, J. Phys. C : Solid State Phys. 12, L 929 (1979)
15. K. Uzelac, R. Jullien and P. Pfeuty, J. Phys. A : Math. Gen. 13, 3735 (1980).
16. A. Berrada, J. Durand, N. Hassanain and B. Loegel, XIV Rare-Earth Research Conf. (Fargo, North Dakota, June 1979).
A. Berrada, private communication (1981).

COMMENTS

STEGLICH: Is there any experimental or theoretical evidence for a "glassylike" γT contribution to the specific heat, originating from the statistical distribution of the crystal field split energies?

COQBLIN: When $\gamma=\Delta_1/\Delta_0$ increases from 0 to 1, one goes from an exponential to a linear behavior of the specific heat at low temperatures. For $\gamma=1$, it behaves therefore as γT at low temperatures, as illustrated by Fig. 4 of the paper. However, the low temperature specific heat of $Pr_xLa_{80-x}Au_{20}$ amorphous alloys has an upward curvature and is closer to the curve B (with $\gamma=1/2$) of Fig. 4 than to the linear curve C (with $\gamma=1$), so that there is presently no experimental verification of this γT law in amorphous magnetic alloys.

MAGNETIC PROPERTIES AND NEUTRON SPECTROSCOPY OF INTERMETALLIC PRASEODYMIUM COMPOUNDS

F.J.A.M. Greidanus*, L.J. de Jongh*, W.J. Huiskamp*,
A. Furrer† and K.H.J. Buschow§

*Kamerlingh Onnes Laboratorium der Rijksuniversiteit Leiden, The Netherlands
†Institut für Reaktortechnik, E.T.H. Zürich, Würenlingen Switzerland
§Philips Research Laboratories, Eindhoven, The Netherlands

INTRODUCTION

Elastic and inelastic neutron scattering, specific heat, susceptibility and resistivity experiments on the series of cubic Laves phase compounds PrX_2 (X = Mg, Al, Co, Ni, Ru, Rh, Pt, Ir) are discussed. In all these compounds, magnetic phase transitions are found. The electronic ground state in the paramagnetic regime is either the singlet Γ_1, or the non-magnetic doublet Γ_3. Furthermore the ratio of the magnetic interaction to the crystal field splitting varies considerably through the series. In $PrNi_2$ the hyperfine interaction determines the nature of the phase transition.

Among the rare earth ions Pr^{3+}-ions exhibit, partly due to the small number of 4f electrons, relatively weak exchange interactions, which leads to several interesting features from a theoretical as well as from an experimental point of view. In intermetallic compounds of Pr, crystal field effects play an important role and will be discussed on the basis of the Lea, Leask and Wolf diagram[1] (LLW). In cubic symmetry the $^3H_4$ ground multiplet of the Pr^{3+} ion is split, leaving a Γ_1(singlet), Γ_3(non-magnetic doublet) or a Γ_5(triplet) as the ground state, as shown in Fig. 1. The possibility of a non-magnetic ground state (Γ_1 or Γ_3) has several important consequences. It was shown by several authors[2-5] that for electronic magnetic order to occur, the dipolar (exchange) interaction should exceed a certain threshold value. For an ensemble of singlet-singlet systems, for

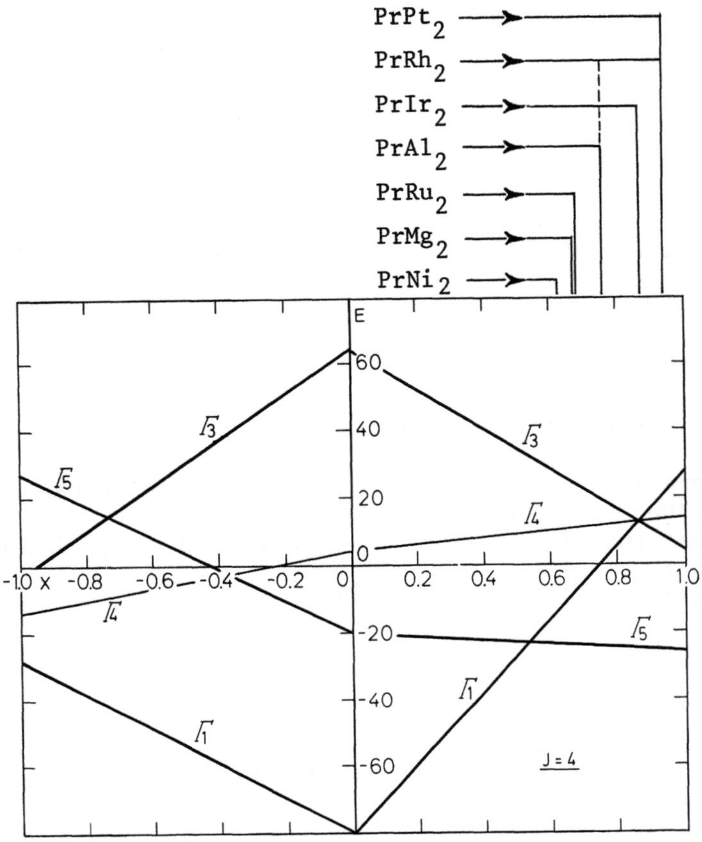

Fig. 1 LLW diagram for Pr^{3+} (after Ref. 1) with a summary of the results as obtained by inelastic neutron scattering.

instance, one can define a parameter $\eta = 4\ K(0)C^2/\Delta$, in which K(0) is the Fourier transform of the interaction constant for zero wave vector, Δ is the energy splitting between the two singlet states, and C is the matrix element of the J_z-operator connecting the two levels. The dependence of the ordering temperature T_c on η can be calculated in the molecular field approximation, yielding $T_c = 0$ for $\eta < 1$. The system then remains paramagnetic to zero temperature. Qualitatively these features are preserved if one goes beyond the molecular field approximation. It may be remarked that the possibility of collective order in singlet systems is of interest also for a wider class of physical problems.

The properties outlined above are not exclusive to singlet ground state systems, but will also apply to systems which have a non-magnetic doublet ground state (Γ_3). However, in that case the situation can become more complicated. Besides the magnetic dipolar exchange interaction, quadrupolar interactions between the (non-vanishing) quadrupole moments within the Γ_3 level can be of importance[6].

The co-existence of these two types of interaction yield the possibility of both dipolar and quadrupolar phase transitions[7].

Another feature of interest is the presence of nuclear magnetic moments. If the system exhibits a phase transition to conventional electronic magnetic order the nuclear sublevels are split by the hyperfine field, giving rise to a Schottky type of anomaly in the specific heat at temperatures well below 1 K. On the other hand, if the dipolar (exchange) interaction is small compared to the crystal field splitting the 4f moment and hence the hyperfine field is quenched. However, the nuclear moments induce a small 4f moment through the hyperfine interaction which can be interpreted as an enhancement of the nuclear moment. Interactions among these, which will involve the RKKY interactions between the induced 4f moments, may lead to phase transitions in the range of 0.1 - 100 mK[8,9]. These enhanced nuclear ordering phenomena have mainly been studied in systems with a singlet Γ_1 ground state[10,11]. Finally it should be mentioned that compounds with enhanced nuclear moments are of considerable importance in obtaining temperatures in the mK range[12].

In view of the above, PrX_2 compounds with X = Co, Ni, Mg, Al, Rh, Ru, Ir, Pt form an interesting series. They all have the C-15 (Laves phase) structure and the Pr^{3+} ions are situated at cubic sites. The lattice constants are listed in Table I. Although the compounds have many common features, their magnetic properties vary in several important respects, as is discussed in the following section.

THERMAL AND MAGNETIC PROPERTIES

The compound $PrMg_2$ was shown to have a ferromagnetic transition at T_c = 9.6 K, from specific heat and magnetization data[13]. $PrAl_2$ has been studied by many workers[14-18], and shown to have a ferromagnetic transition at T_c = 33 K[17]. From single crystal magnetization measurements Purwins et al[17] derive L.L.W. parameters: x = 0.70 and W = -0.38 meV. By adapting these values and using a two-dimensional mean-field model, Bak[18] could explain quite well the specific heat data measured by Deenadas[16].

The specific heat of $PrCo_2$, also measured by Deenadas et al[19] shows a double-peaked structure, with maxima at about T ≃ 39.3 and T ≃ 43.3 K. The origin of this phenomenon is as yet unexplained. The zero field a.c. susceptibility[20] displays only a single peak at T_c = 39.3 K. It should be noted that $PrCo_2$ differs from the other compounds in the series in the sense that the transition metal ion also carries a moment[21].

Although $PrNi_2$ was previously thought to become ferromagnetic at T_c ≃ 8 K[22,23], more recent susceptibility and specific heat data[24-26] show it to remain paramagnetic down to 1.2 K. Recently, peaks in both the ac and dc susceptibility have been found at

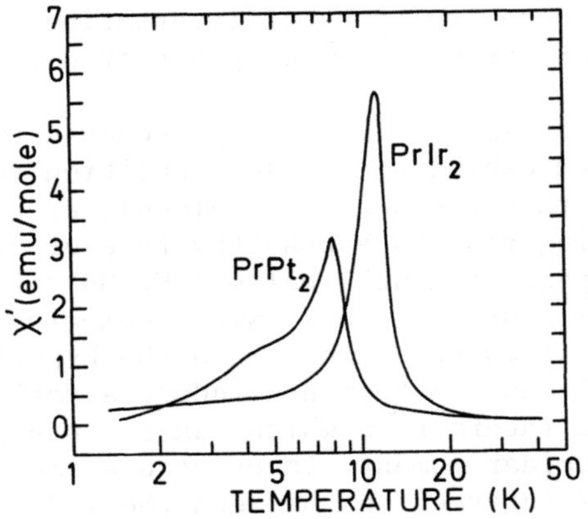

Fig. 2 A.c. susceptibility of $PrPt_2$ and $PrIr_2$ versus temperature.

about 270 mK[27,28]. This was interpreted as a mixed electronic-nuclear transition. The lack of any structure in the specific heat in the same temperature range is remarkable. However it was argued that this is not in contradiction with theoretical calculations on the mixed electronic-nuclear ordering behaviour[28].

The compounds $PrPt_2$, $PrIr_2$, $PrRh_2$ and $PrRu_2$ have nearly equal lattice constants. The zero field ac susceptibility shows a sharp peak at temperatures T_c = 33.9 K, 7.9 K, 11.2 K and 7.7 K for $PrRu_2$, $PrRh_2$, $PrIr_2$ and $PrPt_2$ resp.[29]. These were interpreted as transitions to a ferromagnetic state. Results for $PrPt_2$ and $PrIr_2$ are shown in Fig. 2 as examples. Specific heat data on the same compounds show a more or less pronounced anomaly at the temperature of the susceptibility peak. In addition a broad maximum at about 8 K is observed for $PrIr_2$, and a similar, although less clear-cut, effect is seen for $PrRh_2$ and $PrRu_2$. For $PrPt_2$ the anomaly observed at T = 7.7 K is of quite a different shape. It is considerably sharper and no evidence for the broad secondary maximum is seen. The results for $PrPt_2$ and $PrIr_2$ are shown in Fig. 3.

Measurements of the electrical resistivity ρ show a sharp drop in the ρ versus T curve below T_c. Moreover, plots of dρ/dT versus T correspond qualitatively with those of the specific heat versus temperature. In Fig. 4 results for $PrIr_2$ and $PrPt_2$ are shown as representative examples.

From the above summary it follows that the PrX_2 series exhibits a wide variety of magnetic properties. In order to explain the differences in behaviour of the compounds a detailed knowledge of the crystal field is essential, as this provides the microscopic basis for such an explanation. For the determination of the crystal field splittings neutron spectroscopy has proved to be an indispensable tool. In the next section we will review the inelastic neutron scattering experiments performed on $PrMg_2$[30], on $PrAl_2$[31,32], as well as our own recent neutron scattering experiments, on PrX_2 (X = Pt, Rh, Ir, Ru and Ni) carried out at Würenlingen.

NEUTRON INELASTIC SCATTERING

Experiments on polycrystalline $PrMg_2$ have been performed by Loidl et al[30], who derive values of x = 0.671 and W = -0.345 meV, leaving a Γ_3 ground state. A molecular field calculation based on these numbers predicts a first order phase transition which is in clear contradiction with the specific heat results. This problem was solved by introducing a quadrupolar interaction term by which both the specific heat and the magnetization data could be explained quite well.

In the neutron scattering experiments by Purwins et al[31] on $PrAl_2$ the magnetic excitations in the ferromagnetic state were studied on a single crystal. They inferred values of x = 0.77 and W = -0.29 meV, yielding a Γ_3 ground level. These results differ somewhat from those obtained from magnetization data. Frauenheim et al[32] studied a polycrystalline sample in both the ferromagnetic and the

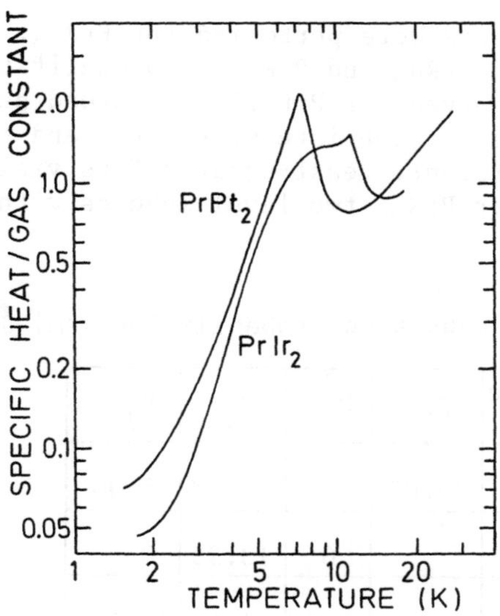

Fig. 3 Specific heat of $PrPt_2$ and $PrIr_2$ versus temperature.

$$\frac{d^2\sigma}{d\Omega d\omega} = N\left(\frac{1.91e^2}{2mc^2} g_J\right)^2 F^2(Q) \frac{k'}{k_o} \exp(-2W) \times \sum_{n,m} \rho_n |\langle m|\vec{J}_\perp|n\rangle|^2 \delta(E_n - E_m - h\delta) \quad (1)$$

the CEF state $|n\rangle$ with energy E_n, $\vec{J}_\perp$ is the component of the total angular momentum operator perpendicular to the scattering vector $\vec{Q}$, exp(-2W) is the Debye-Waller factor and the remaining symbols have their usual meaning. A measured spectrum provides information about the parameters x and W, through the position of the peaks as well as the measured intensities. The calculated transition probabilities $|\langle m|\vec{J}_\perp|n\rangle|^2$ are tabulated in Table I. By studying the relative intensities of the 3.5 meV and 15 meV transitions as well as their temperature dependence, we deduced the LLW parameters $x = 0.93 \pm 0.02$ and $W = -0.38 \pm 0.01$ meV for $PrPt_2$. The peak at 3.5 meV then corresponds to the $\Gamma_1 - \Gamma_4$ ground state transition, whereas the broad peak at 15 meV arises from the $\Gamma_4 - \Gamma_5$ and $\Gamma_3 - \Gamma_5$ excited state transitions. In Fig. 5 the data at 15 K and 40 K are shown together with the least-squares fit to the above LLW parameters. Deviations are strongest at the lowest temperature (15 K), which is probably due to exchange fluctuation effects[36]. It should be noted that by taking only the position of the peaks into account, one may arrive at level schemes which yield the Γ_3 or Γ_5 level as the ground state. However, with those schemes the measured relative intensities could not be explained satisfactorily.

Similar experiments were performed for $PrRu_2$, $PrIr_2$, $PrRh_2$ and $PrNi_2$. The spectra of $PrRu_2$ and $PrRh_2$ show qualitatively the same features as those observed for $PrPt_2$, i.e. broad peaks at $\simeq 3.5$ meV and at $\simeq 15$ meV. However, the intensity ratio and temperature dependence were quite different. Least-squares fits yield a Γ_3 ground state for $PrRu_2$, whereas for $PrRh_2$ two level schemes with either a Γ_1 or

Table I. Transition probabilities $|\langle m|\vec{J}_\perp|n\rangle|^2$.

m \ n	Γ_5	Γ_1	Γ_4	Γ_3
Γ_5	25.00	0	7.00	8.00
Γ_1		0	13.33	0
Γ_4			1.0	18.67
Γ_3				0

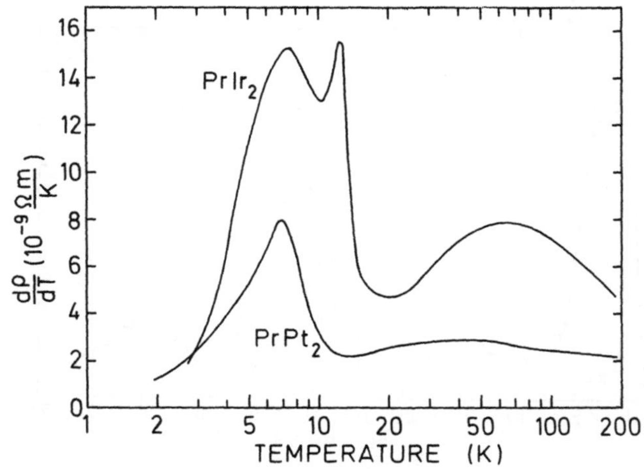

Fig. 4 Temperature derivative of the electrical resistivity $\frac{d\rho}{dT}$, for $PrPt_2$ and $PrIr_2$ versus temperature.

paramagnetic state. They obtained $x = 0.75 \pm 0.02$ and $W = -0.315 \pm 0.01$ meV, which again points to a Γ_3 ground state.

Our own experiments on powders of PrX_2 (X = Pt, Rh, Ru, Ir, Ni) compounds were performed on a triple axis spectrometer at the reactor Saphir in Würenlingen. A doubly bent graphite monochromator and a horizontally bent graphite analyser[33] were used. By adjusting the horizontal curvature to the appropriate neutron energy, the intensity could be increased by an order of magnitude. The analyser energy was kept fixed at 13.7 meV, a pyrolitic filter was used to reduce higher-order contamination. The measurements were carried out at constant Q in the neutron energy-loss configuration at various temperatures. Typical results on $PrPt_2$ are shown in Fig. 5 and will now be discussed briefly.

In the spectrum measured at 15 K, two peaks are clearly visible at energies of $\simeq$ 3.5 meV and 15 meV. By studying both the temperature- and Q-dependence, it could be ascertained that these peaks do not originate from phonon scattering. The assignment of CEF transitions was further confirmed by the absence of these peaks in the energy spectra obtained for the isomorphous diamagnetic lanthanum compound $LaPt_2$.

For a system of N non-interacting ions the differential neutron scattering cross section for CEF transitions is given by[34,35] where F(Q) is the magnetic form factor, k_o and k' are the wavenumbers of the incoming and outgoing neutrons, ρ_n^o denotes the population of

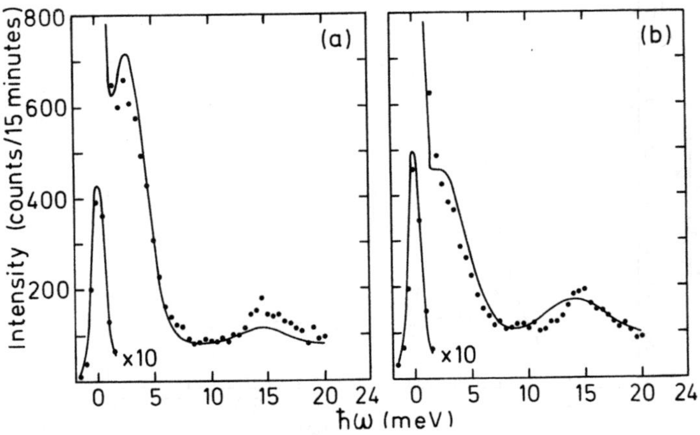

Fig. 5 Energy spectra of neutrons scattered from polycrystalline $PrPt_2$ in the paramagnetic state. (a) T = 15 K. (b) T = 40 K. The curves are the results of the least-squares fitting procedure as explained in the text.

Γ_3 as the ground level can explain the data equally well. In the spectrum of $PrIr_2$, measured at 15 K, only a single transition is observed at around 22 meV, from which we conclude x = 0.86, W < 0. This puts $PrIr_2$ quite close to a point of level crossing in the LLW diagram. The large neutron absorption cross-section of Ir (440 barn) makes data collection rather cumbersome, which resulted in a much more restricted set of experiments, as compared to the other compounds. As a consequence any conclusion about $PrIr_2$ should be regarded with some caution. We also performed measurements in the ordered state on these four compounds, which in general support the above conclusions.

For $PrNi_2$ an enhanced intensity is seen up to ≃ 15 meV, with a peak clearly visible at about 3 meV. This observation is in agreement with unpublished neutron scattering measurements performed by Knorr[37] in the energy range 0 - 6 meV. No set of parameters x and W could fit the data unequivocally, but the best fit was obtained for x = 0.62 and W = -0.19 meV, implying a Γ_3 ground state. Discrepancies between theory and measurements are probably due to dispersion effects or to a non-localized 4f-shell. A summary of the results obtained with inelastic neutron scattering is given in Table II and Fig. 1.

In addition, neutron diffraction experiments have been performed to determine the values for the magnetic moments in the ordered state. These may be compared with moments calculated in a molecular field approximation, on the basis of LLW crystal field states and effective fields derived from the observed transition temperatures. For $PrPt_2$, $PrRu_2$ and $PrIr_2$ the data confirm the crystal field parameters. For $PrRh_2$ the moment measured is just in between the values calculated

Table II. Lattice constants, transition temperatures, LLW parameters and the ground state of the PrX_2 compounds. For $PrRh_2$, two possible sets are given, as outlined in the text.

	a(Å)	T_c(K)	x	W(meV)	ground state
$PrCo_2$	7.312	39.3	unknown	unknown	unknown
$PrNi_2$	7.285	0.27±0.02	0.62±0.02	-0.19±0.1	Γ_3
$PrRh_2$	7.582	7.9 ±0.5	0.93±0.02	-0.35±0.02	Γ_1
			0.75±0.02	-0.33±0.02	Γ_3
$PrIr_2$	7.621	11.2 ±0.5	0.86±0.02	-0.57±0.1	$\Gamma_1\Gamma_3\Gamma_4$
$PrPt_2$	7.709	7.7 ±0.5	0.93±0.02	-0.38±0.01	Γ_1
$PrRu_2$	7.624	33.9 ±0.5	0.68±0.02	-0.33±0.01	Γ_3
$PrAl_2$ [a)]	8.025	33	0.75±0.02	-0.32±0.01	Γ_3
$PrMg_2$ [b)]	8.696	9.6	0.671	-0.345	Γ_3

[a]Ref. 32 [b]Ref. 30

on the basis of the two level schemes, so that a definitive decision between these two possibilities cannot be made yet.

In general the values for the moments measured with neutron diffraction agree reasonably well with those derived from hyperfine specific heat measurements[38]. However, for $PrPt_2$ a difference of a factor two is found, for which an explanation remains to be provided.

DISCUSSION

It has been shown that for the PrX_2 compounds discussed above, the crystal field can be described by LLW parameters $x = +0.8\pm0.2$ and $W < 0$. This region constitutes a very interesting part of the LLW diagram, as it contains a crossing point, the groundstate being the singlet Γ_1 or the non-magnetic doublet Γ_3 for $x > 0.86$ and $x < 0.86$, resp. The available results nicely demonstrate how the magnetic properties are fundamentally influenced by the crystal field. Both single-ion properties and collective behaviour are determined by the character of the ground state and by the relative strength of the crystal field and the magnetic interaction. We note that, with the exception of $PrPt_2$ and possibly $PrRh_2$, all compounds of the series have the Γ_3 doublet as the ground state. The broad maxima observed below T_c in the specific heat for some of the compounds are attributed to the presence of the Γ_3 non-magnetic ground state. One may visualize that this doublet is split by the molecular field at temperatures far enough below T_c (the Γ_3 is split only to 2nd order by a magnetic field). By the splitting process the entropy of the doublet may gradually be dissipated, yielding a Schottky-type specific heat anomaly. The properties of $PrIr_2$ may depend largely on the fact that it is near a point of sixfold degeneracy. The influence of quadrupolar interactions has already been established for $PrMg_2$ and may also play a role in explaining the properties of the Γ_3 ground state compounds. At present we are performing model calculations on the basis of the derived crystal field schemes, in order to explain the magnetic and thermal properties more quantitatively.

Another interesting problem would be the theoretical explanation of the variation in the crystal field parameters, as measured in this series of compounds. Bearing in mind the variation of the lattice constant through the series, one expects that the point-charge model will be (as usual) a poor approximation,and that the band structure needs also to be taken into account.

ACKNOWLEDGEMENT

We would like to thank J.C.M. van Dongen, G.J. Nieuwenhuys and J.A. Mydosh at Leiden, for participating in several experiments, and also W. Bührer and P. Fischer at Würenlingen for their assistance and collaboration. This investigation is part of the research program of the "Stichting F.O.M." with financial support from "Z.W.O.".

REFERENCES

1. K.R. Lea, M.J.M. Leask and W.P. Wolf, J. Phys. Chem. Solids 23, 1381 (1962)
2. G.T. Trammell, J. Appl. Phys. 31, 362S (1960); Phys. Rev. 131, 932 (1963)
3. B. Bleaney, Proc. Roy. Soc.(London) A 276, 19 (1963)
4. B.R. Cooper, Phys. Rev. 163, 444 (1967)
5. Y.L. Wang and B.R. Cooper, Phys. Rev. 172, 539 (1968); Phys. Rev. 185, 696 (1969)
6. P.M. Levy, P. Morin and D. Schmitt, Phys. Rev. Lett. 42, 1417(1979)
7. D.K. Ray and J. Sivardière, Phys. Rev. B 18, 1401 (1978)
8. K. Andres, Phys. Rev. B 7, 4295 (1973)
9. T. Murao, J. Phys. Soc. Jap. 46, 40 (1979)
10. J. Babcock, J. Kiely, T. Manley and W. Weyhmann, Phys. Rev. Lett. 43, 380 (1979)
11. M. Kubota, Thesis, Jülich (1980)

12. K. Andres and S. Darack, Physica 86-88 B, 1071 (1976)
13. K.H.J. Buschow, R.C. Sherwood, F.S.L. Hsu and K. Knorr, J. Appl. Phys. 49, 1510 (1978)
14. K.H.J. Buschow, J.F. Fast, A.M. van Diepen and H.W. de Wijn, Phys. Stat. Sol. 24, 715 (1967)
15. K.H. Mader, E. Segal and W.E. Wallace, J. Phys. Chem. Solids 30, 1 (1969)
16. C. Deenadas, A.W. Thompson, R.S. Craig and W.E. Wallace, J. Phys. Chem. Solids 32, 1853 (1971)
17. H.-G. Purwins, E. Walker, B. Barbara, M.F. Rossignol and P. Bak, J. Phys. C 7, 3573 (1974)
18. P. Bak, Risø Report no. 312 (1974)
19. C. Deenadas, R.S. Craig, N. Marzouk and W.E. Wallace, J. Solid State Chem. 4, 1 (1979)
20. L.J. de Jongh, J. Bartolomé, F.J.A.M. Greidanus, H. de Groot and H.L. Stipdonk, submitted to the J. Magn. Magn. Mater.
21. J. Schweizer, Phys. Lett. 24 A, 739 (1967)
22. E.A. Skrabek and W.E. Wallace, J. Appl. Phys. 34, 1356 (1963)
23. M.J. McDermott and K.K. Marklund, J. Appl. Phys. 40, 1007 (1969)
24. W.E. Wallace and K.H. Mader, Inorg. Chem. 7, 1627 (1968)
25. W.E. Wallace, Rare Earth Intermetallics, Academic Press (1973)
26. W.E. Wallace, R.S. Craig, A. Thompson, C. Deenadas, M. Dixon, M. Aoyagi and N. Marzouk, Les Elements des Terres Rares, Coll. Int. C.N.R.S. No. 180, 427 (1970)
27. H. Mori, T. Fujita, T. Satoh and T. Ohtsuka, Phys. Lett. 79, 121 (1980)
28. F.J.A.M. Greidanus, L.J. de Jongh, W.J. Huiskamp and K.H.J. Buschow, submitted to Physica
29. J.C.M. van Dongen, H.W.M. van der Linden, F.J.A.M. Greidanus, G.J. Nieuwenhuys, J.A. Mydosh and K.H.J. Buschow, J. Magn. Magn. Mater. 15-18, 1245 (1980)
30. A. Loidl, K. Knorr, M. Müllner and K.H.J. Buschow, J. Appl. Phys. 53, 1433 (1981)
31. H.-G. Purwins, W.J.L. Buyers, T.M. Holden and E.C. Svensson, AIP conference Proc., vol. 29, 259 (1975)
32. Th. Frauenheim, W. Matz and G. Feller, Sol. State Comm. 29, 805 (1979)
33. W. Bührer, R. Bührer, A. Isacson, M. Kock and R. Thut, Nucl. Inst. and Meth. 179, 259 (1981)
34. G.T. Trammell, Phys. Rev. 92, 1387 (1953)
35. P.G. de Gennes, Magnetism 3, 115 (1963) (New York and London; Academic Press)
36. A. Furrer and H. Heer, Phys. Rev. Lett. 31, 1350 (1973)
37. K. Knorr, private communication to K.H.J. Buschow
38. F.J.A.M. Greidanus, L.J. de Jongh, W.J. Huiskamp and K.H.J. Buschow, J. Magn. Magn. Mater. 15-18, 1231 (1980)

COMMENTS

COOPER: Is there evidence for nuclear participation in the ordering in the $PrNi_2$ material which has a low ordering temperature? Knight shift measurements would be valuable in this regard.

GREIDANUS: As far as I know Knight shift measurements have not been made on the PrX_2 compounds. These kinds of measurements would be very useful.

ROSSAT-MIGNOD: What is the origin of a large bump in the specific heat of $PrNi_2$?

GREIDANUS: This bump is of nuclear origin and appears below the transition temperature (T_c=0.27 K) as measured with susceptibility.

McEWEN: Knight shift studies have been made on $PrNi_5$ by D.L. Williams et al. (Simon Fraser University, Canada), so there is a good possibility of measuring this effect in $PrNi_2$. But it is necessary to use finely powdered samples.

LÜTHI: Why are there no structural transitions in Laves phase compounds with Γ_3 ground states? Why is the exchange interaction bigger than quadrupolar effects? In addition, I have a remark: in Laves phase compounds one has a Γ_2^g-optic phonon at the Γ-point. One should observe strong magnetoelastic effects of this optic phonon in PrX_2 compounds.

GREIDANUS: It might very well be that the magnetic phase transitions are accompanied by structural transitions, caused by quadrupolar interactions in these Laves phase compounds. Why the quadrupolar interaction is not much larger than the exchange I do not know.

PIERRE: (Comment after a remark of Dr. B. Luthi) Two cubic Pr compounds, $PrMg_3$ and $PrInAg_2$ have a Γ_3 ground state and do not order. Neither have a Jahn-Teller effect until 1.5 K.

PRESSURE INDUCED CHANGES IN THE MAGNETISM OF CRYSTAL FIELD SPLIT SYSTEMS

Robert P. Guertin

Physics Department
Tufts University
Medford, MA 02155, U.S.A.

INTRODUCTION

Crystalline electric field (CEF) effects in f-band materials are often most dramatically evident in that class of systems where the lowest lying sublevel is a singlet. Although this ground state is necessarily non-magnetic, i.e. $\langle 1|J_z|1\rangle = 0$, mixing of this state with excited states within the Hund's rule multiplet is often strong enough to magnetically polarize the ground state, i.e. $\langle 1'|J_z|1'\rangle \neq 0$, where $|1'\rangle$ represents the perturbed ground state[1,2]. This ground state moment can couple with others, giving rise to magnetic order for the system. Generally, the coupling is weak, and these systems (called induced moment systems) tend to order at rather low temperatures. Accompanying the low ordering temperature is a rather small spontaneous moment. In addition the low temperature magnetization falls far short of the maximum saturated free ion value, even in high magnetic fields. In singlet systems which can develop quadrupole moments through an overcritical admixture of ground and excited states, structural transitions can occur. In addition, for metallic systems, aspherical Coulomb scattering provides an additional avenue for quadrupolar coupling. Ott[3] has given a review of these effects, which represent structural analogues to the type of phenomena discussed here.

Hydrostatic pressure and uniaxial stress are expected to be useful diagnostic perturbations for the study of induced moment systems because the rather tenuous ground state moment is likely to be altered by changes in lattice constant. The CEF splittings themselves are expected to be strong functions of lattice constant. For example, in most cubic metallic compounds, the fourth order term B_4, in the CEF Hamiltonian[4], dominates. Consequently, to first approximation, the CEF interaction strength varies as the inverse fifth power of the distance between atoms, $B_4 \sim \langle r^4 \rangle / a^5$. Thus for a 1% change in lattice constant (attained with a pressure of 10 kbar in many rare earth compounds), a 5% increase in CEF interaction strength should result.

Of all the 4f elements, Pr, Tb and Tm are most likely to show CEF-singlet behavior. We will be restricted in this review to Pr-based materials. Pr^{3+}, which has the $4f^2$ configuration, has the J=4 Hund's rule ground state, and a typical CEF level sequence is shown in Fig. 1, which shows the levels for PrSb, which is a non-ordered van Vleck paramagnet[5]. (An unexpected and most likely fortuitous result of this early work was that the observed splittings, observed with inelastic neutron scattering, are closely predicted by a point charge model if Z = -2e is assumed for neighboring charges).

With this background we begin discussion of the Pr-based compounds which fall into four general classes of magnetic behavior, considered more or less in order of complexity as follows: paramagnets, ferromagnets, antiferromagnets and spin glasses. This is followed by a discussion of recent work involving a search for "giant" moments of 4f ions in nearly critical Pr-based singlet systems.

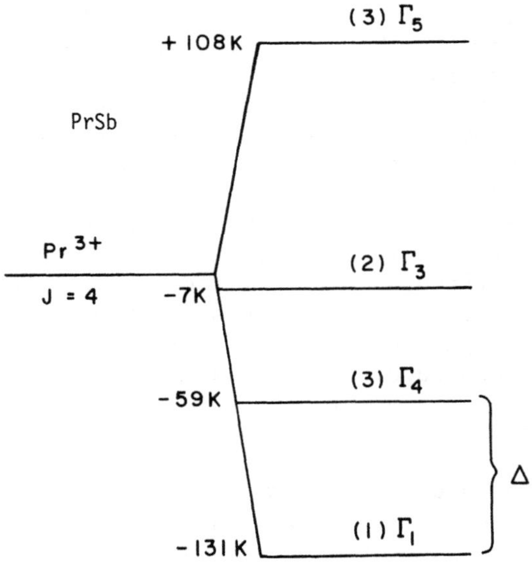

Figure 1. Crystal field levels for PrSb.

SINGLET SYSTEM CLASSES AND PRESSURE EFFECTS

Paramagnets

The Pr- (and Tm-) monopnictides are all well-studied singlet ground state, van Vleck paramagnets. Their measured susceptibility is somewhat larger than that calculated from the CEF-only value, implying an exchange enhancement. The wavelength dependent susceptibility, $\chi(q)$, within the molecular field approximation is given by[6]:

$$\chi(\vec{q}) = \frac{2g^2\mu_B^2\alpha^2}{\Delta-4\alpha^2 J(\vec{q})} \qquad (1)$$

where $J(\vec{q})$ is the Fourier transform of the exchange and α is the matrix element which connects the only thermally occupied state, the Γ_1 singlet, with the triplet state Γ_4. Divergence of $\chi(\vec{q})$ can occur if, for example, the Γ_1-Γ_4 splitting, Δ, becomes small enough.

It has been seven years since the first observations of anomalous behavior under pressure in the low temperature magnetism of Pr- (and Tm-) monopnictides[7,8]. The low temperature susceptibility was found to increase with decreasing lattice constant, not decrease as expected from rather general considerations. (The static magnetic properties of these systems are well accounted for by CEF-only behavior. Therefore, one expects an increase in Δ, and the concomitant decrease in $\chi\sim\Delta^{-1}$, as neighboring ions move nearer the CEF-split ion.) Following thermal expansion measurements[9], the direct observation was made of the decrease under pressure of Δ (and thus the entire level scheme such as that of Fig. 1) using inelastic neutron scattering[10]. Dispersion in the magnetic exciton spectrum and degeneracy lifting in the Γ_4 state of PrSb implies anisotropic exchange interactions. A striking pressure induced paramagnetic to Type I antiferromagnetic transition was found at 30 kbar in PrSb by McWhan et al.[11]. Based on the pressure dependence of Δ, the critical pressure for the divergence of Eq. (1) is about 80 kbar[7] for PrSb. The lower value found experimentally illustrates the breakdown of the molecular field approximation as the level scheme (Fig. 1) begins to close up. In addition, it was found that the X_4 exciton softens substantially but fails to go to zero at and above the critical pressure. The observation of pressure induced antiferromagnetism in PrSb (and presumably in the other Pr-monopnictides) represents an exciting end to the trend in these systems of the reduction of CEF interaction strength with decreasing lattice constant. Devine[12] has advanced a theory based on charge transfer mechanisms to explain the pressure dependence of the CEF parameter in the Pr- and Tm-monopnictides.

Ferromagnets

As discussed above, in magnetically ordered singlet ground state systems, even in the absence of external perturbations such as pressure or magnetic field, the exchange induced mixing of excited CEF levels with the ground state level is sufficient to cause magnetic polarization of the ground state. In ferromagnets the q=0 susceptibility diverges, and such systems have rather low ordering temperatures, T_C, and low spontaneous moments, σ_o.

The magnetism of three Pr-based singlet ground state ferromagnets have been measured as a function of pressure. These are $Pr_3T\ell$ (T_C=12K, σ_o=0.22 σ_s)[13], Pr_3S_4 (T_C=2.3K, σ_o=0.15 σ_s)[14], and Pr_3Se_4 (T_C=15K, σ_o=0.31 σ_s)[14,15] where σ_s (=3.2μ_B) is the maximum free ion moment for the Hund's rule ground state of Pr^{3+}. In Fig. 2 we show the pressure dependence of T_C for each of these ferromagnets. For all data, T_C was measured at very low applied magnetic fields, using the high pressure dc magnetometry techinques described elsewhere[16].

The Γ_1-Γ_4 splitting, Δ, in $Pr_3T\ell$ is 76K[17], and the observed response of T_C to pressure[18], dT_C/dP = -1.5K/kbar, is consistent with the increase of CEF splittings as atomic distances are reduced. For all the Pr-based ferromagnets studied, this is the case, and it

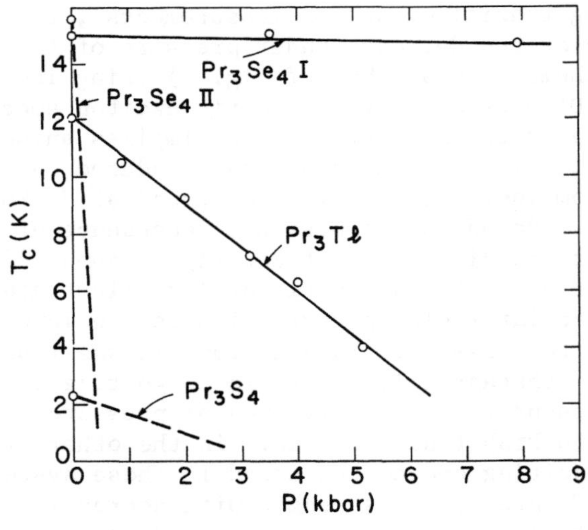

Fig. 2. Pressure dependence of the Curie temperatures for three induced ferromagnets. Pr_3Se_4 shows two-component like behavior, one component being reduced very strongly.

is opposite to the trend noted for the Pr-monopnictides. For Pr_3Tl, substantial softening of the magnetic exciton modes was found not to occur near the critical pressure for induced magnetism[19]. A molecular field treatment, taking into account the pseudo-tetragonal distortion in Pr_3Tl was developed to explain the gross features of this system. Jafarey et al.[20] developed a high temperature series expansion that accounts well for the relation between T_C and σ_0 as pressure reduces both quantities to zero.

For Pr_3S_4 (T_C=2.3K), the results are quite similar to those for Pr_3Tl in that dT_C/dP is large and negative. (We note that for both systems only polycrystalline samples were available.) For Pr_3S_4 the ferromagnetic state is barely overcritical and only small pressures are sufficient to drive it into the enhanced paramagnetic state. This is illustrated in Fig. 3 where we see pressure induced loss of hysteresis in the magnetization of Pr_3S_4 at 1.40K. The ease with which this system can be driven undercritical will be exploited in connection with the Gd moment studies described below.

A particularly complex and interesting example of induced moment ferromagnetism is represented by Pr_3Se_4. The crystal structure for this system (and that of Pr_3S_4) is that of cubic Th_3P_4. Like other light rare earth chalcogenides with this structure, a

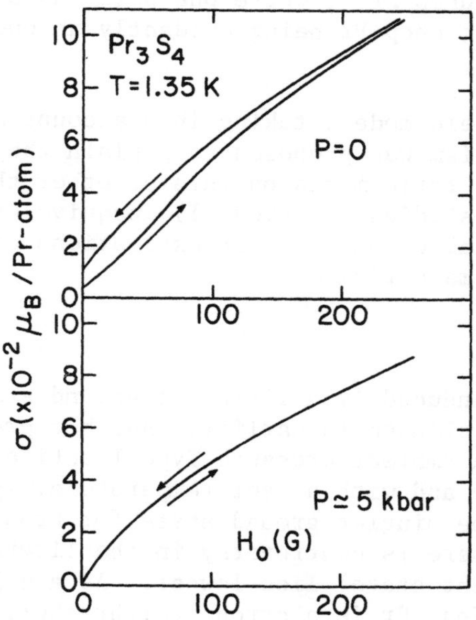

Figure 3. Pressure-induced loss of hysteresis in Pr_3S_4 (T_C = 2.3K).

structural phase transition occurs around 40K whose origin is apparently unrelated to the unfilled 4f shell of the rare earths since it occurs also for La_3Se_4 and La_3S_4[21]. Despite overall cubic symmetry, the local Pr-site symmetry is quite low, leading to a CEF-level splitting involving five singlets and two doublets. Evidence exists for a singlet-first excited state splitting of about 50K[22]. A unique feature of magnetism in this crystal structure is that there are two Pr-sites which, although equivalent in the unstressed, zero magnetic field case, are inequivalent in the presence of an external field (or internal molecular field) and/or a tetragonal [100] distortion.

The low and high field (to 5T) magnetic properties of a single crystal Pr_3Se_4 sample were measured along the principal crystallographic orientations in pressures up to 9.4 kbar[14]. Several pieces of experimental evidence indicate two-component magnetism in this system, thus suggesting a connection with the inequivalence of Pr sites. (Similar results might be observed in Pr_3S_4 were single crystals available.) First, as had been noted previously with ac susceptibility[21], low field dc magnetization studies show two closely spaced steps in the low field σ vs T curve[23]. Secondly, quite small hydrostatic pressures (P<1 kbar) are capable of suppressing one of these transitions, (see dashed line in Fig. 2), and this effect is reversible with pressure. Thirdly, these low pressures also reduce the moderate field ($0 \leq H_o \leq 20$ kOe) magnetization by about 50%, but for higher pressures σ and the other T_C tend to remain unchanged by pressure. Two-site magnetism has been discussed in connection with dhcp Pr[24], where one site has cubic, the other hexagonal, symmetry; dhcp Pr being evidently an undercritical but enhanced paramagnet.

A molecular field model, taking into account all CEF-levels and the two site magnetism was proposed to explain the properties of Pr_3Se_4[25]. However, refinements on this or other theories await neutron scattering studies to reveal 1) inequivalent Pr-site moment and most likely parallel sets of CEF excitations and 2) the kind of ferro- (or ferri-) magnetism.

Antiferromagnets

The pressure induced Type I singlet ground state antiferromagnetism of PrSb was alluded to earlier. Another pressure study is that on $PrSn_3$[26], an ambient pressure Type I antiferromagnet having the Cu_3Au structure and with a Neel temperature, T_N=8.6K. However, unlike for PrSb, the singlet ground state for $PrSn_3$ is not well established, and there is controversy in the literature as to whether Γ_1 or Γ_5 (the triplet state) lies lowest. As a dilute impurity in superconducting $LaSn_3$, Pr is a strong pairbreaker, suppressing rapidly the superconducting transition temperature. Pressure studies[27] reveal CEF singlet-like behavior for the Pr moment, at least in the low concentration range.

Pressure dependent magnetization shows a rapid increase of T_N with pressure, $dT_N/dP=+0.23K/kbar$, the positive pressure derivative being predicted by thermal expansion data[3]. This is illustrated in Fig. 4. A parallel experiment was performed on the Kramers ion-based antiferromagnet $NdSn_3$ ($T_N=5.2K$), which showed $dT_N/dP=+0.05K/kbar$, significantly less than for $PrSn_3$. An important observation in the case of $PrSn_3$ is that no decrease in the σ was found with increasing pressure. Were the magnetism of $PrSn_3$ to be strongly influenced by the Kondo mechanism, a supposition based on dilute Pr in $LaSn_3$ studies, a decrease might be expected. (This is seen, for example, in $CeA\ell_2$[28].) The CEF splittings in $PrSn_3$ are probably quite small[21,29] and so T_N and its pressure dependence may be dominated by the interion exchange. However, the vast amount of data on this and other $RESn_3$ compounds and their dilute impurity superconducting La-based counterparts suggest that isotropic exchange is not nearly sufficient to account for all the data and that other exchange and Coulomb mechanisms may be important. DeLong[30] has provided a recent review of the $RESn_3$ systems.

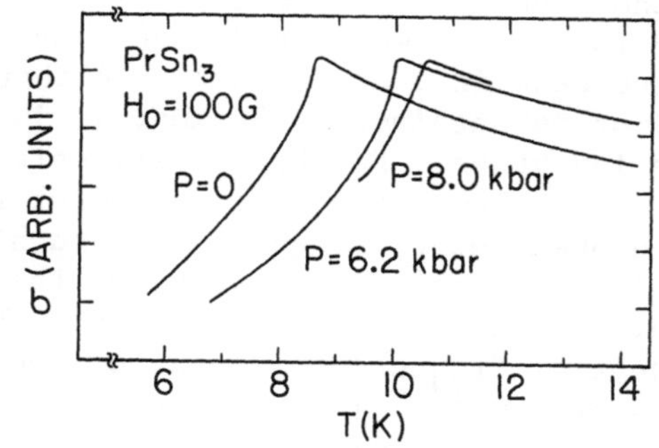

Fig. 4. Pressure dependence of the low field magnetization of antiferromagnetic $PrSn_3$.

Spin Glasses

Returning to the discussion in §A above, we note that divergence of $\chi(\vec{q})$ (see Eq. 1) in paramagnetic systems might also come about through increased exchange as well as decreased Δ. Any influence which, for example, causes an increase in the conduction electron concentration or band overlap might be expected to raise $J(\vec{q})$, which in Eq. (1) represents the Fourier transform of the interion exchange. For PrSb, an initial pressure induced reduction of Δ gave rise to Type I antiferromagnetic order. For PrP, isomorphic to PrSb, the same trend occurs, the CEF level structure for PrP being similar to that in Fig. 1 but with the vertical scale expanded so that Δ=133K[5]. In the system described below, PrP_y ($y \lesssim 1$), it is not clear which mechanism gives rise to the induced moments, although recent inelastic neutron scattering studies[31] suggest Δ for off-stoichiometric, spin glass samples is smaller than in stoichiometric samples. Westerholt and Methfessel[32] first noted that in off-stoichiometric PrP_y, ($1.00 \geq y \geq 0.85$), where the NaCl structure is maintained, spin glass-like behavior was observed, clearly indicating a moment on the Pr sites (an induced moment in view of the neutron results). Hasanain et al.,[33] who performed several dc and ac magnetic measurements, both at ambient and high pressures, point out that most likely only some Pr sites have moments and that these may vary throughout the sample due to variations in local Pr-site symmetry and other environmental effects. This then constitutes an induced moment spin-glass, a theoretical concept introduced by Sherrington[34], who adapts the Edwards-Anderson spin glass model to the singlet ground state case.

In Fig. 5 we show the high field magnetization of two PrP_y samples exhibiting spin glass transitions at 9K and 10K (see the insert of Fig. 5 for the susceptibility data). The σ vs H_o data for stoichiometric PrP is also shown, the dashed line indicating the pressure induced increase in σ at 10 kbar. Although substantial magnetization is attained by 50 kOe, it is far from the $3.2\mu_B$ of the Pr^{3+} free ion value. Further experiments on this system include the detection of low field time dependent remanence effects[35] and a related phenomenon, frequency dependent ac susceptibility measurements which show T_f^{-1}, the inverse of the spin glass transition temperature, varying as log f for $f \gtrsim 1$ Hz. However, for $f \lesssim 1$ Hz, T_f is independent of frequency[36], behavior recently noted in two other spin glass systems[37]. The relevant pressure experiment performed in PrP_y was a measurement of dT_f/dP using a very low frequency (f=0.03 Hz) ac susceptibility technique. The results showed $dT_f/dP \simeq +0.1$K/kbar. This is consistent with the pressure induced decrease in Δ already noted for stoichiometric PrP, T_f being a function of J_{ex} and Δ. The origin of magnetic moments in PrP_y is entirely different from that for more conventional spin glass systems, which rely on the introduction of well-established magnetic impurities into a non-magnetic host. Due to the presence of voids in the P-sublattice, disturbances are introduced in the CEF level structure and magnetic moments are induced at some Pr sites. (These moments presumably vary from site to site.) Accompanying the presence of voids in the P-sublattice is a strong increase in conduction electron concentration[38]. Despite the dissimilar origin of the moments, the

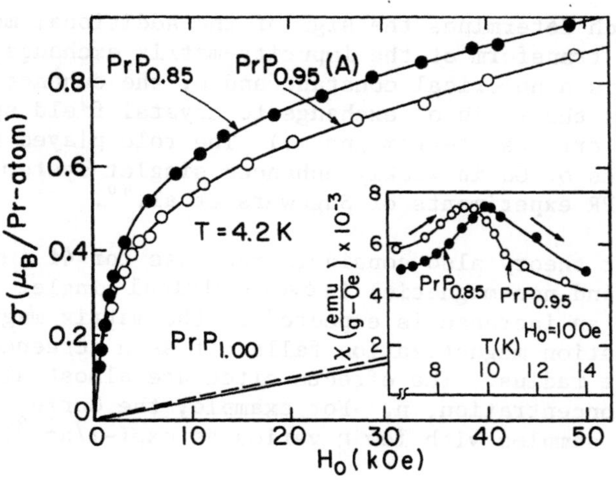

Fig. 5. High field magnetization of two PrP_y samples showing spin glass transitions (see insert) and stoichiometric $PrP_{1.00}$.

macroscopic magnetic properties of PrP_y are quite similar to spin glasses with a random distribution of uniform moments.

MAGNETIC MOMENT OF Gd IN SINGLET SYSTEMS

The work discussed in this section was motivated by the theoretical work of Shender[39], who considered the possibility of observing "giant" moments of well-localized 4f ions (e.g. Gd) in highly enhanced singlet ground state (e.g. Pr-based) systems. The theory considers the change of state of an enhanced or over-critical singlet matrix upon the addition of a small concentration of impurities, both magnetic and non-magnetic. Because a strong magnetic polarization for under-critical matrices can occur near magnetic impurities, this effect is analogous to the "giant moment" Fe in Pd, which involves a highly enhanced d-band system. Indeed, a sufficient number of giant moments could easily "trigger" ferromagnetism through overlap in an otherwise under-critical system.

For singlet matrices where the enhancement of CEF-only behavior is relatively small, the effective moment of, for example, a Gd moment is:

$$\mu_{eff} = \mu_{Gd} \left[1 + \frac{\alpha\mu_h}{\mu_{Gd}} \frac{U(0)}{\Delta(1-\eta)}\right], \tag{2}$$

where U(0), which determines the sign of the additional moment, is the q=0 Fourier transform of the impurity-matrix exchange, μ_h is the host moment, α is a numerical constant and η, the enhancement parameter, signifies the ratio of exchange to crystal field splitting. (η=1 for barely critical ferromagnets.) The role played by U(0) in the spin dynamics of Gd in weakly enhanced singlet systems was explored in the EPR experiments of Sugawara et al.[40].

The Shender theory also considers the case for 4f impurities (both magnetic and non-magnetic) in over-critical singlet ferromagnets. Here, an increase is expected in the matrix magnetization and the perturbation magnetization falls off as a screened Coulomb law with a large radius. The effects cited are almost all non-linear with impurity concentration, n. For example, the Curie temperature for small n for samples with low η varies as $\exp(-1/n^{1/3})$.

The extreme sensitivity of nearly critical singlet matrices to perturbation by 4f impurity addition complements their sensitivity to hydrostatic pressure. The combination of these two perturbations is therefore likely to cause extreme changes in the magnetism of these materials. In view of the discussion in §IIB above, the two most likely candidates for nearly critical singlet host matrices are Pr_3S_4 at P>8 kbar and Pr_3S_4 at $P \gtrsim 1$ kbar. A further possibility is dhcp Pr at P=0[24] but to date this has not been pursued. Pr_3S_4 is particularly appealing because of the ease with which it can be driven subcritical. However, obtaining samples with reproducible properties have plagued work in this system. In Fig. 6 we show the low field susceptibility of Pr_3S_4 with its indicated 2.3K ferromagnetic transition, and a 2% Gd-doped sample. The latter shows a spin-glass-like transition at 9K, below which time dependent remanence was detected. No anomaly was observed near 9K in the temperature dependence of the electrical resistivity (which is quite high, $\rho \sim 300$ μΩ-cm). The dashed line in Fig. 6 shows the effect on the spin glass transition of hydrostatic pressure, the most obvious feature being the large suppression of the susceptibility. At 7 kbar the host, Pr_3S_4, is well under-critical, and the Gd moment given by Eq. 2 may be appropriate. The spin glass freezing temperature, T_f, is lowered, and along with the susceptibility suppression this is consistent with the rapid destruction of ferromagnetism by pressure in the host, Pr_3S_4.

In Fig. 7 we see the low temperature, high field magnetization for the samples discussed above, plus the 7 kbar σ data for the 2% Gd doped sample. The relatively small addition to σ of Pr_3S_4 by the Gd suggests a Gd moment which may be smaller than 7 μ_B, the free ion value. This is supported by 12 kOe magnetic susceptibility studies for $4.2K \leq T \leq 300K$ which suggests a Gd moment of about 5 μ_B. High field, high pressure magnetization studies yield an interesting result: although σ for Pr_3S_4 is suppressed uniformly ($d\sigma/dP$=-0.7%/kbar at 50 kOe), the suppression of σ for the 2% Gd sample is much

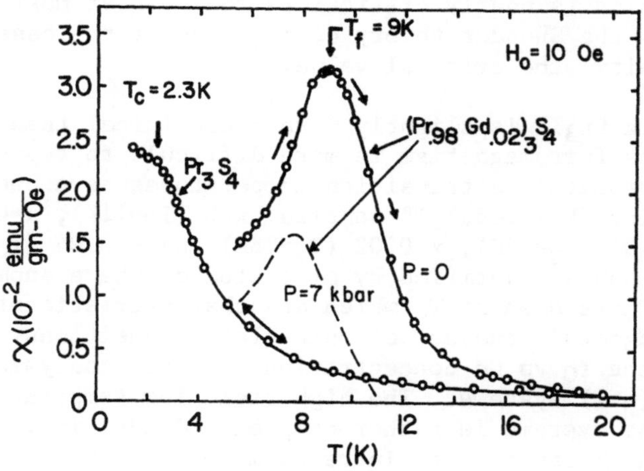

Fig. 6. Low field susceptibility of ferromagnetic Pr_3S_4 and of a 2 at % Gd-doped sample. Dashed line indicates pressure dependence of the spin glass-like transition of the latter.

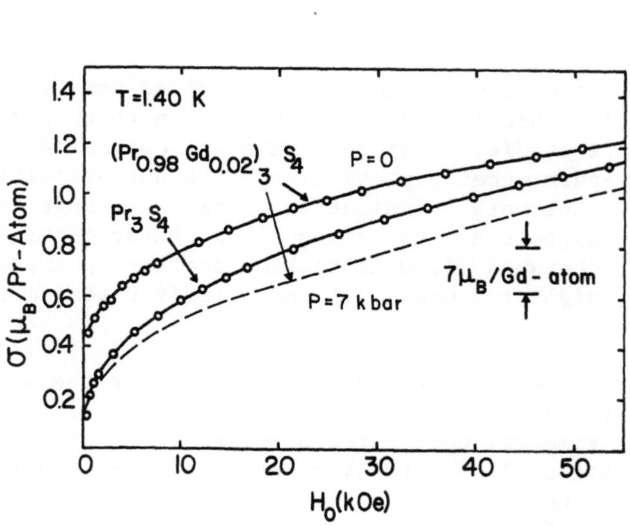

Fig. 7. High field magnetization of two samples of Fig. 6. Note possible decreased Gd moment at high fields and strong pressure dependence of the Gd doped sample.

larger and somewhat non-linear ($\Delta\sigma$=-20% at 5 kbar and decreases more slowly thereafter). This reflects the non-linearity of the Gd moment in a matrix which is nearly critical (a few kbar at most in Pr_3S_4). Unfortunately, the Shender theory does not cover the case where η is close to unity, the critical value.

The matrix $Pr_3T\ell$ is slightly more overcritical than Pr_3S_4 to the extent that its ferromagnetism is more difficult to suppress by pressure. Ferromagnetic transition temperatures have been detected in the following $(Pr_{1-x}Gd_x)_3T\ell$ samples: x=0 (T_C=11K), x=0.005 (T_C=15K), x=0.01 (T_C=22K), x=0.02 (T_C=28K) and x=0.04 (T_C=34K). These transitions are verified by $\rho(T)$ studies where anomalies in $\rho(T)$ and $d\rho/dT$ are seen at T_C which are characteristic of induced moment ferromagnets[41] and not of spin glass transitions. The non-linearity of the T_C vs Gd concentration suggests analysis in terms of overcritical (η>1) systems. The high field low temperature magnetization in these systems is rather complex: Gd addition, even at the level of 0.5 at % causes σ to increase more rapidly at low H, but then tends to attain lower σ than for $Pr_3T\ell$. The data are suggestive of a moment larger that $7\mu_B$ for Gd, at least for x≃0.01. Pressure effects of T_C for high concentration Gd samples are minimal, probably because these materials are well overcritical. For example, T_C(=28K) for x=0.02 is suppressed by only 1K at ≃5 kbar. A more complete presentation of the data and analysis in terms of the Shender theory will be forthcoming.

CONCLUSIONS

It is hoped that the foregoing has provided some insight into why nearly critical singlet systems have proven to be fertile ground for high pressure studies. Future work might emphasize the nature of the magnetic-paramagnetic phase transition and associated critical exponents. Recent work[42] indicating a valence change in Pr at very high pressures suggests new experimental and theoretical efforts. Using the new techniques allowing neutron scattering experiments to be conducted at high pressures will undoubtedly reveal more about singlet systems.

The author is indebted to many colleagues for the work discussed above, in particular, to S. Foner, S.K. Hasanian and J.E. Crow. Much of the experimental work cited here has taken place at the Francis Bitter National Magnet Laboratory at M.I.T., where the author is a Visiting Scientist. He is grateful to the National Science Foundation for financial support for these studies. (DMR 80-10530).

REFERENCES

1. G.T. Trammell, Phys. Rev. 131, 932 (1963). B. Bleaney, Proc. Roy. Soc. A276, 19 (1963).
2. B.R. Cooper and O. Vogt, Phys. Rev. B1, 1211 (1970).
3. H.R. Ott in "Crystalline Electric Field and Structural Effects in f-Electron Systems", Ed. J.E. Crow, R.P. Guertin and T. W. Mihalisin, Plenum, New York, 1980. Pp. 9-19.
4. K.R. Lea, M.J.M. Leask and W.P. Wolf, J. Phys. Chem. Solids 23, 1381 (1962).
5. K.C. Turberfield, L. Passell, R.J. Birgeneau and E. Bucher, J. Appl. Phys. 42, 1746 (1971). R.G. Birgeneau, E. Bucher, J.P. Maita, L. Passell and K.C. Turberfield, Phys. Rev. B8, 5345 (1973).
6. C. Vettier in Ref. 3. Pp. 23-32.
7. R.P. Guertin, J.E. Crow, L.D. Longinotti, E. Bucher, L. Kupferberg and S. Foner, Phys. Rev. B12, 1005 (1975).
8. H.T. Weaver and J.E. Schirber, AIP Conf. Proc. 24, 49 (1975).
9. H.R. Ott and B. Luthi, Phys. Rev. Letters 36, 600 (1976).
10. C. Vettier, D.B. McWhan, E.J. Blount, and G. Shirane, Phys. Rev. Letters 39, 1028 (1977).
11. D.B. McWhan, C. Vettier and R. Youngblood, Phys. Rev. B20, 4612 (1979).
12. R.A.B. Devine in Ref. 3. Pp. 165-171.
13. K. Andres, E. Bucher, S. Darack and J.P. Maita, Phys. Rev. B6, 2716 (1972).
14. R.P. Guertin, S. Foner, B.R. Cooper and R. Siemann, J. Magn. Magn. Mat. 15-18, 11 (1980).
15. E. Bucher, K. Andres, F.J. diSalvo, J.P. Maita, A.C. Gossard, A.S. Cooper and G.W. Hull, Jr., Phys. Rev. B11, 500 (1975).
16. R.P. Guertin in "High Pressure and Low Temperature Physics", Ed. by C.W. Chu and J.W. Woollam, Plenum, New York 1978. Pp. 97-113.
17. R.J. Birgeneau, J. Als-Nielsen and E. Bucher, Phys. Rev. B6, 2724 (1972).
18. R.P. Guertin, J.E. Crow, F.P. Missell and S. Foner, Phys. Rev. B17, 2183 (1978).
19. J.K. Kjems, M. Nielsen, W.J.L. Buyers and J.E. Crow, J. de Physique C5, 132 (1979).
20. S. Jafarey, Y.-L. Wang and K. Rauchwarger, Phys. Rev. B19, 2724 (1979). D. Yang and P.A. Lindgard, J. de Physique C5, 134 (1979).
21. E. Bucher, J.P. Maita, G.W. Hull, Jr., and L.D. Longinotti, Z. Physik B25, 41 (1976).
22. J.G. Houmann, B.D. Rainford and B. Luthi, unpublished result cited in Ref. 21.
23. R.P. Guertin and S. Foner, Phys. Letters 68A, 116 (1978).
24. J. Jensen, J. de Physique C5, 1 (1979).
25. R. Siemann, B.R. Cooper, R.P. Guertin and S. Foner, in Ref. 3. Pp. 457-64.

26. L.E. DeLong, R.P. Guertin and S. Foner, Solid State Commun. 32, 833 (1979).
27. L.E. DeLong, M.B. Maple, R.W. McCallam, L.D. Woolf, R.N. Shelton and D.C. Johnston, J. Low Temp. Phys. 34, 445 (1979).
28. M.C. Croft, R.P. Guertin, L.C. Kupferberg and R.D. Parks, Phys. Rev. B20, 2073 (1979). B. Barbara, M. Cyrot, C. Lacroix-Lyon-Caen and M.F. Rossignol, J. Phys. Colloq. 40, C5, 340 (1979).
29. P. Lethuillier and J. Chaussy, Phys. Rev. B13, 3132 (1976). P.A. Alekseev, I.P. Sadikov, I.A. Morkova and O.D. Chistyakov, Physica Status Solidi (b) 86, K59 (1978).
30. L.E. DeLong in Proc. of the 14th Rare Earth Res. Conf., Fargo, ND, June 1979.
31. S.M. Shapiro, S.K. Hasanain and R.P. Guertin, (to appear).
32. K. Westerholt and S. Methfessel, Physica 86-88B, 1160 (1977).
33. S.K. Hasanain, R.P. Guertin, K. Westerholt, M. Guyot and S. Foner, to be published in Phys. Rev. B.
34. D. Sherrington, J. Phys. C12, L929 (1979).
35. C.N. Guy, J. Phys. F: Metal Phys. 7, 1505 (1977). Ibid. 8, 1309 (1978).
36. M. Guyot, S. Foner, S.K. Hasanian, R.P. Guertin and K. Westerholt, Phys. Lett. 79A, 339 (1980).
37. N. Bontemps, J.C. Rivoal, M. Billardon, J. Rajchenbach and J. Ferré, J. Appl. Phys. 52, 1760 (1981). A.P. Malozemoff and Y. Imry, (to be published).
38. P. Wachter, E. Kaldis and R. Hauger, Phys. Rev. Letters 40, 1404 (1978).
39. E.F. Shender, J. Phys. C: Solid State Phys. 9, 1713 (1976).
40. K. Sugawara, C.Y. Huang and B.R. Cooper, Phys. Rev. B11, 4455 (1975).
41. P. Bossard, S. Bakanowski, J.E. Crow, T. Mihalisin and W.J.L. Buyers, J. Appl. Phys. 50, 1892 (1979).
42. J. Wittig, Phys. Rev. Letters 46, 1431 (1981).

COMMENTS

GREIDANUS: I wonder whether you have ever tried these frequency dependent susceptibility experiments on stoichiometric compounds? We have seen the same effects as you described in $PrNi_2$ where the hyperfine interaction induces the electronic moment.

GUERTIN: No, we have not tried these experiments on any other system.

STEGLICH: As to the off-stoichiometry PrP systems, Wachter's group has observed in one of the Gd chalcogenides or pnictides a large anomaly in the resistivity, i.e. an increase with decreasing temperature. This was attributed to a magnetic defect structure. Did you see similar effects in your samples at temperatures sufficiently high, so that CEF levels are populated?

GUERTIN: Resistivity has not been performed on these samples.

KELLER: Can the increase of T_c in Pr_3Tl by adding Gd be understood as a mean-field calculation using the same value of the exchange interaction for Gd-Pr as for Pr-Pr?

GUERTIN: This has not been done, but we will try.

FULDE: I would like to make a comment on the problem of "giant" moments. The theory for them was worked out a long time ago by Dr. Mais (doctoral thesis). From this work, people at Karlsruhe (Czjzek, Götz, et al.) looked for these effects by adding Eu impurities to Pr metal. What they found were "dwarf" moments due to the antiferromagnetic coupling of Pr and Eu. What is happening is that the polarization induced in the Pr metal cancels to a large extent the magnetic moment of the Eu.

VAN VLECK PARAMAGNETS IN HIGH MAGNETIC FIELDS

E. Leyarovski*, L. Leyarovska*, Chr. Popov* and N. Iliev†

*Institute of Solid State Physics, Bulgarian Academy of Sciences, 1184 Sofia, Bulgaria
†Institute for Low Temperature and Structure Research Polish Academy of Sciences, Wrocław, Poland

INTRODUCTION

As a result of the crystal-field splitting of the lowest multiplet, the electronic ground state of the even 4f-rare-earth ions in many compounds is a nonmagnetic singlet, the excited states being raised by more than 10 K or even 100 K in energy by the crystalline field. Because of the very strong hyperfine coupling between electrons and nuclei in Van Vleck compounds, there is a profound influence of the magnetic field on both the electronic and nuclear low-lying energy level schemes. Besides inducing a large hyperfine enhanced magnetic moment by admixing the excited electronic states into the singlet ground state (SGS) wave function, an external magnetic field also gives rise to an additional electric field gradient which results from the 4f-charge cloud of the rare-earth ion, no matter what the crystal symmetry is. Even in cubic symmetry, a field-induced nuclear quadrupole splitting will be superimposed on the hyperfine pseudo-Zeeman splitting. Experimental evidence for this and some other effects may come from accurate measurements of the hyperfine specific heat of Van Vleck paramagnets in sufficiently high magnetic fields.

THEORETICAL BACKGROUND

In a paramagnetic of the Van Vleck type with an SGS, the interaction with the nuclear magnetic moments (hyperfine interactions-HFI) are considerably enhanced by the hyperfine (HF) magnectic structure. It is well known[1] that such interactions are the magnetic dipole and exchange interactions, the "psuedo-quadrupole" nuclear interactions, the "psuedo-Zeeman" electron-nuclear interactions, and external magnetic field effects ("pure" electron and nuclear Zeeman splitting).

The non-Kramers ions of a rare earth with an even number of electrons in the nonfilled 4f shell, in particular those which have a stable isotope with an uneven nucleus and a large nuclear moment, e.g. ^{141}Pr, are convenient for the study of such interactions.

An adequate analysis of these systems may ordinarily be achieved by employing a comparatively simple Hamiltonian of the kind[1]:

$$\mathcal{H} = V_c + g_J\mu_B(\vec{H}\cdot\vec{J}) + A_J(\vec{J}\cdot\vec{I}) - g_N\mu_N(\vec{H}\cdot\vec{I}) + \mathcal{H}_Q \tag{1}$$

where V_c is the crystal field (CF) potential which causes the splitting of the lowest-lying ion multiplet and the appearance of an SGS.

Besides the ordinary electron and nuclear Zeeman terms and the nuclear quadrupole interaction $\mathcal{H}_Q$, the Hamiltonian contains the term $A_J(\vec{J}\cdot\vec{I})$, corresponding to the magnetic HFI in an ion with a total angular momentum J, I being the nuclear angular momentum.

Bleaney[2] has examined the general case of a system of two singlet-singlet states ($|G\rangle$ and $|E\rangle$). These appear due to the presence of the crystal field V_c, the energy difference is Δ_z, and the assumption is made that a nonzero matrix element exists ($\alpha_z = \langle E|J_z|G\rangle$. It is assumed also that the external magnetic field H is directed along the z-axis, fixed parallel to each ion by the crystal field of the ligands (purely Ising-type interaction).

If the two states of interest ($|G\rangle$ and $|E\rangle$) have energies of $-\Delta_z/2$ and $+\Delta_z/2$, respectively, in the crystal field V_c, then the energy matrix in a magnetic field H_z will be

$$\begin{matrix} E \\ G \end{matrix}\begin{vmatrix} \Delta_z/2 - g_N\mu_N H_z I_z & \alpha_z(g_J\mu_B H_z + A_J I_z) \\ \alpha_z(g_J\mu_B H_z + A_J I_z) & -\Delta_z/2 - g_N\mu_N H_z I_z \end{vmatrix}. \tag{2}$$

After the diagonalization with the help of

$$\begin{vmatrix} \sin(\theta/2) & -\cos(\theta/2) \\ \cos(\theta/2) & \sin(\theta/2) \end{vmatrix} \tag{3}$$

one obtains the expression

$$W_{1,2} = -g_N\mu_N H_z I_z \pm (\Delta_z/2)(\tan^2\theta + 1)^{1/2} + O[(\alpha_z g_J\mu_B H_z)^2] \tag{4}$$

for the energy eigenvalues. Here $\tan\theta=2\alpha_z(g_J\mu_B H_z+A_J I_z)/\Delta_z$. These formulae contain all the parameters of interest in the problem; the CF elements α_z and Δ_z and the constant A_J of the HFI between the nucleus and the 4f-electron shell of the ion.

Eq. (4) can be recast in the equivalent form

$$\omega_{1,2}=-g_N\mu_N H_z I_z\pm(\Delta_z/2)\tan\theta(1+\cot^2\theta)^{1/2}, \quad (5)$$

which is convenient in the case of a strong magnetic field. We have, evidently, $\cot\theta\to 0$ in the strong field limit, i.e.

$$W_{1,2}=-g_N\mu_N H_z I_z\pm\alpha_z(g_J\mu_B H_z+A_J I_z). \quad (6)$$

The splitting of the nuclear sublevels of the singlet ground state is given by

$$W_N=-g_N\mu_N H_z I_z-\alpha_z A_J I_z. \quad (7)$$

Before reaching this limit, the nuclear Zeeman splitting is described by Eqs. (4) or (5). In the limit of Eqs. (6) and (7) the H_{hf}, induced by the electronic moments and experienced by the nuclei, comes together with the HF splitting. This would correspond to achieving the total electronic magnetic moment of the ground state ($\alpha_z=J$).

We are also interested in the behavior of both the electronic and nuclear moments in a magnetic field H. If the temperature is sufficiently low, so that the population of only the ground state need be taken into account, the z component of the magnetic moment would be given by

$$M_z=-dW/dH_z=g_N\mu_N I_z+\alpha_z g_J\mu_B\tan\theta/(\tan^2\theta+1), \quad (8)$$

(see Eq. (4)). The first term here gives the z-component of the "pure" nuclear moment, while the second contains implicitly both the electronic moment induced in the ground state and the HF enhanced nuclear moment. Evidently, in the strong field case, one must use the exact formula (4) before reaching the "very strong field limit". It is evident also how desirable it is to use experimentally such fields for which Eq. (7) could be applied, since the Zeeman and the HF interactions might then be separated.

Knowledge of the energy distribution of the nuclear sublevels of the singlet ground state may be obtained from heat capacity measurements. In the paramagnetic range, when $W_i \ll k_B T$ the T^{-1} power series turns out to be useful[3]. Essential simplification of the formulae for the magnitudes of interest is achieved when the "energy origin" is chosen so that $\langle W\rangle=0$. The principle terms of the heat capacity expansion are then given by

$$C/R=<(W/k_B)^2>T^{-2}-<(W/k_B)^3>T^{-3}+... \qquad (9)$$

The values $<W^n>$ here are evaluated at constant field, so that $C=C_H$.

After the introduction of an expansion of this kind in the spin Hamiltonian and the selection of the HF terms, the high temperature tail of the nuclear Schottky anomaly may be represented as

$$C/R = \sum_{n=2}^{\infty} c_n T^{-n} \qquad (10)$$

where

$$c_2/R=(1/3)a^2 I(I+1)+(1/45)P^2 I(I+1)(2I-1)(2I+3)$$
$$c_3/R=-(1/15)a^2 PI(I+1)(2I-1)(2I+3)$$
$$c_4/R=-(1/30)a^4 I(I+1)(2I^2+2I+1).$$

Some higher-order terms are omitted. The initial condition assumed is that H be parallel to the symmetry axis of an axially symmetric CF gradient. The magnitude a is equal to $A_J<J_z>$, where $<J_z>$ is the SGS magnetic moment induced by the external field. It is only at full saturation that $a=a_o=A_J J$ holds. This value corresponds to a free ion. In Eq. (10), P is the normal nuclear quadrupole interaction:

$$P=\{(-3eQ<r_q^{-3}><J||\alpha||J>)/[4I(2I-1)]\}(<3J_z^2-J(J+1)>) \qquad (11)$$

in the nomenclature of Elliot and Stevens[5]. The "pseudo-quadrupole" nuclear interaction must be added here too.

Two more kinds of interactions exist which participate in the spin Hamiltonian and should be included in the expansion (10); the dipole-dipole and the indirect exchange interaction between the nuclear moments. Both are enhanced by the HFI and most probably by the external magnetic field. In the expression for C_N they are united in a term which is proportional to T^{-2} also[3]:

$$CT^2/R=(1/2)\{[I(I+1)]/3k_B\}^2 \sum_{j>i} (J_{ij}^N)^2. \qquad (12)$$

If we introduce the mean square effective field $<H_i^2>$, so that

$$(g_N\mu_N)^2<H_i^2>=(1/3)I(I+1) \sum_{j>i} (J_{ij}^N)^2$$

then we come to

$$CT^2/R=(\lambda_N/2)<H_i^2>, \qquad (12a)$$

λ_N being the nuclear Curie constant. The dependence of the nuclear heat capacity on the external magnetic field is contained implicitly in the first term of the expansion (11).

HEAT CAPACITY OF $PrIn_3$

$PrIn_3$ is the ideal system for investigations in strong magnetic fields: (i) the fcc lattice ensures cubic crystal symmetry for Pr^{3+}. One may thus assume that the magnetic HFI is isotropic, while quadrupole effects will be absent except in very strong magnetic fields. (The quadrupole moment of In in $PrIn_3$ needs special consideration[6].) (ii) The cubic CF potential for the 4f-ions, represented in equivalent operators, is particularly simple[7]:

$$H_c = W\{x[O_4/F(4)] + (1-|x|)[O_6/F(6)]\}. \qquad (13)$$

The matrix elements of the operator inside the brackets are tabulated[7], and the energy parameter W for $PrIn_3$ is determined by inelastic neutron scattering experiments[8], so that the scheme of the CF levels is well known[6]. (iii) The strong CF effects guarantee a very large splitting of the lowest-lying multiplet of the Pr^{3+} ion, $^3H_4$. The energy difference between the ground state singlet Γ_1 and the first excited triplet Γ_4 is $\Delta_{14} \sim 105$ K[8,9]. For such a large splitting $PrIn_3$ should be an extraordinarily weakly coupled Van Vleck paramagnet and, according to Murao's definition[10], can be examined in a "nuclear regime" for a wide range of magnetic fields.

The magnetic properties of polycrystalline $PrIn_3$ have been studied in fields up to 93.9 kOe[11]. It turned out that the Van Vleck susceptibility, numerically equal to 2.6×10^{-2} emu/mole, is field independent up to 100 kOe. We also measured the specific heat of a polycrystalline sample of $PrIn_3$ in magnetic fields up to 60 kOe and at temperatures higher than 0.3 K, applying our technique described in Ref. 12. The electron and lattice heat capacities were separated in the higher temperature experiments. The HF nuclear specific heats are given in a long-log scale in Fig. 1. It turned out that all the results can be well fitted to a single curve $C_N \sim T^{-2}$ by plotting C_N vs T/Δ; $\Delta = g_N \mu_N H/k_B$. (The H=0 case is discussed below.) Such a result corresponds to a series of Schottky-anomalies (the high temperature part) related to the system of equidistant nuclear levels split by Δ. Such a "corresponding state pattern" would not obtain if the levels were not equidistant, i.e. if second-order effects such as an actual quadrupole effect in a strong magnetic field or a "pseudo-quadrupole" effect were significant. Our C_N results in fields up to 60 kOe confirm the conclusion of Genicon and Tournier[11] for field-independence of the CF effects in $PrIn_3$ up to 100 kOe. In cases like this it should be possible to confine oneself in the description of the experiment to the first term only in the expansion (10) (the one quadratic in H). Fig. 2 shows that this is the case. The enlarged scale of the

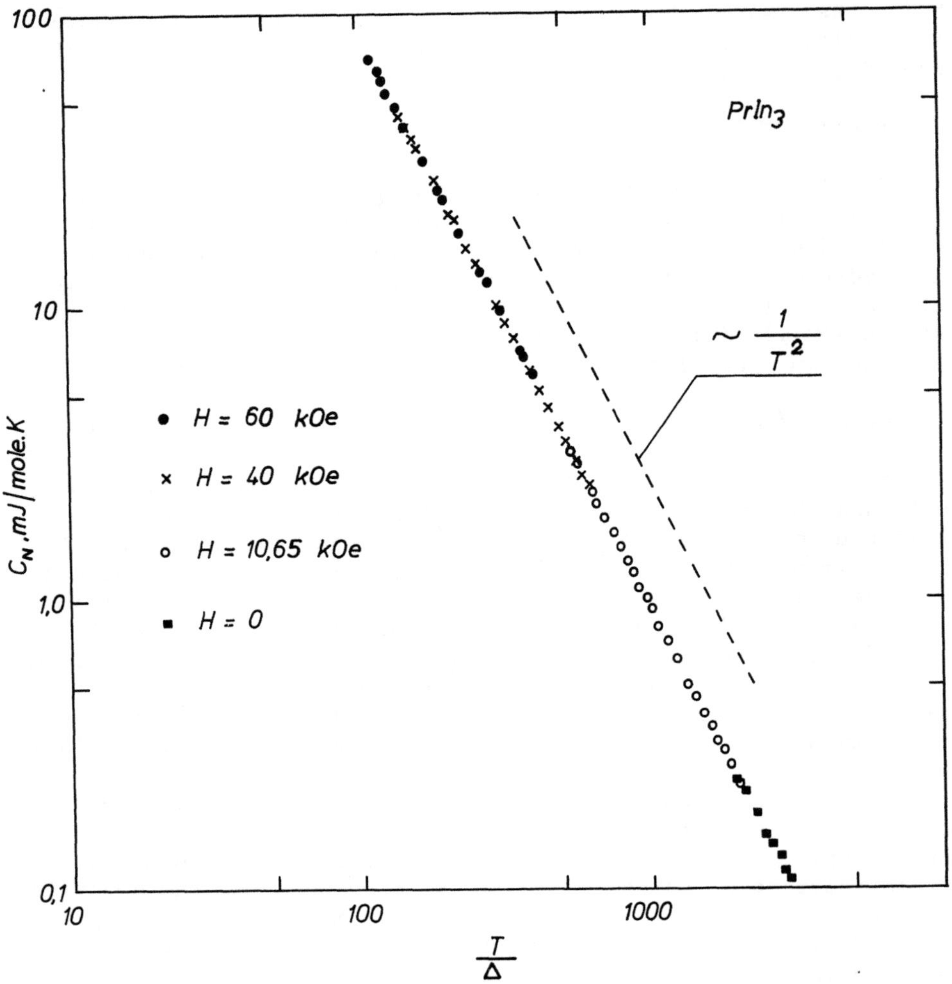

Fig. 1: Hyperfine specific heat of $PrIn_3$ in magnetic fields of different intensities, plotted vs reduced temperatures T/Δ, $\Delta = g_N \mu_N H / k_B$.

insert makes it evident that the straight line $(C_N T^2/R)(H^2)$ does not pass through the origin. A residual field H_o=2920 Oe corresponds to an experimentally measurable heat capacity. It falls remarkably in its proper place on the logarithmic straight line in Fig. 1 (in reduced coordinates T/Δ) in agreement with Eq. (12a). It is most natural to connect this field with the exchange interactions in $PrIn_3$. Having in mind exchange interactions and employing the Murao model[10] for nuclear exchange interactions, Hunik et al.[6] calculated the heat capacity of $PrIn_3$ in a field $H \to 0$, taking into consideration first

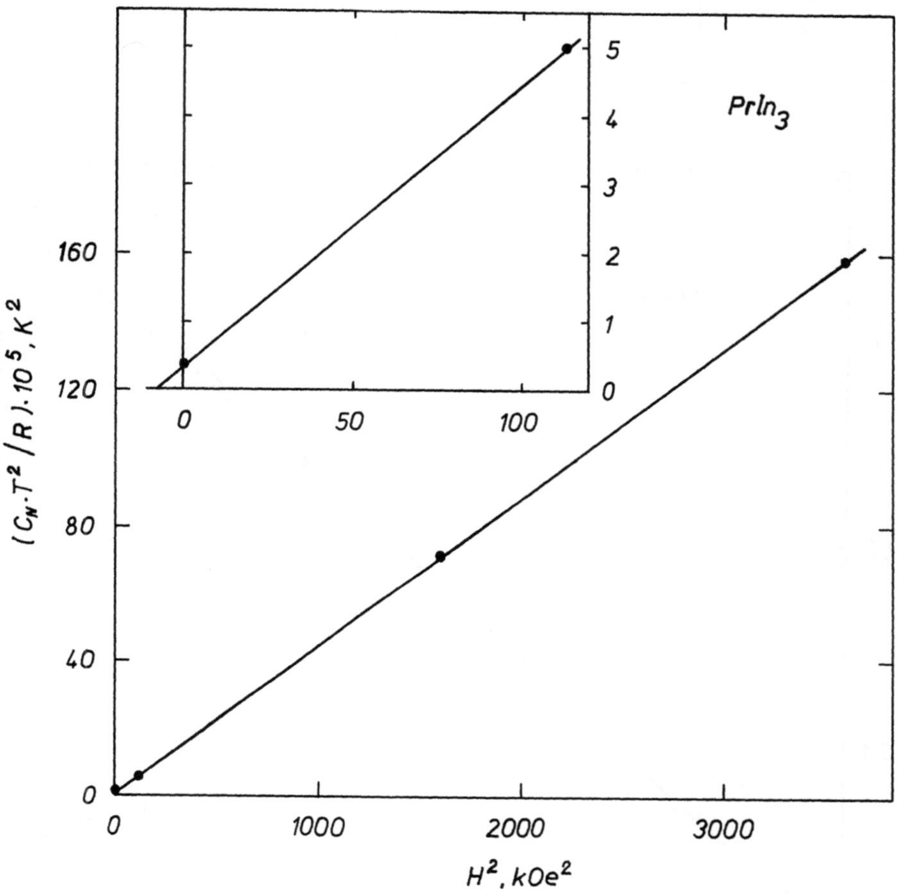

Fig. 2: Reduced hyperfine specific heat of $PrIn_3$ plotted vs H^2.

nearest neighbor exchange only, then next nearest and third nearest neighbor exchange and finally summing over all sites. The exchange energy found in our experiments (H_o=2920 Oe) coincides numerically with the results of Hunik et al. for the nearest and next-nearest neighbor case (H_o=2900 Oe). It is interesting to note that this is precisely the curve which gives the best agreement between theory and their experimental results.

The constant c in the first term of expansion (1) can be determined by the slope of the straight line in Fig. 2 and from c the magnetic moment induced by the external field in the SGS $\langle J_z\rangle = g_J(c/A_J)$. We preferred to calculate $\langle J_z\rangle$ numerically using the experimental data for C_N at a given temperature (0.5 K). The results obtained are shown in Fig. 3. The "zero" point gives the moment induced by the exchange interactions. The dotted line corresponds to

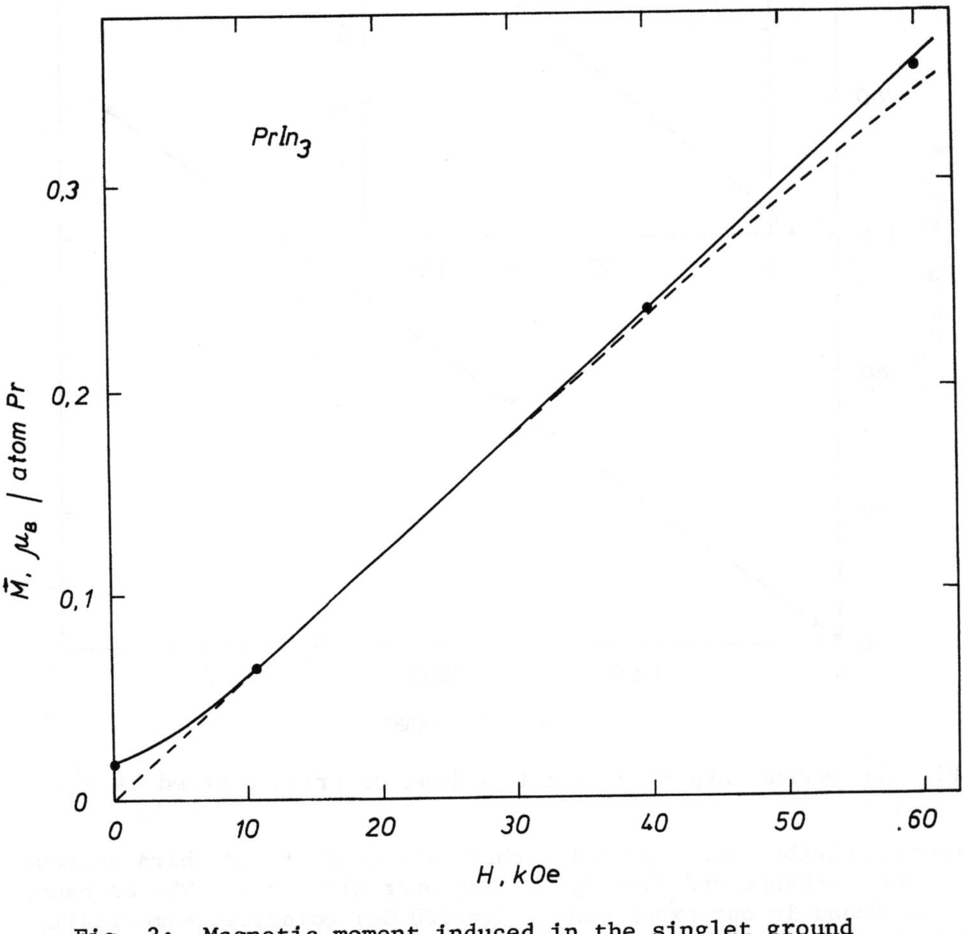

Fig. 3: Magnetic moment induced in the singlet ground state of $PrIn_3$. These data are extracted from measurements of the hyperfine specific heat as a function of the magnetic field strength. The solid line represents our experimental results, whereas the dotted line represents the moments calculated from Equation 8 (see text for details).

the theoretical values of the moment, calculated with the help of Eq. (8). The nonlinearity is clearly observable. The reason for such a disagreement with the experimental data should most probably be found in the inadequacy of the model (singlet-singlet instead of singlet-triplet for the lowest states). It is worth noting, however, that nonlinearity of the induced moment in systems with an SGS has been repeatedly observed[13], the $PrIn_3$ experiments of Genicon and Tournier[11] being among them. We evaluated, nevertheless, the Van Vleck susceptibility by the straight line in Fig. 3 and obtained the result $\chi_{vv}=2.65 \times 10^{-2}$ emu/mole, in fair agreement with their result.

HEAT CAPACITY OF POLYCRYSTALLINE Pr

Among the rare earths Pr is the most convenient element for the study of CF effects in pure metals with an SGS since, to all appearances, it remains paramagnetic down to very low temperatures. It is precisely the problem of the magnetic order in pure Pr that has stimulated numerous experiments on polycrystalline and monocrystalline samples and, in addition, a large number of theoretical investigations. The general discussion of this problem is the subject of the detailed reviews of McEwen[14] (up to 1978) and Jensen[15] (up to 1979). It seems that the latest experiments of McEwen and Stirling[16] on polycrystalline Pr using thermal neutron scattering techniques prove that a long-range magnetically ordered structure really appears in Pr at T<<1 K, the spontaneous magnetic moment being $\sim 0.1\mu_B$. The anomalous increase of the nuclear heat capacity observed at a temperature of about 30 mK by Lindelof et al.[17] in a monocrystalline sample is in agreement with this conclusion.

Metallic Pr is a complex SGS-system. The dhcp crystal structure places half of the Pr^{3+} ions in a hexagonally symmetric CF, and the remaining half in a cubic one. The CF structure itself may be considered to be determined with the help of direct neutron measurements[18]. The splitting insures an SGS, $|J_z\rangle=0$ for both types of point symmetries. The energy separations of the first excited states differ considerably: singlet-doublet, $\Delta_h=41$ K, for hexagonal symmetry and singlet-triplet, $\Delta_c=80$ K, for cubic symmetry. Due to the large splitting at a cubic symmetry site, the influence of the cubic ions on the magnetic and thermal properties of Pr, especially in the paramagnetic range, is appreciably weaker. An external magnetic field, however, essentially changes things, as can be seen from the experiments of Gregers-Hansen et al.[19] on the heat capacity of a Pr polycrystal in fields up to 18.6 kOe.

We have extended the experiments up to 140 kOe fields, at temperatures in the interval 0.3 to 1 K. The results are shown in Fig. 4. Two things may be seen in the insert on an enlarged scale. First, our points are located higher than those of Gregers-Hansen

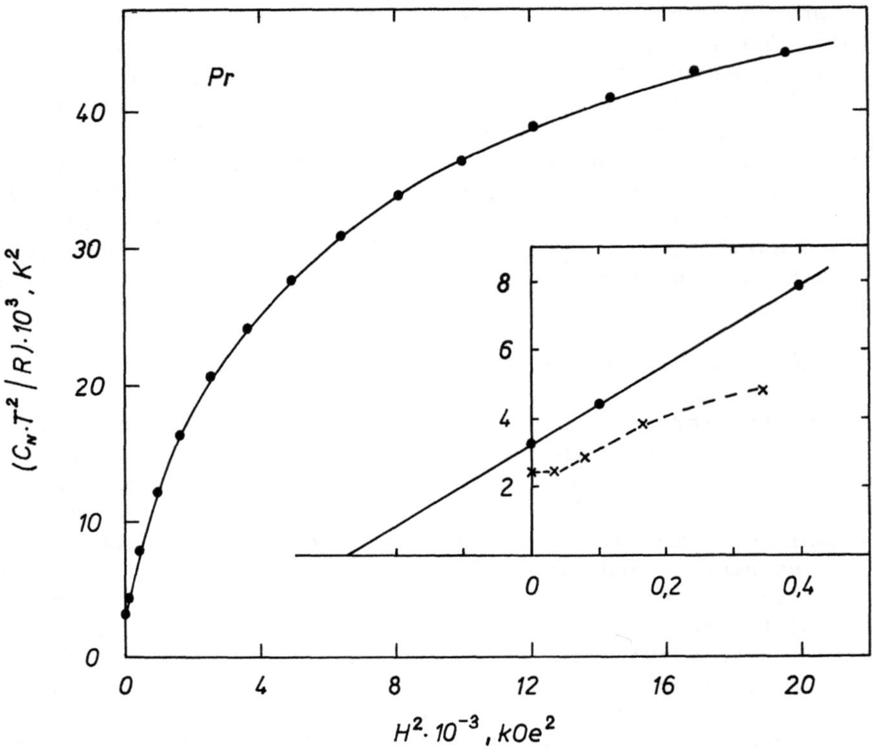

Fig. 4: Reduced hyperfine specific heat of a Pr polycrystal as a function of H^2. Results of Gregers-Hansen et al.[19] are also shown (dotted line in the insert).

et al. (as well as those of Lounasmaa[20]), but coincide with a remarkable precision with those of Holmstrom et al.[21] and Andres et al.[22] on polycrystalline samples. Secondly, at fields of about 20 kOe, $(C_N T^2/R) \sim H^2$, the heat capacity at H=0 corresponding to an energy equivalent to a residual field of 16,300 G.

Various interactions can be connected with such an energy: according to Murao[10], at H=0, in the nuclear regime, the pseudo-nuclear quadrupole interaction part of the effective Hamiltonian for the nuclear spins in Pr can be expressed as

$$H_{pnQ} = -\sum_n \Delta[I_z^2 - (1/3)I(I+1)], \tag{14}$$

where

$$\Delta = (1/2A_J M^a \langle 1/(1-\eta_k^x) \rangle. \tag{15}$$

For the case of $J=4(Pr^{3+})$ and a uniaxial CF with only one singlet state,

$$M^a=J(J+1)/\{[E(\pm1)-E(0)]/k_B\}. \quad (16)$$

The CF parameter $\Delta_h=41$ K is taken from Ref. 18, so that a larger value of Δ is obtained. The averaged exchange-enhancement factor $1/(1-\eta) \simeq 12.5$ to 20. Δ is then equal to 8 to 13 mK. The numerical computation of Murao[10] in reduced coordinates, T/Δ, corresponds in the parametric range to $(C_N T^2/R)=0.4 \times 10^{-3}K^2$ for $\Delta=8$ mK, and to $1 \times 10^{-3}K^2$ for $\Delta=13$ mK. This leads to $H_o=2000$ Oe and 5000 Oe respectively.

It is evident that the CF and HF effects give rise to the "pure" exchange interaction, equivalent to the remaining part of the "residual" field. In energy units, the two Δ values (8 mK and 13 mK) lead to 6.63×10^{-2} meV and 5.23×10^{-2} meV, correspondingly. This agrees with the results of Houmann et al.[18] for uniaxial symmetry along the c-axis (a symmetry which we are forced to assume). Such a result leads, however, to lower η values: evidently, the problem of the HFI between the HF-excited moments in the SGS at H=0 and pseudo-quadrupole effects enhanced through them should be resolved self-consistently.

Let us return to Fig. 4. The clear cut nonproportionality between C_N and H^2 for strong fields is the main feature of the experiment. Obviously it is impossible to fit the experimental data by employing the first term only of the expansion (10). The very possibility of using this expansion seems problematic now.

In Fig. 5, the reduced HF heat capacity is represented as a function of the external field. The curve initially seems to agree qualitatively with Eqs. (4), (5) and (6), but then the dependence becomes manifestly nonlinear. In spite of this, we calculated the field induced moment in the SGS, using the complete expansion in the form (10). In the higher-order terms of the expansion the quadrupole effects lead to corrections which should be taken into account. The quadrupole interaction constant includes both a real and a psuedo-quadrupole term. The latter was introduced above in the Murao formula (14) for J=4. One may assume that the pure quadrupole interaction is primarily due to 4f electrons. Then the strong magnetic field must be taken into consideration, the most essential effect being the admixing of the wave functions of the first excited state, and possibly, of the other excited states, into the SGS. As a result of this, Eq. (11) has to be modified:

$$P_{\|} =\{(-3eQ<r_q^{-3}>)/[4I(2I-1)]\}\sum_{-J_z}^{J_z} n_J^2<J||\alpha||J><+|3J_z^2-J(J+1)|+>, \quad (11a)$$

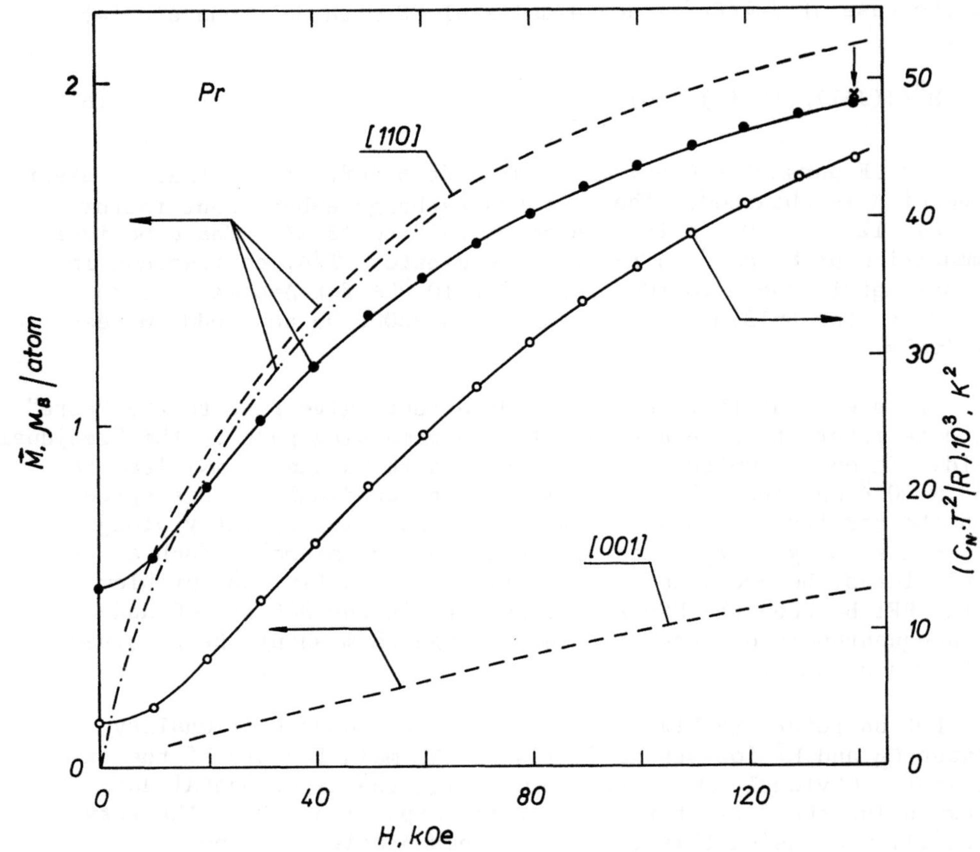

Fig. 5: Reduced hyperfine specific heat and field-induced magnetic moment of a Pr polycrystal plotted vs H. (Note the different axes for ($C_N T^2/R$) and $\vec{M}$).
Dotted lines: magnetization data of McEwen et al.[24] in the easy and hard direction, respectively.
Dash-dotted line: magnetization data of Johansson et al.[23] in the basal plane.
Arrow shows the moment calculated from Eq. (8); see text.

where every term under the summation symbol contains the square of the amplitude of the ground state wave function, and $M^z \neq 0$ in the Murao notation. (The direction of the induced gradient of the electric field coincides with that of H_z.) M^x and M^y surely decrease (down to 0.1 M^a)[18] with the increase of the field generating the uniaxial symmetry. But the very introduction of this symmetry includes the Pr^{3+} from the cubic sites in the quadrupole interactions in a strong field. We have taken into consideration the matrix elements of the $|\pm 1\rangle$ states only.

The course of the obtained field induced ion moment is also shown in Fig. 5. For the sake of comparison, results of the direct measurement of the magnetization of a Pr single crystal are included, both in moderate fields in the basal plane[23] and in extremely high fields (up to 370 kOe) in both the easy and hard directions[24]. The averaging for our polycrystalline sample is automatically effected by the very measuring of an "integral" quantity such as the specific heat. We are not firmly convinced that our fitting of the experimental data with the expansion (10), notwithstanding the corrections introduced, is exact. But in any case, it is no worse than other possible ones. It is well known that the separation of magnetic and quadrupole effects, particularly in systems with a strong HFI, is an old and unsolved problem. On the other hand, the moment induced at 140 kOe, calculated with the help of the exact formula of Bleaney (Eq. (8)), is indicated by the arrow in Fig. 5. Such an agreement is quite significant, the more so bearing in mind that the singlet-singlet model for the hexagonal sites of Pr was proposed by Bleaney[25] as early as 1963. Probably a model like this is not so unreal after all. The coincidence may be fortuitous. In any case, we took Δ_Z=41 K in (8), in agreement with the work of Houmann et al.[18].

DEDICATION

It is pleasure to dedicate this work to Professor Wlodzimierz Trzebiatowski's 75th birthday. The work was performed in the International Laboratory of High Magnetic Fields and Low Temperatures in Wrocław and we much profited from his constant interest and support.

REFERENCES

1. See B.R. Cooper in "Critical Reviews in Solid State Sciences", D. Schuele and R. Hoffman, eds., CRC Press, Cleveland, Ohio (1972), Vol. 3, p.83; B. Bleaney in "Magnetic Properties of Rare-Earth Metals", R.J. Elliot, ed., Plenum Press, London (1972), Ch. 8, p.383.
2. B. Bleaney, Physica 69, 317 (1973).
3. A. Abragam and B. Bleaney, "Electron Paramagnetic Resonance of Transition Ions", Clarendon Press, Oxford (1970), App. A.
4. B. Bleaney and R.W. Hill, Proc. Phys. Soc. (London), LXXVIII, 313 (1961).
5. R.J. Elliot and K.W.H. Stevens, Proc. Roy. Soc. A218, 553 (1953); see also B. Bleaney, J. Appl. Phys. 34, 1024 (1963).
6. R. Hunik, F. Nijhoff and W.J. Huiskamp, Physica 101B, 71 (1980).
7. K.R. Lea, M.J. Leask and W.P. Wolf, J. Phys. Chem. Sol. 23, 1381 (1962).
8. W. Gross, K. Knorr, A.P. Murani and K.H.J. Buschow in "Crystal Field Effects in Metals and Alloys", A. Furrer, ed., Plenum Press, N.Y. (1977), p. 37.

9. D. Davidov, V. Zevin, R. Levin, D. Shaltiel and K. Barberschke, Phys. Rev. B15, 2771 (1977).
10. T. Murao, J. Phys. Soc. Japan 39, 50, 1629 (1975); 46, 40 (1979).
11. J.L. Genicon and R. Tournier, Phys. Rev. Lett. 36, 107 (1976).
12. E. Leyarovski, L. Leyarovska and Chr. Popov in "Proceedings of the Eighth International Cryogenic Engineering Conference", C. Rizzuto, ed., IPC Science and Technology Press, Guilford (1980), p. 466.
13. K. Andres and S. Darak, Phys. Rev. B10, 1967 (1974); K. Andres, E. Bucher, P.H. Schmidt, J.P. Maita and S. Darack, Phys. Rev. B11, 4634 (1975); B.R. Cooper, Phys. Rev. 163, 444 (1967).
14. K.A. McEwen in "Handbook on the Physics and Chemistry of Rare-Earths", K.A. Gschneidner Jr., ed., North-Holland, Amsterdam (1978), Ch. 6, p. 411.
15. J. Jensen, J. de Physique 40, C5, 1 (1979).
16. K.A. McEwen and W.G. Stirling, J. Phys. C14, 157 (1981).
17. P.E. Lindelof, I.E. Miller and G.R. Pickett, Phys. Rev. Lett. 35, 1297 (1975).
18. J.G. Houmann, B.D. Rainford, J. Jensen and A.R. Mackintosh, Phys. Rev. B20, 1105 (1979).
19. P.E. Gregers-Hansen, M. Krusius and G.R. Pickett, Phys. Rev. Lett. 29, 420 (1972).
20. O.V. Lounasmaa, Phys. Rev. 133, A211 (1964).
21. B. Holmström, A.C. Anderson and M. Krusius, Phys. Rev. 188, 888 (1969).
22. K. Andres, E. Bucher, J.P. Maita and L.D. Longinotti, Phys. Rev. B6, 313 (1972).
23. T. Johansson, B. Lebech, M. Nielsen, H. Møller and A.R. Mackintosh, Phys. Rev. Lett. 25, 524 (1970).
24. K.A. McEwen, G.J. Cock, L.W. Roeland and A.R. Mackintosh, Phys. Rev. Lett. 30, 287 (1973).
25. B. Bleaney, Proc. Roy. Soc. A276, 39 (1963).

COMMENTS

McEWEN: You noted that your results could be fitted with Bleaney's crystal field parameters. Perhaps this is not so surprising since these parameters were deduced to fit heat capacity measurements. But, of course, they do not agree with the magnetization and neutron scattering results because the wave functions are incorrect. Did you find any evidence for a larger-than-excpected contribution to the heat capacity from the conduction electrons?

LEYAROVSKI: (Remark) We have calculated the induced moment at 140 kOe with the aid of Bleaney's formula[8], and by employing the singlet-singlet model for the hexagonal part of Pr^{3+}. The crystal field parameters used in this calculation are the very parameters obtained by Houmann et al.[18] in the neutron scattering experiments.

(Answer) Yes, the coefficient in the linear part of the specific heat is γ=28.4 mJ/mol-K^2 for the specimen of polycrystalline Pr under investigation.

GREIDANUS: Could you comment on a possible contribution on to the very low temperature specific heat due to the electric field of the In nuclei, which possess a quadrupolar moment and are at non-cubic sites:

LEYAROVSKI: It is very suspicious indeed, that the "residual" field, extracted from Fig. 2 of the present work (2920 G) is numerically equal to the quadrupole moment interactions in pure In (2900 G). But it is equal as well to the exchange interactions for the nearest and next-nearest neighbors, calculated by Hunik et al.[6]. On the other hand, the electric field gradient can be calculated from the lattice sum over screened ions. In $PrIn_3$ such a calculation should give a zero quadrupole interaction. Since the valences of Pr^{3+} and In^{3+} are equal, this implies an fcc crystal symmetry around the In nuclei. Even if we assumed that a point charge difference of about three units existed, the corresponding contribution to the specific heat should be a factor of 2.5 smaller than that of In metal, according to the point charge model[6].

SINGLET GROUND STATE AND COMBINED

ELECTRON-NUCLEAR MAGNETISM IN PRASEODYMIUM

K.A. McEwen*,†, W.G. Stirling* and C. Vettier*

*Institut Laue-Langevin, 38042 Grenoble cédex, France
†Department of Physics, University of Salford, U.K.

INTRODUCTION

Comprehensive neutron scattering[1] and magnetisation studies[2] have established that the ions in double hcp Pr metal experience a crystalline electric field which splits the $^3H_4$ ground multiplet to produce singlet ground states at both the locally hexagonal and cubic sites. For singlet ground state systems to exhibit magnetic ordering as $T \to 0$, the exchange coupling $J(q)$ must exceed a threshold value, expressed by the condition $\eta = 4\alpha^2 J(q)/\Delta > 1$ where Δ and α denote the energy separation and matrix element between ground and excited states respectively[3]. Pr provides a canonical example of such a singlet ground state system.

A detailed analysis[4] of the magnetic excitations in Pr at 5K confirms that (a) the crystal field ground state on the hexagonal site ions is the pure $|J_z = 0\rangle$ singlet and the first excited state is the $|J_z = \pm 1\rangle$ doublet, and (b) the coupling between these ions is largest, but still subcritical for a wavevector $q \sim 0.25$ Å^{-1} ($\sim 0.125\ \tau_{100}$) along the ΓM direction (equivalent to the real space $\hat{b}$ direction).

From neutron diffraction experiments carried out many years ago, it was deduced that polycrystalline Pr orders antiferromagnetically below $T_N \sim 25$K with a moment of $\sim 1\ \mu_B$/hexagonal site[5]. However, it is very clear that no long range ordering develops in single crystal Pr at such relatively high temperatures. This discrepancy will be explained in the next section.

The low temperature behaviour of singlet ground state systems is modified when interactions between the ions and the nuclei are included. By considering a Hamiltonian of the form

$$H = \sum_i V_i^{CF} - \sum_{ij} J_{ij}\, \underline{J}_i . \underline{J}_j - \sum_i A\underline{I}_i . \underline{J}_i \qquad (1)$$

where the terms describe, in turn, the crystal field potential, the exchange and the hyperfine interactions, it can be seen that at sufficiently low temperatures, a weak magnetic ordering will always be developed via the nuclear spin polarisation, even for subcritical values of η. The ordering of the combined electron-nuclear moment system takes place at a temperature increased by the factor $(1 - \eta)^{-1}$ over that at which the nuclei would order in the $\eta \ll 1$ case[6]. The temperature dependence of the electronic moment $\langle J_z \rangle$ may be very different from the normal Brillouin behaviour, and the saturation moment $\langle J_z \rangle_{T=0}$ is expected to be much less than the free ion moment[6]. Pr is thus an ideal system in which to search for this type of magnetism.

A number of macroscopic experimental techniques (eg. heat capacity, electrical resistivity) are sensitive to the development of magnetic ordering, but the most direct determination of magnetic phase transitions is provided by neutron scattering studies. In a series of experiments using triple-axis spectrometers at the Institut Laue-Langevin, Grenoble, with Pr in a dilution refrigerator, we have detected the onset of magnetic ordering in two crystals of Pr. Some of the results presented in the review below have already been published[7,8]. Details of the crystal preparation and experimental techniques may be found in these papers.

EXPERIMENTAL RESULTS AT LOW TEMPERATURES

Two crystals have so far been investigated. Pr-A (prepared by zone melting) and Pr-B (prepared by solid state electrolysis) contain $\lesssim$ 20 ppm and 6 ppm of Nd, respectively. Crystal Pr-A was mounted with an $\hat{a}$-direction vertical, giving a b-c horizontal scattering plane (orientation I). A complementary orientation (II) was chosen for crystal Pr-B. This was mounted with its $\hat{c}$-direction vertical to yield an a-b scattering plane. Figure 1 shows elastic scattering scans for Pr-A at various temperatures. The scans through the (001) reciprocal lattice point parallel to [100] show a broad central peak around $\underline{Q}_1 = 0.10\ \tau_{100}$ at 4.2K. As the temperature was reduced, the intensity of this peak increased steadily. However, at temperatures below 1K, a second peak appeared at $\underline{Q}_2 = 0.13\ \tau_{100}$ and the intensity of this satellite peak increased more rapidly. It is significant that the two peaks <u>co-exist</u> and that whilst the width of the satellite peak equals that of the experimental resolution, the $\underline{Q}_1$ peak is considerably broader.

This broad peak or "central peak" has been the subject of considerable confusion. In a recent neutron polarisation analysis study[9], we have confirmed that this peak is associated with <u>magnetic</u>

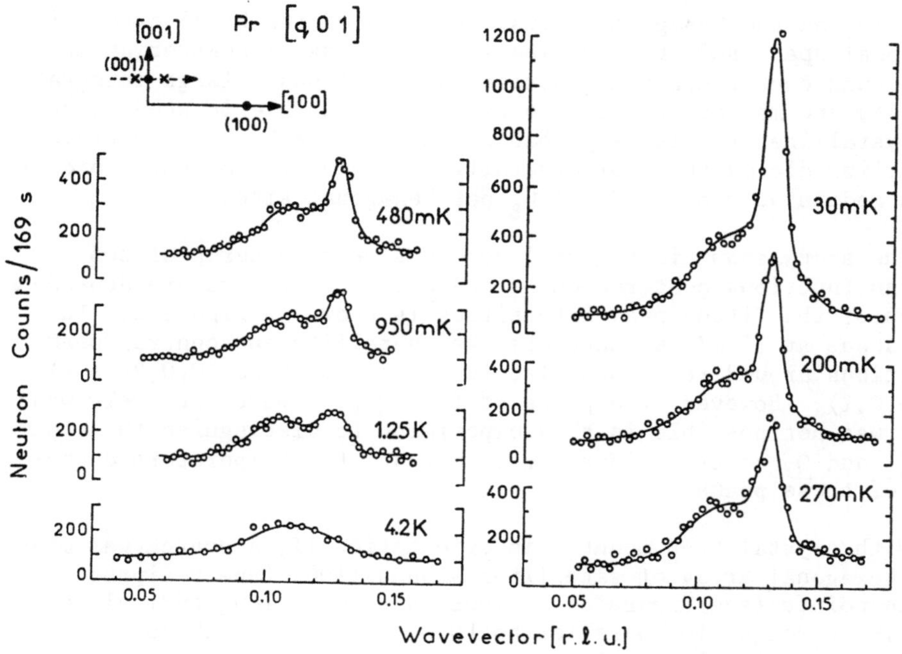

Fig. 1. Elastic scattering scans through the reciprocal lattice position (0.1,0,1) as a function of temperature, showing the central and satellite peaks at wavevectors of 0.11 and 0.13, respectively. The full lines are the sum of 2 Gaussians fitted to the data, as explained in the text. 1 r.l.u. = $4\pi/a\sqrt{3} = 1.976\ \text{Å}^{-1}$.

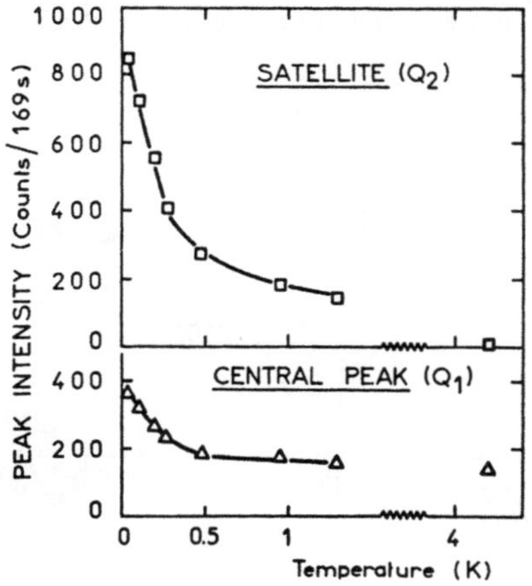

Fig. 2. Temperature dependence of the peak intensities of the central peak and satellite.

scattering on the hexagonal sites. Because of its broad extent in reciprocal space and its distribution as a ring of scattering at $|Q_1|$ around each reciprocal lattice point, it has a large integrated intensity and may be expected to be clearly visible in studies of polycrystalline Pr. It is probable that this peak is responsible for earlier claims that polycrystalline Pr exhibits magnetic ordering, with a modulated moment of $\sim 1\ \mu_B$ per hexagonal site.

The scans shown in Figure 1 were analysed in terms of two Gaussian functions centered on $\underline{Q}_1$ and $\underline{Q}_2$. The temperature dependence of each of the fitted peak intensities is shown in Figure 2. In other scans at 30 mK, we have observed satellite and central peak reflections around reciprocal lattice points of type $(0,0,2\ell + 1)$ and $(h,0,\ell)$. However, the peaks at $(1 \pm Q,0,0)$ were extremely weak and it was not possible in this experiment to distinguish the separate $\underline{Q}_1$ and $\underline{Q}_2$ contributions, or to measure the temperature dependence of these peaks.

With crystal Pr-B mounted in orientation II, a search was made for a hexagonal array of satellites around (100) type reciprocal lattice points (see Figure 3). Figure 4 shows a longitudinal scan, at 36 mK, through (100) parallel to [010]: a broad peak and a satellite peak are clearly resolved. Such peaks were observed at positions A, C, D and F. Peaks were not clearly seen at B and E. Because of geometric factors in the neutron scattering cross-section (equation 5 below), the intensity of the peaks around B and E is expected to be an order of magnitude less than at A, C, D and F.

A transverse scan through the satellite position at A revealed an unexpected splitting which was clearly evident below a temperature $T_2 \approx 200$ mK (Figure 5). The modulation wavevector below T_2 may be described as $\underline{Q}_2 = (Q_x,Q_y,0)$ where $Q_x = 0.13$, $Q_y = 0.007$, representing a tilt of $\underline{Q}_2$ by some 3° away from the symmetry direction.

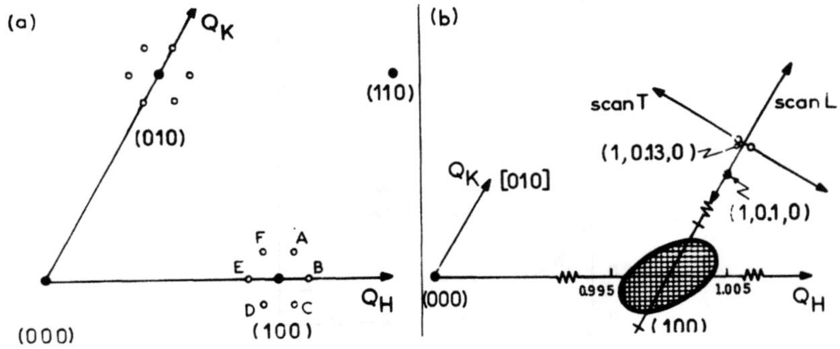

Fig. 3. (a) Orientation II with satellites around (100) and (010). (b) L and T scans through A (scale 5 x (a)), also resolution ellipse around (100) at scale 50 x (a).

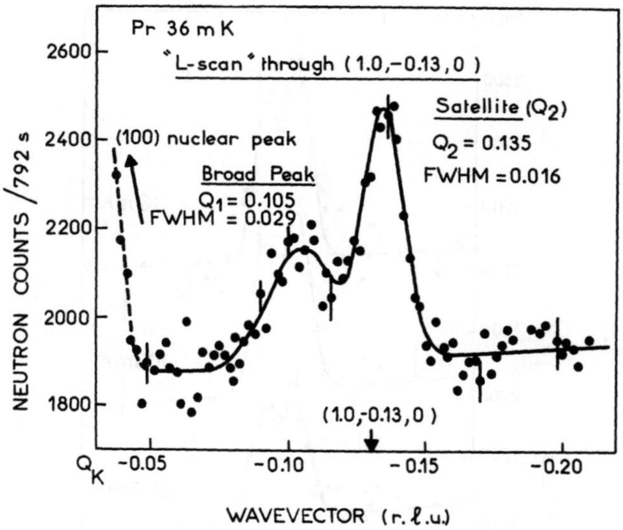

Fig. 4. L-scan through position D, at 36 mK. The centres and widths of the peaks are given in reciprocal lattice units.

MAGNETIC STRUCTURE OF PRASEODYMIUM

We have interpreted the satellite peaks $\underline{Q}_2$ as arising from an incommensurable magnetic structure comprising moments on the hexagonal sites which are sinusoidally modulated according to

$$\underline{m}_h = m_x \hat{b} \sin (\underline{Q}_2 . \underline{R} + \phi_x). \tag{2}$$

In this model, for temperatures between T_N and T_2, we have assumed that both the magnetic moments and the modulation wavevectors are along a $\hat{b}$-direction. The presence of equivalent sets of satellites around each (100) type reciprocal lattice point may be interpreted either in terms of three domains or as a 3-q structure. In the latter case the net magnetic moment at each site is a linear combination of the form

$$\underline{m}_h = \sum_i m_x^i \hat{b}^i \sin (\underline{Q}_2^i . \underline{R} + \phi_x^i); \quad i = 1,2,3 \tag{3}$$

where the phases ϕ_x^i are unknown. In our calculations below we have used the domain model assuming equal domain population, since this is more tractable.

We propose the existence of a second phase transition at the temperature $T_2 < T_N$. At this temperature $\underline{Q}_2$ begins to turn away from [100] as does the magnetic moment. Such a structure may be written as follows

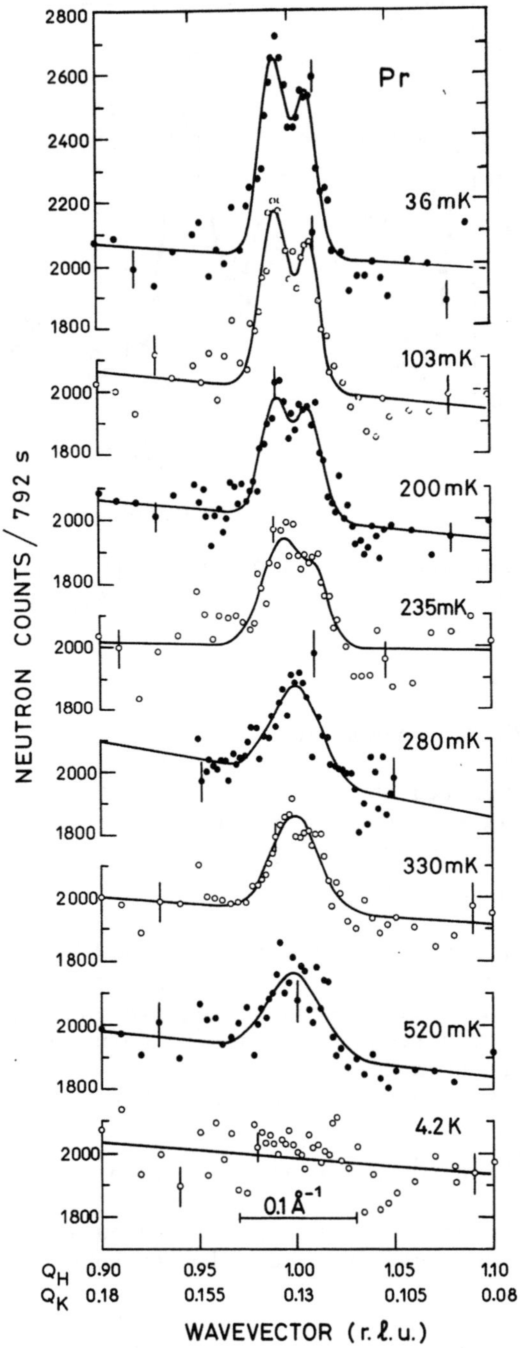

Fig. 5. T-scans through position A at various temperatures, showing the splitting of the (1, Q_2, 0) satellite. The full lines represent fitted Gaussians.

$$\underline{m}_h = m_x\,\hat{b}\,\sin(\underline{Q}_2.\underline{R} + \phi_x) + m_y\,\hat{a}\,\sin(\underline{Q}_2.\underline{R} + \phi_y). \qquad (4)$$

The relative phases of the m_x and m_y components are not known. This structure can be viewed as a freezing of the longitudinal optic exciton together with a contribution from the transverse optic exciton also. Thus, the structure (equation 4) has in general an elliptical polarisation.

The analysis of the data shown in Figure 2 is complicated by the satellite splitting which appears at the lowest temperatures. We have assumed that in orientation I this splitting is not resolved, due to the vertical divergence of the spectrometer, and we have therefore used the following formula to deduce the magnetic moment:

$$\frac{I^{mag}_{ho\ell}}{I^{nuc}_{100}} = \frac{1}{3} \cdot \frac{p_o^2}{b^2_{Pr}} \cdot \frac{m_h^2}{4} \cdot (1-(\hat{\kappa}.\hat{m}_h)^2).|f(\underline{\kappa})|^2.|F(\underline{\tau})|^2. \qquad (5)$$

The ratio $I^{mag}_{ho\ell}/I^{nuc}_{100}$ is related to the experimentally measured intensities, integrated over units of reciprocal space, by the factor[10]

$$(I^{mag}_{ho\ell}/I^{nuc}_{100}) = (I^{mag}_{ho\ell}/I^{nuc}_{100})_{meas} \cdot \frac{|\sin(\theta-\beta)|_{ho\ell}}{|\sin(\theta-\beta)|_{100}} \qquad (6)$$

where θ denotes the Bragg angle and β is the angle between the scattering vector κ and the scan direction, in each case. The factor 3 represents equal domain population. The Lorentz factor expressed in (6) was omitted in our earlier paper[7]. On this basis, the intensity of the $Q_2 01$ satellite at the lowest temperature indicates a hexagonal site moment of 0.13 μ_B/atom.

In our interpretation of the results obtained in orientation II, we have replaced the volume factor of 3 in equation (5) by a factor of 6, to account for the doubling of the number of satellites. On this basis we deduce an average moment of 0.09 μ_B from the satellites in positions A,F,C and D. It may be noted that the split satellites in Figure 5 have unequal intensities. This can be interpreted in terms of an unequal population of two triple-$\underline{Q}$ domains within which the 3 modulation wavevectors are turned away from the symmetry directions by an equal amount. The magnetic structure of Pr at low temperatures has thus a strong resemblance to that of Nd which also exhibits a second phase transition at a temperature slightly below T_N[11].

MEASUREMENTS UNDER UNIAXIAL STRESS

The application of a uniaxial stress along an $\hat{a}$-direction removes the two-fold degeneracy of the excited states on the hexagonal sites and consequently reduces the separation between the ground and first excited states, leading to an increase in the ratio η. Magnetic ordering now occurs at a considerably higher temperature than for zero stress. Figure 6 shows representative elastic scattering scans for a crystal (Pr-C, from the same ingot as Pr-A) mounted in orientation-I, at a stress of 98 MPa. As the temperature is reduced, both the central peak at $\underline{Q}_1$ and a well-defined satellite at $\underline{Q}_2$ are developed around (001) and (003). The intensity at 1.3 K of both peaks is very much greater than the maximum values found under zero stress conditions. Clear evidence for the existence of both a $\underline{Q}_1$ peak and a $\underline{Q}_2$ satellite around the (100) reciprocal lattice point was also found. The temperature dependence of the peak intensities of the central peak ($\underline{Q}_1$), the satellite around (001) and the satellite around (100) is shown in Figure 7. We note that these latter satellites appear at a lower temperature and are much less intense at 1.3 K than those around (001). The $(1-(\hat{\kappa}.\hat{m}_h)^2)$ factor in the neutron scattering cross-section implies that the (1 ± Q,0,0) satellites arise from perpendicular components of the magnetisation (m_y). These observations provide strong support for the existence of a second phase transition at $T_2 < T_N$, as proposed herein. The magnitude of the saturation magnetic moment increases with applied stress and we deduce a value of 0.90 μ_B per hexagonal site at 98 MPa.

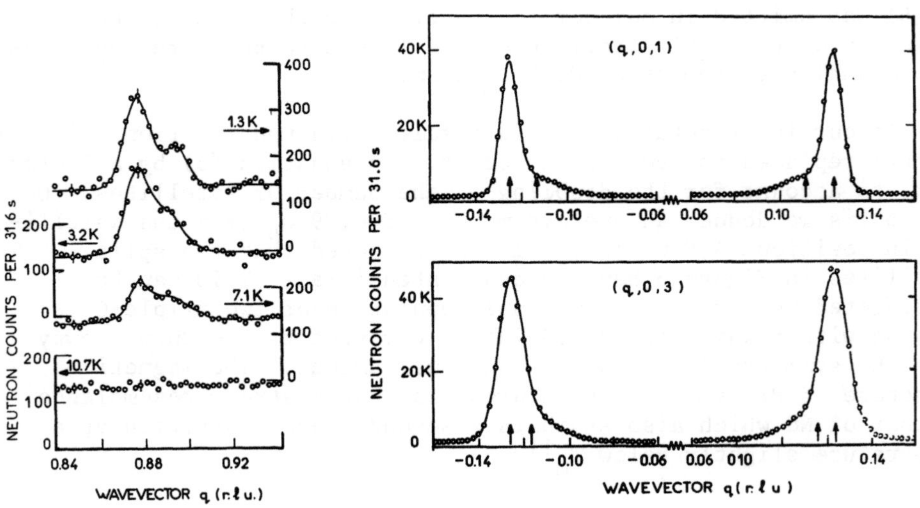

Fig. 6. Scans through (± 0.1,0,1) and (± 0.1,0,3) at 1.3K, together with (q,0,0) scans at various temperatures, under a uniaxial stress of 98 MPa.

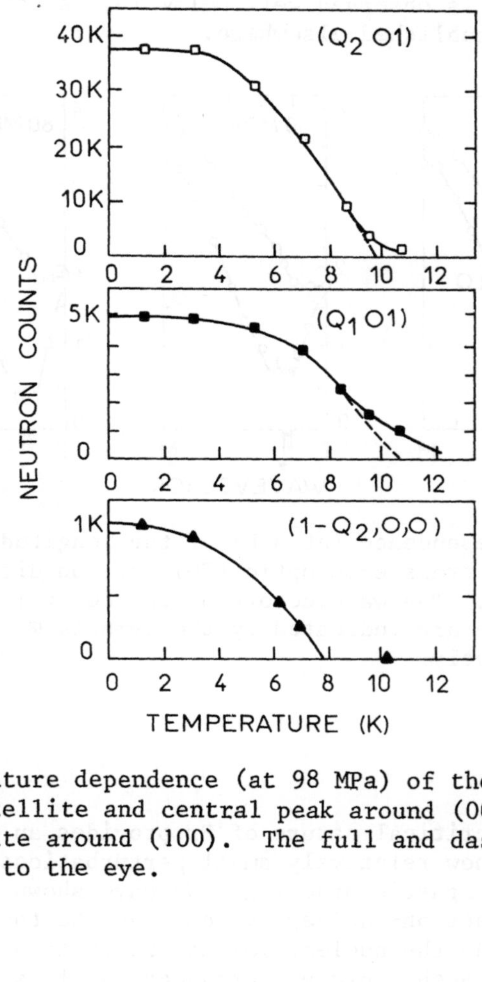

Fig. 7. Temperature dependence (at 98 MPa) of the intensities of the satellite and central peak around (001), and of the satellite around (100). The full and dashed lines are guides to the eye.

Praseodymium also provides a beautiful realisation of a magnetic soft-mode system. In general, there are four branches of magnetic excitations along the ΓM direction: the lowest two have a longitudinal optic and transverse optic character respectively. Figure 8 shows the effect of stress on these excitations at 5 K. There is a clear softening of the longitudinal optic mode which is most pronounced around the wave vector $\underline{Q}_2$. We also note a significant increase in energy of the transverse optic mode. The temperature dependence, at constant stress, of the LO mode is rather complex. For wavevectors some way from $\underline{Q}_2$, the excitation energy falls as the temperature is reduced through T_N (but it is still finite at T_N) until it becomes impossible to resolve from the scattering centered at zero energy. However, at still lower temperatures, the mode energy increases slightly and can then be resolved. For wavevectors very close to $\underline{Q}_2$ the mode becomes heavily over-damped and no well-defined excitation is observed well below T_N. A detailed account of this work will be published elsewhere.

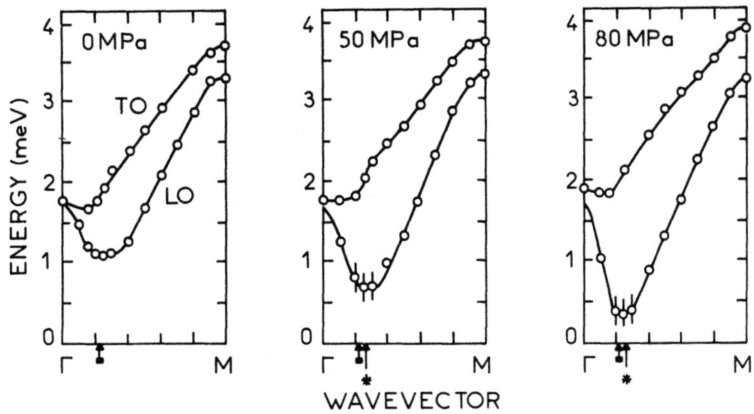

Fig. 8. Stress dependence (at 5 K) of the longitudinal optic (LO) and transverse optic (TO) exciton dispersion curves along ΓM. The wavevectors of the central peak and satellite are indicated by the symbols ■ and ✱, respectively.

CONCLUSION

The close-to-critical nature of Pr provides an excellent opportunity to examine how relatively small perturbations of this SGS system can induce magnetic ordering. We have shown how the coupling of the combined electron-nuclear system leads to the development of magnetic ordering as the nuclear moments begin to polarise at very low temperatures. Such a process produces far less than the full ordered moment of Pr, at the temperatures so far accessible. Alternatively, the application of a suitable uniaxial stress induces phase transitions at much higher temperatures. Concomitant with the development of a relatively large ordered moment, the magnetic excitations exhibit a clear soft-mode behaviour. Evidence for the existence of a second phase transition below T_N has been presented. The magnetic structure of Pr below T_2 can be viewed as a freezing of the LO and TO modes to produce an elliptical modulation. The reason for the existence of the second transition, which is also accompanied by a turning of the modulation wavevector away from symmetry directions, remains to be explained. Moreover, the dynamics of the phase transitions are particularly complex, and will be reported elsewhere.

ACKNOWLEDGEMENTS

Many colleagues at the ILL have contributed to this work, and we wish to thank especially D. Brochier, S. Burke, K. Gobrecht, J.L. Ragazzoni and P. Suttling. We are also grateful to D. Fort and D.W. Jones for the crystal Pr-B. KAM wishes to thank the SERC for financial assistance and the British Council for a travel grant.

REFERENCES

1. J. Jensen, J. Physique 40, C5-1 (1979)
2. K.A. McEwen, in 'Handbook on the Physics and Chemistry of Rare Earths', K A Gschneidner, ed., North-Holland, Amsterdam (1978) p.411.
3. G.T. Trammell, Phys. Rev. 131, 932 (1963).
4. J.G. Houmann, B.D. Rainford, J. Jensen and A.R. Mackintosh, Phys. Rev. B20, 1105 (1979).
5. J.W. Cable, R.M. Moon, W.C. Koehler and E.O. Wollan, Phys. Rev. Lett. 12, 553 (1964).
6. T. Murao, J. Phys. Soc. Japan 46, 40 (1979) and references therein.
7. K.A. McEwen and W.G. Stirling, J. Phys. C14, 157 (1981).
8. K.A. McEwen, W.G. Stirling, C. Vettier, J.L. Ragazzoni, D. Fort and D.W. Jones, Proc. Low Temperature Physics Conf. (LT16), Los Angeles, USA, in press (1981).
9. S.K. Burke, W.G. Stirling and K.A. McEwen, J. Phys. C14, in press (1981).
10. B. Lebech and M. Nielsen, Proc. Neutron Diffraction Conf., Petten, Netherlands (RCN-234), 466 (1975).
11. B. Lebech, J. Appl. Phys. 52, 2019 (1981).

COMMENTS

COQBLIN: When an uniaxial stress is applied on Pr, the upper levels are split. Could you indicate if the mean value of the upper levels remains constant or moves with increasing stress?

McEWEN: The energies of the excitations at the zone boundary (M) provide information on this question. Although the splitting here between the LO and TO modes has increased to approximately 1 meV at 120 MPa, the average energy has changed very little from its zero stress value.

ROSSAT-MIGNOD: Have you any idea about the origin of the broad magnetic peak?

McEWEN: We have recently made a study of the broad central peak (CP) in Pr using polarized neutrons[1]. Its magnetic origin has been confirmed, and it is associated with the hexagonal site ions. The CP coexists with the long range ordering. Under the application of uniaxial stress the intensity becomes too large to be explained by surface or impurity effects. We therefore conclude that it is an intrinsic dynamical effect in Pr, and we hope this experiment will stimulate further theoretical efforts to explain the CP. [Reference 1 is: Burke, Stirling and McEwen, to be published in J. Phys. C (1981)].

COOPER: Are there any Knight shift measurements that can be coordinated with your measurements?

McEWEN: Unfortunately it is technically very difficult to make such measurements on single crystal Pr because of the very small skin depth at high frequencies: it is too good a metal.

STUDY OF THE CRYSTALLINE ELECTRIC FIELD IN PRASEODYMIUM INTERMETALLICS

W. Matz*, B. Lippold§, E.A. Goremychkin†, A. Andreeff†,
H. Greissman† and T. Frauenheim†

*Zentralinstitut für Kernforschung Rossendorf der AdW der DDR, Dresden, GDR
§Karl-Marx-Universität, Sektion Physik, Leipzig, GDR
†Joint Institute for Nuclear Research Dubna, Laboratory of Neutron Physics, Moscow, USSR

INTRODUCTION

The study of crystalline electric fields (CEF) in rare earth intermetallics is of interest for the understanding of the magnetic interaction in metals. The CEF depends on the particular form of the electrical potential in the metal and consequently it can provide knowledge of the electronic structure of a substance. Information about the symmetry and strength of the CEF can be obtained by determining the splitting of the ground state J-multiplet of the 4f electrons into so-called crystal field levels.

In metallic materials crystal field spectroscopy by inelastic neutron scattering seems to be the most direct method to determine the crystal field splitting. But often only comparison with other measurements, e.g. specific heat and magnetic susceptibility, allows measurement of an unambiguous CEF level scheme[1].

In this contribution, we report new results on the crystalline electric field in the cubic Laves phase compound $PrNi_2$ and the hexagonal intermetallic $PrCu_5$. Both substances have been reported to be van Vleck paramagnets[2,3] indicating a nonmagnetic crystal field ground level. Recently, a cooperative nuclear magnetic ordering at T=250 mK was observed on $PrNi_2$[4]. To our knowledge this is the highest nuclear ordering temperature determined so far.

DETERMINATION OF CRYSTAL FIELD PARAMETERS

The phenomenological description of the CEF terms of operator equivalents O_n^m for the case of cubic and hexagonal symmetry is given by the following Hamiltonians:

$$\hat{H}_{cub}=B_4^0(O_4^0+5O_4^4) + B_6^0(O_6^0-21O_6^4), \tag{1}$$

$$H_{hex}=B_2^0O_2^0+B_4^0O_4^0+B_6^0O_6^0+B_6^6O_6^6. \tag{2}$$

The values of the parameters B_n^m determine the relative energies E_i of the crystal field levels and the appropriate wave functions $|\Gamma_i\rangle$. For Pr^{3+} ions the ground multiplet $^3H_4$ splits in the cubic CEF into one singlet Γ_1, one doublet Γ_3, and two triplets Γ_4, Γ_5. For the case of a hexagonal CEF a splitting into three singlets Γ_1, Γ_3, Γ_4 and three doublets Γ_{5A}, Γ_{5B}, Γ_6 occurs.

In the inelastic scattering experiment one can observe transitions between the crystal field levels. The cross section in the dipole approximation is given by

$$\frac{d^2\sigma}{d\Omega d\varepsilon} = \left(\frac{1.91e^2}{2mc^2} g_J\right)^2 \frac{k'}{k} f^2(\kappa) \sum_{i,i'} p_i |\langle\Gamma_{i'}|\hat{J}_\perp|\Gamma_i\rangle|^2$$

$$\times \sqrt{\frac{4\ln 2}{\gamma_{ii'}^2}} \exp\left[-4 \ln 2 \left(\frac{E_i-E_{i'}-\varepsilon}{\gamma_{ii'}}\right)^2\right]. \tag{3}$$

Here, $\vec{k}$ and $\vec{k}'$ are wave vectors of incident and scattered neutrons, respectively, $\kappa=|\vec{k}-\vec{k}'|$ is the momentum transfer, and $f(\kappa)$ the atomic magnetic form factor. A peak at an energy transfer ε indicates the existence of two crystal field levels with energy difference $E_i-E_{i'}$. The peak intensity is proportional to the matrix element $|\langle\Gamma_{i'}|\hat{J}\ |\Gamma_i\rangle|^2$ and contains information about the wave functions of the crystal field levels ($\hat{J}_\perp$ is the component of the total angular momentum operator perpendicular to the scattering vector $\vec{\kappa}$). The temperature dependence of the neutron scattering cross section is described by the thermal population factor of the levels p_i, which is given by Boltzmann statistics.

If there exist several peaks from crystal field transitions in the spectrum of inelastically scattered neutrons, this method provides a unique tool for the determination of the crystalline electric field in metallic substances. Unfortunately, there are cases where an unambiguous evaluation of the CEF parameters from the inelastic neutron scattering experiment is not possible without additional information. The study of the low temperature behavior of different physical properties of the materials can be very helpful because they also depend on the CEF splitting.

The magnetic contribution to the specific heat is given by the following equation:

$$C=\frac{1}{Z^2kT^2}[Z\Sigma_i n_i E_i^2 \exp(-E_i/kT)-\{\Sigma_i n_i E_i \exp(-E_i/kT)\}^2]. \quad (4)$$

This contains only the energies of the CEF levels and their degeneracy n_i. Z is the partition function. The specific heat does not distinguish between different sets of CEF parameters if the energy difference between the levels and their degeneracy is the same. Another aspect is that all specific heat calculations assume ideal levels without any broadening. The effect of broadening can distort the shape of the specific heat considerably.

For the case of the magnetic susceptibility χ not only the energy levels E_i but also the appropriate wave functions $|\Gamma_i>$ must be known.

$$\chi(T)=\frac{N\mu_B^2 g^2}{Z}\{\sum_i \frac{|<\Gamma_i|\hat{J}|\Gamma_i>|^2}{kT} \exp(-E_i|kT) + 2\sum_{i\neq j} \frac{|<\Gamma_i|\hat{J}|\Gamma_j>|^2}{E_i^o-E_j^o} \exp(-E_i|kT)\}. \quad (5)$$

Because the sum in $\chi(T)$ runs over all matrix elements, the magnetic susceptibility is not very sensitive to small changes of the crystal field parameters. This is why it is not selective of the energies and matrix elements like inelastic neutron scattering.

Nevertheless, both properties, specific heat and magnetic susceptibility, depend in a rather direct way on the crystal field parameters and are supplementary to neutron scattering experiments for the determination of the CEF.

Transport properties are also influenced by the crystalline electric field but in a more complicated way. For their evaluation other parameters than the CEF parameters are additionally necessary. Therefore the crystal field parameters are used to compute transport properties rather than using transport property measurements to determine crystal field parameters.

NEUTRON SCATTERING EXPERIMENTS

The inelastic neutron scattering experiments have been performed with a time-of-flight spectrometer in the inverted geometry at the pulsed reactor IBR-30 of JINR Dubna. Spectra have been obtained for sample temperatures of 8 K and 80 K at a scattering angle ϕ=60°. For $PrCu_5$ additional measurements with a triple axis spectrometer at ZfK Rossendorf have been carried out. In order to obtain some information about neutron scattering on phonons the intermetallics $LaNi_2$ and $LaCu_5$ have been studied as the isostructural and nonmagnetic counterparts of $PrNi_2$ and $PrCu_5$, respectively.

For the determination of the crystal field parameters B_n^m from the neutron scattering experiments, first the peaks due to crystal field transitions are to be established in the spectra. After this a computer program[7] evaluates all possible combinations of crystal field parameters which allow for the observed transitions. As additional information the nonmagnetic character of the ground level in both intermetallics was used. For $PrNi_2$ the ground level should be the singlet Γ_1 or the nonmagnetic doublet Γ_3. In the case of $PrCu_5$ a singlet ground level is expected. To obtain the

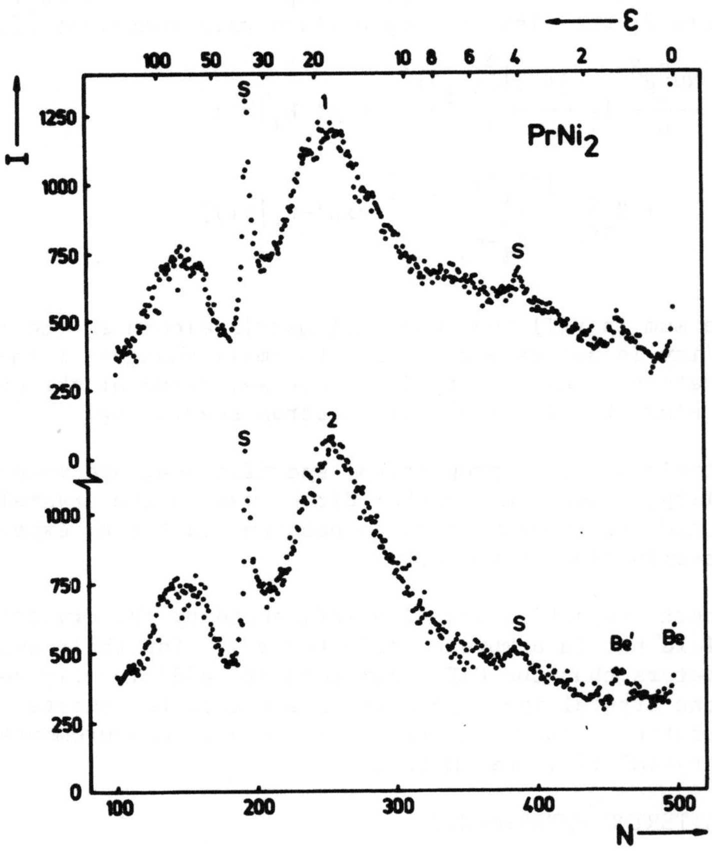

Fig. 1: Time-of-flight spectra of inelastically scattered neutrons on $PrNi_5$ at T=8K (1) and T=80K (2) and a scattering angle of 60°. I - intensity per channel, ε - energy transfer in meV, N - channel number, S - satellites of fast neutrons, Be(Be') - beryllium filter cut-off (precut-off).

best crystal field parameters from all possible combinations, the method of simultaneously modeling phonon and crystal field contributions to the spectra of scattered neutrons was applied.

The spectra for $PrNi_2$ are shown in Fig. 1. One can identify several peaks. The first peak at an energy transfer ε of about 70 meV cannot be unambiguously explained but is probably due to hydrogen adsorbed on the surface of the crystallites. The second broad peak at about 20 meV is due to phonon scattering. This was concluded from the neutron scattering results for $LaNi_2$ and from the temperature dependence of the scattering intensity for $PrNi_2$. With the same arguments it was determined that the two broad peaks at $\varepsilon=(3.0\pm1.0)$ meV and $\varepsilon=(7.0\pm1.5)$ meV should be of magnetic origin. They must be transitions starting from the CEF ground level. Applying the procedure described above the best fit to the spectra was obtained for the following CEF parameters for $PrNi_5$:

$$B_4^0 = (24.5\pm1.8)\cdot10^{-4}\ \text{meV}, \qquad B_6^0 = -(65.7\pm4.9)\cdot10^{-6}\ \text{meV}.$$

In Fig. 2 the comparison between the experimental and calculated neutron scattering spectrum for T=8K is shown. For the line-widths $\gamma_{ii'}$ of the CEF transitions a value of 6 meV was deduced from the

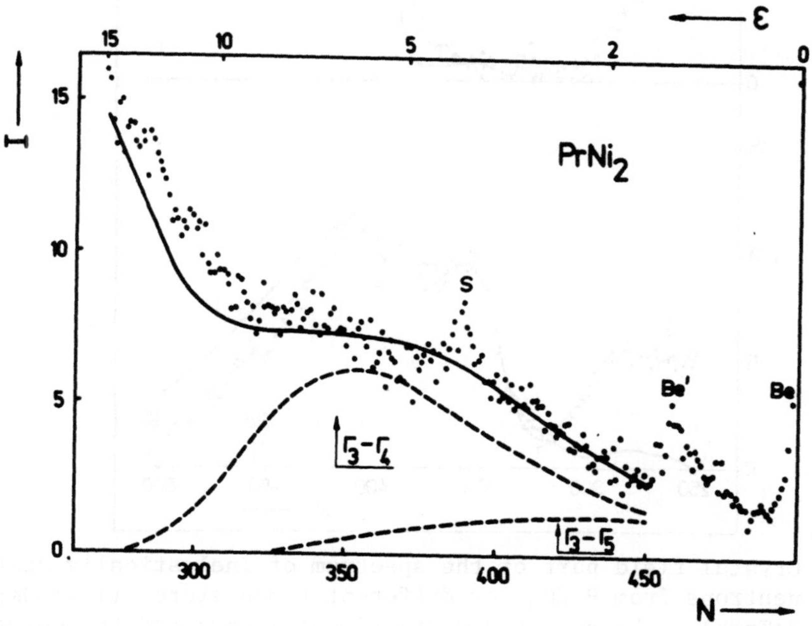

Fig. 2: Computed TOF spectrum compared to the experimental results for $PrNi_2$ at T=8 K. Solid line - sum of the phonon and CEF scattering contributions, broken line - the CEF scattering contribution only.

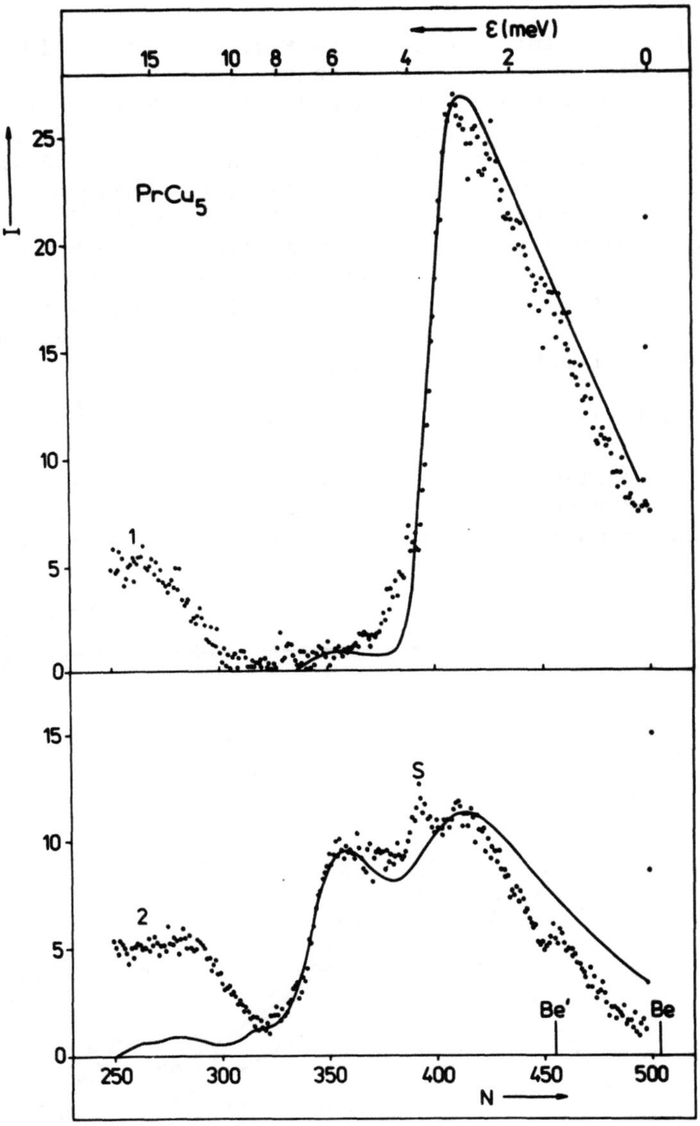

Fig. 3: Crystal field part of the spectrum of inelastically scattered neutrons from $PrCu_5$ for different temperatures (1:T=80K; 2:T=8K). The points are experimental data and the curves represent the fit with the given cyrstal field parameters and line-widths of 0.75 meV for T=80K and 1.5 meV for T=8K.

fit. We also observed such a rather broad line-width in the cubic intermetallics $PrAl_2$ and $PrMg_3$ (Refs.10, 8). A possible explanation for this is an electron-phonon interaction, because such structures have big unit cells with a large number of atoms and consequently many possible phonon branches.

In the spectra of inelastic neutron scattering on $PrCu_5$ one finds only two distinct peaks from crystal field transitions at energies of $\varepsilon=(3.0\pm0.3)$ meV and $\varepsilon=(6.0\pm0.5)$ meV. From the temperature dependence of the peak intensities it was concluded that the transition with $\varepsilon=3$ meV starts from the ground level and the transition with $\varepsilon=6$ meV starts from an excited level. The crystal field transition character of the peaks was confirmed by the experiments at the triple-axis spectrometer operated in the constant momentum transfer mode. The crystal field part of the neutron scattering spectra from $PrCu_5$ as well as the fitted curves are shown in Fig. 3. For the crystal field parameters we obtained for $PrCu_5$:

$B_2^0=(48\pm7)\cdot10^{-2}$ meV $\qquad B_6^0=(49\pm6)\cdot10^{-6}$ meV

$B_4^0=(8.5\pm1.0)\cdot10^{-4}$ meV $\qquad B_6^6=(14.2\pm1.7)\cdot10^{-4}$ meV.

In the analysis the ratio $V=B_6^6/B_6^0$ was restricted as discussed earlier[9] to the value V=29 obtained from a point charge calculation with the real lattice parameters of $PrCu_5$. The splitting schemes of the ground multiplet of Pr^{3+} in both intermetallics are given in Fig. 4.

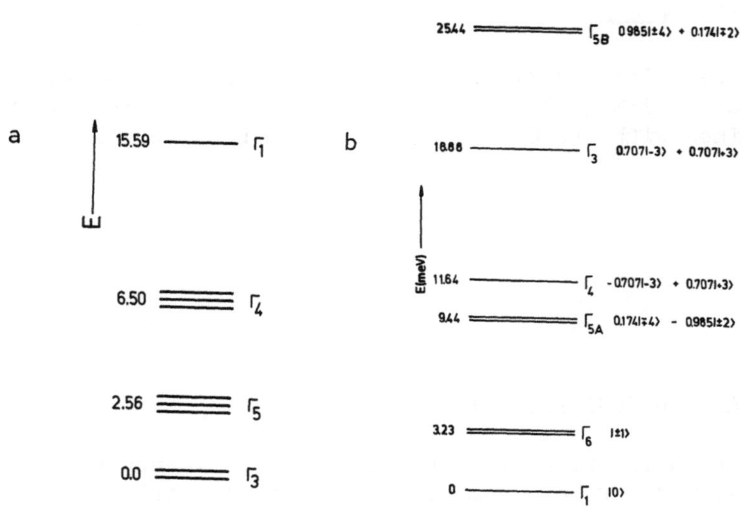

Fig. 4: Splitting scheme (in meV) of the ground multiplet of Pr^{3+} in $PrNi_2$ (a), and $PrCu_5$(b), obtained from the neutron scattering experiments. For $PrCu_5$ the wave functions of the CEF levels are indicated.

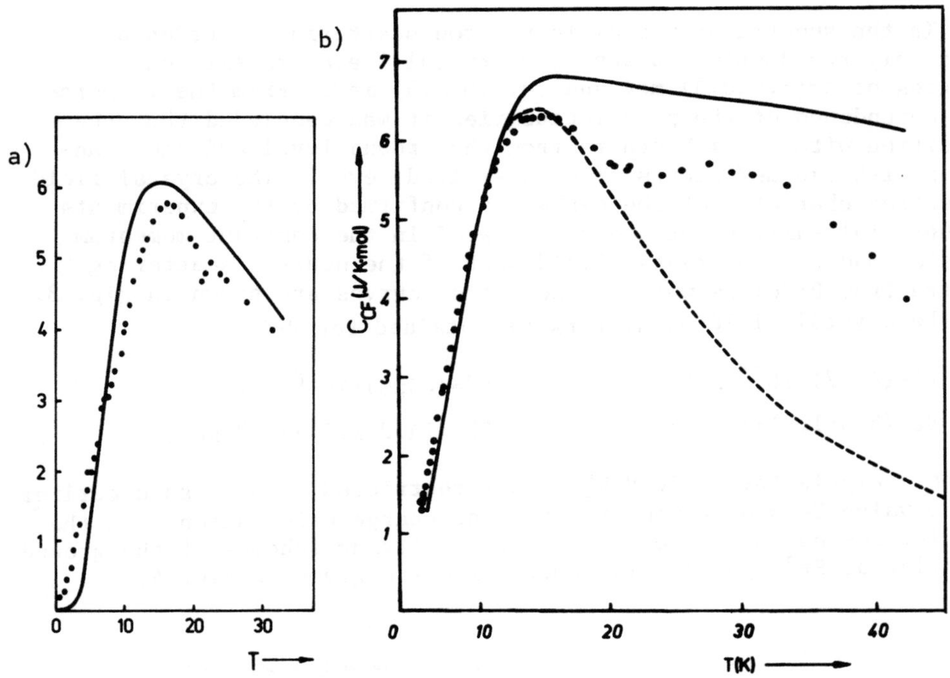

Fig. 5: Crystal field contribution to the temperature dependence of specific heat of $PrNi_2$(a) and $PrCu_5$ (b). The solid line is the theory obtained with the CEF parameters from neutron scattering.

SPECIFIC HEAT AND SUSCEPTIBILITY

In addition to the neutron scattering experiments the specific heat was measured for both samples. The experimental data giving the magnetic part of the specific heat are shown in Fig. 5 together with the calculated curves using the level schemes from Fig. 4. For $PrNi_2$ (Fig. 5a) the coincidence of the experimental and computed results is quite good. The difference in the rise of the low temperature part is presumably due to the great width of the CEF levels, because in the calculation sharp levels were used. In the case of $PrCu_5$ (Fig. 5b) quantitative agreement is obtained for low temperatures. The calculated curve shows only a weak structure (second maximum) compared to the experimental points. A similar result was reported in Ref. 2. From the measured specific heat one can evaluate a splitting of about 3.2 meV between ground and first excited levels. This is in good agreement with $\varepsilon=(3.0\pm0.3)$ meV detected in the neutron scattering experiments.

The magnetic susceptibility of $PrCu_5$ was measured with a SQUID magnetometer in the temperature range from 5K to 80K. A calculation of the inverse susceptibility using the B_n^m and a molecular field parameter of $\lambda=0.72$ mole-cm^{-3} gives a curve which is in good agreement with the experimental points (Fig. 6). But it should be noted that a similar agreement can be achieved with other sets of crystal field parameters also.

From the experimental data[3] of the magnetic field dependence of the induced magnetization of $PrNi_2$ a molecular field constant $\lambda=67$ kOe/μ_B was calculated self-consistently using the Hamiltonian:

$$\hat{H}=\hat{H}_{CEF}-g^2\mu_B^2\lambda\langle J_z\rangle\hat{J}_z+g\mu_B\hat{J}_zH_z. \qquad (6)$$

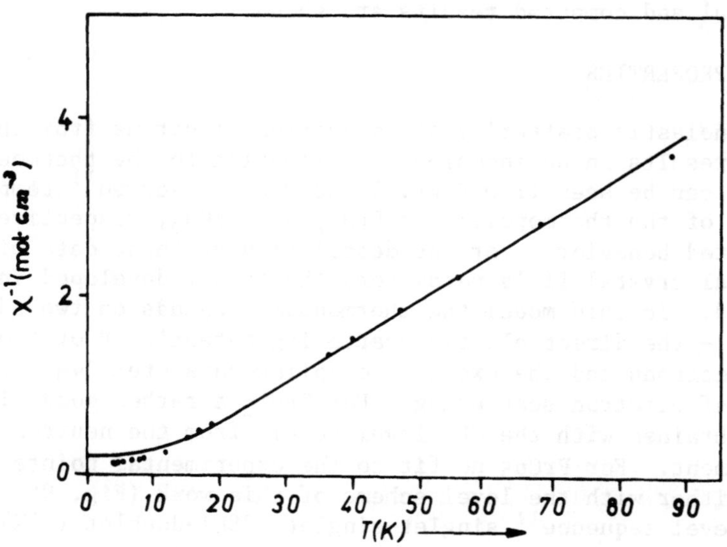

Fig. 6: The inverse susceptibility of $PrCu_5$ vs T. The points are experimental values and the curve is computed with the determined crystal field parameters.

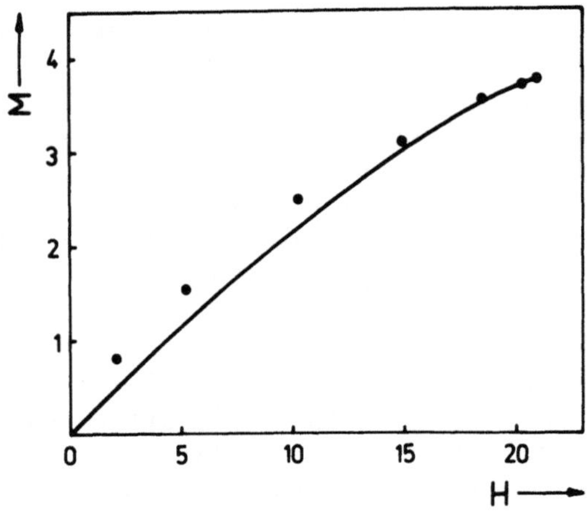

Fig. 7: Magnetic field dependence of the induced magnetization in $PrNi_2$ at T=4.2K: The line is computed using the level scheme from Fig. 4(a) and the molecular field constant λ=67 kOe/μ_B. M and H are given in 1000 erg/(Oe·mole) and kOe, respectively.

This value is very close to the critical one λ_c=86.5 kOe/μ_B for the onset of an induced magnetic ordering in $PrNi_2$. In Fig. 7 the experimental and computed results are shown.

TRANSPORT PROPERTIES

The inelastic scattering of conduction electrons from the f-electrons results in an anomalous contribution to the thermoelectric power. As can be seen from Figs. 8 and 9 the observed[11] temperature dependence of the thermopower for $PrNi_2$ and $PrCu_5$, respectively, shows a complicated behavior. For the description of these data with our experimental crystal field parameters the theory developed in Ref. 12 was used. In this model the thermopower depends on two additional parameters - the direct elastic scattering potential V of the conduction electrons and the exchange coupling parameter J_{MM} for conduction - 4f electron scattering. For $PrNi_2$ a rather good fit could be obtained with the CEF level scheme from the neutron scattering experiment. For $PrCu_5$ no fit to the experimental points was obtained either with the level scheme of this work (Fig. 9b) or with the level sequence[11] singlet-singlet (28K)-doublet (40K). The disagreement between experiment and calculation seems to be a consquence of using the CEF-only-theory, which does not take into account details of the scattering mechanism, the symmetry of the wave functions of the crystal field levels and the influence of phonons on the level scheme[13].

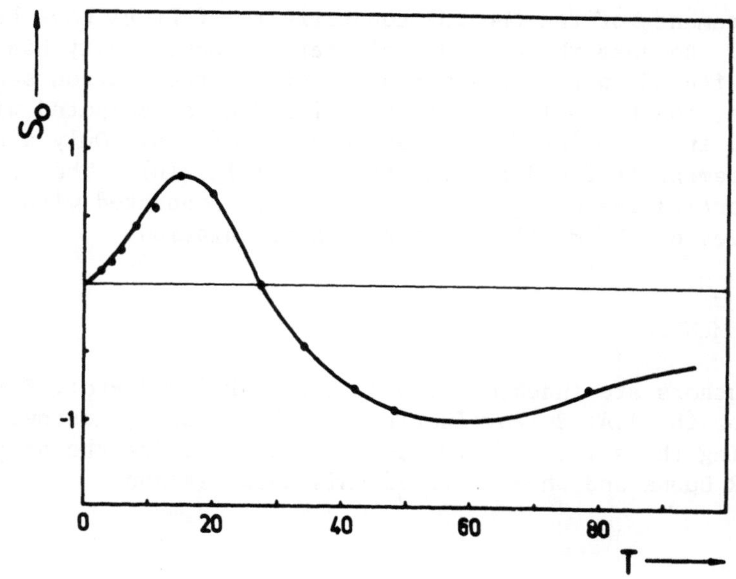

Fig. 8: Temperature dependence of the thermoelectric power S_o (μV/K) of $PrNi_2$. The solid line is a fit to the experimental data (Ref. 11) with our CEF level scheme and the parameters V=-57 meV and J_{11}=-29 meV.

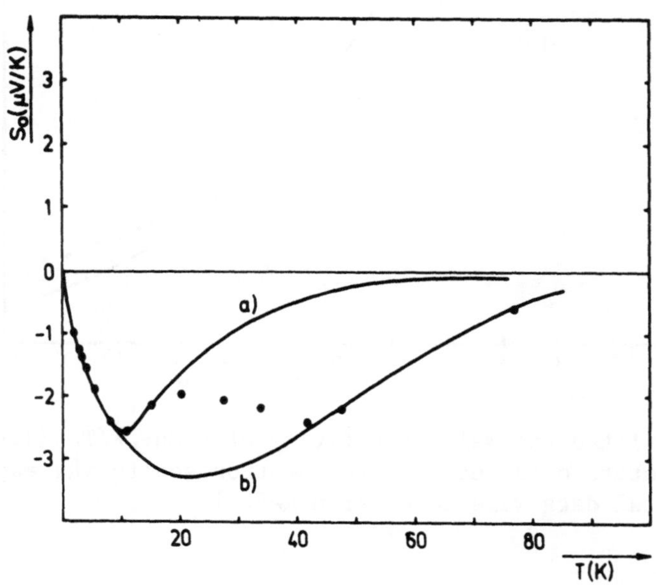

Fig. 9: Experimental (points) and calculated thermoelectric power of $PrCu_5$ using different CEF level schemes (see text). For the adjustable parameters the following values were used: a)V=1 meV, J_{11}=-2.5 meV; b)V=5 meV, J_{11}=-5 meV.

Measurements of the thermal conductivity of $PrCu_5$ have been reported[14]. Because the analysis of thermal conductivity has been performed with CEF parameters which do not fit the neutron scattering experiments, the thermal conductivity of $PrCu_5$ was computed with new parameters, in a similar way as was done in Ref. 14. Only a qualitative agreement is found for the best fit (Fig. 10). The reason for the quantitative disagreement seems to be connected with insufficiencies of the model used for this calculation.

ACKNOWLEDGEMENTS

The authors are indebted to O.D. Chistyakov and Prof. E.M. Savitskii of the A.A. Baikov Institute of Metallurgy, Moscow, for preparing the samples. W.M. and B.L. acknowledge the hospitality of the JINR Dubna and where part of this work was done.

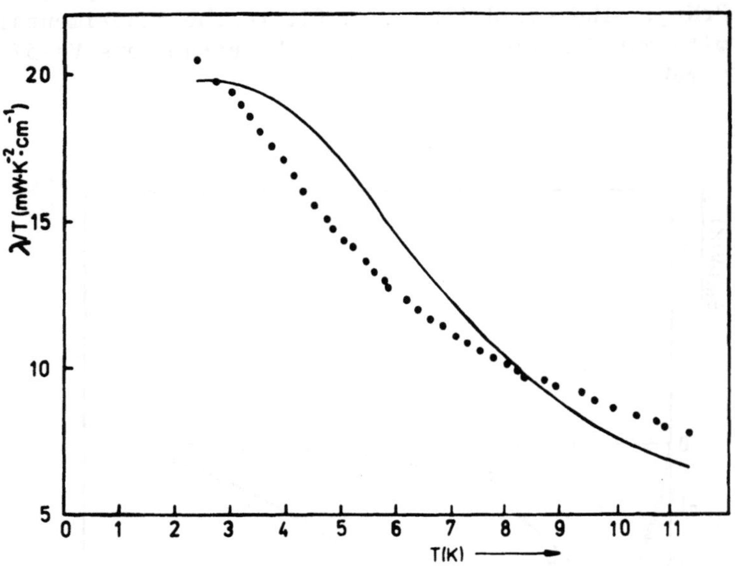

Fig. 10: Calculated thermal conductivity of $PrCu_5$, λ/T, (line) using the CEF parameters obtained from the neutron scattering experiment The experimental data were taken from Ref. 14.

REFERENCES

1. A. Andreeff, L.P. Kaun, T. Frauenheim, B. Lippold and W. Matz, Fiz. Element. Castic i atomn. yadra 12, 277 (1981); for english translation see Soviet Physics: Particles and Nuclei.

2. K. Andres, E. Bucher, P.H. Schmidt, J.P. Maita and S. Darack, Phys. Rev. B11, 4364 (1975).
3. W.E. Wallace and K.H. Mader, Inorg. Chem. 7, 1627 (1968).
4. H. Mori, T. Fujita, T. Satoh and T. Ohtsuka, Phys. Lett. 79A, 121 (1980).
5. W.E. Wallace, Rare Earth Intermetallics, Academic Press, New York and London, 1973, Ch. 3.
6. P.G. de Gennes, in : Magnetism, Vol. III, etc. G.T. Rado and H. Suhl, Academic Press, New York, 1963, p.115.
7. W. Matz, Communications JINR, P14-11111, Dubna 1977.
8. A. Andreeff, E.A. Goremychkin, H. Griessman, B. Lippold, W. Matz, O.D. Chistyakov, and E.M. Savitskii, Phys. Stat. Sol. (b) 98, 283 (1980).
9. P.A. Alekseev, A. Andreeff, H. Greissman, L.P. Kaun, B. Lippold, W. Matz, I.P. Sadikov, O.D. Chistyakov, I.A. Markova and E.M. Savitskii, Phys. Stat. Sol. (b) 97, 87 (1980).
10. K. Hennig, L.P. Kaun, B. Lippold, S. Matthies, W. Matz, W. Drexel, P. Seifert, N.I. Moreva, Solid State Commun. 21, 941 (1977); T. Frauenheim, W. Matz, E. Feller, Solid State Commun. 29, 805 (1979).
11. E. Bucher, J.P. Maita, E.W. Hull, J. Sierro, C.W. Chu and B. Lüthi in: Conf. on Crystal Field Effects in Metals and Alloys, R.A.B. Devine, Ed., University of Montreal, 1974.
12. A.K. Bhattacharjee, B. Coqblin, Phys. Rev. B13, 3441 (1976).
13. V.L. Aksenov, T. Frauenheim, N.M. Plakida and J. Schreiber, J. Phys. F (1981), in press.
14. Y.H. Wong, Phys. Rev. B17, 3899 (1978).

COMMENTS

GREIDANUS: I have a comment concerning $PrNi_2$. In our experiment (on a triple axis spectrometer) we have seen a transition at about 3 meV, and some intensity increase at higher energies (around 8 meV) which one can consider as a second transition. The intensity ratio of these two transitions favors a scheme with a positive x-value, in our opinion. In addition, the neutron spectrum is not consistent with the level scheme Bucher proposed some years ago. This scheme predicts (at a higher temperature) a transition at an energy of 15-20 meV. This transition is not observed.

KNORR: Years ago, I investigated $PrNi_2$ by inelastic neutron scattering (unpublished). I remember that I observed broad, almost overdamped transitions, as you did and that the level scheme was in agreement with that proposed by Bucher. Do you agree with Bucher and how do you explain the large linewidth?

MATZ: A calculation of the linewidth of the Γ_3-Γ_5 excitation in $PrNi_2$ in the BFK theory would give a value of about 1 meV. Also, dispersion averaging cannot explain the observed linewidth of about 6 meV. Maybe the effect discussed in a previous session by Dr. Keller is important here. Our level scheme does not agree with the data of Bucher.

LOEWENHAUPT: What is the origin of the inelastic peak around 100 meV?

MATZ: The peak at about 70-80 meV is due to the scattering of hydrogen. The peak was also found in $LaNi_2$ and can be enhanced in intensity by loading the sample with more hydrogen.

MAGNETIC EXCITATIONS IN TbP UNDER HYDROSTATIC PRESSURE

A. Loidl*, K. Knorr* and C. Vettier†

*Institut für Physik der Universität Mainz, FRG
†Institut Laue-Langevin, Grenoble, France

INTRODUCTION

Singlet ground state magnets still attract much attention. In these systems magnetic ordering occurs via the polarization of the ground state, and noticeable dispersion of well defined excitations is expected in systems where the exchange coupling is about equal to the crystal field energy. In addition, the two singlet-triplet systems PrSb[1] and TbP[2] where single crystals are available and which have been studied in detail showed a pronounced splitting of the Γ_1-Γ_4 magnetic excitations where the triple degeneracy was lifted along [100] and [011] by a strong anisotropy of the exchange.

The evaluation of induced-moment magnetism in TbP was followed below and above the ordering temperature[2-4]. Terbium phosphide crystallizes in the cubic rocksalt structure. Its ground multiplet is split into a singlet Γ_1 crystal field ground state, followed by two triplets Γ_4 and Γ_5 at 1.6 meV and 3.2 meV[2], respectively. The remaining crystal field states are above 10 meV. The critical ratio of the crystal field strength to the exchange coupling is slightly overcritical and at 7.3 K the system undergoes a phase transition to a Type II antiferromagnetic arrangement. The magnetic phase change is accompanied by a slight trigonal distortion of the cubic cell. Above T_N a strong dispersion and a splitting of the Γ_1-Γ_4 magnetic excitons were observed. The evaluation of these results yielded an anisotropic exchange coupling of long range[2]. In the ordered phase we found an accidental coincidence of the first three excited molecular field states. The results could be explained with a magnetic molecular field of H_m=4 T and a quadrupolar field B_2=9μeV.

During the last few years the pressure variable was used in

slightly undercritical singlet ground state systems to induce magnetic order[1,5,6]. In the present study of TbP under hydrostatic pressure we wanted to clarify the following questions: 1) At standard pressure TbP exhibits two soft modes at $\vec{q}_0$=(0.5,0.5,0.5) and at $\vec{q}_1$=(0.25,0,0). The competition of these two modes is best described by the two corresponding theoretical ordering temperatures of T_0=5.5 K and T_1=4.3 K.[2] The actual phase transition is of first order and takes place at 7.3 K at the ordering wave vector $\vec{q}_0$ with an energy gap of 0.5 meV. Can hydrostatic pressure influence the competition between the soft modes and possibly change the ordering wave vector? 2) How does the anisotropic exchange coupling depend on the interion spacing? This information would help to set up a microscopic model.

EXPERIMENTAL DETAILS

The experiments were performed on the triple axis spectrometer IN2 located at a thermal neutron source at the high flux reactor of the Institut Laue-Langevin in Grenoble. Pyrolytic graphite crystals were used as monochromators, analyzer and filter. Most of the data were obtained using the constant-$\vec{Q}$ technique with a fixed incident energy of 14.8 meV. The collimations were chosen to be 60'-30'-30'-40' along the path of the neutrons. The measurements at high pressures and low temperatures were made in a previously developed high pressure apparatus[6] where the oriented single crystal plus pressure transmitting fluid is mounted inside an Al_2O_3 cylinder in an aluminium cell. The TbP crystal was oriented with the [01$\bar{1}$] direction perpendicular to the scattering plane. The crystal volume was approximately 0.3 cm^3 and the volume of the fluid was minimized by filling the space around the crystal with NaCl. This also serves as an independent check on the pressure, as the equation of state of rocksalt is well known. The measurements have been performed at 9 and 16 kbar where we found a lattice spacing of 5.663 and 5.664 Å compared to the value of 5.680 Å at atmospheric pressure.

RESULTS AND DISCUSSION

In a first set of experiments we determined the ordering temperatures at applied hydrostatic pressures. The distinct first order transition at standard pressure[3] is considerably smeared out with a huge hysteresis. The Néel temperature (defined as the midpoint of the hysteresis loop) is almost pressure independent. It is shifted from 7.3 K at standard pressure to 8 and 8.5 K at 9 and 16 kbar, respectively. It seems that the ratio of the exchange to the level splitting stays almost constant over the whole pressure range. Representative scans at 11 K at three pressures are shown in Fig. 1.

The anisotropic exchange splits the Γ_1-Γ_4 exciton into two branches. With increasing pressure this anisotropic splitting increases, resulting in a softening of the lower branch. The

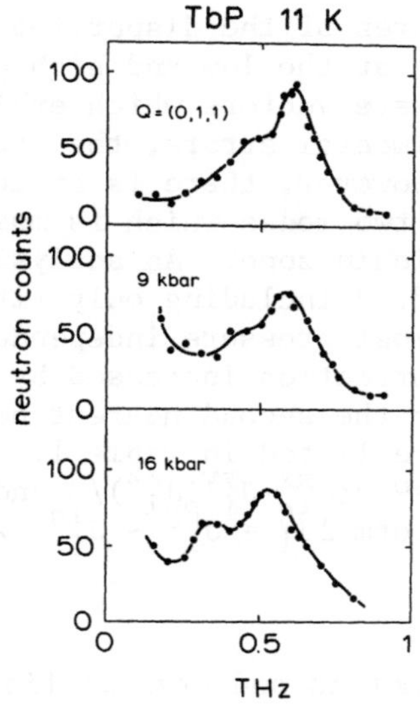

Figure 1: Scans at 11 K at $\vec{Q} = (0,1,1)$ for atmospheric pressure, 9 kbar and 16 kbar.

measured dispersion relation of the magnetic dipole transition between the ground state and the lowest excited state is shown in Fig. 2. Here the data at 16 kbar and 11 K are compared with the model which yielded the best fit of the data at standard pressure.

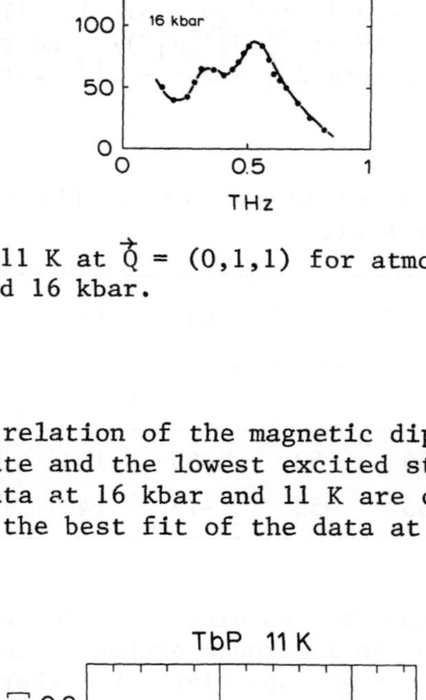

Figure 2: The dispersion of the split Γ_1-Γ_4 magnetic exciton at 11 K and 16 kbar. The solid line represents the best fit of the data at atmospheric pressure using the model described in Ref. 2.

The essential features of the dispersion of the magnetic excitations are the same at the low and high pressure. In particular, at the two wave vectors which exhibit soft mode behavior within the experimental errors, the exciton energies are pressure independent. However, there is an increase in the anisotropic splitting of the two modes which is most significant around the X-point of the Brillouin zone. An analysis using the phenomenological model of Ref. 2 including only first and second nearest neighbors yielded an almost pressure independent isotropic exchange, while the anisotropic correction increased by 40% for the first neighbors and by 30% for the second nearest neighbors. All the parameters of the fit are listed in Table 1. (The isotropic exchange is defined by $J_i^{iso} = (J_i^{xx}+J_i^{yy}+J_i^{zz})/3$ and the anisotropic corrections are of the form $\Delta J_i^{\beta} = J_i^{\beta\beta} - J_i^{iso}$ with $\Delta J_1^y=\Delta J_1^x$ and $\Delta J_2^y=\Delta J_2^z$).

Table 1: Coupling parameters and crystal field splitting in TbP at 0 and 16 kbar.

P	$\Delta(\Gamma_1-\Gamma_4)$	J_1^{iso}	J_2^{iso}	ΔJ_1^x	ΔJ_1^x	ΔJ_2^x	ΔJ_2^x
kbar	meV			µeV			
0	1.6	66	-53	-53	108	29	-14
16	1.6	63	-53	-73	147	38	-17

In calculating the exciton dispersion for even higher anisotropic corrections to the isotropic exchange one finds a soft mode at $\vec{Q}\simeq(0,0.6,0.6)$. One might speculate that therefore at higher pressures the lower branch at this wave vector becomes the new soft mode of the system.

We also studied the dispersion of the magnetic excitations in the ordered phase. In a singlet triplet system one expects two transverse modes originating from J transitions and a longitudinal mode that is split into an acoustic and an optical branch. Only one of these longitudinal branches carries intensity depending on whether the Brillouin zone contains a magnetic or a nuclear Bragg point. In TbP at standard pressure conditions we found an accidental coincidence of the first three excited molecular field states. With increasing pressure this accidental degeneracy is lifted. In Fig. 3 this effect is demonstrated at $\vec{Q}=(0.5,0.5,0.5)$. We believe that hydrostatic pressure changes only the quadrupolar coupling resulting in an increased Néel temperature and in altered transition energies in the ordered phase. A quantitative analysis

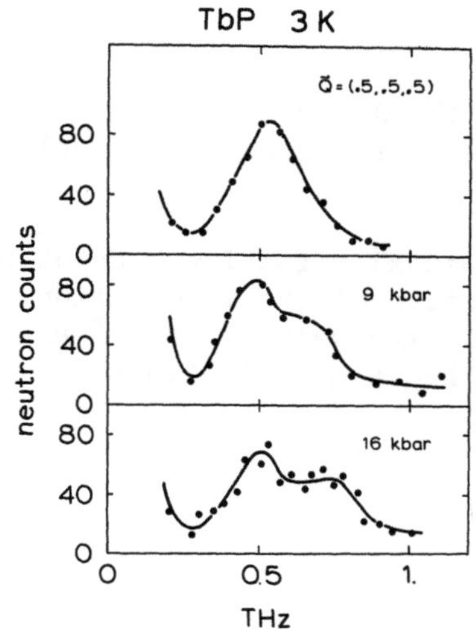

Fig. 3: Scans at $\vec{Q}$ = (0.5, 0.5, 0.5) in the antiferromagnetic phase at 0, 9, and 16 kbar.

of the excitations in the antiferromagnetic state in TbP is in progress.

In conclusion, we found no pressure dependence of the isotropic exchange and of the crystal field splitting in TbP. Relevant changes under pressure occurred only in the anisotropy. In the isomorphic compound PrSb the exchange coupling was nearly pressure independent also, while the crystal field splitting decreased with pressure contrary to the predictions of the point charge model.

This research was supported by the Bundesministerium für Forschung und Technologie.

REFERENCES

1. D.B. McWhan, C. Vettier, R. Youngblood and G. Shirani, Phys. Rev. B20, 4612 (1979).
2. K. Knorr, A. Loidl, J.K. Kjems, in "Crystalline Electric Fields and Structural Effects in f-Electron Systems", p. 141, J.E. Crow, R.P. Guertin and T.W. Mihalisin (eds.) Plenum Press (1980).
3. J. Kötzler, G. Raffius, A. Loidl and C.M.E. Zeyen, Z. Physik B35, 125 (1979).
4. K. Knorr, A. Loidl, J.K. Kjems, Z. Physik B40, 287 (1981).
5. K.A. McEwen, C. Vettier and W.G. Stirling, J. de Physique C5, 26 (1979).
6. D.B. McWhan and C. Vettier, J. de Physique C5, 107 (1979).

COMMENTS

ROSSAT-MIGNOD: Why is the J_i^{xy} component of the exchange tensor equal to zero? This is a very special situation indeed; what does it mean?

LOIDL: Yes, it is a special situation. Unfortunately, the theoretical microscopic picture of the interaction is missing. For pure dipolar interactions, finite values for J_i^{xy} are expected. Higher order terms may cancel these contributions.

TRANSPORT PROPERTIES OF THE INTERMETALLIC $PrAl_3$

H. Müller*, E. Hegenbarth*, W. Matz†, E. Mrosan*
and A. Schmeltzer*

*Technische Universität Dresden, Sektion Physik,
8027 Dresden, GDR
†Zentralinstitut für Kernforschung Rossendorf der AdW
der DDR, 8051 Dresden, GDR

In order to understand the nature of the crystalline electric field in intermetallics containing rare earth ions a variety of properties was studied. In addition to neutron crystal field spectroscopy and investigations of specific heat or susceptibility, which are rather directly influenced by the crystal field, we have recently examined more complicated crystal field effects. Here, we report measurements of the thermal and electrical conductivity of the intermetallic compound $PrAl_3$ at low temperatures. Crystal field spectroscopy by inelastic scattering was performed earlier[1] on the same sample, so we can use these data for analyzing the new experiments.

The polycrystalline samples of $PrAl_3$ and $LaAl_3$ for the measurements of thermal and electrical resistivities were 2x3x15 mm in size. The thermal conductivity (λ) was measured in the temperature range between 4.2 K and 80 K by the usual steady state method. The absolute temperature and the temperature gradient across the sample were recorded simultaneously by gold (0.03% Fe)-chromel thermocouples with an accuracy of about 1 mK. The temperature of the sample holder was kept constant during the experiment by an automatic temperature control device. Fig. 1 shows the results of the measurements on $PrAl_3$ and $LaAl_3$. The thermal resistivity multiplied by the temperature is shown. The experimental accuracy of the thermal resistivity is about 10%.

For the determination of the electrical resistivity (ρ) the standard four probe dc techinques with copper probes were used. Temperature control was as before. The errors are nearly the same

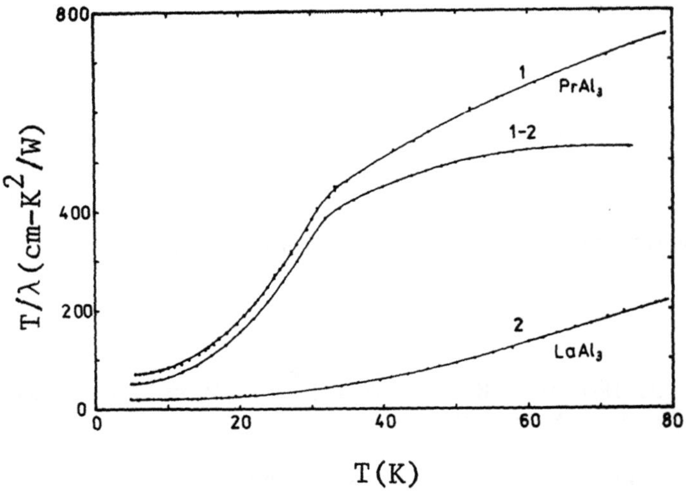

Fig. 1: Thermal resistivity multiplied by the temperature (T/λ) of $PrAl_3$ and $LaAl_3$ vs. temperature. The third curve is the magnetic part of thermal resistivity of $PrAl_3$.

as for the thermal resistivity experiments because the main error is due to geometrical effects in both cases. Fig. 2 displays the absolute electrical resistivity of PrAl3 and the magnetic contribution normalized to the resistivity at T=80K. The crystal field part of the electrical resistivity differs slightly from the data of Mahoney et al.[3].

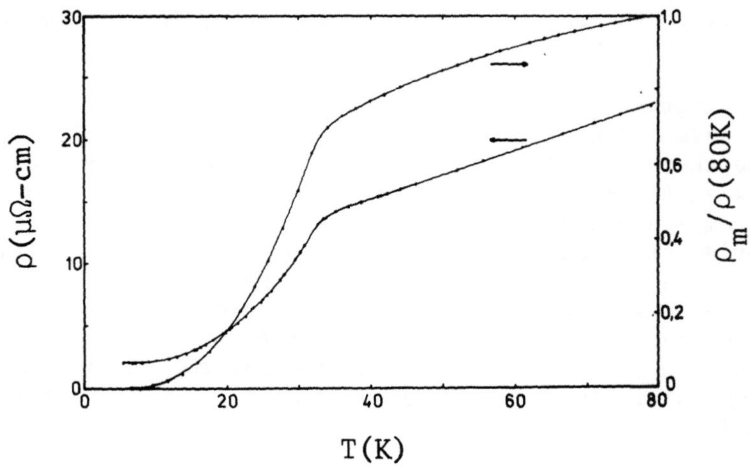

Fig. 2: Electrical resistivity ρ and relative magnetic part $\rho_m/\rho(T = 80\ K)$

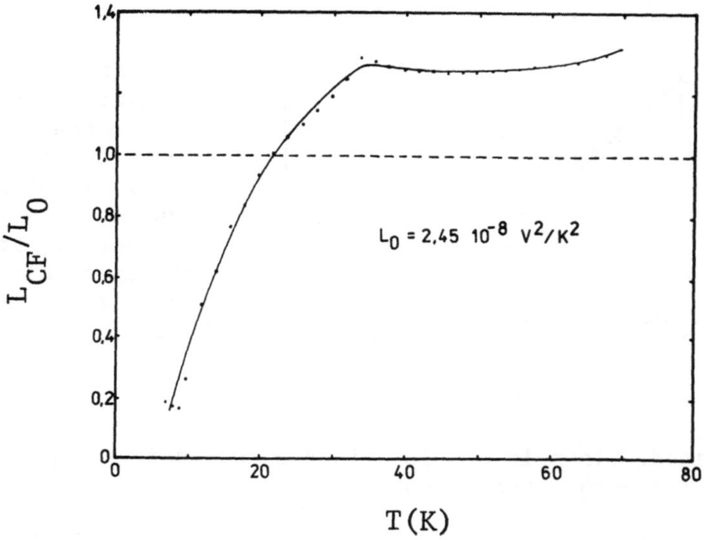

Fig. 3: Temperature dependence of the crystal field Lorenz number for $PrAl_3$.

Because the thermal conductivity in metallic materials is also due to electrons, the thermal and electrical conductivities are a sum of contributions from different scattering processes, according to Matthiesen's rule. Because we are interested in the magnetic part of the resistivities from crystal field effects, we must separate all other contributions such as those from impurity or electron-phonon scattering. For this purpose the isostructural compound $LaAl_3$ is used as the nonmagnetic counterpart of $PrAl_3$.

From the magnetic contributions to the thermal and electrical resistivities a crystal field Lorenz number L_{CF} was computed:

$$L_{CF} = \frac{\rho_{CF}}{W_{CF} \cdot T} \quad . \tag{1}$$

W_{CF} denotes the thermal resistivity. The data are shown in Fig. 3. For metals where only electrons contribute to the thermal conductivity, the relative Lorenz number L_{CF}/L_0 is 1 insofar as there are only elastic scattering processes[2]. Inelastic scattering processes however, have different effects on the thermal and electrical conductivities, leading to a suppression of the Lorenz number. In $PrAl_3$ the crystal field level scheme has a Γ_1 singlet ground state and an excited Γ_6 doublet at 4.6 meV[1]. Because no elastic scattering is possible on the Γ_1 level, for low temperatures there is only inelastic scattering (transition Γ_1-Γ_6) and consequently a low value for L_{CF}/L_0. With increasing temperature the Γ_6 doublet be-

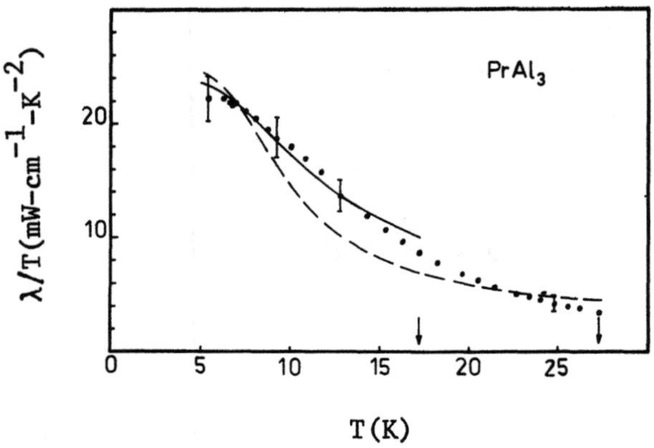

Fig. 4: Experimental and calculated crystal field part of the thermal conductivity of $PrAl_3$. The fits were performed with crystal field parameters from Ref. 1, including data up to different temperatures (marked by the arrows).

comes populated. In this doublet elastic scattering is possible, so that the Lorenz number increases.

The magnetic part of the thermal conductivity of $PrAl_3$ was analyzed within the framework of the equations given by Wong[4] and using the crystal field parameters from the neutron scattering experiment[1]. Several computer runs were done to fit the results of the measurements for different temperature ranges. Fig. 4 shows the experimental points of the thermal conductivity together with the fitted curves. It is seen that a good fit is achieved only if the analyzed region is restricted to a maximum temperature of about 17 K. If the temperature range is expanded it is not possible to get good agreement between calculation and experiment. In order to confirm that the data analyzed are due to crystal field effects, the measured thermal resistivity was fitted by a model which allows for contributions from crystal field, phonon and impurity scattering processes of the electrons. The crystal field part is represented by a simple two level splitting scheme. From this analysis of the temperature dependence over the whole temperature range investigated, it was clarified that more than 90% of the thermal resistivity is due to crystal field effects. It seems therefore that the model of Wong works only over a very small region of low temperatures. On the other hand this model is not very sensitive to different sets of crystal field parameters because it was tested both for $PrAl_3$ and $PrCu_5$.[5] This analysis shows that thermal conductivity is rather insensitive to the crystal field level scheme and information about crystal fields from such measurements is not likely to be too fruitful.

REFERENCES

1. A. Andreef, L.P. Kaun, B. Lippold, W. Matz, N.I. Moreva and K. Walther, Phys. Stat. Sol. (b) 87, 537 (1978).
2. J.M. Ziman, Electrons and Phonons, Clarendon Press, Oxford (1963).
3. J.V. Mahoney, W.E. Wallace and R.S. Craig, J. Appl. Phys. 45, 2733 (1974).
4. Y.H. Wong, Phys. Rev. B17, 3899 (1978).
5. W. Matz, B. Lippold, W.A. Goremychkin, A. Andreeff, H. Greissman and T. Frauenheim, this conference.

EFFECTS OF CRYSTALLINE FIELDS ON THE PHYSICAL PROPERTIES OF PrB_4

M. Kasaya, K. Takegahara, A. Yanase and T. Kasuya

Department of Physics, Tohoku University, Sendai, Japan

INTRODUCTION

Most of the rare earths form tetraborides whose structure is of ThB_4 type. The rare earth ions are in a 3+ state except Ce and Yb.[1] It is interesting that among the metallic rare earth tetraborides which show magnetic ordering, PrB_4 is the only ferromagnet.[2] Furthermore, strong magnetic anisotropy has been observed.[3] In order to investigate the origin of the magnetic anisotropy, we have measured magnetization curves for PrB_4 in magnetic fields up to 85 kOe. Using the crystalline field calculated on the basis of the point charge model, we have tried for the first time to analyze the magnetic anisotropy and the temperature dependent spin-disorder resistivity of PrB_4. The purpose of the present paper is to report and discuss the results of these observations.

EXPERIMENTAL RESULTS

Single crystals of the tetraborides PrB_4 and LaB_4 were prepared from a flux of excess rare earth metal. Powdered boron and the rare earth metal (weight ratio RE/B $\sim$25) were mixed and heated under argon atmosphere at 1700 C for 3 hours. The solution was cooled over a 3 hour period to 900 C and then quenched to room temperature. The typical weight of the samples for which magnetic data are presented is 2.5 mg. We used five pieces of single crystals for the magnetic measurements. Measurements of dc resistivity and Hall effect were carried out from room temperature to 4.2 K.

Experimental results of magnetization at 4.2K, the magnetic field being in the [001] direction, are shown in Fig.1. Berrada et al. have reported the magnetization of PrB_4 is 3.6 μ_B/Pr and nearly independent of external magnetic field.[3] However, the saturation magnetization at low field of the present work is 1.9 μ_B/Pr and increases gradually up to 2.1 μ_B at 85 kOe. The deviation from the free ion value gJ = 3.2 μ_B suggests that crystalline fields should be taken into account to explain the magnetic properties of PrB_4. The magnetization with the magnetic field in the (001) plane is too small to measure even at 85 kOe, namely less than one thirtieth of that along the [001] axis.

It is worthwhile to measure transport properties of PrB_4 to investigate whether or not crystalline fields also play an important role in the electrical resistivity ρ. The temperature dependence of ρ, the current being in the (001) plane, is shown in Fig.2. As a reference system for PrB_4, ρ and the Hall constant R_H of non-magnetic LaB_4 are also shown in the figure. As was expected, both PrB_4 and LaB_4 are metallic. The value of R_H of LaB_4 at room temperature, -2.8×10^{-3} cm^3/C, is nearly the same as that of YB_4 -2.1×10^{-3} cm^3/C.[4] Although the Hall constant of LaB_4 shows weak temperature dependence, this might be due to unequal temperature

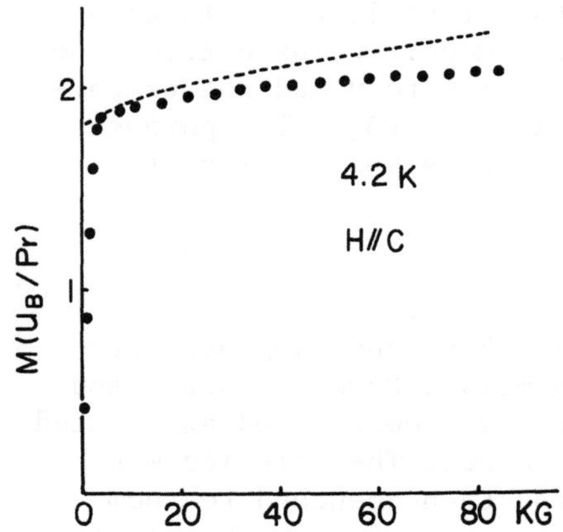

Fig.1 Magnetization of PrB_4 as a function of external magnetic field. The dashed line shows calculated results.

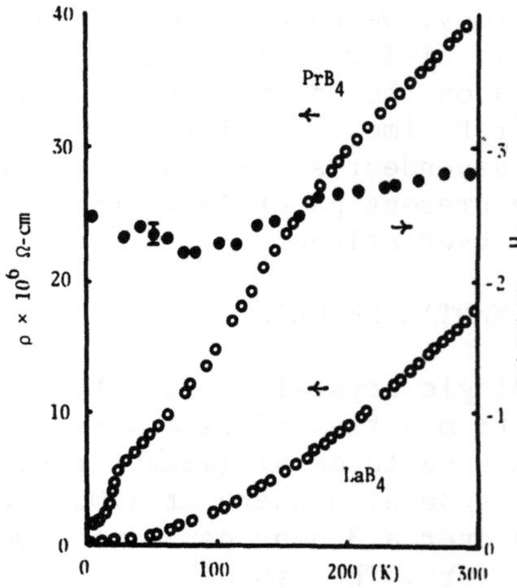

Fig.2 Temperature dependences of ρ of PrB_4 and LaB_4 and R_H of LaB_4. The ordinate axis for R_H is shown on the right side.

dependences of the mobilities of electrons and holes which coexist in the complicated Fermi surfaces. It is beyond the scope of the present study to mention about the band structure of the tetraborides. In the following we assume that the number of carriers is temperature independent. It is apparent from the figure that the ferromagnetic phase transition in PrB_4 gives rise to a sharp bend at 23 K in the ρ vs T curve. Above 100 K, ρ(T) of PrB_4 exhibits a convexity which is clearly absent in LaB_4. Therefore this temperature dependent part can be attributed to the magnetic scattering from Pr^{3+} ions.

CRYSTALLINE FIELD

PrB_4 crystallizes in the tetragonal space group P4/mbm-D^5_{4h} with four formula units/cell occupying the following positions:

Pr	4g	x, 0.5+ x, 0	x= 0.319	
B(1)	4e	0, 0, z	z= 0.196	a= 7.235 A
B(2)	4h	x, 0.5+ x, 0.5	x= 0.086	c= 4.116 A
B(3)	8j	x, y, 0.5	(x= 0.175, y= 0.039	

According to the point charge model, the crystalline electric field potential has the form[5]

$$V = \sum_{\ell=2}^{6} \sum_{m=-\ell}^{\ell} \frac{e^2 \langle r^\ell \rangle}{a^{\ell+1}} A_{\ell m} C_m^{(\ell)} \tag{1}$$

where $C_m^{(\ell)} = (\frac{4\pi}{2\ell+1})^{1/2} Y_{\ell m}(\theta,\phi)$. In the calculation of the coefficient $A_{\ell m}$, the contributions from all Pr^{3+} and B^{3-} ions are considered. Assuming for Pr^{3+}, $\langle r^2\rangle=1.0632$, $\langle r^4\rangle=2.5217$ and $\langle r^6\rangle=11.4920$ in atomic units,[6] we obtain non-zero values of $B_{\ell m}$ $(=e^2\langle r^\ell\rangle A_{\ell m}/a^{\ell+1})$ as follows: $B_{20}=4400.17$, $B_{22}=3892.42i$, $B_{40}= -1060.95$, $B_{42}=0.25i$, $B_{44}= -169.41$, $B_{60}= -20.09$, $B_{62}=8.83i$, $B_{64}=5.06$ and $B_{66}= -9.24i$ K. The nine-fold degenerate $^3H_4$ term splits in the crystalline field with the potential (1) into nine singlets. The energy scheme for splitting of the $^3H_4$ term, along with the eigenvectors, is shown in Fig. 3.

DISCUSSION

Firstly, we will discuss the effect of crystalline fields on the magnetic anisotropy of PrB_4. To calculate the saturation magnetization for the ferromagnetic state, it is necessary to investi-

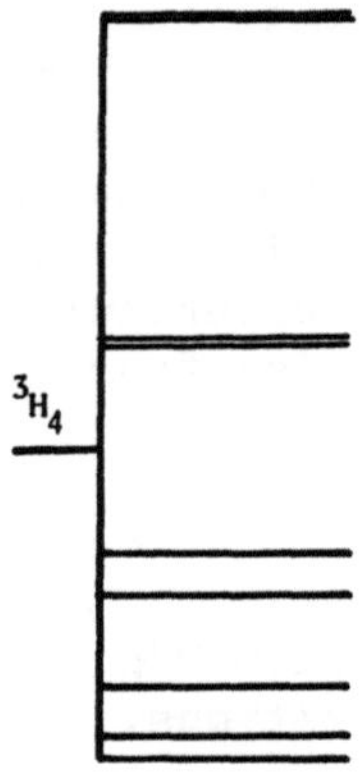

E(K)	
2079.9	$-0.066\|-4\rangle+0.434i\|-2\rangle+0.784\|0\rangle-0.434i\|2\rangle-0.066\|4\rangle$
2077.3	$0.208i\|-3\rangle+0.676\|-1\rangle-0.676i\|1\rangle-0.208\|3\rangle$
525.7	$0.456\|-3\rangle-0.541i\|-1\rangle+0.541\|1\rangle-0.456i\|3\rangle$
491.3	$0.210i\|-4\rangle+0.675\|-2\rangle+0.675\|2\rangle-0.210i\|4\rangle$
-492.0	$0.350i\|-4\rangle+0.463\|-2\rangle+0.571i\|0\rangle-0.463\|2\rangle+0.350i\|4\rangle$
-684.3	$0.676\|-3\rangle+0.208i\|-1\rangle+0.208\|1\rangle+0.676i\|3\rangle$
-1131.4	$0.541i\|-3\rangle-0.456\|-1\rangle-0.456i\|1\rangle+0.541\|3\rangle$
-1377.9	$-0.675\|-4\rangle-0.210i\|-2\rangle-0.210i\|2\rangle+0.675\|4\rangle$
-1488.6	$0.611\|-4\rangle+0.313i\|-2\rangle-0.242\|0\rangle-0.313i\|2\rangle+0.611\|4\rangle$

Fig.3 Crystalline-field scheme for Pr^{3+} in the orthorhombic C_{2v} point symmetry.

gate the effects of the exchange field on the crystalline field scheme presented in Fig.3. From the experimental value of T_C =23 K, we estimated the value of the exchange field as 140 kOe ($\sim$7.5 K). The calculated result for the magnetization is shown by a dashed line in Fig.1. Reasonable agreement between the experimental result and the calculated one was obtained, provided that energy splitting shown in Fig. 3 is reduced by a factor of 1/2. The magnetization with the magnetic field being in the (001) plane, was also calculated and found to be about 0.1 μ_B/Pr in an external field of 80 kOe, which is consistent with the experimental results.

Next we will discuss the temperature dependent spin-disorder resistivity. The elastic and inelastic contributions to ρ from s-f exchange coupling have been investigated by Hirst and Andersen.[7] The resistivity can be expressed as $\rho_{mag} = \rho_0 \mathrm{Tr}(PQ)$, where the matrices P and Q are

$$P_{ij} = \frac{\exp(-E_i/kT)}{\Sigma_k \exp(-E_k/kT)} \cdot \frac{(E_i-E_j)/kT}{1-\exp[-(E_i-E_j)/kT]} \tag{2}$$

and
$$Q_{ij} = |\langle i|J_z|j\rangle|^2 + \frac{1}{2}|\langle i|J_+|j\rangle|^2 + \frac{1}{2}|\langle i|J_-|j\rangle|^2,$$

respectively. Q_{ij} is calculated using the eigenvectors shown in Fig.3. Calculated results are shown by a dashed line in Fig. 4. In the calculation, we used the 1/2-reduced energy splitting and the calculated curve was fit to the experimental value at room temperature. The asymptotic value of ρ_{mag} is $\rho_o J(J+1)=20\rho_o$. The tendency to saturation of ρ_{mag}, i.e., $\rho(PrB_4)-\rho(LaB_4)$, above 200 K can be reproduced by the present calculation. ρ_o is $1.3\times10^{-6}\,\Omega$-cm. However, the discrepancy between the experimental result and calculated one is large below 200 K.

In conclusion, magnetic anisotropy of PrB_4 was explained qualitatively by the effects of crystalline field. More detailed investigations are necessary to clarify the transport properties.

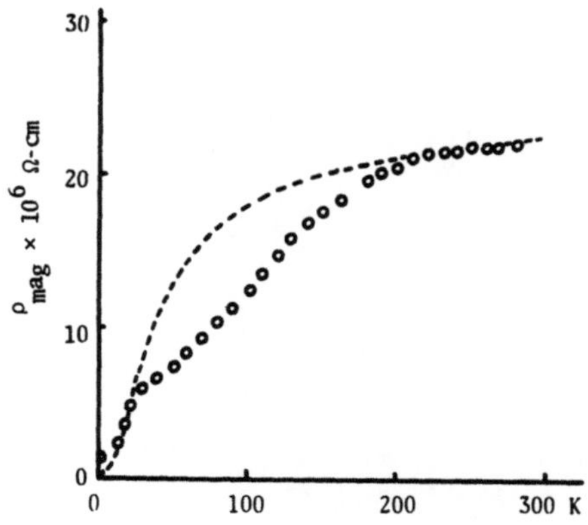

Fig.4 ρ_{mag} as a function of temperature. o : $\rho(PrB_4)-\rho(LaB_4)$. --- : calculated result.

REFERENCES

1. Z. Fisk, A.S. Cooper, P.H. Schmidt and R.N. Castellano, Mat. Res. Bull., 7,285(1972).

2. K.H.J. Buschow and J.H.N. Creyghton, J. Chem. Phys., 57,3910(1972).

3. A. Berrada, J.P. Mercurio, B. Chevalier, J. Etourneau, P. Hagenmuller, M. Lalanne, J.C. Gianduzzo and R. Georges, Mat. Res. Bull., 11,1519(1976).

4. R.W. Johnson and A.H. Daane, J. Chem. Phys., 38,425(1963).

5. B.G. Wybourne, "Spectroscopic Properties of Rare Earths," Interscience Pub., New York, (1965).

6. S. Fraga, J. Karwowski and K.M.S. Saxena, "Handbook of Atomic Data," Elsevier, Amsterdam, (1976).

7. L.L. Hirst, Solid State Commun., 5,751(1967).
N.H. Andersen, P.E. Gregers-Hansen, E. Holm, H. Smith and O. Vogt, Phy. Rev. Lett., 32,1321(1974).

CRYSTAL ELECTRIC FIELD EFFECTS IN THE ESR OF DILUTE ALLOYS

Klaus Baberschke

Institut für Atom- und Festkörperphysik, Freie Universität Berlin, D-1000 Berlin 33, FRG

INTRODUCTION

Typical microwave frequences for paramagnetic resonance (ESR) are 1, 3, 9 or 35 GHz. This corresponds to an energy of 50 to 1500 mK·k_B, respectively, 4 to 140 μeV. The energy resolution of the ESR is limited by the linewidth of two adjacent transitions; in metals for dilute alloys typically a few Gauss or 10^{-2} to 10^{-4} of the Zeeman energy. Inelastic neutron scattering (INS) as another microscopic technique covers a range of energy transfer of the order of 1 to 100 K·k_B with an energy resolution of 1% to 10%. The lowest concentration of rare earth (RE) ions in dilute alloys to be detected is 10 to 100 ppm for ESR and the order of 1% for INS.

This comparison demonstrates the applicability of ESR for questions of the crystalline electric field (CEF):

(1) The low concentration favors the investigation of non-interacting RE ions.

(2) The energy range limits the ESR to the investigation of the magnetic groundstate. Experiments on thermally excited levels of the resonance ion[1] or in Van Vleck host materials[2] will not be discussed here.

(3) The high energy resolution enables the investigation of small CEF splittings ∿1 K, this is to say the CEF of the S-state ions Gd^{3+} and Eu^{2+}. The importance of this is twofold: (a) All experiments on S-state ions in the millikelvin regime (resistivity, spin glass, etc.) have to consider the finite S-state splitting, and (b) the absence of the largest contribution to the CEF originated from the angular momentum L, keeps small contributions (i.e. from the conduction electrons) more "visible". This argument is in complete analogy to the immense theoretical and experimental work on the core polarization in the hyperfine field for S-state ions[3].

In the following section we discuss the $\Gamma 8$ resonance especially for $(YEr)Al_2$ and compare the ESR with INS. Next we review the CEF of the S-state ions and the influence of the conduction electrons (CE) on the $4f^7$ configuration. Then we briefly present ESR and CEF effects in the superconducting phase. Finally we will show how higher order contributions (Kondo) in the ESR yield information on CE and CEF splitting.

At the CEF Conference in 1979 K.W. Stevens[4] pointed out that there exist objections to conventional crystal field theory. For simplicity and because of lack of another more appropriate theory, we will continue to parametrize the CEF effect on the ESR in the conventional way. For notation, see the standard literature[5,6].

$\Gamma 8$-RESONANCE OF $(Y_{1-c}Er_c)Al_2$

The CEF of RE-ions in the dialuminide Laves phases still remains a puzzle. All four quadrants in the LLW diagram have been proposed[7a,7b], $W \lessgtr 0$ and $x \lessgtr 0$. A strong concentration dependence has been proposed[8]. However, very early magnetization measurements[9], INS on c=.20[10] and c=.02[11], magnetic anisotropy experiments[12] and ESR for c=1500 ppm[13] as well as the result of the present work with c=500 ppm all lead to roughly the same W, x parameters, namely $W \sim -0.3$ K and $x \sim -0.26$ to -0.32, corresponding to $A_4>0$ and $A_6<0$. In the following we will discuss our results on very dilute samples[14] and demonstrate how the ESR technique can contribute to the solution of CEF. Similar arguments would work if one replaces Er by Dy, etc. or Y by Sc, La or Lu.

From ESR on polycrystalline samples a $\Gamma 7$ ground state was deduced[15]; more recent experiments on single crystals yield a $\Gamma 8$ ground state but two solutions for $x \lessgtr 0$ fit the experiments. In the case of an isolated $\Gamma 8 (W \to \infty)$ only two magnetic dipole transitions are expected within the $\Gamma 8$ quartet[5]. The transitions for $\pm 1/2 \leftrightarrow \pm 3/2$ coincide. For finite splitting and (for purposes of discussion $W<0$, $x<0$) the first excited Γ_8^2 would mix via the Zeeman energy with the Γ_8^3 ground state. The energy separation between the $|+1/2\rangle$ and $|+3/2\rangle$ states and $|-1/2\rangle$ and $|-3/2\rangle$ are no longer equal, resulting in different resonance transitions. Our first results (Fig. 1 in Ref. 13) suggest such a possibility. The line shape changes with orientation (superposition of more than one line), and the linewidths are inconsistent (local low symmetry distortions - inhomogeneous broadening). To clarify this question we prepared more dilute samples. After various thermal treatments we succeeded in narrowing the individual linewidths and suddenly a third line appeared (Fig. 1). The angular dependence in the {110} plane is shown in Fig. 2. In the |001| direction we attribute the $-3/2 \leftrightarrow -1/2$ transition to the high field line, the $1/2 \leftrightarrow -1/2$ to the low field and the $+3/2 \leftrightarrow +1/2$ to the weak line. The level diagram in Fig. 3

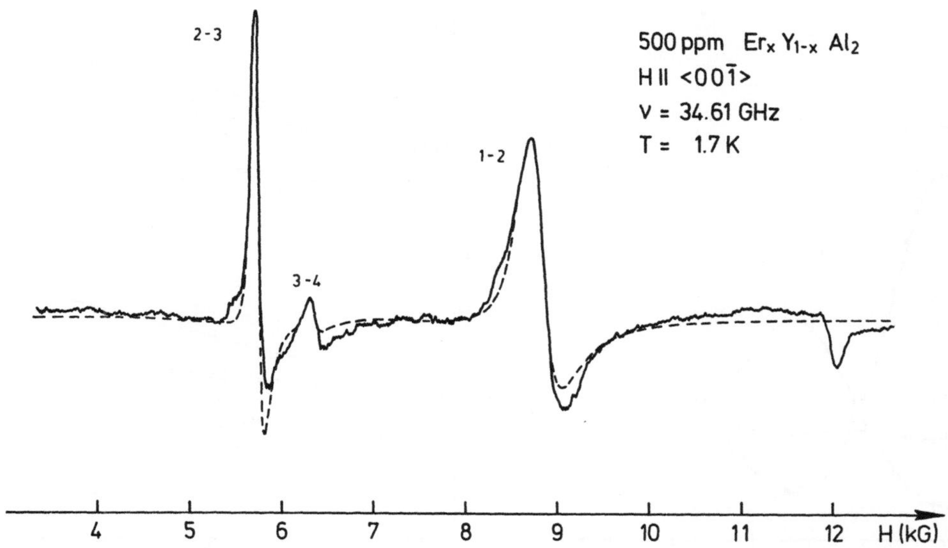

Fig. 1: ESR spectrum of a (YEr)Al_2 single crystal. The peak at 12 kG is a background signal of the empty cavity. The dashed line is the theoretical fit, see text. The hyperfine lines of the Er isotope can be clearly seen on the high and low field sides of each of the transitions.

shows the 3 transitions for 9 and 35 GHz. From this complete set of information only one set of parameters could be deduced, namely W=-0.29(2) K$\cdot k_B$ and x=-0.32(2). A fit for x=+0.23 is impossible.

So far we have not included exchange interactions with the CE. Assuming $\mathcal{H}_{ex}=J_{eff}\underline{S}\cdot\underline{\sigma}$ and the validity of projecting $\underline{L}$ or $\underline{S}$ on $\underline{J}$, the full Hamiltonian is given by:

$$\mathcal{H}=W|x|\frac{O_4(J)}{F_4}+(1-|x|)\ \frac{O_6(J)}{F_6}\ +g_J\mu_B\underline{H}\cdot\underline{J}+J_{eff}(g_J-1)\underline{J}\cdot\underline{\sigma}. \qquad (1)$$

The magnetization of the CE $<\sigma_z>$ acts as an internal field and shifts H_{res} for the individual transitions. This effect has clearly been detected yielding a coupling strength between the Er ion and the CE of $N(E_F)\cdot J_{eff}$=+0.09. This number is in fair agreement with the value for (YGd)Al_2, $N(E_F)J_{eff}$=+0.07[16]. Within our experimental error bar we were not able to detect any contribution of anisotropic exchange. Finally we compare in Table 1 some selected other results from various techniques and using data ranging from 100% Er to our lowest concentration. From this comparison, we are tempted to

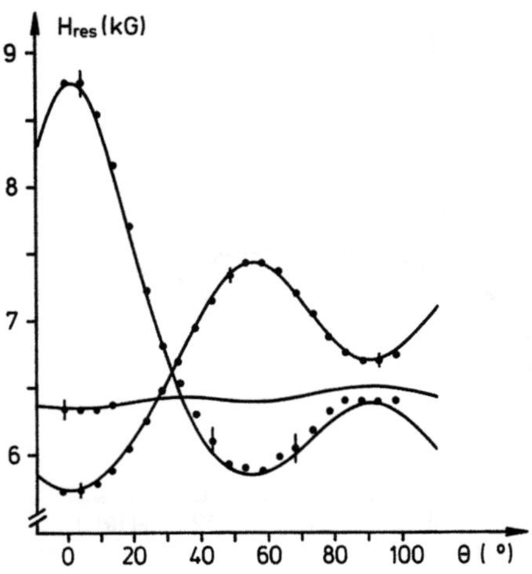

Fig.2: Angular dependence of H_{res} as a function of $\theta = ∢[(001),H]$. The g-shift corresponding to $N(E_F)J_{eff}$ is in the theoretical fit (solid lines) included.

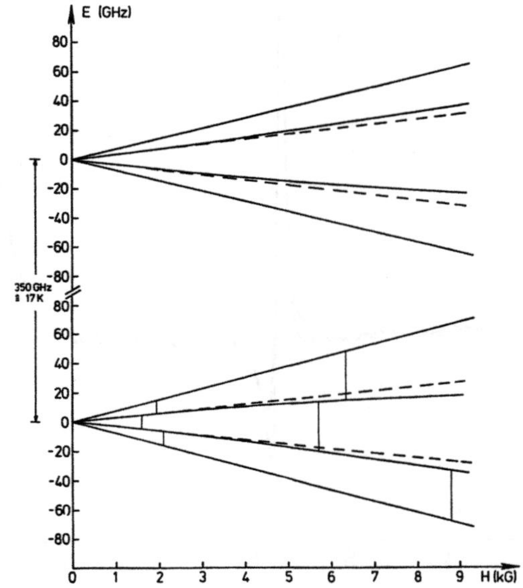

Fig. 3: Level scheme for Γ_8^3 ground state and Γ_8^2 first excited state. Full lines are the best ESR fit, dashed lines INS data, Ref. 11, see Table 1.

Table 1. CEF Parameters from several experiments

$(Y_{1-c}Er_c)Al_2$	$W/k_B \cdot K$	x	Comment
c = 1.00	-0.29	-0.26	magnetization Ref. 9
0.20	-0.25	-0.3	INS Ref. 10
0.02	-.25(10)	-0.31(5)	INS Ref. 11
0.005	-.25	-0.3	Param. anisotr., Ref. 12
0.0005	-.29(2)	-.032(2)	ESR, this work

conclude that no clear concentration dependence of W,x can be seen. It is more likely that a "pseudo concentration dependence" is due to incomplete data sets and analysis and/or different sample preparations[7b,8,12,15].

In summary: a long list of CEF level schemes has been published for $(YEr)Al_2$. Many mistakes were caused by incomplete data sets: missing transitions in ESR, INS and other techniques as well. In the present example we believe that different techniques finally yield the same CEF parameters and ESR, if it works, is quite precise and sensitive to small effects, i.e. distortions, CE effects, etc.

CEF OF S-STATE IONS Eu^{2+} AND Gd^{3+}

There exists a fair number of ESR results on lanthanide S-state ions in metals and intermetallic compounds. At first sight the CEF seems to be as difficult to interpret as in insulators[17]. Direct, relativistic and intra-atomic configuration mixing effects interfere with each other[17,18]. However, in a detailed view, characteristic anomalies appear for metals. Remarkable examples are: (1) the large B_2^0 of Gd in Sc compared to Y and Lu, which may be attributed to the position and width of the 5d band in Sc[19], (2) the enormous CEF splitting and different sign of the g-shift for Eu^{2+} and Gd^3 in some Chevrel phases[20], (3) similar effects observed for Gd^{3+} and Eu^{2+} in $BaAl_4$[21]. We will discuss these "metallic" contributions to the CEF for Gd^{3+} vs Eu^{2+} in $LaAl_2$ single crystals.

In insulators there are essentially three contributions to the CEF: the direct contribution, via spin-orbit mixing into the $|^8S_{7/2}\rangle$ ground state, relativistic contributions, and intra-atomic configuration mixing such as $4f^6$ 6p or $4f^6$ 5f. For these three processes it is very difficult to explain the CEF splitting of 480 mK for Eu^{2+} whereas no splitting at all ($\lesssim$40 mK) was detected for the Gd^{3+} ion. The essential differences of the two ions in $LaAl_2$ are not the mechanisms mentioned above, but the position of the $4f^7$ level with respect to the conduction band. The interatomic covalent contributions to the exchange interaction mix $|4f^6\text{+val}; J,J_zks_z\rangle$ into the $|4f^7;S_z\rangle$. The hybridization between the conduction band and the $4f^8$ state has to be discussed as well. In Fig. 4 we show the proposed mechanism.[24] The contribution to the CEF is given to first order by[18,24]:

$$\frac{B_4^0}{A_4^0\langle r^4\rangle} \simeq \frac{10}{63}\left(\frac{\zeta}{E_-}\right)^4 \cdot \frac{N(E_F)|V_{kf}|^2}{E_-} \qquad (2)$$

For Eu^{2+} $E_-\simeq 1$ eV, $\zeta\simeq 0.16$ eV, in contrast to Gd^{3+} where the $4f^7$ level is far below E_F ($E_-\simeq 8$ eV). In the Schrieffer-Wolff limit the second term can be rewritten $N(E_F)|V_{kf}|^2/E_-=-N(E_F)J_{c.m.}$. This quantity in turn is accessible in an ESR experiment; $\Delta g_{cov.mix.}=N(E_F)J_{c.m.}$. In fact, the experimental g-shifts for Gd^{3+} and Eu^{2+} in $LaAl_2$ are different, namely 0.11 or 0.7 respectively, yielding $N(E_F)J_{c.m.}\simeq -.04$.

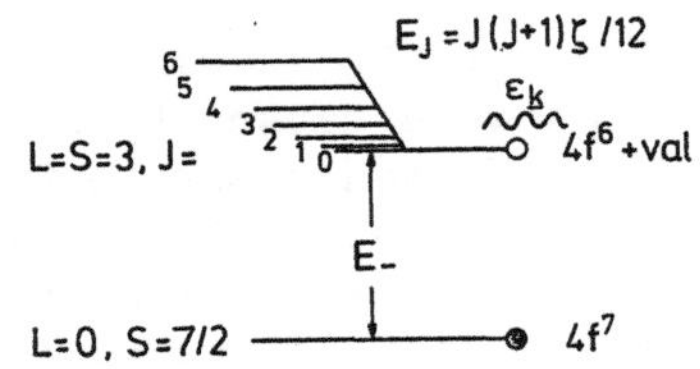

Fig. 4: Schematic energy scheme for the $4f^7$ and $4f^6$ state of Eu in $LaAl_2$.

Similar results can be drawn from a band structure calculation:

$$|\psi> \simeq \sqrt{1-\alpha^2}|4f^7>+\alpha|4f^6> \tag{3}$$
$$|\alpha|^2 \simeq \sum_k \frac{|V_{kf}|}{(\varepsilon_d+U-\varepsilon_k)} + \frac{N(E_F)|V_{kf}|^2}{E_-} \to N(E_F)J_{c.m.} \simeq -0.04.$$

We return to the main question: are real conduction electron effects to the CEF small? Many experiments are discussed without mentioning at all that the host material is a conductor. In small S-state crystal fields we believe that those effects are "visible". Unfortunately, all CEF effects enter via higher order perturbation theory, depending on many parameters, e.g. ζ/E_-. At present only the trend can be seen. Nevertheless, the problem is of general interest. For $E_-\to 0$ one can discuss the CEF in valence fluctuating compounds, i.e., for isomorphic Eu $Pt_{2-x}Rh_x$.

ESR IN THE SUPERCONDUCTING PHASE

In this section we will demonstrate another very interesting application of ESR, namely the determination of the relaxation rate or the order parameter in the superconducting phase. The local properties (at the impurity) of superconducting dilute alloys are of current interest. Very recently the first INS experiment[24] in $(La_{1-c}Tb_c)Al_2$ has been published. Mossbauer experiments[25], NMR[26], etc. are known as well. For ESR essentially two experiments are known involving CEF and relaxation in the superconducting phase. R. Odermatt[20] investigated Gd and Eu in the Chevrel phase materials (polycrystalline). We detected the Gd resonance in single crystals of $LaOs_2$[27,28]. In Fig. 5 we briefly recall the impetus of the experiment. The CEF splitting of S=7/2 in cubic symmetry into a Γ_6, Γ_7 and Γ_8 causes an unequal spacing of the 8 Zeeman levels for a fixed magnetic field. This is reflected in the so-called fine-structure, and its angular dependence in an ESR spectrum. Depending on temperature and microwave frequency some of the dipole transitions may not be visible, because of thermal depopulation. The largest splitting appears for H parallel to the |001| and |111| direction. If, however, the relaxation rate of the individual transitions is large or comparable to the energy separation of adjacent lines (roughly speaking) the spectrum collapses to an exchange narrowed

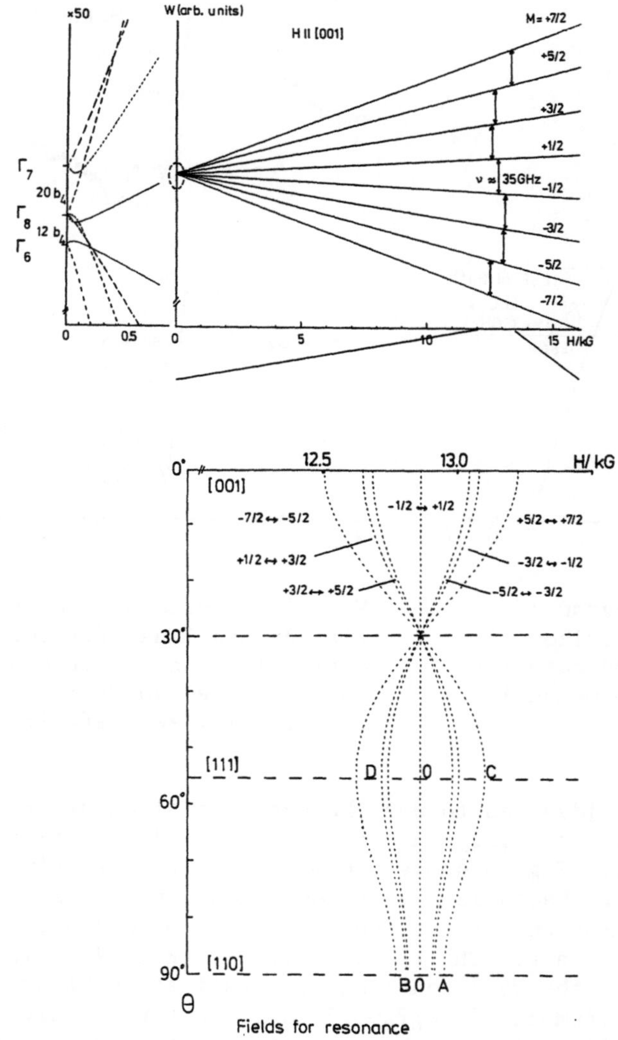

Fig. 5. Level diagram for Gd^{3+} in $LaOs_2$. The zero field splitting (CEF) equals $\Delta \simeq 32\ B_4 \simeq 75mK$. The lower part shows the angular dependence. Note that the $-7/2 \leftrightarrow 5/2$ transition is always on the other side of the center of gravity from the $-5/2 \leftrightarrow -3/2$ and $-3/2 \leftrightarrow 1/2$ transitions.

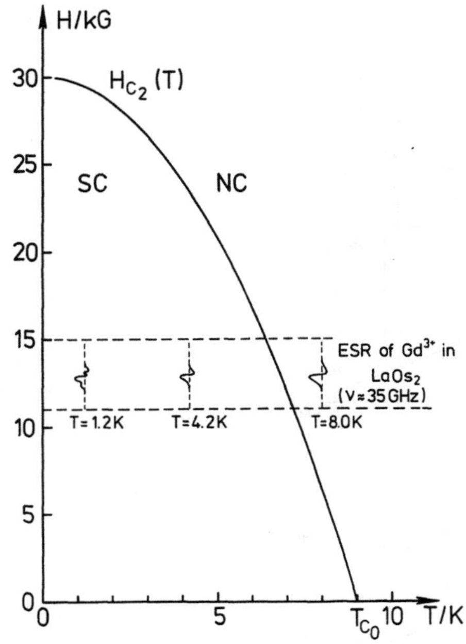

Fig. 6: Phase diagram of (LaGd)Os_2. The horizontal (dashed) lines indicate the scan range of the magnetic field.

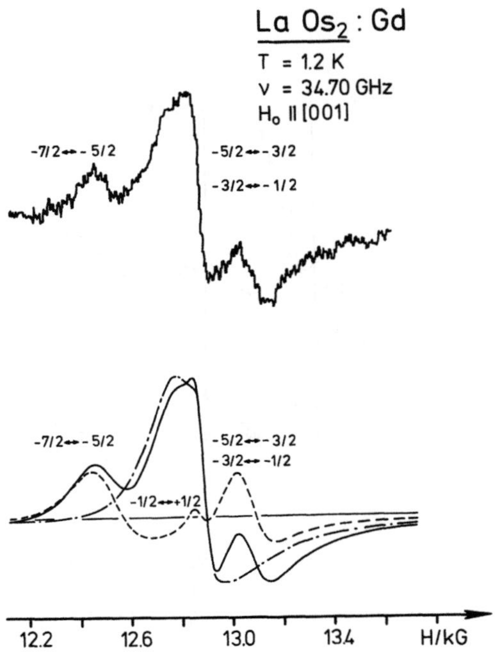

Fig. 7: Experiment and theoretical fit: dashed line: fit for non-interacting ions and superconducting phase, dashed-dotted: cluster line. For details, see Ref. 28.

single line. Our idea was to use the exponential decay of the relaxation rate in the s.c. phase as a switch and to detect a resolved spectrum. Fig. 6 shows schematically the experiment and Fig. 7 the result. $LaOs_2$ is a very convenient Type II superconductor with high T_c and H_{c2} but it was chosen because of our principal interest in Laves phase. The theoretical fit (Fig. 7) using the full Plefka-Barnes theory[29] confirms our qualitative picture and yields an order parameter $2\Delta/k_BT_c$=3.5 to 5. Unfortunately, at present the experimental error bars in most of the systems[20,24,25,27] are so large that the very interesting question of the local superconducting order parameter in the vicinity of a magnetic impurity cannot be treated.

DEGENERACY OF THE CONDUCTION ELECTRONS AND THE CEF

One fundamental interaction between localized moments and the CEF is the exchange $J(\underline{k},\underline{k}')$. Caroli[30] has stated, since usually one makes the crude approximation that $J(\underline{k},\underline{k}')$ is a constant J, this leads to a J-value depending on the experiment, alloy, etc... More realistically, one should take the angular dependence into consideration, yielding a degeneracy parameter $(2\ell+1)=d$ in various formulae. In ESR this enters in the following way:

$$\Delta g \propto N(E_F)J_1 \rightarrow \sum_{\nu=1}^{d} N_\nu J_1 \rightarrow dN_\nu J_1, \tag{4}$$

$$\tau^{-1}T^{-1} \propto (N(E_F)J_2)^2 \rightarrow \sum_{\nu=1}^{d}(N_\nu J_2)^2 \rightarrow d(N_\nu J_2)^2 \tag{5}$$

and the scaling temperature[31] for higher order effects (Kondo) becomes

$$k_B T_K = D\ \exp(d/N_\nu J_o). \tag{6}$$

These equations imply many crude assumptions[30,31,32]. But we did include an angular momentum ℓ and consequently the CEF comes into play. Here we want to focus on the following aspect: To introduce a degeneracy, i.e. $2\ell+1=5$, may be as arbitrary as $2\ell+1=1$, as long as we cannot measure it explicitly. It just changes the J constant. Is it possible to measure d? If yes, we should keep the number variable because the CEF will partially lift the degeneracy.

In a recent ESR experiment[32] on <u>AuYb</u> we believe we have seen such an effect. To include higher order effects into the formulae for ESR we replace the bare exchange <u>integral</u> J_o by the temperature dependent exchange <u>parameter</u> $J_{eff}(T)$[33]:

$$N_\nu J_o \rightarrow N_\nu J_{eff}(T) = (\ln\ T_K/T)^{-1} \tag{7}$$

yielding $\Delta g \propto d(\ln\ T_K/T)^{-1}$ and $h\tau^{-1} \propto k_B T \cdot d(\ln\ T_K/T)^2$. If this is experimentally verified the fit will yield <u>two</u> parameters, the degeneracy d from the slope and $\ln T_K$ from the intercept. In our experiment[32] we found $d \sim 3$. This is a very reasonable number because the cubic CEF of Au (1) splits d-bands into Γ_3 and Γ_5 ($J_{atomic} \rightarrow J_{4f-5d}$ VBS), (2) splits f-bands into Γ_2, Γ_4, Γ_5 ($J_{cov.mix} \rightarrow J_{4f-|k,f\rangle}$) and (3) the p-band remains 3-fold degenerated (contributes to $J_{cov.mix}$, also).

We do not want to overemphasize our data analysis; more detailed experiments are under current investigation. But we feel that it is correct to account for the angular momentum of the C.E.. Moreover, since in our experiment[32] we were able to detect the Kondo effect in the ESR, we have an (almost) independent measure of d (the slope) and the coupling strength T_K (the intercept $\ln T_K$).

REFERENCES

1. D. Davidov, C. Rettori, A. Dixon, K. Baberschke, E.P. Chock, R. Orbach, Phys. Rev. B8, 3563 (1973).
2. D. Davidov, V. Zevin, R. Levin, D. Shaltiel, K. Baberschke, Phys. Rev. B15, 2771 (1977) and V. Zevin, D. Davidov, R. Levin, D. Shaltiel, K. Baberschke, J. Phys. F7, 2193 (1977).
3. Hyperfine Interactions, A.J. Freeman and R.B. Frankel, eds., Academic Press, New York (1967).
4. K.W.H. Stevens in "Crystalline Electric Field and Structural Effects in f-Electron Systems", J.E. Crow, R.P. Guertin and T. Mihalisin, eds., Plenum Press, New York (1980).
5. A. Abragam, B. Bleaney, ESR of Transition Metals, Clarendon Press, Oxford (1970).
6. K.R. Lea, M.J.M. Leask, W.P. Wolf, Phys. Chem. Solids 23 1381 (1962).
7. a) See various contributions in Ref. 4; b) Discussion on p.489ff in Ref. 7a.
8. R. Levin, A. Grayevsky, D. Shaltiel, V. Zevin, Sol. St. Com. 37, 69 (1981).
9. H-G. Purwins, E. Walker, B. Barvara, M.F. Rossignol, A. Furrer, J. Phys. C9, 1025 (1976).
10. M. Loewenhaupt, K. Baberschke, H. Scheuer, Sol. St. Com. 33, 175 (1980).
11. M. Loewenhaupt, S. Horn, B. Frick, this conference.
12. H.E. Hoenig, R. Voitmann, Sol. St. Com. 33, 43 (1980).
13. K. Baberschke, B. Bachor, H. Luft, J. Pellison, J. de Physique C5, 51 (1979).
14. U. Doebler, K. Baberschke, details to be published.
15. R.A.B. Devine, M. Poirier, T.J. Cry, J. Phys. F5, 1407 (1975).
16. W. Schäfer, H.K. Schmidt, B. Elschner, K.H. Buschow, Z. Phys. 254, 1 (1972).
17. D.J. Newman, W. Urban, Adv. Phys. 24, 793 (1975).
18. S.E. Barnes, K. Baberschke, M. Hardiman, Phys. Rev. B18, 2409 (1978).
19. For detailed discussion and original reference, see Ref. 18.
20. R. Odermatt, Sol. St. Com. 32, 1227 (1980) and R. Odermatt, Thèse, Genève 1981, Helv. Phys. Acta 54, (1981).
21. H.H. Luft, Thesis, Darmstadt, 1977, unpublished.
22. K. Baberschke, B. Bachor, S.E. Barnes, Phys. Rev. B21, 2666 (1980).
23. E. Müller-Hartmann, Verh. DPG 3, 240 (1980).
24. R. Feile, M. Loewenhaupt, J.K. Kjems, H.E. Hoenig, Phys. Rev. Lett. 47, 610 (1981).
25. W. Wagner, G.M. Kalvius, V.D. Gorobschenko, J. Magn. Magn. Mat. 15-18, 626 (1980).
26. For review of standard techniques, see e.g. "Superconductivity", Vols. I, II, R. Parks, ed, Marcel Dekker Press.
27. H. Luft, K. Baberschke, J. Appl. Phys. 52, 2095 (1981).

28. For review on polycrystalline results, see K. Baberschke, Z. Physik B24, 53 (1976).
29. For details see review article by S.E. Barnes, Adv. in Phys., to appear.
30. B. Caroli, J. Phys. F5, 1399 (1975).
31. P.W. Anderson, Comments, Sol. St. Phys. 1, 31 and 190 (1968), and S.E. Barnes, J. Phys. F6, 1713 (1976).
32. K. Baberschke, E. Tsang, Phys. Rev. Lett. 45, 1512 (1980) and E. Tsang, K. Baberschke, J. Appl. Phys. 52, 2208 (1981).
33. Similar arguments have been used for NMR in CuMn, see H. Alloul, F. Hippert, H. Iskii, J. Phys. F9, 725 (1979).

COMMENTS

COQBLIN: Concerning the problem of your d value, I think that the 4f electrons are mixed with the f(ℓ=3) part of the conduction electrons in Au(Yb) alloys and that this k-f mixing mechanism is certainly dominant in these alloys, as shown by the presence of the Kondo effect.

BABERSCHKE: Yes, that is what I meant. The question is: does this lead to $2\ell+1=7$ or is this split in the cubic symmetry into Γ_2, Γ_4 and Γ_5?

LÜTHI: Why does one observe a dynamical Jahn-Teller effect for the Γ_8 ground state ion? Has one observed this effect for rare earth ions?

BABERSCHKE: I agree with this possibility. But it would not show up in the resonance position. The contributions to the (residual) linewidth would be difficult to extract from our data.

SHENOY: The observation of a dynamic Jahn-Teller effect for the Γ_8 state will depend on the frequency used in your experiment. Have you varied the frequency?

BABERSCHKE: Since we do not know the frequency spectrum of a Jahn-Teller fluctuation, we could not follow it experimentally. We did see the hfs of Er^{167}, but the signals are too weak to find Jahn-Teller contributions.

KASUYA: The anisotropy mechanism considered by you for the S-state is the same as the ferromagnetic exchange mechanism considered by us more than ten years ago for Eu chalcogenides, in which the f-d mixing matrices are anisotropic. A Kondo state in real materials is much more complicated because usually the 5d conduction bands are of multivalleys with a complicated Fermi surface. For example CeB_6.

BABERSCHKE: Yes, I agree with your first comment. I also believe that "real" Kondo systems are more complicated. We think that the determination of $(2\ell+1)=d$ from an experiment will provide us with more information concerning the Fermi surface, I think Au is a little bit simpler than CeB_6.

ZIPPER: Do you assume that 5d electrons in $LaAl_2$:Gd form a virtual bound state?

BABERSCHKE: No, we do not [see Phys. Rev. B21, 2666 (1980)]. But here we focus more on the anomalous Eu^{2+} and so we cannot exclude contributions in the 5d VBS.

WOHLLEBEN: Covalent mixing is thought to occur for Eu impurities but not for Gd. Therefore, the difference of CEF splittings and g-shifts is found by subtracting Gd from Eu. Why should Gd and Eu be the same even if both were configurationally stable, in view of our lack of knowledge of why the parameters are what they are even for Gd?

BABERSCHKE: g-shifts and the CEF are not so simply related. Compare for example the g-values and B_4 parameters of Gd in Sc, Y and Lu. But I agree with you on the difference between Gd and Eu. This in turn justifies my arguments in the third section of my talk.

CRYSTAL FIELD INTERACTION IN RARE EARTH HYDRIDES:

EPR AND LOW TEMPERATURE SPECIFIC HEAT MEASUREMENTS

Henryk Drulis

Institute for Low Temperature and Structure Research
Polish Academy of Sciences, 50-950 Wrocław, Poland

INTRODUCTION

Rare earth metal hydrides represent an interesting group of interstitial compounds, especially with respect to their electronic properties and chemical bond mechanism. A remarkable feature is that they form phases which exist over considerable composition ranges. They are based on stoichiometries RH_2 so-called dihydrides, and RH_3, so-called trihydrides. The dihydrides RH_2 exhibit the CaF_2 - type structure, where the hydrogen atoms occupy nominally tetrahedral sites, whereas the trihydrides of the light rare earth elements crystallize with the BiF_3 - type structure, where nominally octahedral sites are also occupied. In the ideal CaF_2 and BiF_3 structures the nearest neighbours of the metal ion are eight hydrogen atoms situated on the tetrahedral sites at the corners of the cube. The next - nearest neighbours are hydrogen atoms on the octahedral sites at the corners of an octahedron.

The problem of characterizing the hydrogen electric charge can be elucidated by crystal field effect investigation. Much experimental evidence suggests that hydrogen in rare earth dihydrides is anionic, i.e. H^- in character[1]. Early studies of the magnetic properties of the rare earth hydrides showed that the rare earth paramagnetism is not modified by hydrogenation, suggesting that the hydrogen acquires its electron to form H^- not from the core but from the conduction band.

In this review we present the results obtained in this laboratory on the crystal field splitting studies using heat capacity and EPR measurements in some hydrides of the light lanthanide elements.

SCHOTTKY ANOMALY IN SPECIFIC HEAT

The crystal field interactions influence the heat capacity of the system as well as the magnetic susceptibility. At the lowest temperatures the ions are in the ground crystal field state or states.

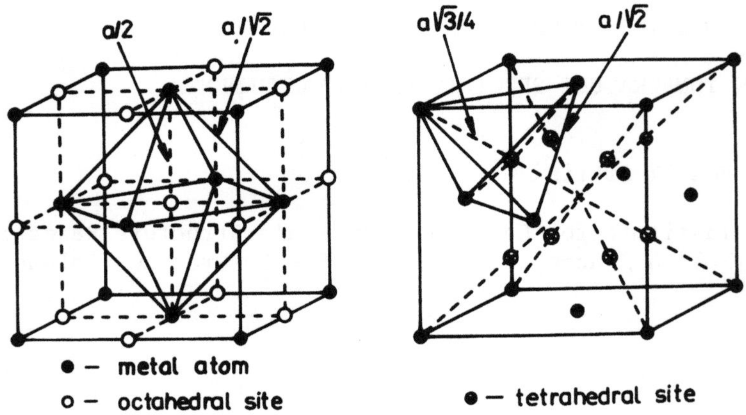

Fig. 1. Crystal structure of the light rare earth hydrides exhibiting the tetrahedral and octahedral hydrogen sites.

With increasing temperature the excited states become gradually occupied giving rise to some contribution to the heat capacity. A general problem with this kind of excitation was first considered by Schottky. According to theory, the transitions between m crystal field levels separated from the ground state by energies δ_1, δ_2, .. δ_m and with degeneracies g_1, g_2, .. g_m respectively, bring about a large change in the average internal energy E equal to

$$\tilde{E} = \frac{N \sum^{m} \delta_i \, g_i \exp(-\delta_i/T)}{\sum_{i=0}^{m} g_i \exp(-\delta_i/T)} \qquad (1)$$

The analysis of Schottky specific heats gives us often a possibility of determining the splitting energies, the positions of the electronic levels and their degeneracies.

Crystal field splitting in the light rare earth dihydrides

In this section we raport the early published results on the specific heat for compounds CeH_2^2 and PrH_2^2, where CEFs have cubic symmetry. In order to obtain the Schottky contributions to the total specific heats of Ce and Pr dihydrides it has been assumed that the specific heat of the nonmagnetic and isostructural lanthanum dihydride represents the lattice - vibrational and electronic contributions to the specific heat.

$\underline{CeH_2}$. Fig. 2 shows experimental results on the heat capacities for cerium dihydride. Under the influence of the cubic crystal field the six-fold degenerate $F_{5/2}$ state splits into the Γ_8 and Γ_7 states, with degeneracies 4 and 2, respectively. Such a splitting is schematically shown in Fig. 2 for the protonic, H^+, and hydridic, H^-, model.

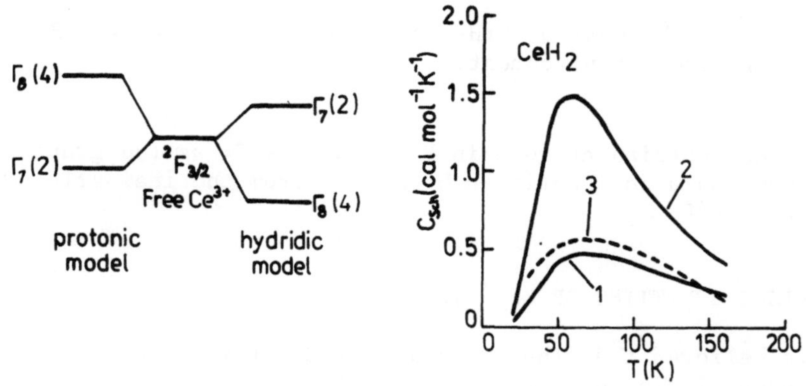

Fig. 2. Schottky specific heat of cerium dihydride and splitting pattern for Ce^{3+}.

On assuming that Γ_8 is the lowest lying level for the theoretical values of the Schottky specific heats, one obtains curve 1 compared to curve 3, which represents the experimental one. The best fit could be obtained for a splitting energy δ = 156 K. For a reversed level scheme and for the same splitting energy one obtains curve 2.

$\underline{PrH_2}$. The results for PrH_2 are presented in Fig. 3. In this case the nine-fold degenerate ground state, $^3H_4$, of the free Pr^{3+} splits into Γ_1 , Γ_4 , Γ_3 and Γ_5 with degeneracies 1, 3, 2, 3 respectively. We proceed now to make the following analysis for PrH_2:
If we assume the triple Γ_5 to be the lowest lying energy level and that the transitions occur between the triplet Γ_5 and doublet Γ_3 only, the Schottky specific heat is given by curve 1 in Fig. 3. The best fit could be obtained for a splitting energy δ_1 = 217 K. If we also include the triplet Γ_4 , and splitting energies δ_2 = 272 K, curve 2 is obtained. By assuming that the singlet Γ_1 is the lowest lying level and that the transitions occur between the singlet Γ_1 and the

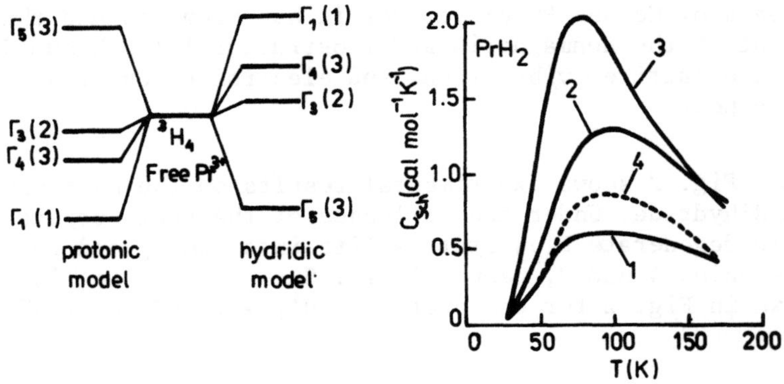

Fig. 3. Schottky heat of PrH_2 and splitting pattern for Pr^{3+} in cubic environment.

triplet Γ_4 one obtains curve 3 in Fig. 3. The Γ_5 as the ground state of Pr^{3+} ions was also identified by Knorr[3] from the inelastic neutron scattering in PrD_2 .

CRYSTAL FIELD SPLITTING OF Er^{3+} IN LuH_2

It is believed that the experimental Schottky specific heats are estimated within an error of about 15 %, which results mainly from the high lattice contribution to the total specific heat. The discrepancies between experimental and theoretical data for cerium and praseodymium dihydrides cannot be explained by this error and some other effects should be taken into account. Many other physical measurements indicate that the dihydrides, RH_2 , do not have the ideal fluorite structure, since a few percent of tetrahedral sites are vacant whereas some octahedral interstices are already occupied. This disturbs the cubic crystal field symmetry. For example, when more than 12 % of the octahedral sites are occupied, most of the metal ions at least have one octahedral hydrogen atom in their surrounding.

Indeed, EPR studies on Er doped ScH_2[4] , YH_2[5] , and LuH_2[6] indicate that in addition to expected cubic symmetry, the Er ions occupy at least one of two types of sites with axial symmetry. Fig. 4 shows a typical low temperature EPR spectrum of the Er^{3+} ion in $LuH_{2-\delta}$. This displays one intense line among a few weaker satellite lines. The intense line originates from Er ions in cubic symmetry and the observed g value of 6.68 is very close to $g = 6.80$ expected for an isotropic Γ_7 doublet. The satellite line may be attributed to $g_{\parallel}$ and $g_{\perp}$ from the powder pattern of Er^{3+} ions in axial symmetry. We have found that decreasing H/Me from 1.98 to 1.75 brought about a decrease in the intensities of this axial line.

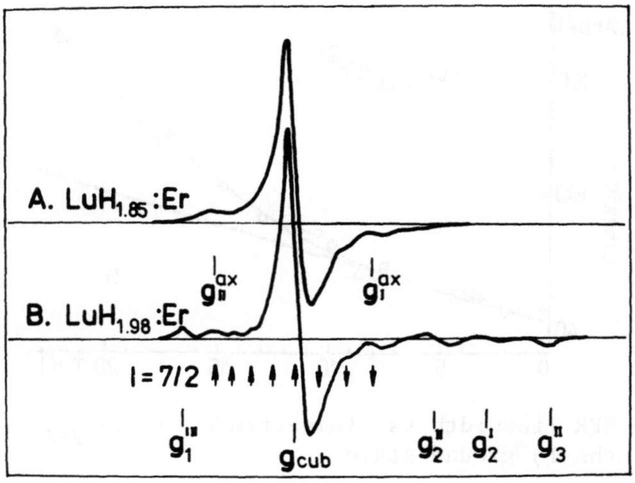

Fig. 4. The observed EPR derivative signal of LuH_y: Er for $y = 1.98$ and 1.85, at $T = 4.2$ K.

LuH_2 : Er . In order to determine the crystal field splitting of an Er ion in lutetium dihydride using the EPR method, we have analysed the Γ_7 resonance linewidth as a function of temperature.

In metallic substances it is usual to consider the linewidth as the sum of temperature dependent and independent parts. The former may be written as follows:

$$\Delta H(T) = \frac{g(g_J - 1)^2}{g_J^2 \mu_B} k_B \pi \left[J N(E_F) \right]^2 \cdot T + \sum_i \frac{C_i \Delta_i}{\exp(\Delta_i / k_B T) - 1} \quad . \quad (2)$$

Here g denotes the g factor of the ground state, g_J is the Lande g factor, J is the local moment-conduction electron exchange constant, and $N(E_F)$ the density of states at the Fermi level. Δ_i is the energy difference between ground state and the i-th excited state. The second term of Eq. (2) is due to the relaxation through excited states which are mixed by isotropic conduction electrons to the local moment interaction.

The measured Γ_7 EPR linewidth as a function of temperature for Er^{3+} in $LuH_{2-\delta}$ is presented in Fig. 5. Above about 10 K the linewidth deviates from the low temperature linear behaviour, according to the second term in Eq. (2). Inspection of Fig. 6 indicates that the next excited state for the Er ion with a Γ_7 ground state is either the $\Gamma_8^{(1)}$ or Γ_6 depending on the value of x . However only the Γ_8 state contributes to the linewidth. We have computed the linewidth using

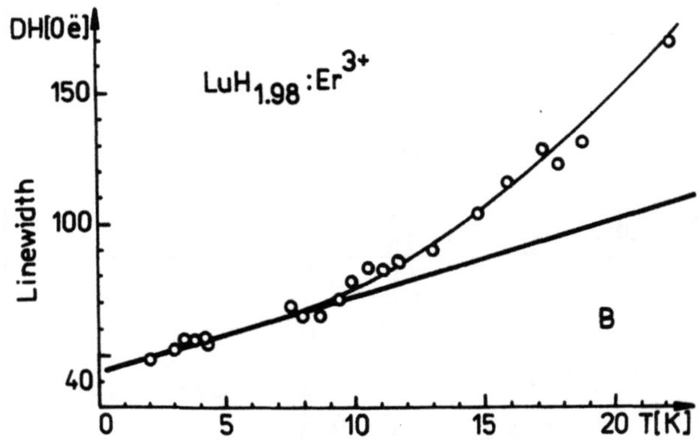

Fig. 5. EPR linewidth vs. temperature for $LuH_{2-\delta}$: Er for the Γ_7 ground state.

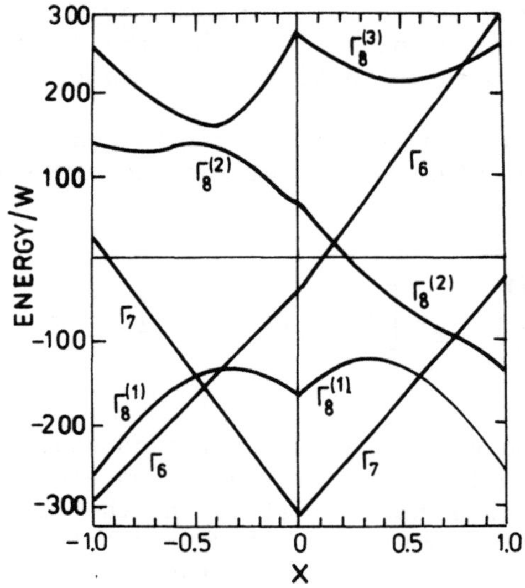

Fig. 6. The energy level splitting E/W for Er^{3+}, J = 15/2, in a cubic CEF as a function of x , after LLW[7].

Eq. (2) and found that the best fit to the data is given by $\Delta(\Gamma_8^{(1)} - \Gamma_7)$ = 57 ± 10 K. Unfortunately in this case we could not determine precise values of the x parameter.

SCHOTTKY SPECIFIC HEAT OF THE NON-STOICHIOMETRIC RARE EARTH HYDRIDES

As was mentioned above, for intermediate H - concentrations, between 2 and 3, the octahedral sites are only partially occupied. The total CF experienced by the magnetic ion can be considered as a sum of the fields created by the shells of the first and second (octahedral sites) depending on the way in which these sites are occupied.

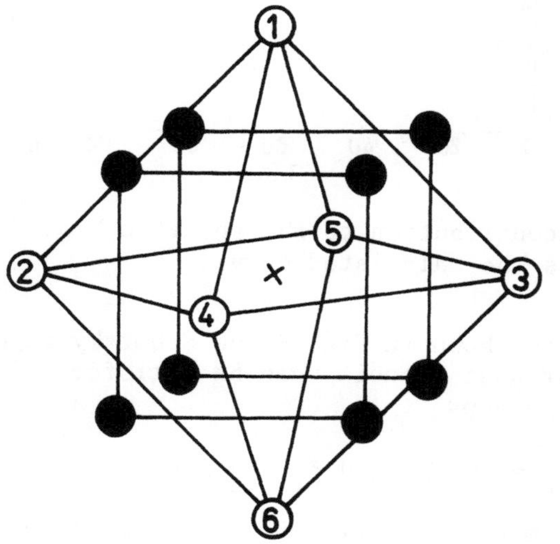

Fig. 7. The first and second neighbours of the magnetic ion occupy tetrahedral and octahedral sites, respectively.

Moreover, previous neutron diffraction experiments on the cerium deuterides $CeD_{2.15}$ and $CeD_{2.75}$[8] have shown that in the low temperature range a long range order of the octahedral hydrogen atoms exists. This means that in all cases when H/Me > 2 the CF symmetry at the magnetic ion is different from cubic, (see Fig.7).

$CeH_{2.56}$: Fig. 8 presents experimental and theoretical curves of the Schottky specific heat as a function of temperature. The theoretical curve was obtained assuming that some lowering of crystal field symmetry takes place and by considering three doublets separated by δ_1 = 80 K and δ_2 = 210 K.

We have also assumed that all cerium ions have only one type of crystal field symmetry. The neutron scattering measurements in praseodymium hydride[3] with composition close to H/Me = 2.5 have shown that occupation of the octahedral interstitials occurs in a well defined way. The hydrogen atoms probably occupy sites 1, 2, 3 as shown in Fig. 7. This leads to an orthorhombic crystal field at all magnetic ion sites. This interpretation confirms the results which were obtained for neodyium hydride $NdH_{2.48}$.

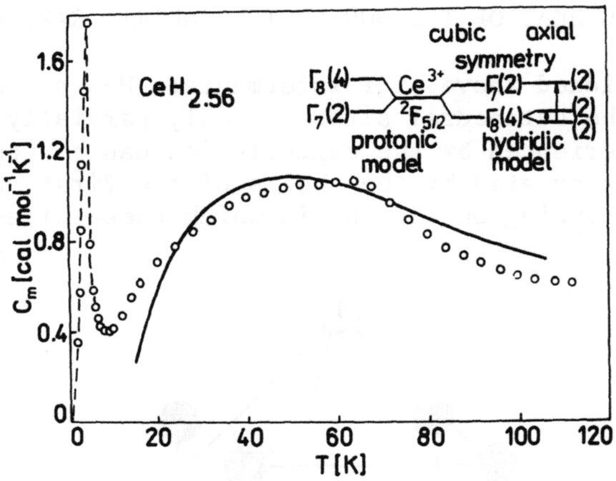

Fig. 8. Schottky contribution to the specific heat for $CeH_{2.56}$ as well as the calculated curve.

$\underline{NdH_{2.48}}$: As is shown in Fig. 9 the Schottky magnetic contribution of the molar heat capacity can be satisfactorily described by one set of energy gaps

$$\delta_1 = 48\ K,\ \delta_2 = 200\ K,\ \text{and}\ \delta_3 = 350\ K.$$

This uses the system of the four doublets of Nd^{3+} ions in orthorhombic symmetry.

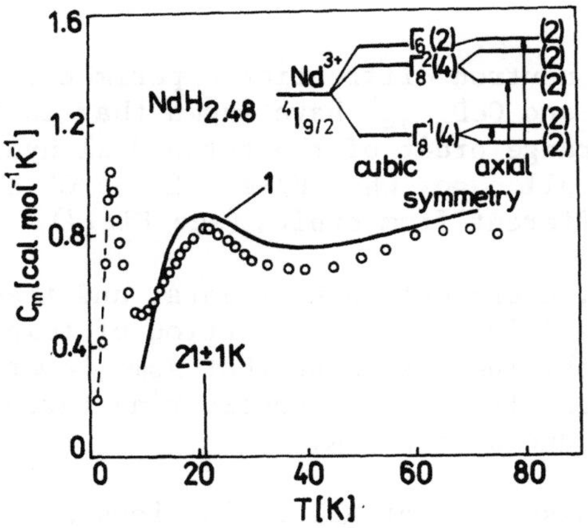

Fig. 9. Experimental Schottky specific heat for $NdH_{2.48}$ compared with the theoretical curves.

$\underline{NdH_{2.70}}$[9] : In Fig. 10 the specific heat of neodymium hydride with compositions $NdH_{2.70}$ is given. If we consider a wide tempera-

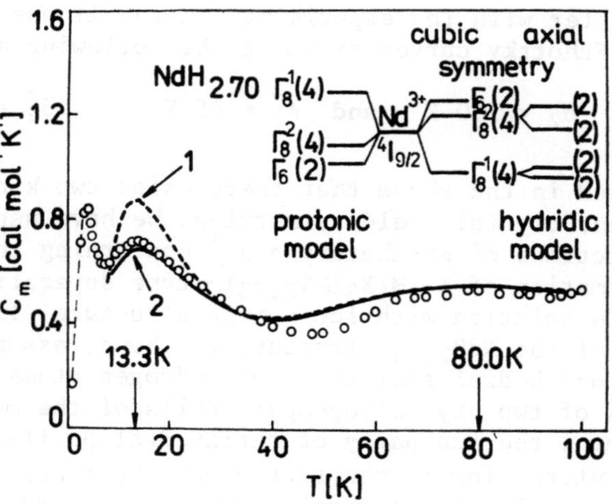

Fig. 10. Schottky anomaly in the specific heat of $NdH_{2.70}$ as well as the curves calculated for models described in text.

ture range up to 100 K, we notice that the Schottky specific heat reveals two anomalies in the vicinity of 13 K and 80 K. The theoretical curve 1 has been obtained assuming thermal excitation from the doublet ground state into two other doublets located at $\delta_1 = 32$ K and $\delta_2 = 220$ K. However, using this approach the theoretical Schottky specific heat considerably exceeds the experimental values.

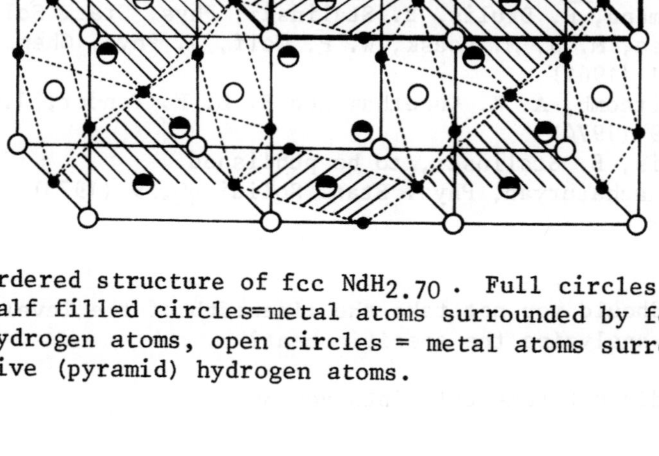

Fig. 11. Ordered structure of fcc $NdH_{2.70}$. Full circles = H atoms, half filled circles=metal atoms surrounded by four hydrogen atoms, open circles = metal atoms surrounded by five (pyramid) hydrogen atoms.

Curve 2 agrees better with the experimental data. It was obtained by composing two Schottky curves assuming the following splittings

$$\delta_1 \doteq 19\ K, \quad \delta_2 = 220\ K \quad \text{and} \quad \delta_1^* = 45\ K, \quad \delta_2^* = 230\ K .$$

It was assumed in the above that there exist two kinds of Nd^{3+} ions with different crystal field symmetries. We have used here the theoretical predictions of Khachaturyan's[10] concerning the possibility of the formation of a $M_4X_3(MX_{2.73})$ - type superstructure in interstitial solid solution with fcc - type structure. In Fig. 11 the ideal scheme of the $NdH_{2.75}$ structure is shown, assuming superstructure in the octahedral sublattice of hydrogen atoms realized within the limits of two crystallographic cells of the metal. To clarify the picture, the occupancy of tetrahedral positions is not indicated. Most interesting is the fact that there exist two kinds of the metal atoms. Half of the metal atoms is surrounded by four (square) hydrogen atoms and another half by five (pyramid).

By using this model we can explain not only the Schottky specific heat anomaly of $NdH_{2.70}$ but also the high temperature order -disorder anomaly in specific heat of the light lanthanide hydrides with composition near $LnH_{2.75}$. The hypothesis about the existence of a long range ordered superstructure predicted by Khachaturyan for interstitial solid solution with composition M_4X_3 requires some additional studies using other physical methods than specific heats. We believe that EPR would be most promising for this purpose.

REFERENCES

1. A. Fujimori and N. Tsuda, J. Phys. C , 14, 1427 (1981)
2. Z. Biegański, Ber. Bunsenges. Phys. Chem. 76, 1183 (1972)
3. K. Knorr, B. E. F. Fender, W. Drexel, Z. Physik B30, 265 (1978)
4. E. L. Venturini, J. Appl. Phys. 50, 2053 (1979)
5. H. Drulis, J. Pyter, B. Staliński, Bull. Acad. Polon. Sci., Sér. Chim., 27, 903 (1979)
6. K. Hoffmann, H. Drulis, B. Staliński, Phys. Stat. Sol., in press.
7. K. R. Lea, M. J. M. Leask, W. P. Wolf, J. Phys. Chem. Solids, 23, 1381 (1962)
8. C. G. Titcomb, A. K. Cheetham and B. E. F. Fender, J. Phys. C7, 2409 (1974)
9. M. Drulis, B. Staliński, to be published.
10. A. G. Khachaturyan, Phys. Stat. Solidi 60, 9 (1973)

COMMENTS

GREIDANUS: Should you not take the linewidth of the levels into account when analyzing the specific heat?

DRULIS: We did not take this into account.

BERTHIER: Did you observe any deviation in the hyperfine constant that you can measure in the $LuH_{1.85}$:Er EPR spectrum compared to the value obtained in other metallic systems?

DRULIS: The value of the hyperfine constant is $A=74\pm1$ Oe and is similar to that observed in other metallic systems. It is slightly higher than the value observed in a nonmetallic host.

CRYSTAL FIELDS AND CONDUCTION ELECTRON EFFECTS IN INTERMETALLIC COMPOUNDS AND ALLOYS

M. Loewenhaupt,* S. Horn† and B. Frick *

*Institut für Festkörperforschung, KFA-Jülich
D-5170 Jülich
†Institut für Festkörperphysik, TH Darmstadt
D-6100 Darmstadt

INTRODUCTION

This paper is intended to continue the work on "neutron studies of CEF excitations and linewidths of dilute systems" presented at the preceding conference on CEFSE in Philadelphia, 1979.[1] Here, however, we restrict ourselves to a more detailed description of inelastic neutron scattering experiments performed on Er and Tm ions diluted in the cubic Laves-phase compounds RAl_2 with R=Sc,Y, and La. A large number of publications already exists for these systems, including all kinds of macroscopic measurements (susceptibility, magnetization, specific heat, etc.) and microscopic measurements (ESR, neutron scattering, etc.). The interpretation of the data yielded a great but inconsistent variety of CF parameters for the different systems. This (mis-)led several authors to speculate about a drastic dependence of the CF parameters on concentration and diluent matrix[2]. Our experiments as presented in this paper are at variance to this speculation and yield a consistent set of CF parameters for both Er and Tm ions in RAl_2 diluent matrices, including the full compounds.

SOME EXPERIMENTAL DETAILS

We report on neutron scattering experiments performed on different time-of-flight spectrometers at the HFR in Grenoble (D7 and IN5) and at the FRJ-2 in Jülich (SV5). We used neutrons with incident energies ranging from E_0=3.6 meV (D7, IN5, SV5) to E_0=14.6 meV (SV5, using the 2nd order reflection of the monochromator). The relative experimental resolution $\Delta E/E_0$ for elastic scattering varied from 1-3 % for SV5 and IN5 to 10% for D7. Because of the high neutron flux most experiments were performed on D7; the other instruments were used only in a few special cases where a better resolution or a higher incident neutron energy was needed. A typical experiment included the measurement of the magnetically doped samples, of the undoped reference compounds RAl_2 (R=Sc,Y,La), of the empty sample holder, of the sample holder with a Cd plate in place of the sample and of a Vanadium standard for absolute intensity calibration. All spectra were corrected for angle and energy dependent absorption and transmission dependent background. Inelastic magnetic scattering could always be separated from phonon scattering in all investigated systems by comparison with the spectra of the undoped reference compounds and by using the different Q and temperature dependence of magnetic and phonon scattering.

To fit the spectra we used a least-squares program allowing for nuclear and magnetic scattering contributions. The nuclear scattering was approximated by a Q and temperature independent elastic line, its width being determined only by the experimental resolution and by inelastic phonon lines corresponding to the different matrices. The magnetic scattering was calculated from the x and W parameters including the diagonalization of the CF Hamiltonian determination of the eigenstates and their eigenvales. We then calculate the dynamical susceptibility by either

(i) replacing the δ-functions by Lorentzians modified by a factor to retain the detailed balance condition or

(ii) using the theory of Becker, Fulde and Keller [3,4].

The fitting parameters are thus x and W and the linewidths, (i), or the effective coupling constant between local moment and conduction electrons, (ii).

EXPERIMENTAL RESULTS

Crystal Field Parameters

At the beginning of this section we present the results for the system $Er_{0.02}R_{0.98}Al_2$ with R = Sc, Y and La. Fig. 1 shows the low temperature spectra measured on D7 with an incoming neutron energy of E_0=3.6 meV. All spectra are dominated by a large elastic peak which contains the incoherent elastic nuclear scattering originating mainly from the matrix and the quasielastic magnetic scattering from the groundstate of the Er ions. Both contributions cannot be separated because the linewidth of the quasielastic magnetic scattering in this low temperature region is considerably smaller than the experimental resolution. The nuclear scattering, however, is temperature independent and its value can be determined from the high temperature spectra. Magnetic scattering can be clearly identified in the inelastic part of the spectra because the matrices show only a flat background in this energy region. We observe one inelastic magnetic

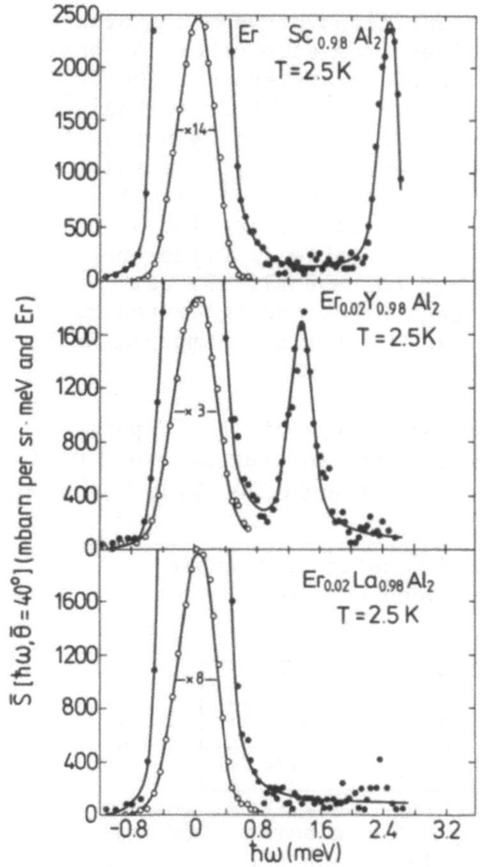

Fig. 1 Scattering law for $Er_{0.02}$ $R_{0.98}$ Al_2 at T = 2.5 K. (D7, E_o= 3.6 meV)

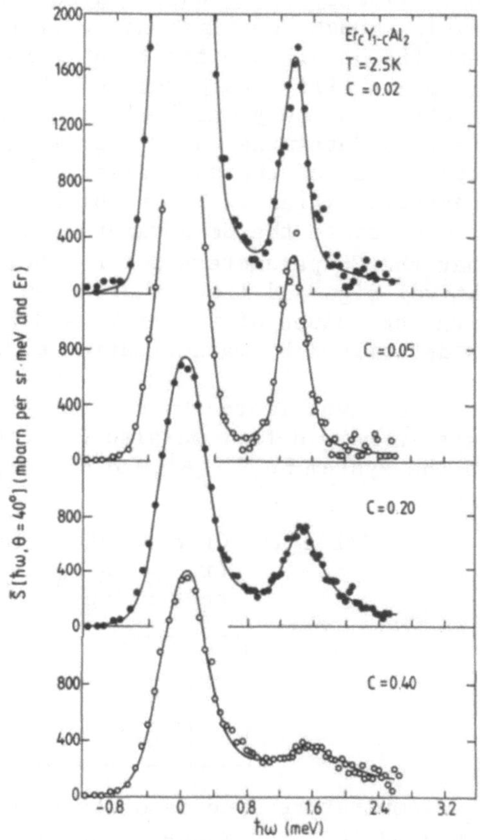

Fig. 2 Scattering law for Er_c Y_{1-c} Al_2 at T = 2.5 K. (D7, E_o= 3.6 meV)

line originating from the excitation of the Er ions from the ground state to the first excited state. In the case of $Er_{0.02}$ $La_{0.98}$ Al_2 this inelastic line is not resolved from the elastic scattering because the first excited state lies only a few K above the ground state as will be shown later. For $Er_{0.02}$ $Sc_{0.98}$ Al_2 and $Er_{0.02}$ $Y_{0.98}$ Al_2 the separation between the two lowest states is (28 ± 1)K and (16 ± 1)K, respectively.

To determine the CF parameters unambiguously, however, more information is necessary. We therefore also performed measurements

(i) with higher incoming neutron energy,
(ii) at elevated temperatures and
(iii) with better experimental resolution.

This allowed us to observe also the transitions between the other levels, though sometimes not well resolved from each other. A consistent fit of all spectra then yields the CF parameters for the $Er_{0.02}R_{0.98}Al_2$ alloys as given in Table 1. In the case of Er in $LaAl_2$ the spectra show only one inelastic line with $\Delta \approx 55K$ which is not sufficient to determine two CF parameters. However taking the absolute intensities of the quasielastic line including the $\Gamma_8^{(3)}-\Gamma_8^{(2)}$ transition which is not resolved and of the inelastic line at 55 K we can locate the CF parameters as given in Table 1. Please note that the CF parameters for R = Sc, Y and La have all the same (negative) sign and lie rather close to each other and agree roughly with the values of x = -0.26 and W = -0.29 K for the compound $ErAl_2$ as determined by magnetization measurements [5].

To investigate the concentration dependence of the CF parameters in more detail we also performed neutron scattering experiments on the system $Er_cY_{1-c}Al_2$ for c=0.02, 0.05, 0.20, and 0.40. Fig. 2 shows the

Table 1. CF parameters in LLW notation for cubic $Er_{0.02}R_{0.98}Al_2$ as deduced from neutron scattering.

R =	Sc	Y	La
x	-0.20(2)	-0.30(2)	-0.48(5)
W	-0.26(2)	-0.25(2)	-0.18(2)

low temperature spectra of these alloys. The elastic line again has a nuclear and magnetic scattering which cannot be separated at low temperatures. The inelastic part, on the other hand, is of purely magnetic origin, showing the transition from the $\Gamma_8^{(3)}$ ground state to the $\Gamma_8^{(2)}$ first excited state. The increasing concentration does not change the inelastic line drastically. Increasing concentration however yields an increase of linewidth, a decrease of intensity (per Er ion) and a slight shift to higher energies.

The effect of an increasing concentration in $Er_cY_{1-c}Al_2$ on the inelastic transition between the ground state and the first excited state is qualitatively similar to the effect of an external magnetic field on this transition. This is shown in Fig. 3 for the alloy $Er_{0.05}Y_{0.95}Al_2$ and an external field of 15 kOe. The inelastic transition is split by the field (broadened if the resolution would be worse), its intensity is reduced (note the different scales!) and its mean position is shifted to higher energies. This similarity demonstrates the presence of internal magnetic fields (originating from spin-spin interactions) in the more concentrated alloys. The CF parameters themselves, however, depend only very weakly on concentration. We obtain for $Er_cY_{1-c}Al_2$ for all investigated alloys up to c= 0.40 the same CF parameters within experimental error as those given

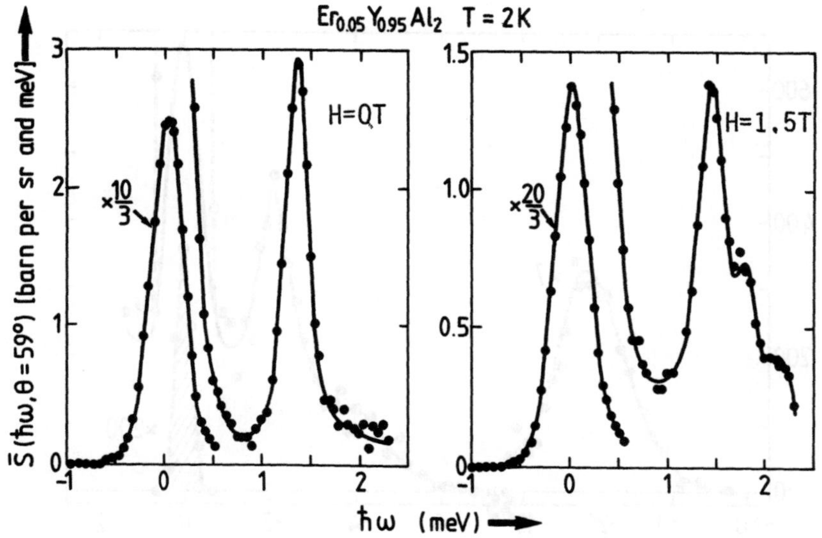

Fig. 3 Scattering law for $Er_{0.05}$ $Y_{0.95}$ Al_2 at T = 2 K in zero external field and with H = 15 kOe perpendicular to the scattering plane measured on D7 with E_o = 3.6 meV but with increased resolution compared to Figs. 1 and 2.

in Table 1 for c=0.20. The parameters were deduced from fits to the neutron spectra at elevated temperatures where spin-spin interactions are of minor importance. We note however that for c=0.20 and 0.40 the intensity ratios for the different transitions deviated from the single ion CF-only calculations. This was even more pronounced for the low temperature spectra where we observed a transfer of intensity from the inelastic (Δ=16K) line to the quasielastic line. At the same time a Q dependence of the quasielastic intensity developed which deviated considerably from the $4f^{11}$ form factor and which peaked at $\underline{Q}$ = (1/2, 1/2, 1/2), indicating antiferromagnetic short range order.

We now proceed and present the results for the system $Tm_{0.02}$ $R_{0.98}$ Al_2 with R = Sc, Y and La. As example we show in Fig. 4 a high temperature spectrum of $Tm_{0.02}$ $Sc_{0.98}$ Al_2. The different contributions to the scattering intensity are shown separately:
(i) elastic nuclear scattering (dashed line and hatched area),
(ii) phonon scattering around 20 meV (dash-dotted line) and
(iii) inelastic and quasielastic magnetic scattering (full line).
From a consistent evaluation of the spectra measured on different instruments (D7, SV5 and IN5) and at different temperatures between 2K and 150K we obtained the CF parameters for the (Tm,R)Al_2 alloys as given in Table 2. Again the CF values for any R have the same sign and lie not very far from each other and agree roughly with

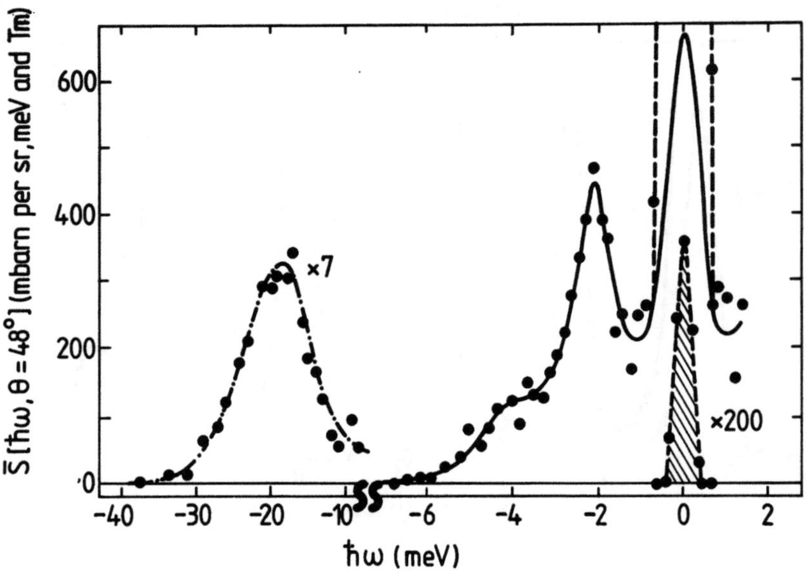

Fig. 4: Scattering for $Tm_{0.02}$ $Sc_{0.98}$ Al_2 at T = 150 K (D7, E_o = 3.6 meV).

Table 2. CF parameters for cubic $Tm_{0.02}$ $R_{0.98}$ Al_2 as deduced from neutron scattering

R =	Sc	Y	La
x	0.41(2)	0.52(2)	0.5 (10)
W	0.46(2)	0.42(4)	0.3 (10)

the values x = 0.52 and W = 0.5 K for the compound $TmAl_2$ as also determined by inelastic neutron scattering [6].

An investigation of the concentration dependence of the neutron spectra of the alloy system Tm_c La_{1-c} Al_2 yields results[7] which are not as simply interpreted as in the case of Er_c Y_{1-c} Al_2. As seen in Fig. 5 the spectra look different for different Tm concentrations. Three magnetic lines denoted by I, II and III are visible for c = 0.15 and 0.07. For c=0.02 line II, which already showed a shift to lower energies for c=0.07 compared to 0.15, has merged with line III to

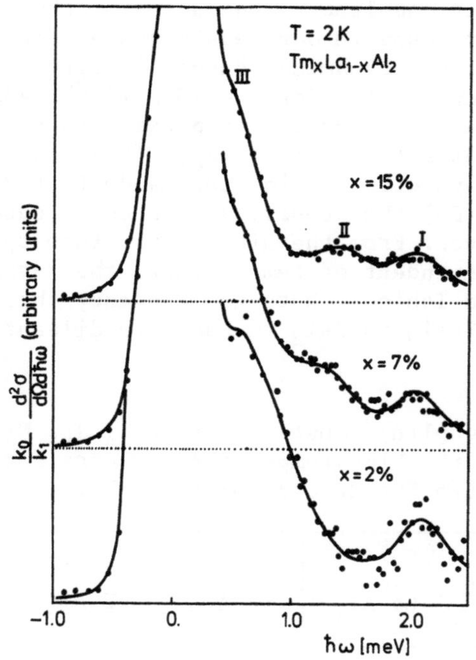

Fig.5: Scattering for $Tm_cLa_{1-c}Al_2$ at T=2.0 K (D7, E_0=3.6 meV).

just one inelastic contribution with Δ = (8±1) K. The position of line I (although the data points show strong scattering in this region) is obviously unchanged compared to the higher concentrations. It is not possible to explain the spectra for c=0.07 and 0.15 (including the high temperature spectra) by a cubic CF operator. A reasonable fit of the data can only be achieved if an additional term The present state of the experiment, however, does not allow giving preference to any of the above mentioned explanations.
(i) phonon exchange between ions (Jahn-Teller effect),
(ii) quadrupole-quadrupole interactions or
(iii) biquadratic exchange.
The present state of the experiment, however, does not allow giving preference to any of the above mentioned explanations.

Linewidths

To obtain reasonable fits of the measured neutron spectra we had to introduce to the experimental resolution an additional intrinsic linewidth of the CF transitions. Typical values for the intrinsic linewidths were 1 K to 5 K (HWHM) for a Lorentzian type of lineshape, increasing with increasing temperature. This suggests a Korringa type of mechanism for the relaxation of the stable local

4f moments due to their coupling to the spins of the conduction electrons. We therefore used in our least-squares fitting program for the calculation of the dynamical susceptibility, the theory of Becker, Fulde and Keller adopted for J = 15/2 (Er^{3+}) and J = 6 (Tm^{3+}). The only free parameter, besides the CF parameters x and W, is the effective coupling constant $\rho = (g-1)\cdot J_{ex}\cdot N(E_F)$ between local moment and conduction electrons, with g being the Landé factor, J_{ex} the exchange integral and $N(E_F)$ the density of states of conduction electrons at the Fermi level. From the high temperature spectra (50 K to 150 K) we obtain (independent of temperature) the ρ-values for the c=0.02 alloys as given Table 3. Note the tendency that ρ is increasing when going from $ScAl_2$ to YAl_2 to $LaAl_2$ as diluent matrix for

Table 3. Effective coupling constant $\rho = (g-1)\ J_{ex}\ N(E_f)$ as obtained from the high temperature spectra for the c=0.02 alloys. g=6/5 for Er and g=7/6 for Tm.

Er in			Tm in		
Sc Al_2	Y Al_2	La Al_2	Sc Al_2	Y Al_2	La Al_2
0.008(2)	0.008(2)	0.012(2)	0.008(2)	0.010(2)	0.015(5)

both Er and Tm ions. Fits of the low temperature spectra yield different (larger) values for ρ (especially for T=2K). This indicates that there are additional mechanisms for line broadening which are not included in the Becker, Fulde, Keller theory. These additional mechanisms are most likely spin-spin interactions which give a practically temperature independent contribution to the linewidth. Our analysis shows that they are even present in the low concentration c=0.02 alloys.

DISCUSSION

The aim of this investigation was to find a consistent set of CF parameters for Er and Tm ions diluted in RAl_2 (R=Sc, Y, La) matrices. Fig. 6 shows the positions of the parameter x for all investigated systems in the LLW diagram for Er^{3+} and Tm^{3+} in cubic symmetry corresponding to the values given in Tables 1 and 2. We see that for the Er systems the x values range from -0.2 to -0.5 and that the ground state is always a $\Gamma_8^{(3)}$ as indicated by the thick

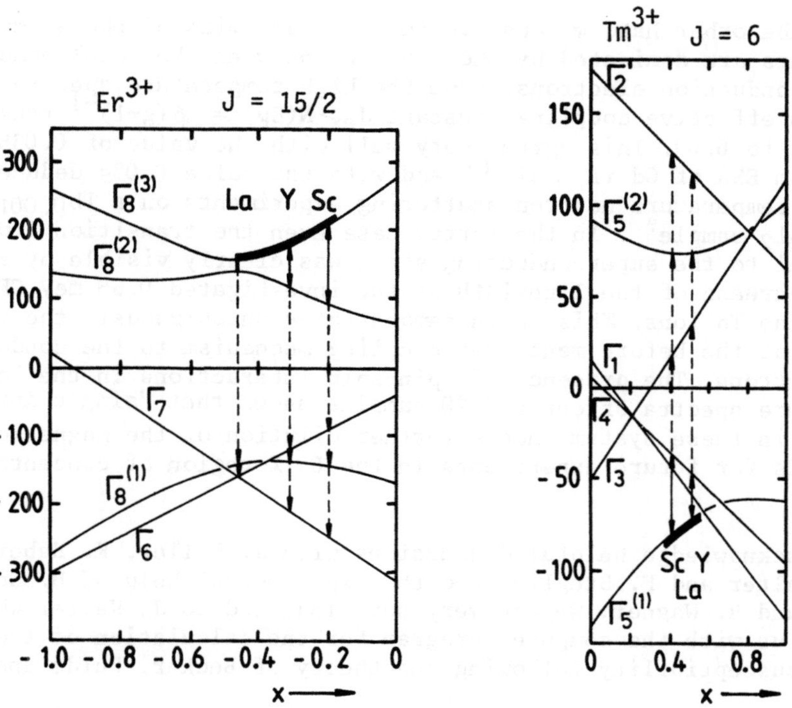

Fig. 6 LLW diagrams for J = 15/2 (x<0) and J = 6 (x>0). Also shown are the positions of x for the investigated systems Er RAl$_2$ and Tm RAl$_2$ as given in Tables 1 and 2.

curve. The Tm systems have x values around + 0.45 and a $\Gamma_5^{(1)}$ as ground state. Calculating from the x and W values the A_4 and A_6 "true" CF parameters we obtain, for both Er and Tm in all investigated matrices and for the full compounds, a positive A_4 parameter and a negative (5 to 10 times smaller) A_6 parameter. This is at variance with the speculation that the CF parameters in RAl$_2$ have a drastic dependence on concentration and diluent matrix which was attributed to the influence of the conduction electrons. Our conclusion however, is supported by recent neutron studies on Pr, Nd, Tb, Dy and Ho ions diluted in YAl$_2$[8] and on Tb in LaAl$_2$[9] which all give a positive A_4 and a negative A_6 as CF parameters. Thus the influence of the conduction electrons on the CF parameters in RAl$_2$ systems is only of minor importance. There may be some connection between the details of the band structure and the systematic increase of /x/ and decrease of /W/ when going from ScAl$_2$ to YAl$_2$ to LaAl$_2$. This however needs further clarification.

On the other hand we observe that the dynamics of the 4f magnetic moments are dominated by the coupling between the local moments and the conduction electrons. From the high temperature spectra we obtain an effective coupling constant $J_{ex} \cdot N(E_F) = \rho(g-1)^{-1}$ ranging from 0.04 to 0.09. This agrees very well with the value of 0.035 deduced from ESR of Gd in YAl_2[10] and with the value 0.034 deduced from low temperature neutron scattering experiments on a $Tb_{0.001}La_{0.999}Al_2$ sample[9]. In the latter case even the transition from the normal to the superconducting state was clearly visible by an abrupt decrease of the linewidth of the investigated 0.68 meV CF line of the Tb ions. This again demonstrates unambiguously the importance of the before mentioned coupling mechanism to the conduction electrons. The presence of spin-spin interactions in the low temperature spectra of our c=0.20 samples shows that "single ion behaviour" in these systems needs further dilution of the magnetic ions. This calls for future experiments in the 0.1% region of concentration.

We acknowledge helpful discussions with J. Keller, K. Baberschke, U. Walter and F. Steglich and the experimental help of H. Scheuer, W. Just and R. Wagner. We are very much indebted to J. Keller who supplied us with the computer program for the calculation of the dynamical susceptibility following the theory of Becker, Fulde and Keller.

REFERENCES

+Work performed within the research programme of the Sonderforschungsbereich 125

1. M. Loewenhaupt in "Crystalline Electric Field and Structural Effects in f-Electron Systems" edited by J.E. Crow, R.P. Guertin, T.W. Mihalisin (Plenum, New York, 1980) p. 589
2. B. Barbara, M.F. Rossignol, R.A.B. Devine, Y. Berthier and A.P. Murani in same as Ref. 1 p. 481.
3. K. Becker, P. Fulde, J. Keller, Z. Phys. B28, 9 (1977).
4. J. Keller, this conference
5. H.G. Purwins, E. Walker, B. Barbara, M.F. Rossignol, A. Furrer, J. Phys. C 9, 1025 (1976).
6. S. Horn, M. Loewenhaupt, unpublished
7. S. Horn, M. Loewenhaupt, H. Scheuer, F. Steglich, J. Magn. Magn. Mat. 14, 239 (1979)
8. U. Walter, E. Holland-Moritz to appear in Z. Phys..
9. R. Feile, M. Loewenhaupt, J.K. Kjems, H.E. Hoenig, Phys. Rev. Lett. 47, 610 (1981)
10. W. Schäfer, K.H. Schmidt, B. Elschner, K.H.J. Buschow, Z. Phys. 254, 1 (1972)

COMMENTS

BERTHIER: Did you consider the opportunity for the next neighbors pair-pair interaction in order to explain the extra line you obtain in the $Tm_xLa_{1-x}Al_2$ spectrum? Also, do you know other determinations of ω and u parameters for $TmAl_2$; for example, from magnetization curves, specific heat measurements, etc.?

LOEWENHAUPT: Yes, to the first question, in the sense that I called it an interaction with the other ions (quadrupole-quadrupole or biquadratic exchange). A detailed study of pair interactions was not done. No to the second question.

LUTHI: The CEF effects for elastic constants can be explained with the CEF parameters given by Loewenhaupt. The magnetoelastic and quadrupole coupling constants are, however, too small to explain the unexpected neutron peak in $Tm_xLa_{1-x}Al_2$ for x=0.07, 0.15 and 0.40.

deJONG: If you formulate your crystal field parameters not as x and W, but as B_4/β_J and B_6/γ_J, do you then get consistent results for Tb, Er and Tm?

LOEWENHAUPT: Yes, in all the cases we get the same signs (A_4+, A_6-), and the values differ by not more than a factor of two.

POSITIVE MUONS AS LOCAL PROBES IN PARAMAGNETIC RARE EARTH SYSTEMS

H. Wehr*, K. Knorr*, F.N. Gygax†, A. Schenck† and
W. Studer†

*Institut für Physik, Universität Mainz, D-65 Mainz FRG
†LHE/ETH Zurich, CH-5324 Villigen, Switzerland

INTRODUCTION

In recent years a new research method has been developed in solid state physics which is based on the asymmetry of muon decay[1]. A positive muon at rest decays after a mean lifetime of 2.2 μsec into a positron and two neutrinos, the positron being emitted preferentially in the direction of the muon spin. Hence if spin polarized muons are implanted in a solid, the time evolution of the muon spin polarization can be determined by a measurement of the angular distribution of the positrons. In an external magnetic field transverse to the initial spin direction of the muon one can study the muon spin precession frequency, which is determined by the actual field at the muon, and τ_2-relaxation effects. (It is this experimental arrangement from which the most commonly used name "muon spin rotation" [μSR] is derived.) A longitudinal field arrangement allows study of τ_1-relaxation effects. The time scale which can be covered by (μSR) is determined by the magnetic field at the muon, the field fluctuations and the muon lifetime. In nonmagnetic substances the time scale ranges from 10^{-8} sec to 10^{-5} sec, in magnetic substances it can be extended down to 10^{-11} sec.

In the present paper we will be mainly interested in the spin polarization of the conduction electrons in the paramagnetic state of metals which contribute to the local magnetic field at the muon site via the Fermi contact field. We will investigate three rare earth (RE) systems in which the conduction electron density can be varied either by alloying ($LaAg_{1-x}In_x$) or by a valence change of the RE atoms (Ce, Ce-Th; $CeSn_3$).

THEORETICAL BACKGROUND: THE MUON KNIGHT SHIFT IN PARAMAGNETIC METALS

In a metal the muons occupy interstitial sites or possibly vacancies. The actual precession frequency ω_μ of the muon spin in the metal due to an externally applied magnetic field (corrected for demagnetization fields) will in general be different from its value in vacuum ω_o. One defines the Knight shift constant K as: $K \doteq (\omega_\mu/\omega_o)-1$. Apart from atomic and nuclear dipole fields the value of K is determined by the electronic spin density at the muon. It is common practice to split K up into contributions of different origin in order to simplify the discussion. This procedure can be justified by experiment. We now list the various contributions to K:
1. Conduction electrons in wide bands which are spin polarized and have a nonzero density at the muon contribute a positive term to K via the Fermi contact field:

$$K_+^s=(8\pi/3)<|\psi|^2>\chi_s/N\equiv(B_s/\mu_B)\cdot\chi_s. \qquad (1)$$

$<|\psi|^2>$ is the average conduction electron density at the muon, which may be different from the unperturbed case when the muon is absent, χ_s is the conduction electron spin susceptibility per atom, N is the number of conduction electrons per atom, B_s represents the induced contact hyperfine field. In most simple metals K_+^s is of the order of $1:10^6$.[2] 2. A negative term K_{dia} results from a diamagnetic shielding of the muon charge by electrons which have bonding character with respect to the muon. The value of K_{dia} is about -20 to -30 ppm in most metals[3]. Quite naturally one expects that the two temperature independent terms $K_+^s+K_{dia}$ should describe the experimental Knight shift values in simple metals. Experimentally, however, it was found that other contributions can exist which become dominant in more complicated electron systems[4]. 3. In most transition metals studied so far negative K values were obtained which scale with the d-electron susceptibility $\chi_d(T)$.[5] It is believed that the negative electron spin polarization at the muon originates from low lying μ-metal bonding states which can be core polarized by Fermi surface d-electrons[6]. Formally, we write:

$$K_d(T)=(B_d/\mu_B)\cdot\chi_d(T), \qquad (2)$$

B_d being the induced hyperfine field arising from the d-electrons.
4. In RE-systems the 4f-moments may contribute to the muon Knight shift either directly or indirectly depending on the degree of localization of the 4f-shell. This has not been experimentally verified for the paramagnetic regime. In the indirect mechanism a local 4f-moment causes an additional conduction electron polarization via the RKKY-exchange interaction. If the conduction electrons at the muon have only free electron character and a uniform polarization model is adopted, the additional term may be written[7]

$$K_{4f}(T)=K^s+\frac{(g_J-1)}{g_J\mu_B^2}\,J_{sf}\cdot\chi_{4f}(T)\equiv(B_{4f}/\mu_B)\cdot\chi_{4f}(T). \qquad (3)$$

$\chi_{4f}(T)$ is the susceptibility of the 4f-ions, J_{sf} is the phenomenological exchange constant, B_{4f} is the induced hyperfine field which arises from the local 4f moments, and the other symbols have their usual meaning.

On the basis of these considerations the Knight shift results for some RE systems will now be discussed. The application of μSR to magnetic problems in solids is a rather new field. The interpretation of the results must take into account two delicate points whose understanding is still at an early stage of development, 1) the effect of the muon charge on the conduction electrons in its environment and 2) possible trapping effects of the muon at irregular sites like voids or grain boundaries. The measurements were carried out at the Schweizer Institut für Nuklearforschung-Villigen in a transverse field of 7.4 kG. In order to obtain the high precision which is necessary for muon Knight shift measurements (ppm!) in a reasonable time limit, a stroboscopic method was used, which allows a high data collection rate[8].

RESULTS AND DISCUSSION

$LaAg_{1-x}In_x$

Since the muon Knight shift is susceptible to both conduction electrons and local 4f electrons, for a study of RE systems it is of paramount importance to obtain first an understanding of the role of the 6s,5d-electrons before turning to specific effects of the 4f-electrons. In particular the 5d electrons may give rise to a temperature dependence of K. We have chosen the system LaAg(In) which crystallizes in the cubic CsCl structure and undergoes a crystallographic transition to a lower symmetry structure at temperatures which strongly depend on the In concentration x.[9] Magnetic susceptibility measurements[10] and band structure calculations[11] suggest that the Fermi level in LaAg lies just below a region of a high density of states which originate from the 5d-e_g states of La. This region can be gradually populated when Ag is replaced by the higher valent In. Hence one can change the number and possibly the type of the conduction electrons by alloying with In and then study the influence on the Knight shift.

The measurements were performed in the cubic phase of polycrystalline LaAg(In). As will be shown the muon performs a site average over the sample and K must be regarded as a mean value. Fig. 1a shows K as a function of x at room temperature. K is slightly negative in LaAg and shifts to more negative values when x is increased. In Fig. 1b we present K as a function of tempera-

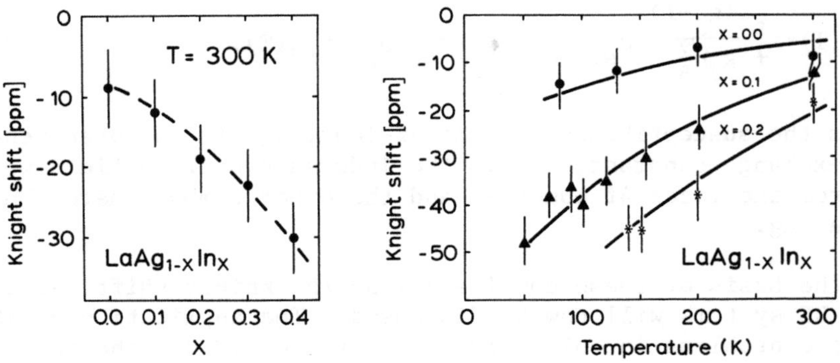

a) b)

Fig. 1: (a) the muon Knight shift in $LaAg_{1-x}In_x$ as a function of x at room temperature. The dotted line is a guide to the eye.
(b) the muon Knight shift in $LaAg_{1-x}In_x$ as a function of temperature. The solid lines result from a fit to the d-electron susceptibility as discussed in the text.

ture for the cubic phase of three different samples. We try to analyze the temperature dependence of K in terms of the expression

$$K(T)=(K_+^s+K_{dia})+(B_d/\mu_B)\cdot\chi_d(T) \tag{4}$$

where $\chi_d(T)$ is the temperature dependent part of the bulk susceptibility (taken from Ref. 10) which was attributed to the d-electron spin susceptibility. The lines in Fig. 1(b) represent the best fits to the experimental data using the values for $(K_+^s+K_{dia})$, T and B_d as given in Table 1. According to the fit both the temperature independent part of K and the coupling to d-like states is increased when Ag is replaced by In. If we assume that the valence electrons of In enter band-like states, the results suggest that: additional electrons enter both states in wide bands contributing to K_+^s [Eq.(1)], and narrow bands which are responsible for the negative temperature dependent Knight shift, [Eq. (2)]. The values of B_d are comparable to the numbers derived for typical d-band metals[5]. So far the muon Knight shift confirms the importance of d-like contributions to the density of states in LaAgIn.

From muon Knight shift measurements at interstitial sites one cannot discriminate between d-electron contributions from the La site and d- or p-like contributions from the In site. A negative Knight shift contribution may also arise from the electrons which screen the excess charge of In if these states have p- or mixed p and d-character. In fact indications for a short screening distance of the In charge was found in a recent crystalline electric field study of CeAgIn[12].

Table 1: Results from a fit of $K(T)-\chi_d(T)$ in $LaAg_{1-x}In_x$ (see text and Fig. 1(b) for the temperature independent part of the muon Knight shift ($K_+^s+K_{dia}$) and the coupling to d-like states B_d.)

x	0.0	0.1	0.2
$(K_+^s+K_{dia})$ (ppm)	+15(±10)	+40(±5)	+70(±8)
B_d (kG/μ_B)	-3.6(±2.0)	-7.2(±0.5)	-11.0(±1.0)

Finally we comment on the intrinsic linewidth of the muon signal in LaAgIn which represents the τ_2-relaxation rate of the muon spin. The linewidth contains information about trapping and site change effects of the muon which could mask the temperature dependence of the Knight shift. In LaAgIn line broadening arises first of all from the nuclear moments of La and In. The results, which will be discussed in more detail elsewhere, show that above 80 K the muon is diffusing most probably between both octahedral and tetrahedral interstitial sites. There is no indication for trapping of the muon at irregularities like In clusters, etc. Hence the temperature dependence of K is really an intrinsic effect of the electronic system of LaAgIn.

γ-to-α Transition in Ce and $Ce_{0.74}Th_{0.26}$

The purpose of the muon Knight shift investigations in polycrystalline fcc Ce and fcc Ce-Th was to find out how the electron spin density midway between atoms is influenced if the local $4f^1$ electron is converted into a delocalized band electron. The isostructural transition from a high volume γ phase to a low volume α phase can be forced in Ce when a high external pressure is applied, and in Ce-Th when the temperature is reduced below 150 K.[13,14] Concomitant with the lattice contraction is a delocalization of the $4f^1$ electron, the exact nature of which is still a matter of discussion: the electron band may either be promoted to the (s,d) conduction band or attain promotion on its own via a Mott transition[13]. The muon Knight shift results are shown in Figs. 2(a) and 2(b). In the 4f-free reference system La-Th [see Fig. 2(a)] the shift is almost zero and temperature independent. Since a temperature dependence is lacking we believe that a negative d-electron contribution to K, as discussed in LaAgIn, is negligible, and we attribute most of the cancellation of the positive term K_+^s to diamagnetic shielding of the muon charge.

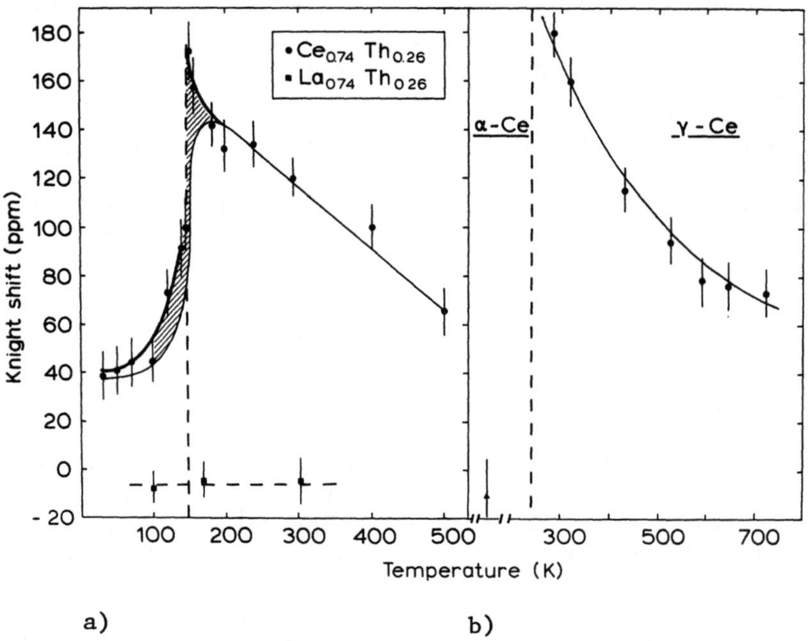

Fig. 2: (a) the muon Knight shift in $Ce_{0.74}Th_{0.26}$ and $La_{0.74}Th_{0.26}$ as a function of temperature. The dotted line and the heavy solid lines are guides to the eye. The light solid line gives the bulk susceptibility taken from Ref. 20 and scaled to fit the data at high temperatures.
(b) The muon Knight shift in γ Ce and α Ce as a function of temperature. The light solid line gives the bulk susceptibility taken from Ref. 21 after appropriate scaling to fit the Knight shift data. For experimental details, see Ref. 22.

In γ Ce and in Ce-Th positive values of K are observed over the whole temperature range for the present study [Figs. 2(a),(b)]. In the γ phase well above the phase transition the data follow the Curie type behavior of the bulk susceptibility $\chi_B(T)$. This result may be interpreted in terms of an indirect polarization of the conduction electrons at the muon site via the RKKY exchange interaction [Eq.(3)]. From the linear relation between K(T) and $\chi_B(T)$ (Fig. 3) one obtains the induced hyperfine field per $4f^1$ moment in the γ phase: $B_{4f}/\gamma=0.35$ kG/μ_B and $B_{4f}/\gamma=0.46$ kG/μ_B for Ce-Th and Ce respectively.

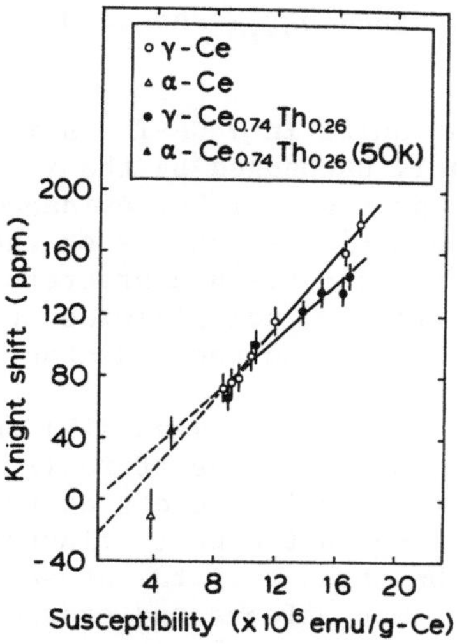

Fig. 3: Demonstration of the linear relation between the muon Knight shift and the bulk susceptibility in γ Ce and γ $Ce_{0.74}Th_{0.26}$ for temperatures above 180 K. The α phase values are also shown.

We have identified $\chi_{4f}(T)$ with $\chi_B(T)$. It is interesting to compare the relatively low values of B_{4f}/γ with the values of B_d, which are negative and about ten times higher in LaAg(In) or with values of B_s in simple metals which are typically of the order of 50 kG/μ_B. The magnitude of the induced hyperfine fields can be regarded as a measure of the degree of localization of the corresponding electron spin density. In simple metals with quasifree and hence delocalized conduction electrons the muon experiences a local spin density which often is enhanced compared to the average polarization [Eq. (1)]. Hence the values of B_s are high. In lanthanide metals the excess polarization at an interstitial site due to the 4f moments is established by an indirect interaction. Hence the values of B_{4f} are low.

In the α phase of Ce and Ce-Th it is generally assumed that the formation of the local $4f^1$ moment is suppressed or at least significantly reduced. Hence the bulk susceptibility χ_B is interpreted as an enhanced Pauli spin susceptibility. If we write $\chi_B \equiv \Delta\chi + \chi_s$, where $\Delta\chi$ refers to the enhanced part of the susceptibility, the Knight shift in the α phase may be written in analogy to Eq. (3):

$$K_\alpha = (K_+^s + K_{dia}) + B_{4f,\alpha}\Delta\chi. \tag{5}$$

If $\Delta\chi \gg \chi_s$, $\Delta\chi$ can be approximated by χ_B and again a linear relation between K and χ_B is established.

One might speculate that the delocalization of the $4f^1$ electrons is large enough to increase the direct Fermi contact field [Eq. (1)], or, according to the observations from LaAg(In), one could expect that larger negative contributions to K arise if the $4f^1$ electron is converted into a d-electron. In both cases $B_{4f,\alpha}$ should be different from its value in the γ phase.

The experiments show that the Knight shift in α Ce-Th is indeed above the value for La-Th but B_{4f} is nearly unchanged by the γ-to-α transition [Figs. 2(a), 3]. The results provide neither evidence for a large increase of the number of conduction electrons at the muon site nor for a significant change from $4f^1$-to-5d character. The results suggest that the conduction electron polarization at the muon may still be described in terms of an indirect interaction.

In Ce-Th deviations from the linear relation between K(T) and $\chi_B(T)$ were observed around the γ-to-α transition temperature T_s [Fig. 2(a)]. In the same temperature range the line width of the muon signal increased anomalously suggesting that density fluctuations are effective at the phase transition. Unfortunately the present experiment cannot distinguish between dynamic effects and static effects (lattice inhomogeneities during the large volume collapse). Hence it cannot be decided whether the Knight shift is increased near T_s by an increased exchange coupling J_{sf} or by a site change effect of the muon. In any case trapping effects of the muon were absent for temperatures well different from T_s in the γ and in the α phase.

$\underline{CeSn_3}$

$CeSn_3$, which crystallizes in the cubic Cu_3Au structure, is regarded as an intermediate valence system displaying valence fluctuations in a wide temperature range[15]. There is an anomalous thermal expansion[16] and a broad maximum in the magnetic susceptibility near 140 K (taken from Ref. 17 and shown as an inset in Fig. 4). This suggests a partial conversion of magnetic Ce^{3+} ($4f^1$) ions to non-magnetic Ce^{4+} ($4f^0$) ions. Whereas an Sn Mossbauer study did not reveal any change in the isomer shift or in the electric field gradient of Sn in the temperature range 1.6 to 293 K,[18] the ^{119}Sn Knight shift (K_{Sn}) exhibited a small temperature dependence similar to $\chi_B(T)$. However, a linear relation between K_{Sn} and χ_B was lacking[17].

These observations provided the motivation to study the muon Knight shift at interstitial sties. The first, yet limited, results are presented in Fig. 4. The Knight shift values are positive both in $LaSn_3$ and in $CeSn_3$. The large positive shift in $LaSn_3$ compared to the value in La-Th reflects the fact that the non-transition metal Sn contributes a large proportion of s-electrons to the muon site (Eq. 1). Despite the limited data the temperature behavior of $LaSn_3$ indicates that $LaSn_3$ is far from being a simple $4f^0$ reference[19].

The Knight shift curve in $CeSn_3$ also shows a maximum below 200 K as it is found in the bulk susceptibility and the ^{119}Sn-Knight shift. However, at low temperatures K decreases much more strongly than χ_B

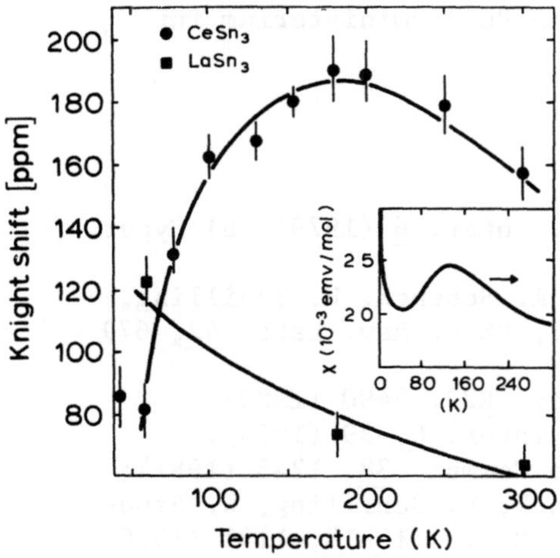

Fig. 4: The muon Knight shift in $CeSn_3$ and $LaSn_3$ as a function of temperature. The solid lines are guides to the eye. For comparison the bulk magnetic susceptibility of $CeSn_3$ taken from Ref. 17 is shown (inset).

(or K_{Sn}) in the same temperature range (Fig. 4). The most interesting feature perhaps is the crossing of the Knight shift curve for $CeSn_3$ and $LaSn_3$ below 100 K. Obviously the electronic environment of a muon at an interstitial site is more sensitive to a change in the 4f state than the electrons at the Sn site. We recall that the muon Knight shift in α Ce-Th did not decrease below the value of La-Th. However, the electronic environment of the muon in $CeSn_3$ and Ce-Th may be quite different; hence the electronic states may be influenced differently if the $4f^1$ electron assumes more band character.

OUTLOOK

In the previous sections we presented the first muon Knight shift measurements performed to date in paramagnetic RE systems. The results are promising and give rise to hope that the μSR method can be successfully used as a complement to existing techniques such as NMR, Mössbauer, etc. as a probe of the local site. We plan to extend our muon Knight shift investigations to other RE-compounds exhibiting valence changes since we believe that these systems provide a good test of the sensitivity of the method. In order to obtain a self consistent picture of the behavior of positive muons in RE systems we will also intensify muon diffusion studies. This aspect of the muon experiments was only mentioned in this paper. The diffusion behavior of a light interstitial in a softening lattice (Ce, Ce-Th, $CeSn_3$, etc.) is in itself a fascinating problem which should deserve more attention.

This research was supported by the Bundesministerium für Forschung und Technologie, F.R.G.

REFERENCES

1. For recent reviews, see a) Hyper. Inter. 6 (1979); b) Hyper. Inter. 8 (1980/81).
2. M. Camani, F.N. Gygax, W. Rüegg, A. Schenck, H. Schilling, E. Klempt, R. Schulze and H. Wolf, Phys. Rev. Lett. 42, 679 (1979).
3. E. Zaremba and D. Zobin, Phys. Rev. B22, 5490 (1980).
4. J. Keller and A. Schenck, Hyper. Inter. 6, 39 (1979).
5. a) F.N. Gygax et al., Solid State Commun. 38, 1245 (1981); b) F.N. Gygax, W. Rüegg, A. Schenck, H. Schilling, W. Studer and R. Schulze, ICM 79 (Munich),J.M.M.M. 15-18, 1191 (1980); c) H. Yasuoka, R.S. Hayano, Nishida, N. Nagamine, T. Yamazaki and Y. Ishikawa, Solid State Commun. 26, 745 (1978).
6. J. Kanamori et al., Hyper. Inter. 8, 573 (1980/81).
7. K. Yosida, Phys. Rev. 106, 893 (1957).
8. M. Camani, F.N. Gygax, E. Klempt, W. Rüegg, A. Schenck, H. Schilling, R. Schulze and H. Wolf, Phys. Lett. 77B, 326 (1978).
9. H. Balster, H. Ihrig, A. Koekel and S. Methfessel, Z. Physik B21, 241 (1975).
10. H. Ihrig, D.T. Uigren, J. Kübler and S. Methfessel, Phys. Rev. B8, 4525 (1973).
11. A. Hasegawa, B. Bremicker and J. Kübler, Z. Physik B22, 231 (1975).
12. H. Wehr, K. Knorr, A.P. Murani and W. Assmus, this conference.
13. D.C. Koskimaki and K.A. Gschneider, Jr., in "Handbook on the Physics and Chemistry of Rare Earths", K.A. Gschneider and L. Eyring, eds., (North-Holland, Amsterdam, 1978).
14. J.M. Lawrence, M.C. Croft and R.D. Parks, Phys. Rev. Lett. 35, 289 (1975).
15. T. Tsuchida and W.E. Wallace, J. Chem. Phys. 43, 3811 (1965).
16. E. Umlauf, P. Sütsch and E. Hess, in "Crystalline Electric Field and Structural Effects in f-Electron Systems", (Plenum, New York, 1980), J.E. Crow, R.P. Guertin and T.Mihalisin, eds.,341.
17. S.K. Malik, V. Vijaraghavan, S.K. Garg and R.J. Ripmeester, Pur. Appl. Chem. 40, 223 (1974).
18. G.K. Shenoy, B.D. Dunlap, G.M. Kalvius, A.M. Toxen and R.J. Gambino, J. Appl. Phys. 41, 1317 (1970).
19. See, for example, a) L.B. Welsh, A.M. Toxen and R.J. Gambino, Phys. Rev. B4, 2921 (1971); b) A.M. Toxen and R.J. Gambino, Phys. Lett. 28A, 214 (1968).
20. J.M. Lawrence and R.D. Parks, J. Phys. (Paris) 37, 64 (1976).
21. C.R. Burr and S. Ehara, Phys. Rev. 149, 551 (1966).
22. H. Wehr et al., to be published in Phys. Rev. B (1981).

COMMENTS

LOEWENHAUPT: How sensitive is μSR to the quality of samples (e.g. Sn precipitation in $CeSn_3$)?

WEHR: In a certain temperature range, which depends on a specific metallic system, the muon may be trapped at impurities. In the temperature range where our experiments have been performed, we find that the muon diffuses between different interstitial sites. For example, in $RESn_3$ no indication was obtained for trapping at Sn clusters.

FELLER: The common J_{sf} depends on the distance from the RE ion. Can you use this method if the polarization of the conduction electrons is not homogeneous?

WEHR: We measure the local spin density at the muon site, which in fact depends critically on the distance between the muon and the RE ions.

WOHLLEBEN: Can you put error bars to the statement that in the α-γ phase transition of CeTh alloys you cannot see any 4f-5d charge transfer?

WEHR: The statement is based on the assumption that the spin susceptibility of the 4f electrons in the α-phase may be identified with the bulk susceptibility. This identification is correct within the error of about 10 to 20%.

MÖSSBAUER STUDIES OF THE $Er_6(Fe_{1-y}Mn_y)_{23}H_x$ SYSTEM

J. Zukrowski[+], G.A. Stewart, G. Kalkowski,
G. Wortmann and G. Wiesinger*

Institut für Atom- and Festkörperphysik
Freie Universität Berlin, Berlin
*Technische Universitat Wien, Austria

INTRODUCTION

The $R_6(Fe_{1-y}Mn_y)_{23}H_x$ compounds (R=rare earth,Y) are especially interesting for two reasons:

(1) x=0. The high magnetic ordering temperatures observed for the boundary phases of the host materials plunge dramatically in the intermediate composition region[1] (for example see Table 1).

(2) x≠0. The effect of loading with hydrogen varies wildly[2-4]. It can result in either suppression or enhancement of the magnetic order, or alteration of the magnetic structure itself.

The crystal structure is the cubic Th_6Mn_{23} type with 4 non-equivalent lattice sites occupied by Fe or Mn and a single R site. The Y compounds have been studied most extensively because they are not complicated by a magnetic R sublattice. In order to determine what role a magnetic R plays we have carried out a microscopic investigation of the $Er_6(Fe_{1-y}Mn_y)_{23}H_x$ system using the ^{166}Er and ^{57}Fe Mössbauer resonances. Parallel measurements were also performed on some of the related Y-compounds. Table 1 summarizes the existing magnetic data for R=Er,Y.

SAMPLE PREPARATION AND EXPERIMENTAL DETAILS

All of the intermetallic compounds were prepared by induction melting on a water cooled copper finger under an argon atmosphere. Hydrogenation was carried out by exposing coarsely ground host

[+] On leave from Solid State Physics Dept., AGH, Cracow, Poland.

Table 1. Existing magnetic parameters for $R_6(Fe_{1-y}Mn_y)_{23}H_x$ with R=Er,Y. F(ferromagnetic), AF(antiferromagnetic) and FI(ferrimagnetic) refer to the alignment of the 3d moments. Parenthesised information refers to the 4f-3d magnetization alignment.

	y=0	y=.5	y=1
$Er_6(Fe_{1-y}Mn_y)_{23}$	F, T_c=480 K[1] (A, T_{comp}=100 K)[1]	T_c=60 K[1]	FI[5] T_c=420 K[1] (F)[5]
$Y_6(Fe_{1-y}Mn_y)_{23}$	F, T_c=484 K[12]	No long range order[8]	FI[5,8] T_c=486 K[12]
$Y_6(Fe_{1-y}Mn_y)_{23}H_x$ x=23-26	$T_c \simeq 640$ K[3]	$T_c \simeq 150$ K[3]	AF[4] T_N=160-180 K[6,7]

material to H_2 gas in a stainless steel chamber. The number of hydrogen atoms, x, absorbed per formula unit was determined by monitoring the pressure change in the known volume. Since a range of x values is possible, a single phase hydride is defined by a unique lattice expansion of the cubic host material. The Mn-rich compounds absorb hydrogen more readily. Therefore, single-phase $Er_6Mn_{23}H_{21}$ and $Y_6Mn_{23}H_{26}$ could be prepared at ambient temperature using a hydrogen pressure of 6 atmospheres. However, the same conditions yielded mixed phase hydrides ($x \simeq 14$) for the Er compounds having more substantial Fe content. Single phase hydrides ($x \simeq 21$) of these compounds were prepared using a pressure of 30 atmospheres and heating slowly to 300°C.

The ^{166}Er 80.6 keV resonance experiments were performed using a source of ^{166}Ho in $Ho_{0.4}Y_{0.6}H_2$ held at 4.2 K. For the ^{57}Fe experiments, a source of ^{57}Co in Rh at room temperature was used.

RESULTS AND DISCUSSION

Fig. 1 shows the ^{166}Er spectra recorded at 4.2 K for $Er_6(Fe_{1-y}Mn_y)H_{21}$ with y=0,0.5 and 1.0. Because the ^{166}Er ground state has zero spin, the positions of the resonance lines describe directly the energy positions of the excited 80.6 keV level having spin I=2. The spectrum for y=0 was fitted with five identical Lorentzian profiles assuming an effective hyperfine field B_{eff} and an aligned axial electric field gradient (EFG) to act at the nucleus. Such a spectrum is usually associated with a magnetically

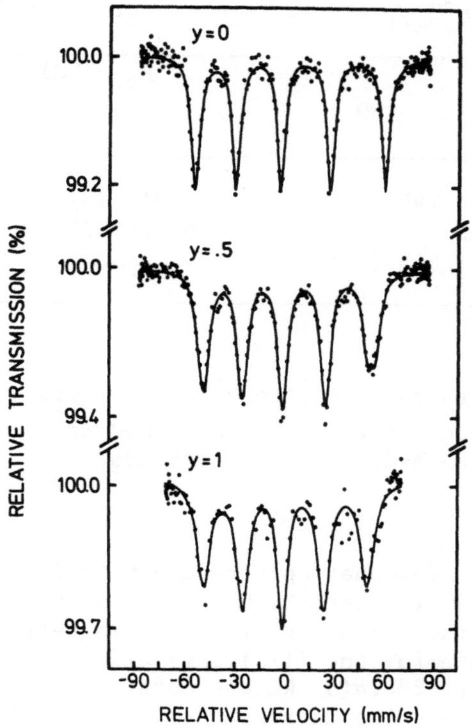

Fig. 1. ^{166}Er Mössbauer spectra for the hydrides $Er_6(Fe_{1-y}Mn_y)_{23}H_{21}$.

ordered system and as expected the spectra measured for the three ordered host materials (not shown) were also of this form. However, the ^{166}Er spectrum for the hydride with y=.5 could be fitted best by superimposing two quintets of identical Lorentzians and the Mn-rich hydride (y=1) required a symmetric variation of the Lorentzian linewidths and amplitudes. The latter is a typical relaxation effect but could also result from a superposition of two quintets with identical EFG's and slightly different B_{eff} values. The average B_{eff} values are collected in Table 2 where the results for the mixed phase hydrides (x≃14) have also been included. For comparison, the "free ion" effective field originating solely from a fully polarized 4f shell where $\langle J_z \rangle = J = 15/2$ is 8102 kG[10]. The approximately free ion B_{eff} value measured for Er_6Fe_{23} is typical for R-Fe intermetallic compounds. In the other two host materials the weaker Er-Mn exchange interaction evidently enables the crystal field (CF) interactions to mix the Zeeman-split 4f-levels of Er^{3+} thus lowering the $\langle J_z \rangle$ ground state expectation value and B_{eff}. As supporting evidence, the quadrupole interaction at ^{166}Er in these compounds was observed nearly to disappear. Hydrogenation also leads to a reduction in B_{eff} and this can again be explained in terms of a weaker exchange interaction.

Table 2. Average B_{eff} at 4.2 K for ^{166}Er in $Er_6(Fe_{1-y}Mn_y)_{23}H_x$.

y	B_{eff}(kG) x=0	x=14	x=21
0	7988(126)	7424(118)	7462(119)
0.5	6762(136)	6468(102)	6581(104)
1.0	6633(136)	———	6423(136)

The essentially itinerant electrons, which mediate the exchange between the 3d-lattice and the well-localized Er^{3+} 4f shell, are partially removed to hydrogen-derived bands.

^{57}Fe Measurements

y=0. The ^{57}Fe Mössbauer spectra measured for Er_6Fe_{23} concurred with those reported elsewhere[14]. Excellent statistics permitted a detailed analysis incorporating all seven magnetically inequivalent Fe sites appropriate for the (111) easy axis. A weighted average effective field B_{eff}=286.6(3) kG was obtained at 4.2 K. Upon hydrogenation (x=21) this value increased to B_{eff}=319(1) kG. Coupled with the ^{166}Er results it can be concluded that Er_6Fe_{23} behaves similarly

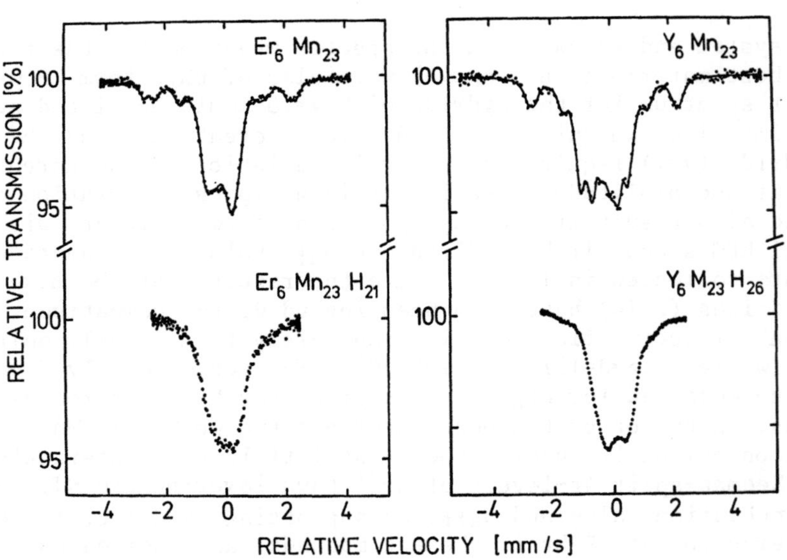

Fig. 2. Mössbauer spectra obtained at 4.2K for 0.5 at% ^{57}Fe impurity in R_6Mn_{23} (R=Er,Y) before and after hydrogenation.

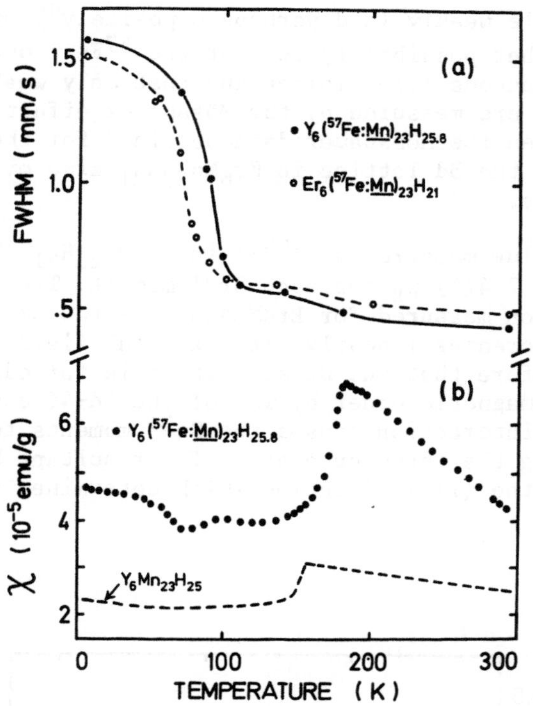

Fig. 3. Temperature dependence of (a) the ^{57}Fe Mössbauer spectrum full width at half maximum absorption (FWHM) and (b) the dc-field susceptibility (χ). (The $Y_6Mn_{23}H_{25}$ χ curve is taken from Ref. 7.)

to Y_6Fe_{23} where loading with hydrogen brings about a 20% increase in the magnetization at 4.2 K as well as an increase in the Curie temperature[11,3].

y=1. 0.5 at% ^{57}Fe was incorporated into the Er_6Mn_{23} and Y_6Mn_{23} host materials and the resulting ^{57}Fe Mössbauer spectra at 4.2 K both before and after hydrogenation are shown in Fig. 2. By comparison the magnetic splittings are much smaller than those observed for the Fe-rich compounds. The effect of hydrogenation is seen to be very similar for both the Er and Y cases. This is even more clearly demonstrated by the curves in Fig. 3 (a). The sudden increase of spectrum widths as the temperature falls is evidence for a magnetic transition in both hydrides. It is now well established by neutron-scattering measurements[4] that $Y_6Mn_{23}D_{23}$ is antiferromagnetic and structurally distorted at 4.2 K and this result is also in agreement with the form of the susceptibility curves (Fig. 3(b)) which give T_N=160–180 K depending on hydrogen (and/or ^{57}Fe) content. The fact that the rapid increase in the ^{57}Fe spectrum width for $Y_6Mn_{23}H_{26}$ starts at a temperature lower than T_N suggests that the

^{57}Fe probes couple weakly (and perhaps oppositely[6]) with their Mn neighbours. Another possibility is that the ^{57}Fe probes themselves support no spontaneous local moment and that only weak transferred hyperfine fields are measured by the Mössbauer effect. Due to the similarity between the Mössbauer data obtained for the two hydrides we conclude that the 3d lattice in $Er_6Mn_{23}H_{21}$ also orders antiferromagnetically.

The B_{eff} value measured at ^{166}Er in $Er_6Mn_{23}H_{21}$ (Table 2) sets a lower limit of 7.4(1) μ_B for each Er^{3+} moment. The low-temperature bulk magnetization measured for $Er_6Mn_{23}H_{23}$ is too small to accommodate this and increases linearly with external field up to 24 kG [2]. It appears therefore that the Er sublattice is forced to participate in the antiferromagnetic order by way of the 3d-4f exchange. In Er_6Mn_{23}, the CF interaction causes the Er^{3+} moments to point noncollinearly along the three cube axes. The resultant Er magnetization is thus in the (111) direction which determines the alignment

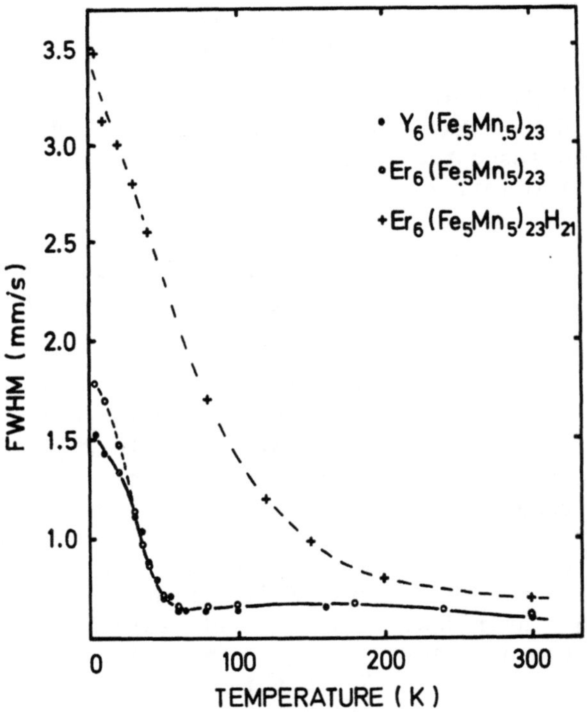

Fig. 4. Temperature-dependence of the ^{57}Fe Mössbauer spectrum full width at half maximum absorption (FWHM).

of the ferrimagnetic Mn sublattice[5]. Because of this, Er_6Mn_{23} has a strong magnetoanisotropy whereas Y_6Mn_{23} is easily saturated in an applied field of 2 kG. The antiferromagnetic alignment of Er^{3+} moments predicted above for $Er_6Mn_{23}H_{21}$ matches in with this model in the sense that the structural distortion which accompanies the magnetic transition would single out a unique axis for the Er^{3+} CF interaction.

y=0.5. The ^{57}Fe spectra obtained for $Er_6(Fe_{.5}Mn_{.5})_{23}$ and its Y counterpart were narrow and featureless. The latter is expected since different combinations of Mn and Fe neighbours result in a distribution of magnetic hyperfine fields at the ^{57}Fe nuclei. Spectrum widths were monitored as a function of temperature and the results are presented in Fig. 4. Again, the Y and Er compounds yield almost identical curves. The sharp increase in the spectrum width at 60 K is in excellent agreement with the bulk magnetization result for $Er_6(Fe_{.5}Mn_{.5})_{23}$ (Table 1).

That $Y_6(Fe_{.5}Mn_{.5})_{23}$ is magnetically ordered is most surprising. Bulk magnetization and neutron scattering measurements[8] show that no long range order exists for this compound in the region $0.4<y<0.7$ (for small external field). Nevertheless, for y=.6, neutron scattering results[13] indicate that a short range order persists up to 180 K. On the basis of the ^{57}Fe results, we propose that the 3d short range magnetic order in both $Y_6(Fe_{.5}Mn_{.5})_{23}$ and $Er_6(Fe_{.5}Mn_{.5})_{23}$ is essentially the same. However, in the Er compound, the Er^{3+} moments are directed along specific crystallographic axes as a result of the CF interactions. This effect, in turn, superimposes a long range order on the 3d lattice magnetization by means of the 3d-4f exchange.

The spectrum width has also been measured as a function of temperature for $Er_6(Fe_{.5}Mn_{.5})_{23}H_{21}$. Both the magnetic broadening and the ordering temperature are increased (see Fig. 4) by hydrogenation. The ordering temperature of about 180 K is of similar magnitude to that measured macroscopically[3] for $Y_6(Fe_{.5}Mn_{.5})_{23}H_{22}$.

CONCLUSIONS

In conclusion, the microscopic behaviour of the 3d magnetization in the $Er_6(Fe_{1-y}Mn_y)_{23}H_x$ system is almost identical to that for the corresponding Y system. The magnetic Er^{3+} ion primarily enables CF information to be transmitted to the Fe-Mn sublattice via the 3d-4f exchange interaction.

ACKNOWLEDGEMENTS

This work was supported by the Sonderforschungsbereich 161 of the Deutsche Forschungsgemeinschaft and the Kernforschungszentrum Karlsruhe GmbH.

REFERENCES

1. G. Hilscher and H. Rais, J. Phys. F8, 511 (1978).
2. F. Pourarian, E.B. Boltich, W.E. Wallace, R.S. Craig and S.K. Malik, J. Magn. Magn. Mater. 21, 128 (1980).
3. H. Oesterreicher and H. Bittner, Phys. Stat. Sol. (a) 41, K101 (1977).
4. K. Hardman, J.J. Rhyne, H.K. Smith, W.E. Wallace and S.K. Malik, J. Appl. Phys. 52 2070 (1981).
5. A. Delapalme, J. Déportes, R. Lemaire, K. Hardman and W.J. James, J. Appl. Phys. 50, 1987 (1979).
6. G.A. Stewart, J. Zukrowski and G. Wortmann, J. Magn. Magn. Mater. (in press).
7. S.K. Malik, T. Takeshita, and W.E. Wallace, Solid State Commun. 23, 599 (1977).
8. K. Hardman, J.J. Rhyne and W.J. James, J. Appl. Phys. 52, 2049 (1981).
9. J.A. Hodges, G. Jehanno and A. Schuhl, Hyperfine Interactions 11, 29 (1981).
10. B.D. Dunlap, 1971, in: "Mössbauer Effect Methodology", Vol. 7, I.J. Gruvermand, ed., Plenum Press, New York.
11. K.H.J. Buschow and R.C. Sherwood, J. Appl. Phys. 49, 1480 (1978).
12. H.R. Kirchmayr and W. Steiner, J. de Physique 32, C1-665 (1971).
13. P.V. Blanckenhagen, C. Lin and G. Hilscher, presented at the Frühjahrstagung der Deutschen Physikalischen Gesellschaft, Münster 1980 (unpublished).
14. G. Wiesinger, Dissertation, Tech. Univ. Wien (1974).

^{141}Pr-NMR INVESTIGATIONS ON THE DIELECTRIC VAN VLECK PARAMAGNETIC COMPOUND $Pr_2(SO_4)_3 \cdot 8H_2O$ AT VERY LOW TEMPERATURES

G. Feller*, M. Staudte* and M.A. Teplov†

*Karl Marx Universität, Leipzig, GDR
†Uljanov-Lenin University, Kazan, USSR

INTRODUCTION

Measurements of the longitudinal relaxation time τ_1 of $Pr_2(SO_4)_3 \cdot 8H_2O$ at high, mean and zero magnetic induction fields in the temperature range between 40 mK and 1 K are reported. At B=0 a short $\tau_1 \propto T^{-3}$ is observed, depending on the crystal surface-to-volume ratio. At B=0.2 T the behavior known for relaxation through impurities is studied, and at high fields, B=4 T, a shortening of τ_1 could be detected.

In 1966 Altschuler[1] suggested for magnetic cooling the use of substances containing rare earth ions with an even number of 4f-electrons and with a nonmagnetic singlet ground state due to the crystalline electric field (CEF). At low temperatures all 4f-electrons occupy the singlet ground state and in zero magnetic field the magnetic moment of the ion is zero. In an external field B a magnetic moment of the 4f-shell proportional to B can be induced by the Zeeman interaction $\mathcal{H}_Z = g_J\mu_B\vec{B}\cdot\vec{J}$, which causes the admixture of wave functions of excited CEF levels with those of the singlet ground state. A similar hyperfine interaction $\mathcal{H}_{hfs} = A_J\vec{I}\cdot\vec{J}$ between the 4f moment $\vec{J}$ and the nuclear spin $\vec{I}$ results in an additional moment given by $A\alpha_1^2/\Delta_1$, where Δ_1 is the energy difference between the singlet ground state $|0\rangle$ and the first excited state $|e_1\rangle$, and for which nonvanishing matrix elements $\langle 0|J_z|e_1\rangle = \alpha_1$ exist. The suggested cooling method has been tested over the last 13 years[2,3], where intermetallic compounds were used due to the short relaxation times caused by the sf exchange and the hyperfine interaction. The exchange interaction leads also to an additional enhanced nuclear moment[4] and to higher ordering temperatures. The lowest temperature obtained is 0.2 mK.[3]

We have measured the longitudinal relaxation time τ_1 of the ^{141}Pr nucleus in a dielectric Van Vleck paramagnetic compound with the aim of proving some suppositions about their use for magnetic cooling. This requires fast relaxation at high, mean and low magnetic fields if other substances, e.g. ^3He, should be cooled.

In $Pr_2(SO_4)_3 \cdot 8H_2O$, the Van Vleck behavior was first observed in 1932 by Gorter and De Haas[6]. The magnetic (mixed electron-nuclear) resonance[7] has also been seen. The low symmetry (monoclinic) causes a CEF singlet ground state of the $^3H_4$ multiplet. The magnetic resonance signals of ^{141}Pr may be analyzed by a spin Hamiltonian for rhombic symmetry[7]:

$$H = \sum_{i=1}^{3}\gamma_i B_i I_i + D[I_z^2 - \frac{1}{3}I(I+1)] + E[I_x^2 - I_y^2],\ i = x,y,z, \tag{1}$$

where

$$D = P + A_J[(\Lambda_x + \Lambda_y)/2 - \Lambda_z]$$

and

$$E = P\eta/3 + A_J(\Lambda_y - \Lambda_x)/2,$$

$$P = 3e^2qQ/[4I(2I-1)],$$

$$\gamma_i = \gamma(1+2g_J\mu_B\Lambda_i/\gamma\hbar),$$

$$\Lambda_i = \alpha_J \sum_n \alpha_n^2/\Delta_n.$$

Here γ = 1.31 kHz/Oe is the gyromagnetic ratio of ^{141}Pr, q the CEF gradient component in the z-direction, η the asymmetry parameter of the CEF gradient, Q the quadrupole moment, and Σ is the sum over states $|e_n\rangle$ that have nonvanishing matrix elements $\alpha_n = \langle 0|J_z|e_n\rangle$ with the singlet ground state $|0\rangle$. Finally, Δ_n is the energy difference between states $|0\rangle$ and $|e_n\rangle$.

EXPERIMENTAL DETAILS

The experiments have been carried out in a home made dilution refrigerator with cw and pulse NMR-spectrometers at a frequency of 13.4 MHz. The $Pr_2(SO_4)_3 \cdot 8H_2O$ monocrystal is placed in the mixing chamber of the cryostat together with the rf-coil such that the probe is exposed to the diffusing ^3He stream. The maximum repetition pulse sequence is chosen so that heating of the probe unit is prevented. The temperature T_b of the surrounding $^3He/^4He$ mixture is measured by a calibrated Germanium resistor. The temperature T_s of the ^{141}Pr moments is measured by the amplitude of the NMR signal following a Curie law and calibrated before each experiment. The sample used is the same one for which τ_1 measurements in zero field are reported[8].

RESULTS AND DISCUSSION

Relaxation at B=0

In zero magnetic field the singlet ground state is split into three levels $m_I = \pm 5/2$, $\pm 3/2$ and $+1/2$ due to the quadrupole interaction. Exciting the $\pm 3/2 - \pm 1/2$ transition (energy difference E/h=13.4 MHz) by short rf-pulses ($\sim 1 \mu s$ is equivalent to a $\pi/2$ pulse in NMR) we observed a quadrupole echo whose amplitude A_q is proportional to the population difference of $\pm 3/2$ and $\pm 1/2$ levels and hence proportional to T^{-1}. Thus a plot of A_q over T_b^{-1} reveals that the nuclear spin system can be cooled by the $^3He/^4He$ bath at T_b. The experiment shows that the spin temperature of the quadrupole system T_s follows T_b down to 35 mK. The relaxation τ_1 is measured by the observation of the echo amplitude at the time τ after a "$\pi/2$" pulse. We noticed two relaxation times for the recovery of the polarization of the three-level system. The short one, τ_{1s}, corresponds to the process of establishment of equilibrium in the three-level system, and the long time $\tau_{1\ell}$ to the relaxation to the bath temperature. The measured values for $\tau_{1\ell}$ are plotted in Fig. 1. There we observed a T^{-3} behavior which depends on the surface to volume ratio of the probe. The results can be expressed by $T_{1\ell}^{-1}=T^3/\nu$, where $\nu=\nu_m=400$ ms for the monocrystal as seen earlier, whereas for crystal discs of 0.5 mm thickness we obtained $\nu_\alpha = 8.3$ ms, being 48 times smaller than ν_m. We conclude that the main thermal resistance at B=0 is caused by Kapitza resistance as in intermetallic compounds. ν_m/ν_d is 8 times larger than the relation of volume to surface ratios of the probes whereas the quality of the monocrystal and disc surfaces is different. If we calculate with Kapitza resistance we expect relaxation times short enough to cool ^3He in the region below 1 mK if fine powder is used.

Relaxation at B=0.2 T

In the magnetic field the three quadrupole doublets $\pm m_I$ split off, and owing to the low symmetry and large hyperfine enhanced quadrupole moment many lines can be observed depending on the crystal orientation. We chose several intensive lines and noticed the recovery of magnetization after a $\pi/2$ pulse by producing a spin echo at a variable time τ from the pulse. The recovery of magnetization can be described by several relaxation times τ_1. From the largest time constant $\tau_{1,th}$ selected, we measured the temperature dependence. $\tau_{1,th}$ is responsible for the energy exchange with the lattice, whereas the other time constants stand for the establishment of the equilibrium of the excited moments with other reservoirs[9]. The same values of $\tau_{1,th}$ have been measured by cw NMR at low temperatures by observing the velocity of line growth after adjusting a lower bath temperature T_b. The results of our measurements are shown in Fig. 2. We observed the same relaxation times for both the monocrystal and the crystal discs.

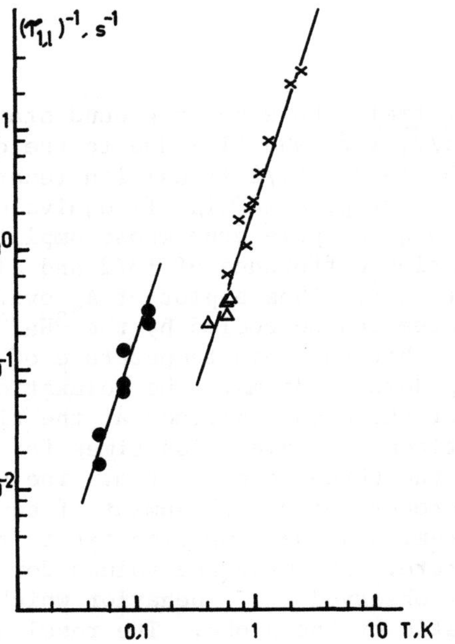

Fig. 1: Temperature dependence for τ_1: x - monocrystal[6]; Δ - monocrystal and ● - crystal discs of this paper.

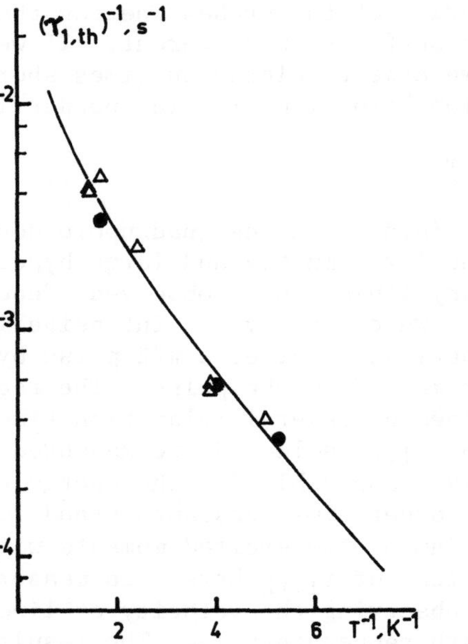

Fig. 2: Temperature dependence of $T_{1,th}$: Δ - monocrystal; ● - crystal discs.

The results can be fitted by the relation

$$(\tau_{1,th})^{-1} = \tau_B^{-1} \text{ ctnh } x(1 - \tanh^2 x), \tag{2}$$

$$x = g\mu_B B/2kT, \quad \tau_B \simeq 5 \times 10^3 \text{ s}, \quad g \simeq 4.$$

From τ_1 measurements above 1 K[10] we know that the paramagnetic impurity is responsible for the relaxation behavior of the enhanced spin system in dielectric compounds. Here a temperature dependence of $\tau_1^{-1} = a \cdot T$ is found.

From the theory of nuclear relaxation in dielectric crystals with impurities we are aware[9] for the case of fast spin diffusion that τ_{1n} behaves like

$$\tau_{1n}^{-1} = (1-P_o)^2 \cdot \tau^{-1} \tag{3}$$

where P_o (here P_o=tanh x) is the polarization of the impurity spins which have the g-factor g [see Eq. (2)], and $\tau^{-1} \propto$ctnh x is an effective relaxation rate depending on the special case of spin diffusion and the impurity relaxation times measurable by EPR. A rough analysis shows that the experimental values can be explained by nuclear relaxation over the dipole-dipole reservoir of the Nd^{3+} and Gd^{3+} impurity ions due to the lattice. A detailed analysis, however, is only possible with the knowledge of relaxation times of the crystal impurities and that is complicated due to the low crystal symmetry which leads to asymmetric spin diffusion parameters.

Relaxation at B=4 T

Owing to the very low temperatures the relaxation rate over the paramagnetic impurity is proportional to e^{-2x}. At higher fields, τ_1^{-1} approaches zero and the nuclear spin system is expected to be isolated.

We measured τ_1^{-1} by an indirect method. The probe is cooled in the presence of the resonance field B by lowering the ^{3}He-^{4}He bath temperature to T_b=40 mK. For $T_b \simeq$(110-130) mK the relaxation time corresponding to (2) becomes very large. The polarization of the enhanced moment freezes in and is consistent with a value of spin temperature near 130 mK verified by cw NMR amplitude measurements. (Almost the same spin temperature is obtained by cooling in zero field to 40 mK and then switching on the resonance field.) The fast relaxation in zero field guarantees that the "spin temperature" of the three-level quadrupole system T_s is also 40 mK. By switching on the field B_r, adiabatic magnetization starts at a field B' where the relaxation time becomes too long, B' $\simeq$ (40 mK/130 mK) x $B_r \simeq$ 25 mT.

At constant temperature the field is enlarged from B_r to 4 T and after 1 hour reduced to B_r. Then the spin temperature is measured by NMR. Repeating this procedure several times, a successive growth of the NMR amplitude is detected with a time constant of the order of 10 hours which we cannot explain by relaxation through impurities. Possibly, the direct relaxation process of the enhanced moment spin system expected to be proportional to B^4, is responsible for the relaxation at high fields[11].

CONCLUSION

The short relaxation times in zero field and their dependence on the crystal size show that the fine powdered dielectric Van Vleck compound has good thermal contact to the bath and can be used for quadrupole resonance thermometry to very low temperatures. The possible cooling of the moments in high fields at 50 mK can be used to precool the spin system which needs a long time due to the limited cooling power of the cryostat. Reversible cooling of other substances can be done only if shortening τ_1 in fine powders at mean magnetic fields takes place as experienced with other probes[12].

REFERENCES

1. S.A. Altschuler, Zh. Eksp. Teor. Fiz. Pisma 3, 177 (1966).
2. K. Andres and E. Bucher, Phys. Rev. Lett. 21, 1221 (1968).
3. H.R. Folle, M. Kubota, Ch. Buchal, R.M. Mueller and F. Pobell, Z. Phys. B - Condensed Matter 41, 223 (1981).
4. S.L. Tsarevskij, Fiz. Tverd. Tela 12, 2047 (1970).
5. S.A. Altschuler and M.A. Teplov in: Problemy Magninogo Resonansa, Nauka Moskva (1978).
6. W.G. Penney and R. Schlapp, Phys. Rev. 41, 194 (1932).
7. M.A. Teplov, Zh. Eksp. Teor. Fiz. 53, 1510 (1967).
8. F.L. Auchadeev, V.A. Grevcev, I.S. Konev, M.S. Tagirov and M.A. Teplov, Fiz. Tverd. Tela 18, 2107 (1976).
9. G.P. Chuzishvili, Usp. Fiz. Nauk 96, 441 (1968).
10. J.S. Konev and M.A. Teplov, Fiz. Tverd. Tela 18, 853 (1976).
11. S.A. Altschuler, private communication.
12. J.V. Gates and W.H. Potter, Phys. Rev. B13, 8 (1976).

CRYSTAL FIELD EFFECTS ON THE ESR SPECTRA OF RARE EARTHS IN $CePd_3$

C. Rettori*, E. Weber*, G.E. Barberis†, J.P. Donoso†
and F.C. Gandra†

*II. Physikalisches Institut, Universität Köln,
5000 Köln 41, FRG
†Instituto de Fisica UNICAMP, Campinas 13100 (SP), Brazil

INTRODUCTION

Crystal field effects on the J-multiplet ground state of rare earths (RE) in the mixed valent host of $CePd_3$ have been investigated recently by neutron scattering[1], ESR[2,3] and susceptibility[4] experiments. However it seems to be difficult to obtain the level scheme for the various non-S state RE ions in $CePd_3$ from the neutron and susceptibility[4] data. Since the ESR spectra give a clear identification of the low lying crystal field levels we performed ESR experiments at low temperatures in the limit of very diluted Dy, Er and Yb in polycrystalline samples of $CePd_3$.

EXPERIMENTAL DETAILS

The ESR experiments were carried out in a conventional ESR X-band spectrometer. Only the sample was cooled by using either a stainless steel dewar with a quartz tail or a helium gas flow system which fits into the microwave cavity kept at room temperature and modulated at 100 kHz.

The samples were prepared in argon atmosphere by arc-melting stoichiometric amounts of the constituents (Ce 4n's, Pd 5n's and REs 3n's) for the corresponding compounds $RE_xCe_{1-x}Pd_3$ (RE = Dy, Er, Yb). X-ray analysis indicated the presence of a single phase (Cu_3Au structure) for all the samples. The various series of diluted alloys were prepared from master alloys of x=0.02, x=0.01 and x=0.005. The ESR experiments were carried out on fine powders of the dilute compounds. In most cases the powdered samples were sealed in a vacuum and annealed at temperatures ranging between

100 and 900 C. A small decrease (∿10%) of the ESR residual linewidth was observed due to this annealing treatment.

RESULTS AND ANALYSIS

$Dy_xCe_{1-x}Pd_3$

Diluted alloys of concentrations varying from 100 to 1000 ppm were run between 2 and 30 K. Fig. 1 shows the temperature dependence of the spectra for the 1000 ppm sample. At high temperature, the spectra show clearly the presence of a resonance with a g-value of 7.50±0.05 corresponding to a Γ_7 level. Lowering the temperature, this resonance narrows and the relative intensity grows, reaching a maximum around 7 K. At lower temperatures, the relative intensity decreases, indicating that this resonance corresponds to an excited level. On the other hand, the relative intensities of the two broad and undefined resonances centered around 700 and 1450 G increase when the temperature is lowered. We believe that these resonances correspond to the powder spectra of a $\Gamma_8^{(1)}$ ground state. Obviously experiments on single crystals are needed in order to clarify this point. The resonance at 1000 G corresponds to the Γ_7 ground state of natural impurities of Er (∿10 ppm) present in all our $CePd_3$ samples (see below). Also the resonance of natural impurities (∿50 ppm) of Gd^{3+} was present.

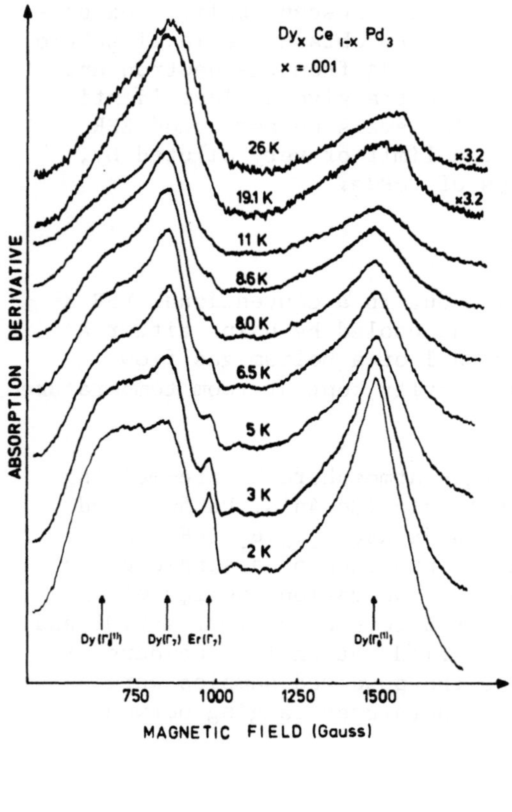

Fig. 1: X-band ESR spectra of 1000 ppm Dy^{3+} in $CePd_3$ for various temperatures.

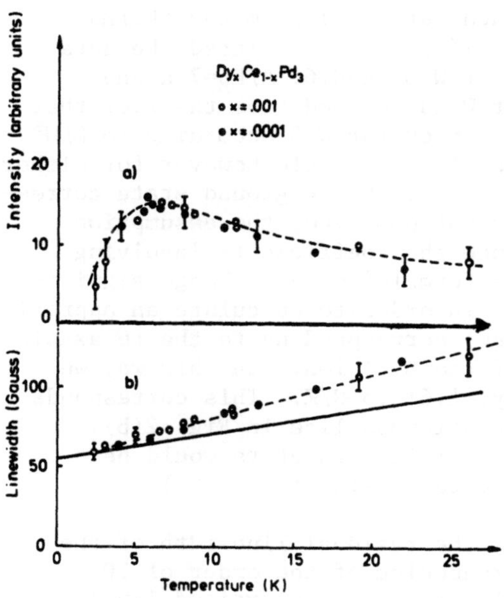

Fig. 2 (a): Temperature dependence of the Γ_7 ESR signal intensity of Dy^{3+} in $CePd_3$. (b): Temperature dependence of the Γ_7 ESR linewidth of Dy^{3+} in $CePd_3$. The continuous line corresponds to the Korringa rate b_K=1.6±0.5 G/K.

Fig. 2(a) shows the temperature dependence of the relative intensity of the resonance for the excited Γ_7 level. We took the intensity to be proportional to the peak height times the square of the linewidth, correcting for the dispersive part of the resonance. The dashed line is the fitting of the data to the theoretical intensity expected for an excited doublet with a $\Gamma_8^{(1)}$ quadruplet as a ground state. This can be written as:

$$I \propto \frac{\Delta_o}{k_B T} \frac{\exp(-\Delta/k_B T)}{2+\exp(-\Delta/k_B T)} \tag{1}$$

where Δ_o is the Zeeman splitting of the Γ_7 doublet, Δ the energy separation between the $\Gamma_8^{(1)}$ ground state and the Γ_7 excited state, and k_B is the Boltzmann constant. In Eq. (1) we have used $\Delta_o<k_B T$. From this fit we obtain Δ/k_B=7±2 K.

To the best of our knowledge, Fig. 2(b) shows the first careful measurement of the temperature dependence for the linewidth associated with the resonance of an excited doublet (Γ_7 in our case) in a metallic host. Hirst's formula[5] predicts a small departure from linear behavior at low temperatures ($\Delta>k_B T$). On the other hand, for high temperatures ($\Delta \lesssim k_B T$) the linewidth should broaden almost linearly with temperature. The rate of this broadening includes the contribution from the relaxation of the $\Gamma_8^{(1)}$ ground state due to the coupling between the Γ_7 and $\Gamma_8^{(1)}$ via the exchange interaction with the conduction electrons[5].

In order to extract the Korringa rate (b_K) from the thermal broadening of the linewidth of Fig. 2(b), we have fitted the data to Hirst's formula[5] (dashed line) for W>0, x=0.6, Δ/k_B=7 K and Δ_o/k_B=0.45 K. A positive value for W is assumed from the fact that a low lying Γ_7 level can be only observed for W>0 according to LLW[6]. The observation of an excited Γ_7 level is possible however for either positive or negative values of x. But as the Γ_6 ground state corresponding to values of -1<x<-0.45 was not observed, the assumption of x>0.48 seems to be correct. Since the coefficients involving various matrix elements in Hirst's formula[5] do not change significantly for 0.48<x<1, we chose x=0.6 in order to calculate an approximate value for the Korringa rate (b_K) corresponding to the relaxation of the isolated magnetic doublet of the Dy^{3+} ion. In this way we found the Korringa rate to be b_K(Dy)=1.6±0.5 G/K. This corresponds to the initial slope shown by the continuous line in Fig. 2(b). Obviously a knowledge of the crystal field parameters would be needed in order to obtain a more accurate value for b_K(Dy).

It is interesting to note that the residual linewidth of the resonance includes a homogeneous broadening of the order of 10 G for Δ/k_B=7 K, associated with the lifetime of the excited level.

$Er_xCe_{1-x}Pd_3$

Various series of dilute compounds for concentrations between 100 to 5000 ppm were analyzed between 1.5 to 50 K. Fig. 3 shows a typical spectrum for a 200 ppm sample at 8 K. Fig. 4 shows that the relative intensity of the resonance increases at lower temperatures. This and the measured g-values of 6.72±0.05 clearly identify the ground state of Er^{3+} in $CePd_3$ as Γ_7. Within the experimental accuracy the linewidth is linearly dependent on temperature. The measured residual width varies between 25 and 50 G depending on the concentration and thermal treatments of the sample. The thermal broadening of the linewidth was found to be 1.2±0.3 G/K. The hyperfine constant extracted from the hyperfine structure of the spectra, using the Breit-Rabi formula, is A(^{167}Er)=75.4±0.5 G.

Following van der Linden et al.[7] we plot in Fig. 4 (continuous line) the expected temperature dependence of the intensity of the ESR resonance for an isolated doublet ground state. We normalized this plot to the experimental data at 7 K. It can be seen that the experimental intensity decreases faster than the theoretical one as the temperature is raised. This is an indication of the presence of low lying excited crystal field levels. As estimate of the position of these excited states from this plot is not possible because of the uncertainty and scatter of data from sample to sample. The presence of nearby crystal field levels is also supported by neutron data[1], where a clear inelastic peak is observed on the energy gain side of the spectrum at about 0.7 meV (∿8.1 K).

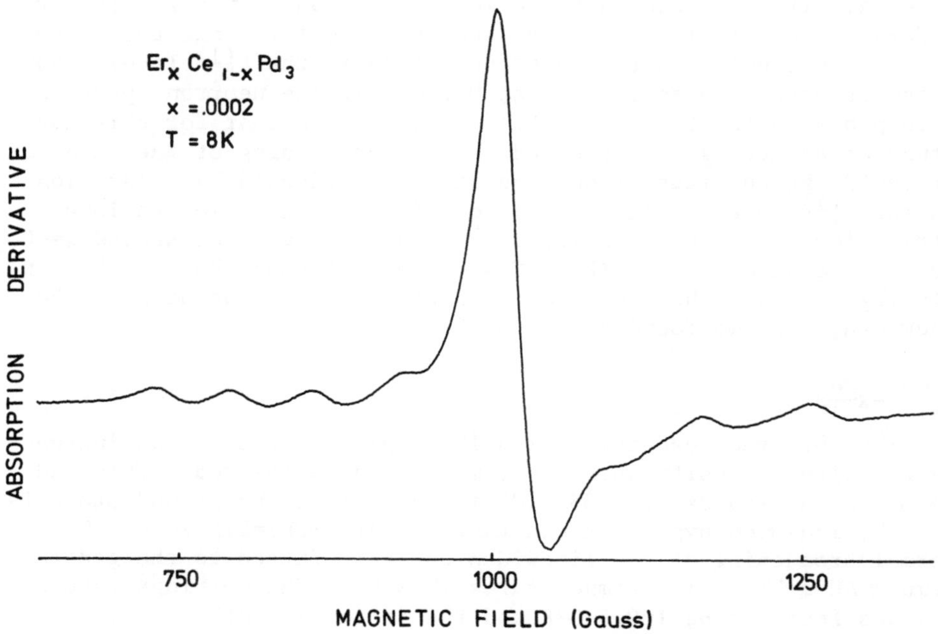

Fig. 3: X-band ESR spectrum of 200 ppm Er^{3+} in $CePd_3$ at T=8 K.

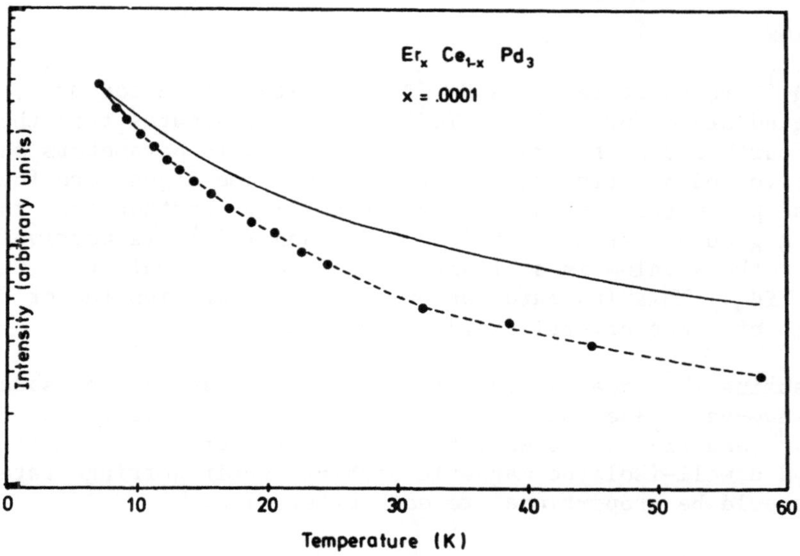

Fig. 4: Temperature dependence of the relative ESR signal intensity for 100 ppm Er^{3+} in $CePd_3$.

Assuming the same crystal field parameters for Er^{3+} as for Dy^{3+} in $CePd_3$, with a Γ_7 ground state we have W>0 and $-0.45<x<0$ for Er^{3+} in $CePd_3$. Then from LLW[6] we can see that the first excited state that can be coupled to the Γ_7 ground state is the $\Gamma_8^{(1)}$ level. Now if the observed inelastic peak at 0.7 meV in the neutron spectrum corresponds to the transition $\Gamma_8^{(1)}\leftrightarrow\Gamma_7$, we know that for this magnitude of splitting the measured thermal broadening of the linewidth for the Γ_7 ground state should be strongly affected by relaxation via the $\Gamma_8^{(1)}$ level[5]. Again, the coefficients involving various matrix elements in Hirst's formula[5] do not change much around x=-0.3. For this value of x and $\Delta/k_B=8$ K we have estimated that the Korringa rate (b_K) is about half of the measured thermal broadening of the linewidth, i.e. we found $b_K(Er)=0.60\pm0.2$ G/K.

$Yb_xCe_{1-x}Pd_3$

The observed resonance for a 1000 ppm sample shows an increasing relative intensity when the temperature is lowered. This and the measured g-value of 3.33±0.05 indicate that the ground state is Γ_7. The measured hyperfine constant was $A(^{171}Yb)=575\pm20$ G. The thermal broadening of the linewidth could be fitted to the general formula $\Delta H=a+bT$ at low temperatures (T<9 K). The Korringa rate (b_K) was found to be 1±0.5 G/K and the residual width a=50±10 G. At temperatures above 9 K the linewidth broadens exponentially indicating the effects of relaxation via excited crystal field levels. Our data were found to be in general agreement with those of Gambke et al.[2].

DISCUSSION

A $\Gamma_8^{(1)}$ ground state and a Γ_7 first excited state for Dy^{3+}, and a Γ_7 ground state for Er^{3+} and Yb^{3+} in $CePd_3$ indicates that the hosts' fourth and sixth order cubic crystal field parameters should be negative and positive respectively. The same signs were found for these parameters in the case of YPd_3[1,8,9]. Furthermore, the Γ_7 and Γ_6 ground states of Er^{3+} in $CePd_3$ and YPd_3[9], respectively, show that the x value is at least a factor of two smaller in $CePd_3$ than in YPd_3. This indicates an even larger reduction factor for the ratio of these crystal field parameters.

Assuming that the average on the Fermi surface of the square of the wave-vector-dependent exchange integral J(q) is the same for Gd^{3+}, Dy^{3+} and Er^{3+} it is easy to show[5] that, for a given host metal and a well-isolated magnetic doublet, their Korringa rate values should be proportional to each other, i.e.

$$\frac{b_K(Dy)}{b_K(Er)} = 2.5; \qquad \frac{b_K(Gd)}{b_K(Er)} = 2.6. \tag{2}$$

Table 1: ESR parameters of Gd^{3+}, Dy^{3+}, Er^{3+} and Yb^{3+} in $CePd_3$ and YPd_3

	$CePd_3$			YPd_3		
RE	b_K (G/K)	g-value	A (Gauss)	b_K (G/K)	g-value	A (Gauss)
Gd	1.7±0.2	2.048±0.005		3.2 ±0.2[a]	2.021±0.005[a]	
Dy	1.6±0.5	7.50 ±0.05				
Er	0.6±0.2	6.72 ±0.05	75.4±0.5	1.25±0.5[b]	6.77±0.05[b]	
Yb	0.5±0.3	3.34 ±0.05	580 ±10	1.0 ±0.5[b]	3.34±0.05[b]	575±20[b]

[a] Ref. 2.
[b] Ref. 9. The value of $b_K(Er)$ in YPd_3 has been calculated using the method described in the text for the crystal field parameters given in Ref. 8.

Table 1 shows that the values, calculated in the above section, for the Korringa ratios of Dy^{3+} and Er^{3+} give a ratio which is in reasonable agreement to that expected from Eq. (2). For Gd^{3+} we also found that the Korringa rate above 200 K scales with that of Dy^{3+} and Er^{3+}, following Eq. (2). A comparison between the Korringa rates for the various RE ions in these two hosts gives a reduction factor of the order of two upon going from YPd_3 to $CePd_3$.

In conclusion, we believe that as far as the crystal field effects are concerned this work and that of Gambke et al.[2] show that the ESR data of non S-state and S-state RE ions in the mixed valent compound $CePd_3$ can be analyzed in much the same way as is usually done in non-mixed valent compounds.

ACKNOWLEDGEMENTS

We thank Profs. D. Davidov and D. Wohlleben for many stimulating discussions. This work was partially supported by CNPQ, FAPESP (Brazil) and the Deutsche Forschungsgemeinschaft. One of us (C.R.) is grateful to the Alexander von Humboldt Foundation for granting a fellowship.

REFERENCES

1. E. Holland-Morlitz, D. Wohlleben, H. Scheuer and M. Loewenhaupt, J. Magn. Magn. Mat. 14, 234 (1979).
2. T. Gambke, B. Elschner and L.L. Hirst, Phys. Rev. Letters 40, 1290 (1978); T. Gambke and B. Elschner, J. Physique C5 40, 331 (1979).

3. J.P. Donoso, F.G. Gandra, C. Rettori, E. Weber and G.E. Barberis, Verh. DPG (VI) 16, 313 (1981).
4. H.W. Ludwigs, Dissertation, Köln 1981. H. Borchardt, R.F. Hoyt, A.C. Mota and D. Wohlleben, Proceedings of the 16th International Conference on Low Temperature Physics, Los Angeles, U.S.A., 1981.
5. L.L. Hirst, Phys. Rev. 181, 597 (1969); D. Davidov, C. Rettori, A. Dixon, K. Baberschke, E.P. Chock and R. Orbach, Phys. Rev. B8, 3563 (1973).
6. M.R. Lea, M.J.M. Leask and W.P. Wolf, J. Phys. Chem. Solids 23, 1381 (1969).
7. H.W.M. van der Linden, G.J. Nieuwenhuys, H.D. Dokter, D. Davidov and I. Felner, J. Magn. Magn. Mat. 15-18, 42 (1980).
8. U. Walter, diploma thesis, Köln (1980).
9. C. Rettori, E. Weber, J.P. Donoso, G.F.C. Gandra and G.E. Barberis, Solid State Commun. 39, 1025 (1981).

^{27}Al-NMR INVESTIGATIONS ON $PrAl_2$ AND $Pr_xLa_{1-x}Al_3$, x=1.0, 0.5, 0.25

G. Feller, A. Frieser, B. Lippold and M. Mühle

Sektion Physik, Karl Marx Universität, Leipzig, GDR

INTRODUCTION

Intermetallic compounds of rare earths exhibit an interesting magnetic behavior because of the large conduction electron spin polarization described by RKKY theory and its influence on properties measured by nuclear magnetic resonance[1-4]. Moreover, substances with singlet crystal field ground states and f-f exchange energies below a critical value (Van Vleck paramagnetic compounds) are considered for nuclear cooling.

This paper deals with NMR measurements on $PrAl_2$ and Van Vleck paramagnetic $Pr_xLa_{1-x}Al_3$. $PrAl_2$ has a cubic Laves structure with cubic point symmetry for the Pr and a tetrahedral one for the Al. NMR investigations[1-4], crystal field parameters[5], and magnetic properties[6,7] have been reported by many authors.

$Pr_xLa_{1-x}Al_3$ has a hexagonal Ni_3Sn structure with orthorhombic point symmetry for the Al and has been studied by NMR[8], neutron scattering , and magnetic measurements. We measured the ^{27}Al Knight shift down to low temperatures and compared the results directly with the crystal field parameters.

According to the molecular field approximation of the RKKY theory the conduction electrons are polarized by the s-f exchange interaction $H_{sf} = -J_{sf}\vec{s}\cdot\vec{S}$ in an external field B. In general, this polarization changes with the distance from the 4f ion, and Jsf corresponds to the value of the polarization at the site of the resonance nucleus. By way of the Fermi contact interaction, the ^{27}Al nucleus "sees" an additional field $\delta B_f=K_oJ_{sf}\langle S_z\rangle/g\mu_B=\langle S_z\rangle B_{sf}$, where $B_{sf}=K_oJ_{sf}/g\mu_B$ is the s-f hyperfine field, g=2, and

$K=\mu_B N(E_F)B_{hfs}$ is the Knight shift without consideration of the 4f spins. B_{hfs} is the contact hyperfine field and $N(E_F)$ is the density of states at the Fermi surface. By adding δB_f and $\delta B_o=K_o \cdot B$, we obtain[1]

$$K = K_o + K_f = K_o[1 - J_{sf} \frac{g_J - 1}{g g_J \mu_B^2} \frac{\chi_f^o}{N_f}] = K_o(1+\lambda_{sj}\chi_f^o) \quad (1)$$

where $\chi_f^o = N_f g_J \mu_B \langle J_z \rangle / B = N_f g_J \mu_B \langle S_z \rangle / (g_J - 1)B$ is the crystal field susceptibility, g_J the Landé factor, and N_f the number of spins per volume unit.

EXPERIMENTAL DETAILS

The samples were prepared from 99.9% pure praseodymium and 99.99% pure aluminium in the State Institute for Rare Earth Metals, Moscow. For NMR measurements the polycrystalline pieces were pulverized to grain sizes of 40-100μ in a He atmosphere and put in glass tubes at a 60% He pressure atmosphere. The measurements were repeated with the same probes after annealing them in a vacuum for 70-100 hours at 830 C ($PrAl_3$) and 600 C ($PrAl_2$). X-ray and neutron diffraction analyses do not show the presence of other phases within an accuracy of <5%.

The measurements were mainly performed at 17 MHz with a NMR pulse spectrometer either by Fourier transformation of the free induction decay or by spin-echo techniques. For $PrAl_2$ only the latter has been used. τ_1 is measured by variation of the time distance between a saturation pulse (or a $\pi/2$-pulse for $PrAl_3$) and a pair of pulses (1 to 3 μs) with a fixed distance. For τ_2 measurements only, the pair is used and the distance between the pulses is varied. The reference signal is obtained from an Al-sulfate solution or from Al powder with consideration of the ^{27}Al Knight shift of 0.162%. Measurements were carried out down to temperatures of T=4.2 K ($PrAl_3$) or 37.5 K ($PrAl_2$). We ascertained that the temperature of the sample in the Helium gas-filled glass cell was equal to that of the surrounding He or N_2 gas.

RESULTS AND DISCUSSION

Knight Shift

Fig. 1 shows the results of Knight shift measurements on $PrAl_2$ and $Pr_xLa_{1-x}Al_3$. The respective values of the Knight shift K are corrected by the ^{27}Al Knight shift of lanthanum compounds, K=0.06% for $LaAl_2$ and $K_o=(0.095\pm0.01)\%$ measured by us for $LaAl_2$ in the temperature range between 4.2 K and 300 K. Our value is substantially higher than that obtained by other authors[8]; $K_o=0.056\%$. All the results can be described by

$$K_f = K - K_o = \nu/(T-\theta), \quad \nu = K_o \lambda_{sf} C_o \quad (2)$$

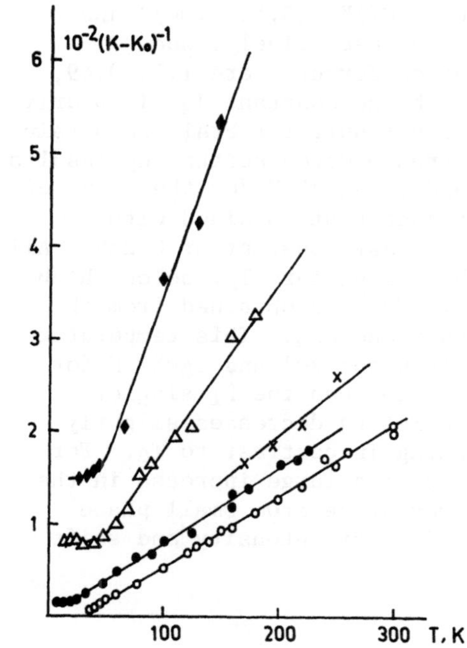

Fig. 1: $(K-K_o)^{-1}$ vs T. o=$PrAl_2$. For $Pr_xLa_{1-x}Al_3$ the symbols are • for x=1, x for x=0.75, Δ for x=0.5 and ♦ for x=0.25.

for $T \gtrsim T_\Delta$, the temperature at which K remains constant for Van Vleck paramagnetic $Pr_xLa_{1-x}Al_3$.

$PrAl_2$

The measured values can be fitted by $\theta=(31\pm1)$K and $\nu=\lambda_{sf}K_oC_o$ $=(1.31\pm0.11)$K. The highest measured value of K is 17.5% at 37.5K. The signal to noise ratio in this case was 5:1. At lower temperatures the signal disappears.

With the measured value $\lambda_{sf}K_oC_o$ and the susceptibility calculated from neutron scattering results[5] ($C_o\simeq1.5$ K-cm^3/mole) we obtain $K_oJ_{sf}=-2.26\cdot10^4$eV, which is in good agreement with earlier values[3] of $-2.37\cdot10^4$eV. With $K_o=0.06\%$ we find $J_{sf}=-0.38$eV, which is somewhat lower than $J_{sf}=-0.43$eV of other authors[3]. From $\theta=31$ K we calculate by use of the Weiss molecular field theory a molecular field constant of $\lambda_W=20.7$ mole/cm^3, which is in reasonable agreement with the value 18.2 mole/cm^3.

$PrAl_3$

The results shown in Fig. 1 can be fitted in the high-temperature region by (2), where ν (taken from a log-log plot similar to

Fig. 1) amounts to (1.26±0.06)K, (1.02±0.07)K, (0.55±0.04)K and (0.28±0.03)K for x=1, 0.75, 0.5 and 0.25 respectively, and θ=(0±5)K. The relation of ν values for different x are 1.1, 0.49, 0.89 and 0.25, which shows that the exchange constant J_{sf} is nearly the same for all values of x. Using the result for $PrAl_3$ we obtain (using the susceptibility calculated from neutron scattering results) $\chi^o=C_o/(T-\theta)$ with C_o=1.645 Kcm^3/mole and θ =-6.75 K for the value of $K_oJ_{sf}=-1.85\cdot10^{-4}$eV, which is 6% higher than that obtained with K_o=0.056%. With K_o=0.056% we evaluate a J_{sf} constant of 0.2eV. All probes reveal Van Vleck behavior. The temperature T_Δ, below which the shift remains constant, gives K_{VV}. This is obtained from the crossing point of the $\nu(T-\theta)^{-1}$ asymptote and K_{VV}. This temperature increases with decreasing x, e.g. T_Δ=18 K for x=1 and T_Δ=51 K for x=0.5 and x=0.25. However, the distance between the Γ_1 singlet ground state and the first excited doublet Γ_3 decreases slightly by several percent from x=1 to x=0, being in contrast to T_Δ. For the compounds x=0.5 and x=0.25 we observed a large increase in the susceptibility at 14 K and 6 K which may arise from small phase inhomogeneities and which may cause a drop of intensity and shift of the ^{27}Al line starting at T$\sim$20 K.

Line Widths

Both compounds possess a noncubic point symmetry of ^{27}Al (with I=5/2), and the high quadrupole moment of Al ($Q/e=0.149\cdot10^{-24}cm^{-2}$) causes a splitting in a central line ($m_I=\pm1/2$) and two satellite pairs ($m_I=\pm3/2$, $\pm5/2$). Here the width of the central line and the distance to the first satellite (for powder) are given by $\delta\nu_{1/2}$= (25/9)A and $\nu_{3/2}-\nu_o=\nu_Q/2$, respectively, with $A=(\nu_Q^2/16)[I(I+1)-3/4]=(1/2)(\nu_Q^2/\nu)$, and $\nu_Q=3e^2qQ/h2I(2I-1)=(3/20)e^2qQ/h$, with eq the strength of the electric field gradient. To first order, the central line is unshifted; to second order its resonance frequency is shifted by $\Delta\nu=(7/18)A$ to lower frequencies. With e^2qQ/h=4.54 MHz[10] for $PrAl_2$ we obtain $\delta\nu_{1/2}$=37.8 kHz, and $\nu_Q/2$=340 kHz, as observed from cw NMR measurements at room temperature. The shift $\Delta\nu/\nu_o$ amounts to 0.03% for ν_o=16 kHz, and must be corrected in the room temperature region, whereas for $PrAl_3$ the second order shift is neglected. Because of the quadrupole splitting in $PrAl_2$, we saturated only the central line by the first pulse. We observed an echo by starting the second pulse at the time τ after the first so that the free induction decay vanished. The amplitude of the echo is proportional to $\exp(-2\tau/\tau_2^*)$ and the relaxation time τ_2^* measured by variation of τ does not contain the quadrupole interaction. We obtained $\tau_2^{*-1}=\Delta\nu_o^*[1+\phi^*/(T-\theta^*)]$, $\Delta\nu_o^*$=(4.1±0.2) kHz, ϕ^*=(20.1±2)K and θ^*=(35±5)K.

For $PrAl_3$ we observed one single (Lorentzian) line with a width of 12.4G at room temperature and ν=17 MHz independent of macroscopic demagnetization factors (cylindrical or spherical samples). The width at room temperature showed a frequency dependence of $(\delta\nu)_1=(\delta\nu_o^2+\alpha\nu^2)^{1/2}$, with $\delta\nu_o$=9.1 kHz and $\alpha=5.1\times10^{-7}$.

For x=0.5 and 0.25 we observed one central line with two satellite pairs with a distance of $\nu_{\pm3/2}-\nu_{\pm1/2}=\gamma(12\pm1)$Oe and a width of $\delta\nu_{1/2}=\gamma\cdot6.8$ Oe at 16 MHz. The results of line width measurements can be described by $\delta\nu=\delta\nu_o[1+\phi/(T-\theta)]$ with $\delta\nu_o=(6\pm1.5)$kHz, which is of the order of the $LaAl_3$ line width. $\phi=(0.66\pm0.1)$ K for x=1, and $\phi=(2.1\pm0.3)$ K for x=0.5, 0.25 (including satellites). The central line also shows the same dependence. All lines have Lorentzian shape. From the satellite distance we calculate $e^2qQ/h=(173\pm20)$kHz, which is much less than the reported value 0.8 MHz. Quadrupole effects therefore can be neglected for the Knight shift calculations in the Al_3 compounds, even for $LaAl_3$. The temperature dependence of the width can be explained by the interaction of the Al nuclear moment with the mean praseodymium moment $\vec{\mu}=\chi_f B/N_f$ by demagnetization fields, which are difficult to estimate, because of the irregular particle form. $\delta\phi_o$ is determined by the quadropole interaction and the Al-Al-dipole-dipole interaction is determined as in $LaAl_2$.

T_1-measurements

In $PrAl_2$ we noted two relaxation times for the recovery of magnetization, $M_z(t)$. Here, $M_o-M_z(t)=A\exp(-t/\tau_{1s})+(1-A)\exp(-t/\tau_{1\ell})$ with A=0.66, since we saturated only the ±1/2 central line. Both τ_{1s} and $\tau_{1\ell}$ follow the temperature dependence of $\chi=\tau_1 T=\tau(T-\theta)$ where $\tau_s^{-1}=(3.51\pm0.25)$kHz is close to $\Delta\nu_o^*$ (see τ_2^* measurements), and $\tau_\ell^{-1}=(830\pm80)$Hz and $\theta\simeq31$ K. The short time describes the approach to equilibrium of the ±1/2 spin system with the other spins at $m_I=\pm3/2$, ±5/2 and $\tau_{1\ell}$ the relaxation to the bath temperature. Our $\tau_{1\ell}$ values are 13 times shorter than those reported. A second sample also verified for phase homogeneity within 5% error, showed the same τ_s but a shorter $\tau_\ell=(1.2\pm0.2)$ms. For $PrAl_3$ we obtained a corresponding temperature dependence for the Korringa relation $(\tau_1 T)^{-1}=(\chi)^{-1}=\chi_o^{-1}+(\tau T)^{-1}$ with $\chi_o=(7\pm2)$ sK and $\tau\simeq15$ ms.

REFERENCES

1. V. Jaccarino, J. Appl. Phys. Suppl. 3, 102 S(1961).
2. B.G. Silbernagel, V. Jaccarino, P. Pincus and J.H. Wernick, Phys. Rev. Lett. 20, 1091 (1968).
3. B.G. Barnes and E.D. Jones, Sol. State Commun. 5, 285 (1967).
4. S. Tanigawa, T. Tsuchida and Y. Nakamura, J. Phys. Soc. Jap. 46, 1488 (1979).
5. T. Frauenheim, W. Matz and G. Feller, Sol. State Commun. 29, 805 (1979).
6. K.H. Mader, E. Segal and W.E. Wallace, J. Phys. Chem. Sol. 30, 1 (1969).
7. M.G. Purwins, E. Walker, B. Barbara, M.F. Rossignol and P. Bak, J. Phys. C7, 3573 (1974).

8. A.M. von Diepen, W.H. de Wijn and K.H.J.Buschow, J.Chem.Phys. 46, 3439 (1967).
9. A. Andreeff, L.P. Kaun, B. Lippold, W. Matz, N.I. Moreva and K. Walther, Phys. State Sol. (b) 87, 535 (1978). See also literature cited therein.
10. G.C. Carter, L.H. Bennett and D.J. Kahan, Metallic Shift in NMR, Progress in Material Science 20, Pergamon Press (1977).

MAGNETIC BEHAVIOR OF TbF_3

M. Piotrowski

Institute for Nuclear Research
05-400 Swierk, Poland

INTRODUCTION

Relatively little is known about the magnetic properties of rare earth fluorides, in contrast to other rare earth compounds, which have been extensively studied over the past two decades. Presumably, one of the reasons for this is that the crystal structures of the rare earth fluorides are in general much more complicated, resulting in a low symmetry point group for the rare earth site. This makes a quantitative analysis of magnetic properties very difficult. On the other hand, among the known rare earth trifluorides only TbF_3 was reported as undergoing a magnetic phase transition at 3.95 K.[1-3] Usually, the most probable magnetic coupling between paramagnetic ions in the non-conducting compounds will be a superexchange interaction via the anions. In TbF_3, however, the ordering temperature is low enough that other mechanisms in addition to superexchange should be considered as a possible cause of the magnetic ordering.

Another implication of a low Néel temperature is that the magnetic interaction, whatever its origin, will be a small perturbation compared to the crystalline electric field. The crystal field splits the free ion ground state multiplet $^7F_6$ into singlets, of which the lowest two lie close together and are well separated from the next levels[1]. Therefore, TbF_3 represents in fact a two-level system and can be described to a good approximation as a spin 1/2 Ising magnet. These properties are extremely interesting from the point of view of the magnetic structure behavior.[4]

EXPERIMENTAL DETAILS

The magnetic structure of TbF_3 was studied on two powder samples. Both samples were prepared from Tb_4O_7 by the wet method described in detail by Zalkin and Templeton[5]. The only difference between the two samples was that sample II was additionally heated in an electric furnace in vacuum at 600 C for about 20 hours. After heating, the sample changed slightly in color from white to gray. Such an observation was also made by Zalkin and Templeton but no explanation was given. No difference was found in the crystal structure after this additional heat treatment.

The neutron diffraction experiments, which form the subject of this paper, were carried out at the EWA reactor in Swierk using a standard two-axis diffractometer. The wavelength of 1.324 Å from a Zn monochromator was used throughout the experiment. In the temperature range 1.5-4.2 K, measurements were performed using either the bath cryostat or the cryostat where the sample was attached to a cold finger. The latter was used together with a conventional electromagnet allowing the collection of data with a magnetic field up to 1 T.

For each sample, four neutron diffraction runs were made: at 1.5 K and 4.2 K without magnetic field and with a magnetic field of 0.4 T parallel to $\vec{e}$. A few additional measurements on both samples were also made; namely the temperature dependence of the strongest magnetic peak (sample I) as the cryostat was warmed up slowly, and the dependence of peak intensities on the magnetic field (samples I and II).

RESULTS

Figure 1 shows illustratively the magnetic behavior of both samples of TbF_3. It is seen (Figs. 1(a) and 1(b)) that down to 4.2 K they remain paramagnetic, i.e., on the neutron diffraction patterns only nuclear peaks are present and strong paramagnetic incoherent scattering is observed. However, when the temperature is lowered to 1.5 K, additional peaks appear and their positions provide experimental evidence for each type of magnetic ordering.

For sample I (Fig. 1(a)) the magnetic reflections cannot be adjusted to any sequence of Miller indices h_M, k_M, l_M. The structure can be satisfactorily described only as a spiral spin structure with the spiral axis parallel to the modulation vector $\vec{\tau}=\{0,y/b,0\}$. The model generated in this fashion is shown on the left side of Fig. 2. While the structure seems rather formidable because of the large magnetic unit cell (about 12 times as large as the chemical one) it can be described in quite a simple way: there are two simple spirals and both are generated by the same vector $\vec{\tau}=\{0,y/b,0\}$ with y=0.082 ±0.005. Between two spirals there is a phase shift of 60°. In Fig. 2,

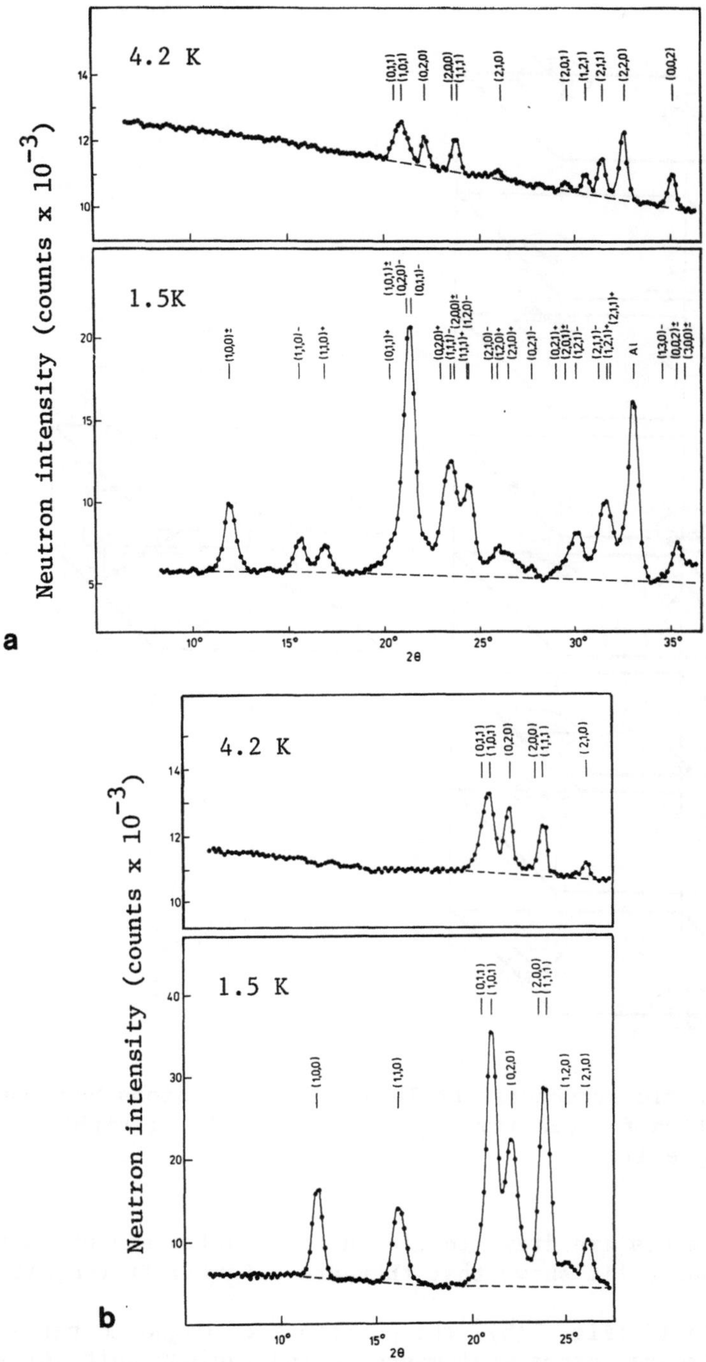

Fig. 1: Neutron diffraction patterns of TbF_3 at 4.2 K and 1.5 K. (a) sample I; (b) sample II.

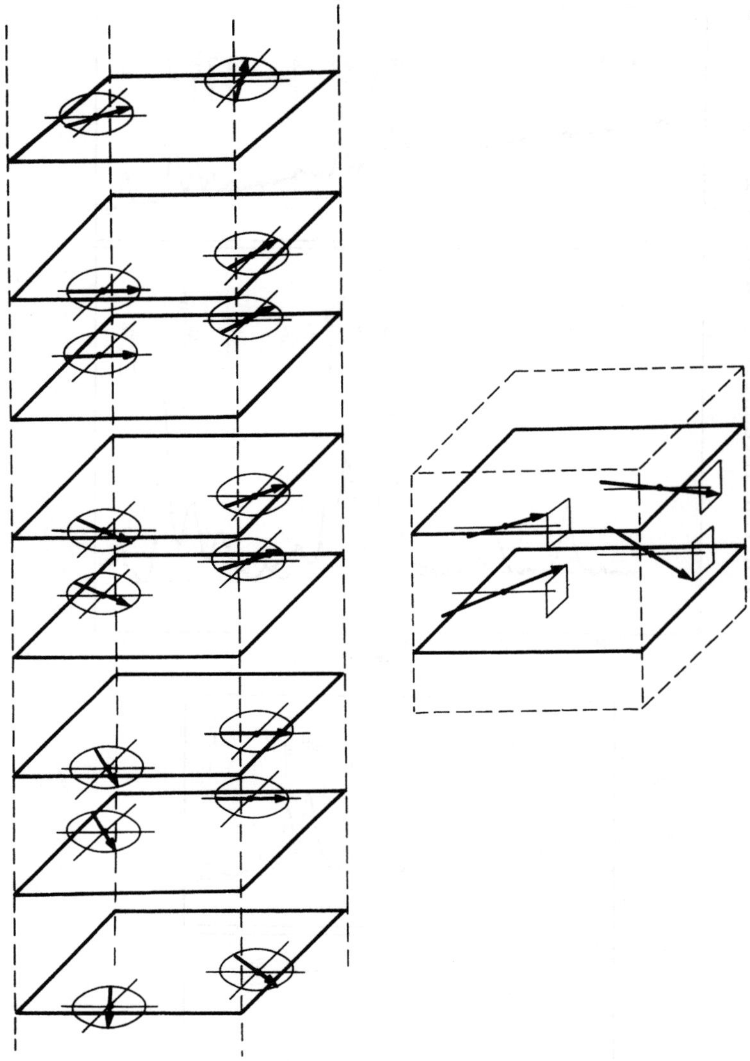

Fig. 2: Magnetic structures of TbF_3. Fluorine atoms have been omitted for clarity. Left side: sample I; right side: sample II.

for clarity moments are drawn to lie in the basal plane while the best least-squares fit shows that they are slightly tilted, $\sim 5°$.

For sample II (Fig. 1(b)) the positions of magnetic reflections provide experimental proof that magnetic and nuclear unit cells coincide. The analysis of neutron data is rather straightforward. From the fact that the magnetic reflections (001), (010) and (200) are zero, whereas (021) does not vanish, the following relations between the magnetic moments of Tb ions can be deduced:

Tb(1):	$(x, \frac{1}{4}, z)$	$\mu(1) = (\mu_x, \mu_y, \mu_z)$
Tb(2):	$(x+\frac{1}{2}, \frac{1}{4}, \frac{1}{2}-z)$	$\mu(2) = (\mu_x, -\mu_y, -\mu_z)$
Tb(3):	$(-x, -\frac{1}{4}, -z)$	$\mu(3) = (\mu_x, -\mu_y, \mu_z)$
Tb(4):	$(-x-\frac{1}{2}, -\frac{1}{4}, z+\frac{1}{2})$	$\mu(4) = (\mu_x, \mu_y, -\mu_z)$.

Notice that the sums $\vec{\mu}(1)+\vec{\mu}(2)$ and $\vec{\mu}(3)+\vec{\mu}(4)$, respectively, have resultant moments pointed along the axis in agreement with the prediction of Holmes et al.[1]. In order to determine the moment directions, a least-squares fitting procedure was applied to the observed spectrum. The best agreement was obtained for the following set of μ_x,μ_y, and μ_z: μ_x=7.4, μ_y=-2.4, and μ_z=-3.4 (all in units of μ_B). The resulting magnetic structure is shown for comparison on the right side of Fig. 2. For both samples, calculations were based on structural parameters determined in Ref. 6, and magnetic form factors after Ref. 7. From these experiments, it follows: μ_{sat}=9.0±0.5 μ_B for sample I and μ_{sat}=8.5±0.5 μ_B for sample II.

We report here also on the unusual behavior of TbF_3 in an external magnetic field. In the absence of a magnetic field, the magnetic phase transition takes place at 3.95 K, in excellent agreement with bulk measurements reported earlier[1]. Apparently, the ordering process proceeds smoothly towards saturation and no short range ordering persists at temperatures above T_c (see Fig. 3(a)). At 4.2 K, i.e. slightly above T_c and in the presence of a magnetic field of 0.4 T applied along the scattering vector $\vec{e}$, long range ordering is induced as seen in Fig. 3(b). This effect was observed in both samples. To determine how the external field perturbs the spiral structure of sample I, a set of measurements was carried out in which the magnetic field dependence of $(110)^{\pm}$ satellites was studied. The results are shown in Fig. 3(c). It is seen that as the magnetic field increases from 0 to about 0.3 T, both satellite lines gradually disappear and at the same time the intensity of the central peak (110) gradually increases. The positions of satellite reflections seem to be unaffected. This conclusion may be inferred from the least squares fit of calculated intensities where widths and positions of three reflections $(110)^{\pm}$ and (110) were varied. The solid line through the experimental points of Fig. 3 (c) represents theoretical calculations.

DISCUSSION AND CONCLUSIONS

The present neutron diffraction investigation confirmed the fact that in TbF_3 at about 4 K, long range ordering occurs. The most striking feature found in these experiments is the observation of two structures - which indicates that the ground state wave

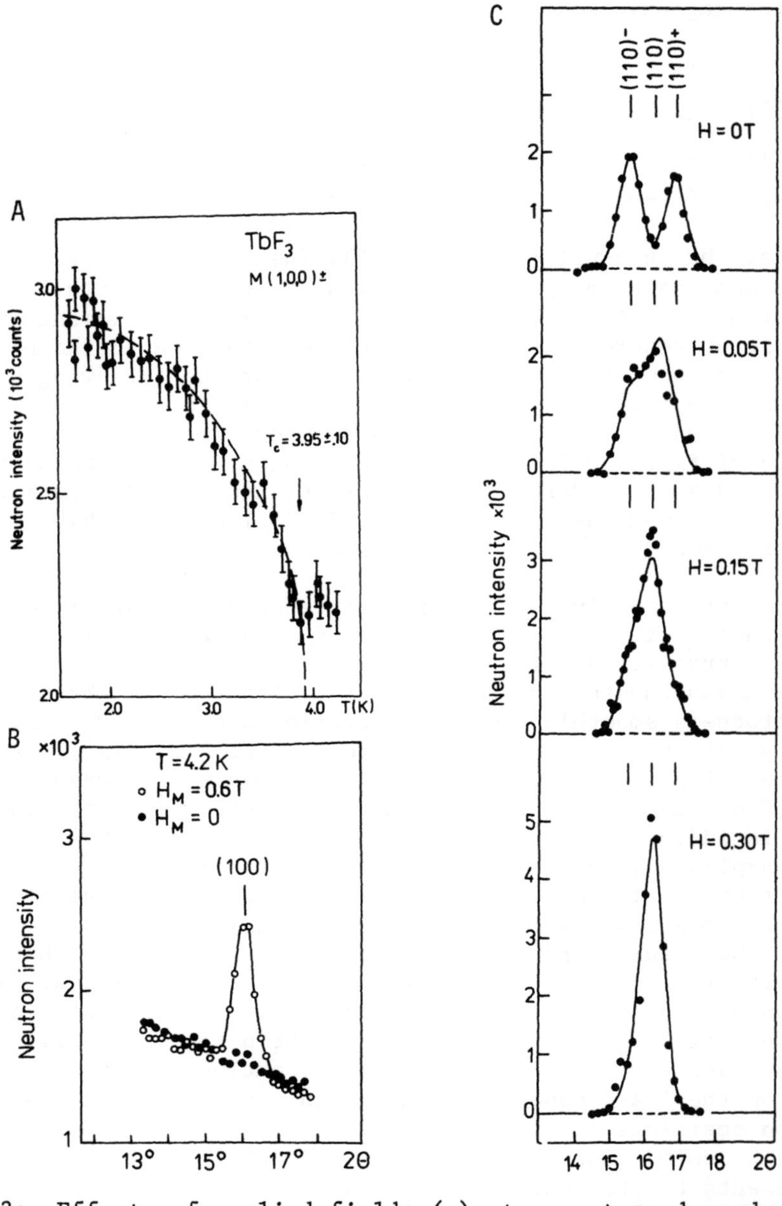

Fig. 3: Effects of applied field; (a): temperature dependence of the height of the magnetic peak $(100)^{\pm}$ in the absence of magnetic field (sample I), (b): effect of applied field observed for both samples at T=4.2 K (open circles), (c) behavior of the $(110)^{\pm}$ satellites in a magnetic field applied along $\vec{e}$ at T = 1.5 K (sample I).

function is strongly dependent on the strength of magnetic interactions which in turn can be easily perturbed by the method of sample preparation. The exchange energy necessary for cooperative magnetic effects is supposed to arise from Tb ion - F anion - Tb ion interactions, as the metal ion to metal ion distances in TbF_3 are too large for direct interaction. As an additional complication we should note that there is strong evidence indicating that magnetic dipole interaction may be important[2].

If magnetic interactions are weak, as in the case of TbF_3, this can lead to unusual magnetic behavior as a function of temperature or magnetic field. In particular, when low lying states are closely spaced, an applied field can easily change the degree of admixture of the crystal field states. This effect in addition to the weak magnetic interaction, means that the structure can be easily influenced by applied field. Indeed, as shown in Fig. 3(c) an external field of 0.3 T completely destroys the spiral structure of sample I and both samples exhibit the same behavior. In a similar way, we can ascribe the onset of magnetic ordering in the presence of a magnetic field at 4.2 K, i.e. slightly above T_c. An external field mixes the ground state singlet with the closely spaced first excited level to produce an energetically lower-lying polarized ground state. The magnetic moment which appears in such a case is essentially an induced moment, which in turn self-consistently determines an exchange field, and yields the long range magnetic order. This effect seems to strongly support the assumption that in TbF_3 the crystal field ground state is a pseudo-doublet[4].

REFERENCES

1. L. Holmes, H.J. Guggenheim and G.W. Hull, Solid State Commun. 8, 2005 (1970).
2. J. Brinkmann, R. Courtis, S. Hüfner and H.J. Guggenheim, J. Magn. Magn. Mat. 6, 279 (1977).
3. M. Piotrowski and A. Murasik, Phys. Stat. Sol. (a) 60, K195 (1980).
4. L.A. Bumagina, B.N. Kazakov, B.Z. Malin and A.L. Stolov, Soviet Phys. - Solid State 19, 1073 (1977).
5. A. Zalkin and D.H. Templeton, J. Amer. Chem. Soc. 75, 2453 (1953).
6. M. Piotrowski, H. Ptasiewicz-Bak and A. Murasik, Phys. Stat. Sol. (a) 55, K163 (1979).
7. G.H. Lander and T.O. Brun, J. Chem. Phys. 53, 1387 (1970).

CRYSTAL FIELD SPLITTING IN LIGHT RARE EARTH DICARBIDES STUDIED BY NEUTRON SPECTROSCOPY

W. Wegener*, A. Furrer†, W. Buhrer† and S. Hautecler*

(*) S.C.K./C.E.N., Mol, Belgium

(†) ETH Zürich, Institut für Reaktortechnik, Würenlingen, Switzerland

INTRODUCTION

The REC_2 compounds (RE=Ce,Pr,Nd,...) possess the CaC_2-type structure with a tetragonal site symmetry for the RE ions[1]. At low temperatures they order antiferromagnetically with the spins along the c-axis[2]. The ordered moments at 4.2K are appreciably smaller than the free ion values (see Table 1); this has been ascribed to the crystalline electric field (CEF) splitting of the RE^{3+} ground state[2]. Because of the metallic character of these systems neutron spectroscopy is an appropriate method to investigate their CEF splittings. A full account of the CeC_2 study has already been published elsewhere[3].

Table 1. Some relevant information on light rare-earth dicarbides.

	T_N (K)	μ (μ_B)	μ_{free} (μ_B)	free ion gr.state	level scheme configuration (paramagnetic phase)
CeC_2	33 ± 2	1.74 ± 0.05	2.14	$^2F_{5/2}$	$\Gamma_{t6} + 2\Gamma_{t7}$
PrC_2	14 ± 2	1.41 ± 0.05	3.20	$^3H_4$	$2\Gamma_{t1}+\Gamma_{t2}+\Gamma_{t3}+\Gamma_{t4}+\Gamma_{t5}$
NdC_2	29 ± 2	2.95 ± 0.05	3.27	$^4I_{9/2}$	$3\Gamma_{t6} + 2\Gamma_{t7}$

EXPERIMENTAL PROCEDURE AND RESULTS

The powder samples (prepared by arc melting) were sealed in cylindrical Al containers; a neutron diffraction analysis did not indicate any impurity. The inelastic neutron experiments were performed on the triple-axis spectrometers at the reactor SAPHIR (Würenlingen). The spectra were obtained in the constant-Q mode of operation with $1.9\ \text{Å}^{-1} \leq Q \leq 5.5\ \text{Å}^{-1}$ (Q is the scattering vector) and in the neutron energy loss configuration. The sample temperature was varied from 8 to 250K. By observing the variation of the peak intensities one could identify CEF transitions in the observed spectra. Typical results are shown in Fig. 1 for the paramagnetic as well as for the antiferromagnetic state; at energies higher than those shown in Fig. 1 only phonon peaks appeared. For NdC_2 no CEF peaks could be found in the spectra up to energy transfers of 50 meV; therefore these data were not included in Fig. 1.

ANALYSIS OF THE RESULTS

For $T > T_N$ the Hamiltonian of a RE^{3+} ion in the tetragonal CEF (with its polar axis in the c direction) is

$$\hat{H}=B_2^0\hat{O}_2^0 + B_4^0(\hat{O}_4^0+u\hat{O}_4^4) + B_6^0(\hat{O}_6^0+v\hat{O}_6^4) \tag{1}$$

with B_n^m being the CEF parameters, $\hat{O}_n^m$ the Stevens operator equivalents[4], and $u = B_4^4/B_4^0$, $v = B_6^4/B_6^0$.

For $T < T_N$ one has to add the exchange interaction $\hat{H}_{ex}$, which in the molecular field approximation is given by

$$\hat{H}_{ex}= -g_J\ \mu_B\ \vec{H}_{mf}\cdot \hat{\vec{J}}, \tag{2}$$

$$H_{mf}= g_J\ \mu_B\ \lambda <\hat{\vec{J}}>,$$

λ being the molecular field constant.

The eigenstates $|n\rangle$ of $\hat{H}$ with the energies E_n determine the differential neutron cross-section according to

$$d^2\sigma /d\omega\, d\Omega \sim F^2(Q) \cdot (k_o/k')\cdot \sum_{n,m} \exp(-\hbar\omega /k_BT)\,|\langle m|\hat{J}_\perp|n\rangle|^2\, \delta(E_n-E_m-\hbar\omega) \tag{3}$$

where F(Q) is the magnetic form factor; k_o,k' are the neutron wave numbers before and after scattering and $J_\perp$ the component of total angular momentum perpendicular to $\vec{Q}$.

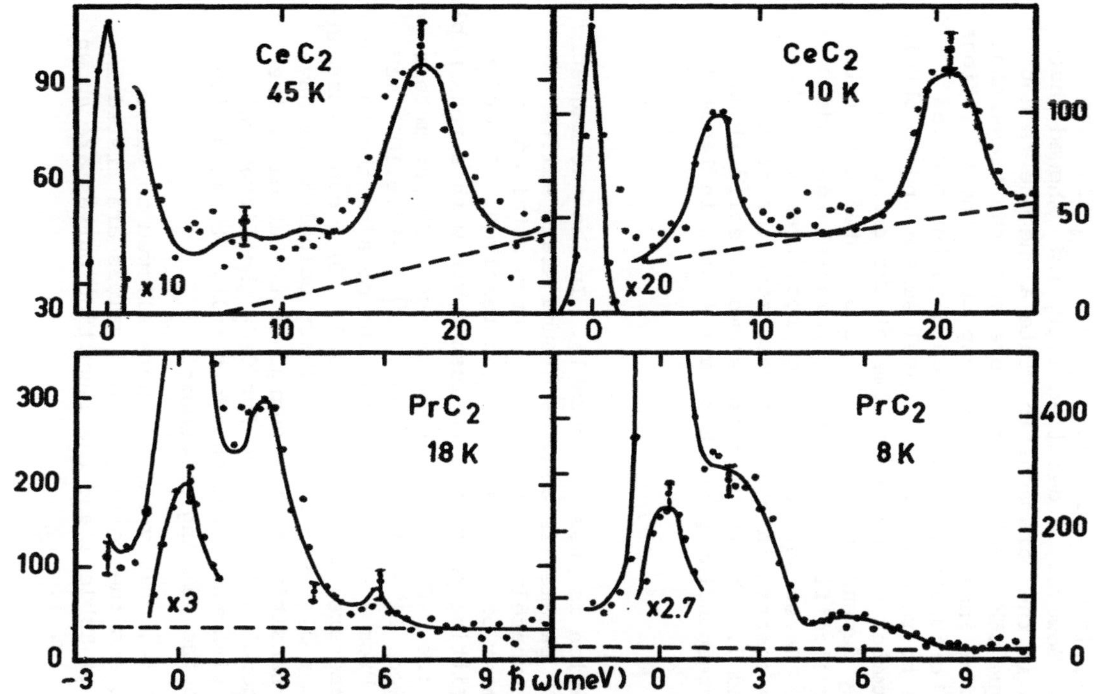

Fig. 1. Energy spectra of neutrons scattered from polycrystalline CeC_2 and PrC_2 samples. The left hand side curves correspond to the paramagnetic, the right hand side curves to the antiferromagnetic state. The ordinate scales correspond to counts in 30 minutes. The CeC_2 data have been obtained with $Q=2.5$ Å^{-1}, the PrC_2 data with $Q=1.8$ Å^{-1}. Dots represent experimental points, solid curves results of least-squares fits of theoretical neutron cross-sections replacing the δ-function by the Gaussian.[3]

CeC_2

At first sight, it would seem that the analysis of the CeC_2 results is rather easy. First, the sixth order terms do not enter in the Hamiltonian (1) and so one is left with three CEF parameters. In addition the level scheme configuration is relatively simple: the ground state splits into three doublets for $T > T_N$ and into six levels for $T < T_N$. However, a detailed analysis[3] showed that from the neutron data in the paramagnetic state alone one is not able to derive a unique CEF level scheme. Moreover, the transition probabilities are rather insensitive with respect to the variations of the parameter u. Fortunately, from the data in the ordered state it was possible to retain only one level scheme and to obtain the value of the parameter u since then some of the transition probabilities strongly depend on it. It is to be noted that this value does not differ significantly from the point-charge model (PCM) result u = 6.3. Least-squares fitting procedures were then applied to the energy spectra in order to determine the remaining parameters B_2^0, B_4^0, and λ. The results are shown as solid lines in Fig. 1 and the corresponding parameters are listed in Table 2.

PrC_2

For PrC_2 the situation is less comfortable. In the paramagnetic state one has five CEF parameters and the ground state splits into 7 levels. Unfortunately the neutron spectrum obtained at 18 K discloses that most of the peaks are not resolved. In order to succeed with the fitting procedure we must begin with a reasonable set of values. From the CeC_2 results some reduced CEF parameters for PrC_2 can be derived: $B_2^0 = -6.1.10^{-2}$ meV, $B_4^0 = -14.6.10^{-4}$ meV, $B_6^0 = 10.4.10^{-6}$ meV. From a PCM calculation one obtains the values of the remaining parameters: u = 6.6, v = -3.4. In a cubic surrounding one would have u = 5, v = -21. Preliminary calculations have shown that v is very sensitive to the charge distribution of the C_2^{2-}-ions on the tetragonal axis. Since for CeC_2 the fitted value of u is in agreement with the point charge result, it was decided to keep u constant in the fitting procedure.

In the ordered state, only v and H_{mf} were allowed to vary. The fitted parameters and the corresponding values are listed in the last two columns of Table 2 and the resulting curves are shown in the lower part of Fig. 1.

NdC_2

As in the case of PrC_2 one has to include the sixth order terms in $\hat{H}$. The ground state splits up into five doublets for $T > T_N$ and into ten singlets for $T<T_N$. But under the experimental conditions of this study no CEF transitions could be found in the NdC_2

neutron spectra. This could be explained if the observed decrease of the overall level scheme splitting from CeC_2 to PrC_2 is extrapolated to NdC_2. This explanation is supported by a calculation of the NdC_2 level scheme using approximate CEF parameters derived from those of CeC_2 and PrC_2.

Table 2. CEF and molecular field parameters determined by fits to the neutron spectra. P: paramagnetic, A: antiferromagnetic state. '*' means that for A the same values have been used as for P. For PrC_2, u=6.6 was not fitted, but taken as a constant. The quoted errors are the statistical errors of the fits.

	CeC_2 P	CeC_2 A	PrC_2 P	PrC_2 A
B_2^0 (10^2 meV)	-0.36 ± 0.04	-0.34 ± 0.04	-2.51 ± 0.2	*
B_4^0 (10^4 meV)	-3.5 ± 0.2	-3.2 ± 0.2	-16.1 ± 1.2	*
B_6^0 (10^6 meV)	-	-	66.9 ± 3.9	*
u	*	6.7 ± 0.7	6.6	*
v	-	-	1.2 ± 0.4	1.8 ± 0.3
λ ($10^{25}Oe^2$/erg)	-	2.2 ± 0.1	-	0.65 ± 0.13

DISCUSSION

One has to consider the significance of the parameter values derived from these analyses. Only in the spectra of CeC_2 did the CEF peaks appear clearly visible and well separated. Since the number of parameters is small, their determination may be considered trustfully. In addition a self-consistent calculation of the zero-field magnetization as a function of T has been carried out using the values of Table 2. The resulting $\mu(0)=1.91\ \mu_B$ and $T_N=40$ K are in reasonable agreement with the experimental data of Table 1. The deviations can be ascribed to J-admixture effects which reduce the magnetic moment by 5-10%[5] and to inadequacy of the molecular field theory at the phase transition.

For PrC_2 on one hand, the number of parameters and of CEF levels is higher; on the other hand the peaks in the neutron spectra were not resolved. One could be afraid that the results of the analysis are not as reliable as those for CeC_2. Nevertheless, they are supported by the fact that the quality of the fits is very good and that the B values obtained were only weakly sensitive to the details of the fit method (choice of background, etc.). On the other hand the value of v does not appear so well defined; but it seems to be sure that $|v|$ is small in comparison with the cubic value 21. $\mu(0)=2.5\ \mu_B$ and $T_N=11$ K have been calculated in the same way as for CeC_2; again T_N is in reasonable agreement with the experimental value (Table 1), whereas the relative deviations are slightly higher for $\mu(0)$.

ACKNOWLEDGEMENT

We are grateful to Mr. H. Ooms (Materials Science Department, S.C.K./C.E.N.).

REFERENCES

1. M. Atoji, J. Chem. Phys. 35, 1950 (1961).
2. M. Atoji, J. Chem. Phys. 46, 1891 (1967).
3. W. Wegener, A. Furrer, W. Buhrer and S. Hautecler, J. Phys. C14, 2387 (1981).
4. K.R. Lea, M.J.M. Leask and W.P. Wolf, Phys. Chem. Solids 23, 1381 (1962).
5. A. Furrer and U. Tellenbach, Helv. Phys. Acta 48, 451 (1975).

COMMENTS

GREIDANUS: Are these compounds metals or insulators?

WEGENER: They are metallic. I do not know if conductivity measurements have been performed, but the crystals show typical metallic shine.

LINEWIDTH OF CRYSTAL FIELD EXCITATIONS IN METALLIC RARE EARTH SYSTEMS

K.W. Becker and J. Keller

Universität Regensburg
8400 Regensburg, Federal Republic of Germany

INTRODUCTION

In a series of papers[1-3] P. Fulde and we at Regensburg have studied theoretically the influence of conduction electrons on the linewidth of elastic and inelastic crystalline electric field (CEF) transitions of rare earth (RE) alloys and compounds. The theory has been applied to explain the temperature dependence of the linewidth for dilute alloys[4] and also the temperature dependence of the quasi-elastic line close to the ferromagnetic transition in the magnetic compound Pr_3Tl.[5]

Our calculations are based on a Hamiltonian of the form

$$H = H_{CF} + H_{el} - j\sum_i \vec{J}_i \cdot \vec{\sigma}_i \qquad (1)$$

where we assume an isotropic exchange interaction between the local moments $\vec{J}_i$ of the RE ions and the conduction electron spin density $\vec{\sigma}_i/2$ at the RE site. H_{CF} describes the CEF splitting of the RE 4f-states and H_{el} the conduction electrons.

This Hamiltonian has been used to calculate the dynamical susceptibility $\chi(q,\omega)$. Peaks in its imaginary part, which yield the inelastic neutron scattering cross section $S(q,\omega) = \chi''(q,\omega)/(1-\exp(-\beta\omega))$, reflect the CEF splitting of the RE ions. Its low frequency behaviour is also of interest: it determines the ultrasonic attenuation[6] and other relaxation phenomena.

The main results for the influence of conduction electrons on the linewidth can be summarized as follows (we do not consider Kondo and intermediate valence effects):

Single ions: The linewidth of elastic transitions is proportional to $(jN(0))^2T$, where N(0) is the conduction electron density of states at the Fermi surface (Korringa law). For well separated inelastic transitions at low temperatures the linewidth is prop. to $(jN(0))^2\Delta$, where Δ is the excitation energy. At higher temperatures, when the inelastic lines overlap, they form a comparatively narrow quasi-elastic peak which becomes broad again at very high temperatures (Korringa law for the degenerate level scheme). Such a narrowing effect which is common for ESR may have been seen also in recent experiments on perturbed γ-γ correlations in CeBa[7].

In this communication we will report on new experimental results on the linewidth of very dilute alloys in the superconducting state[8] and their theoretical explanation[9].

Magnetic compounds: Here the conduction electrons give rise to an effective coupling (RKKY interaction) between different RE moments, which leads to a q-dependence of the excitation energies. In the RPA they are given by the poles of

$$\chi(q,\omega) = u(\omega)/(1-J_q u(\omega)) \tag{2}$$

where $u(\omega)$ is the dynamic single ion susceptibility and J_q is the Fourier transform of the effective coupling, which is prop. to the conduction electron susceptibility.

The influence of the conduction electrons on the linewidth can be split approximately into a Korringa type damping entering the single ion susceptibility $u(\omega)$ and a damping due to particle hole excitations (Landau damping) which is prop. to q^{-1}. The latter effect can be incorporated into a frequency dependent effective coupling $J(q,\omega)$.

Another damping mechanism discussed by several authors[10-14] results from the effective coupling between different local moments. Here one starts from a Hamiltonian of the form

$$H = H_{CF} - \frac{1}{2} \sum_{ik} J_{ik} \vec{J}_i \vec{J}_k . \tag{3}$$

In the case of a 2-level system, which is discussed the most (Ising model in a transverse field), the inelastic line can be rather broad, particularly for the case of strong coupling $J_{ik} \simeq \Delta$. At the ferromagnetic transition temperature and for small q the inelastic line merges into a quasielastic peak[12].

Concerning the experimental situation, one finds well resolved inelastic transitions for weak coupling systems at low temperatures[15-17]. Their q-dependence and mode-splitting indicates that besides an isotropic exchange one also needs anisotropic and directional dependent interactions[10,16]. One of the most simple systems from a theoretical point of view is Pr_3Tl. Here the coupling is strong enough to produce ferromagnetic order. On the other hand the transition temperature is low enough to make the resolution of inelastic transitions possible in the paramagnetic state[17,5]. From the CEF splitting of the Pr ions (assumed level scheme: Γ_1 ground state, Γ_4 triplet at 77 K) one expects a strongly q-dependent mode which becomes (nearly) soft at the ferromagnetic transition. However, in experiment the predicted temperature dependence is not observed. Unfortunately no single crystals are available. Therefore the inelastic line is broad because of an overlap of lines with different dispersion. This makes a comparison with theoretical results difficult.

So far no complete theory of the linewidth for realistic systems is available which includes a full level scheme, anisotropic interactions and the combined effect of damping by electronic excitations and damping due to the RKKY interaction between local moments. As both damping mechanisms are based on the exchange interaction between local moments and conduction electrons, it is desirable to derive both within the same theoretical approach and to compare their influence on the linewidth.[18] This will be done in the last section; but there we will restrict ourselves to a simple level scheme and an isotropic interaction.

LINEWIDTH OF CEF EXCITATIONS OF RE IONS IN SUPERCONDUCTORS

In recent neutron scattering experiments Feile et al [6] have measured the linewidth of CEF excitations in very dilute alloys of $Tb_xLa_{1-x}Al_2$ (x=0.001 and 0.003) and found a pronounced decrease of the linewidth when the sample becomes superconducting. $LaAl_2$ is a superconductor with a transition temperature of T_{co}=3.24 K. In the cubic field of $LaAl_2$ the Tb^{3+} ions have a nonmagnetic ground state Γ_1 followed by a Γ_4 triplet at an excitation energy Δ_{14}=8 K. In Fig. 1 the half width of the inelastic transition Γ_1-Γ_4 is shown for the sample with x=0.001. At the superconducting transition temperature T_c=3.1 K of the alloy a drop in the linewidth can be observed. This effect which has been calculated by us a long time ago[9] can be explained by the reduced ability of conduction electrons to absorb energy in the superconducting state.

For simplicity, if we approximate the level scheme of Tb ions by a 2-level system (a generalization to the full level scheme is straightforward) the linewidth of the inelastic transition is given by

$$\Gamma = j^2M^2\Phi''(\omega)\coth\frac{\beta\omega}{2}\Big|_{\omega=\Delta_{14}} \qquad (4)$$

where M is the matrix element of the transition and $\Phi(\omega)$ is the local conduction electron susceptibility. Its absorptive part $\Phi''(\omega)$ depends on the number of states in the electron gas which can be excited at the energy ω. In the normal state $\Phi''(\omega)=\pi N^2(0)\omega$, which leads to a finite linewidth at T=0. In the superconducting state $\Phi''(\omega)$ contains contributions from scattering of thermally excited quasi particles and from processes which break the Cooper pairs. The first contribution vanishes for $T\to 0$, while the second is possible only for $\omega > 2\Delta_s(T)$, where $\Delta_s(T)$ is the superconducting energy gap. As $\Delta_{14} < 2\Delta_s(0)$ in the present case, one expects the linewidth to vanish at low temperatures.

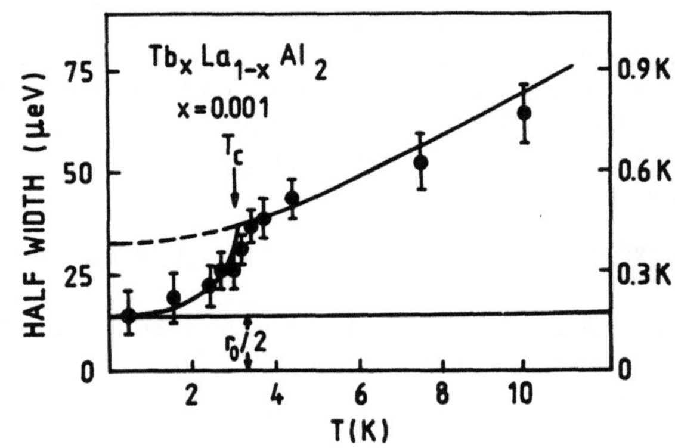

Fig. 1. Half width of the inelastic transition of Tb ions in $LaAl_2$ measured by neutron scattering (from Ref. 8). T_c denotes the superconducting transition temperature of the sample.

In the experiment, however, a finite linewidth is observed at T=0, whose origin is unclear at present. In Fig. 1 this residual linewidth has been added to the theoretical result calculated from Eq. (2) and a numerical evaluation of the local conduction electron susceptibility in the superconducting state[9]. Then good agreement between theory and experiment is obtained.

LINEWIDTH OF VAN VLECK MAGNETS WITH CONDUCTION ELECTRONS

In this section we want to present a new theory for the linewidth of van Vleck magnets with conduction electrons which combines the damping effect of electronic excitations with the damping due to the RKKY interaction between local moments. For the purpose of demonstrating our method we will restrict ourselves to a 2-level system described by the Hamiltonian

$$H = -\Delta \sum_i S_i^x - j \sum_i \sigma_i^z S_i^z + H_{el}. \tag{5}$$

Here $\vec{S}$ is an effective spin with S=1/2 and Δ is the energy separation. Its dynamical behaviour has to be compared with an effective spin Hamiltonian where the conduction electrons as dynamical variables are eliminated:

$$H_{eff} = -\Delta \sum_i S_i^x - \frac{1}{2} \sum_{ik} J_{ik} S_i^z S_k^z . \tag{6}$$

Here J_{ik} is the RKKY interaction which can be expressed by

$$J_q = j^2(\chi_{el}(q) - \frac{1}{N} \sum_k \chi_{el}(k)) \tag{7}$$

where $\chi_{el}(q)$ is the static free conduction electron susceptibility (2 x Lindhard function summed over reciprocal lattice vectors[2]).

The static magnetic properties are described within RPA by a set of static susceptibilities* (per ion)

$$\begin{aligned} \chi_{ss}^z(q) &\equiv \beta(S_q^z|S_q^z) = u/(1-J_q u) \\ \chi_{ss}^y(q) &\equiv \beta(S_q^y|S_q^y) = u = \frac{1}{2\Delta} \tanh\left(\frac{\beta\Delta}{2}\right) \\ \chi_{s\sigma}^z(q) &\equiv \beta(S_q^z|\sigma_q^z) = -j\chi_{el}(q)u/(1-J_q u) \end{aligned} \tag{8}$$

where u is the single ion susceptibility. Note that in the present model S^z is the magnetic moment, S^y is prop. to the time derivative of S^z, i.e. $LS^z \equiv [H,S^z] = i\Delta S^y$, and S^x describes the occupation number difference of the two levels.

---

*The scalar products are defined by

$$(A|B) = \frac{1}{\beta}\int_0^\beta d\lambda \langle e^{\lambda H} A^+ e^{-\lambda H} B\rangle - \langle A^+\rangle\langle B\rangle$$

For the calculation of the dynamical properties we use a projection formalism described in Ref. 2. Expanding the relaxation function

$$\Phi_{ss}(q,\omega) \equiv \frac{\chi^z_{ss}(q)-\chi^z_{ss}(q,\omega)}{\omega} = \beta(S^z_q| \frac{1}{\omega+i\eta-L} S^z_q) \tag{9}$$

into a continued fraction we obtain

$$\Phi_{ss}(q,\omega) = \frac{\chi^z_{ss}(q)}{\omega-\frac{\delta_1(q)}{\omega-\Sigma(q,\omega)}} \tag{10}$$

with $\delta_1(q)=\Delta^2(S^y_q|S^y_q)/(S^z_q|S^z_q)=\Delta^2(1-J_q u)\equiv\Delta^2_q$. Here Δ_q is the inelastic excitation energy in RPA. Damping effects are contained in $\Sigma(q,\omega)$ which is the relaxation function of the orthogonal part of LS^y_q of the original dynamical variables S^z_q and S^y_q:

$$\Sigma(q,\omega) = (QLS^y_q|\frac{1}{\omega+i\eta-QLQ}QLS^y_q)/\chi^y_{ss}(q) \tag{11}$$

$$QLS^y_q = i\frac{1}{\sqrt{N}}j\sum_k Q(\sigma^z_k S^x_{q-k}) . \tag{12}$$

In order to show how the coupling of a local spin mode to free conduction electrons and to other local spin modes enters simultaneously into the dynamical properties let us first orthogonalize the conduction electron spin density with respect to S^z_q and S^y_q:

$$Q\sigma^z_q \equiv \sigma^z_{\perp q} = \sigma^z_q - \frac{(S^z_q|\sigma^z_q)}{(S^z_q|S^z_q)} S^z_q . \tag{13}$$

The variable $\sigma^z_{\perp q}$ can be considered as that part of the conduction electron spin density that is not coupled to the local spin mode S^z_q and hence can be approximated by free conduction electrons (its static susceptibility contains no critical enhancement factor: $\chi_{\sigma^z_\perp\sigma^z_\perp}(q) = \chi_{el}(q)$).

Substituting Eq.(10) for Eq.(9) and writing $S^x_q=\langle S^x\rangle\sqrt{N}\delta_{q,0}+\delta S^x_q$ the variable QLS^y_q can be decomposed into three independent terms

$$-iQLS^y_q = \frac{1}{\sqrt{N}}\sum_k J_k S^z_k \delta S^x_{q-k} + j\langle S^x\rangle\sigma^z_{\perp q} + \frac{1}{\sqrt{N}} j \sum_k \frac{1}{2}\{\sigma^z_{\perp k},\delta S^x_{q-k}\}_+ . \tag{14}$$

The first term describes the coupling between local spin modes and has the same form as that obtained from the effective spin

Hamiltonian Eq. (6). The second and third terms contain the coupling to "free" conduction electrons. The third gives rise to a small Korringa-like damping, which will be neglected in the following discussion, the second leads to a q-dependent Landau damping. Thus $\Sigma(q,\omega)$ can be written as

$$\Sigma(q,\omega) = \Sigma_{eff}(q,\omega)+\Sigma_{el}(q,\omega). \qquad (15)$$

The relaxation due to the spin-spin interaction can be further approximated by

$$\Sigma_{eff}(q,\omega) = \frac{\delta_2^s}{\omega+i\Gamma_s}$$

with (16)

$$\delta_2^s = \frac{1}{2N}\sum_k \frac{J_k}{\Delta_k \tanh(\beta\Delta_k/2)}\ , \quad \Gamma_s = (\tfrac{\pi}{2}\,\delta_2^s)^{1/2}.$$

Here we have used a method-of-moment approximation similar to Ref. 11. It may be compared with a kind of mode-coupling approximation used in Ref. 10 in which, however, the relaxation of δS^x (occupation number fluctuations) have been neglected giving unreliable results at low temperatures and for $\omega\simeq\Delta$). With a more sophisticated mode-coupling treatment one can show that (i) the conventional theory of critical dynamics can be applied and (ii) Eq. (16) is a reasonable approximation for the frequency dependence of the relaxation due to spin-spin interactions. For Σ_{el} which describes the relaxation due to electronic excitations we will use the following approximation

$$\Sigma_{el}(q,\omega) = \frac{\delta_2^\sigma}{\omega+i\Gamma_\sigma}$$

with (17)

$$\delta_2^\sigma = (j\langle S^x\rangle)^2\chi_{el}(q)/u\simeq\Delta^2 uJ_0\ , \quad \Gamma_\sigma = \frac{2}{\pi}\,qv_F\ .$$

For the numerical evaluation we have used a spherical BZ with radius q_0 and an effective interaction J_q of the form $J_q=J_0(1-x(1-\cos\pi q/q_0))$ with x=0.62 chosen such that $\Sigma J_q=0$, i.e. $J_{ii}=0$. Typical results are shown in Fig. 2. In Fig. 2a the strong effect of the spin-spin interaction on the linewidth is clearly seen if one compares with Fig. 2b where Σ_{eff} has been dropped and only the damping due to particle-hole excitations is kept. This is remarkable since the relaxation rate due to spin-spin interaction is formally of higher order (j^4) in the exchange interaction than the electronic relaxation rate (j^2). The electronic damping mechanism leads to a linewidth prop. to q^{-1} (Fig. 2b). It will be important only for systems with weak coupling $J\ll\Delta$ and a high density of states N(0). This may be the case for dhcp Pr where a linewidth prop. to q^{-1} has been observed[15].

Note, that in Fig. 2a the inelastic peak follows roughly the RPA result Δ_q (compare the peak positions with Fig. 2b) but stays above Δ_q for small q before it merges into the quasi-elastic peak. Until now, however, it is not clear whether this frequency shift is sufficient to explain the non-softening of the inelastic mode observed in Pr_3Tl.[17]

In this paper we have presented a theory for the dynamics of van Vleck magnets with 2 singlet levels including for the first time the damping due to the effective spin-spin interaction as well as due to electronic particle hole excitations. For realistic level systems resulting from cubic crystalline fields the theory is different. However, the dynamical characteristics will be similar as long as the ground state is nonmagnetic and the first excited

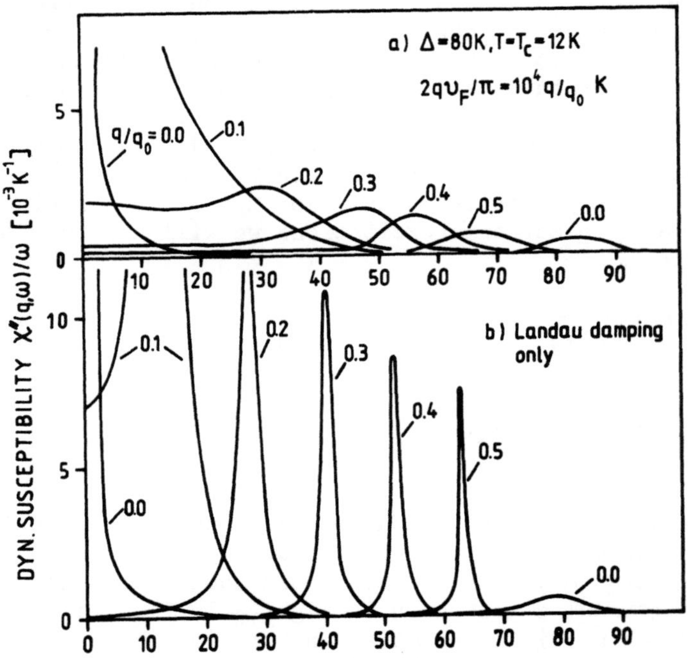

Fig. 2. Imaginary part of the dynamic susceptibility as function of energy ω for a 2-level system with energy separation Δ at the ferromagnetic transition temperature T_c. Fig. 2a shows the damping effect of spin-spin interactions and Landau-damping, Fig. 2b only the effect of Landau-damping.

state has a sufficiently high energy. Let us mention finally that for electronic Heisenberg systems described by a Hamiltonian Eq. (1) with vanishing CEF, the presented mechanism leads to an interesting cross-over between a relaxation rate prop. to $q^{5/2}$ for large q and prop. to q^2 for small q at $T=T_c$.[19]

REFERENCES

1. P. Fulde and I. Peschel, Adv. Phys. 21, 1 (1972); K.W. Becker and P. Fulde, in:'Crystal Field Effects in Metals and Alloys', A. Furrer, ed., Plenum Press, New York (1977) p. 284
2. K.W. Becker, P. Fulde and J. Keller, Z. Physik B28, 9 (1979)
3. J. Keller and K.W. Becker in: 'Crystalline electric field and structural effects in f-electron systems', J.E. Crow, R.P. Guertin, and T.W. Mihalisin, eds., Plenum Press, New York (1980) p. 173
4. H. Loewenhaupt, S. Horn, B. Frick, this volume H.E. Hoenig, private communication
5. J.K. Kjems, M. Nielsen, W.J.L. Buyers, and J.E. Crow, J. de Physique 40, C5-132 (1979)
6. K.M. Leung, P.L. Huber, and B. Lüthi, J. Appl. Phys. 50, 1831 (1979), D.L. Huber and K.M. Leung, in: 'Crystalline electric field and structural effects in f-electron systems', J.E. Crow, R.P. Guertin, and T.M. Mihalisin, eds., Plenum Press, New York (1980) p. 39
7. H.J. Barth, G. Netz, K. Nishijama, and D. Riegel, Phys. Rev. Letters 45, 1015 (1980)
8. R. Feile, J. Kjems, M. Loewenhaupt, and H.E. Hoenig, Phys. Rev. Letters
9. J. Keller and P. Holzer, Proceedings LT 14, M. Krusius and M. Vuorio, eds., North-Holland (1975) Vol 3, 438
10. P. Bak, Phys. Rev. B12, 5203 (1975)
11. T.N. Tommet and D.L. Huber, Phys. Rev. B11, 1971 (1971)
12. T.H. Cheung, Z. Physik 267, 251 (1974)
13. P. Prelovsek and I. Sega in: 'Lattice Dynamics', M. Balkanski, ed., Flammarion, Paris (1977) p. 154
14. P.-A. Lindgård, in: 'Crystalline electric field and structural effects in f-electron systems, J.E. Crow, R.P. Guertin, and T.W. Mihalisin, eds., Plenum Press, New York (1980) p. 153
15. J.G. Houmann, B.D. Rainford, J. Jensen, and A.R. Mackintosh, Phys. Rev. B20, 1105 (1979)
16. A. Loidl, K. Knorr, J.K. Kjems, and B. Lüthi, Z. Physik B35, 253 (1979)
17. R.J. Birgeneau, J. Als-Nielsen, and E. Bucher, Phys. Rev. Letters 27, 1530 (1971) and Phys. Rev. B6, 2724 (1971); J. Als-Nielsen, J.K. Kjems, W.J.L. Buyers, and R.J. Birgeneau, J. Phys. C10, 2673 (1977)
18. V. Zevin and E. Barboy, Z. Physik B39, 173 (1980)
19. K.W. Becker, to be published

COMMENTS

WOHLLEBEN: When the interaction temperatures of RE compounds are low, we face the possibility of quadrupole-quadrupole interactions becoming as important as spin-spin interactions, or even more so. What would this do to the T and q dependence of the linewidths? How and where would this modify the results of the calculations based on spin-spin only?

KELLER: A quadrupole interaction will also contribute to the linewidth of the magnetic susceptibility. If one starts from an RPA expression for the dynamical susceptibility, it will add to the relaxation in the single ion susceptibility and will be q-independent in this approximation.

KASUYA: Can you say something about possible mechanisms for the remaining linewidth in the superconducting state?

KELLER: Not seriously. One could speculate about the relaxation from spin-spin interactions, but I do not know if this is still effective at such low concentrations. The theory also assumes the electron to be in a BCS state. If there is a sufficient pair breaking by magnetic impurities with a magnetic ground state, one also obtains a contribution to the linewidth from the quasiparticle scattering.

LOEWENHAUPT: The residual linewidth of Tb in $LaAl_2$ seems to be the same for c=0.001 and 0.003 (i.e. concentration independent). One possible explanation is that it originates from defects in the samples. But note that the residual linewidth is just a little above the experimental error bars.

STEGLICH: A comment concerning the residual linewidth in the superconducting state of $La(Tb)Al_2$: previous experiments on nominally pure $LaAl_2$ showed a considerable excess of specific heat, well below the transition temperature. This was attributed to some 100 ppm spurious Ce. Thus, I would like to suggest performing low temperature specific heat measurements on the best $LaAl_2$ sample in order to check whether the residual linewidth could arise from such impurities.

BABERSCHKE: Does the neutron experiment in $La(Tb)Al_2$ probe the local property at the Tb site? What about the local order parameter (Kümmel, 1972)?

KELLER: We did not consider this effect. I am not sure if the superconducting order parameter is much depressed locally by a nonmangetic ground state ion and I do not know how much the quasiparticles which may exist in this region contribute to the linewidth.

ELECTRONIC STRUCTURE OF $LaIn_3$ AND $LaSn_3$

Akira Hasegawa

General Education Department
Niigata University
Niigata 950-21, Japan

INTRODUCTION

A series of the lanthanum intermetallic compounds with the $AuCu_3$ crystal structure, LaX_3, in which X means In, Sn, Tl or Pb, was the subject of extensive study, because of their interesting superconducting and magnetic properties. Many experimental works were made for these compounds, notably the work of Grobman (1972) and Toxen et al (1973). Among these compounds, $LaIn_3$ and $LaSn_3$ have recently attracted renewed interest as a proper reference material for the study of the electronic structure of the valence-fluctuating state in $CeSn_3$, $CeIn_{3-x}Sn_x$ ($0 \leq x \leq 3$), etc. One of the most interesting recent works is the Fermi surface measurement of $CeSn_3$ by Johanson et al (1981); other important references are cited in their paper.

A step toward a better, quantitative understanding of the physics of these La compounds and various Ce compounds with the same crystal structure may be to investigate the one-electron energy band structure of $LaIn_3$ and $LaSn_3$ by band theory. In contrast to many experimental works, however, only a few semiquantitative band structure calculations have been carried out for these compounds (Gray and Meisel, 1972; Freeman and Koelling, 1972; Hackenbracht and Kübler, 1979). This paper thus intends to report some results of an improved band calculation.

METHOD OF CALCULATION

This calculation is based on a self-consistent symmetrised APW method with the muffin-tin approximation. The relativistic effects are taken into account by the method of Koelling and Harmon (1977).

This method eliminates the spin-orbit interaction term from the Hamiltonian, and takes relativistic energy shifts into account. To construct the exchange and correlation potential, the local-spin-density approximation of Gunnarsson and Lundqvist (1976) is used. The starting crystal charge density is a superposition of self-consistent charge densities for neutral atoms. The frozen core approximation is used.

After the self-consistent iteration processes, energies for many bands are calculated at 86 points in the irreducible 1/48 of the simple cubic Brillouin zone. These energies are used to construct the interpolation formula for each band by the Fourier series expansion. The interpolation formula is carefully constructed so as to fit each band with sufficient accuracy, especially near E_F. It is then used to generate energies at finer mesh points, from which the density of states, the Fermi surface and the cyclotron mass is calculated. For details of the procedure, the reader is referred to Hasegawa and Yanase (1977).

RESULT FOR $LaIn_3$

Fig. 1 shows the result for the energy band structure for $LaIn_3$ together with the density of states. According to the charge distribution analysis, the Bloch states in the lowest three bands consist dominantly of the In s state. The Bloch states in an upper part of the valence band consist mainly of the In p and La d states. The number of La d electrons in the APW sphere amounts to 1.05; that of the In s and p electrons amounts to 3.93 and 3.45, respectively. Many narrow bands lying above E_F originate from the La f state. By the hybridisation with the In p bands, the f electrons mix into the valence band, but do not distort it seriously; the number of La f electrons in the APW sphere is only 0.08. The Fermi energy lies in the vicinity of a sharp peak in the density of states. The density of states at E_F is 32.1 states/Ryd. cell. The calculated value is much smaller than an experimental value, but the discrepancy can be attributed to the enhancement due to the electron-phonon interaction (Hackenbracht and Kübler, 1979).

The Fermi surface consists of two sheets of the sixth and seventh bands. Fig. 2 shows the seventh-band sheet centred at the R point. It is essentially a sphere, which bulges slightly toward the M point, and contains 0.38 electrons/cell. Its extremal cross sectional area in units of $(2\pi/a)^2$ is 0.42, 0.40 and 0.36, and its cyclotron mass is $0.42m$, $0.36m$ and $0.35m$ in the <100>, <110>, <111> directions of the magnetic field, respectively, m being the free-electron mass. The sixth-band surface is complex, having heavier masses. Transport properties reflect these electronic structures. Grobman (1972) found that the temperature dependence of the electrical resistivity and thermopower is similar to that of a typical transition metal rather than that of a simple metal. Details of the calculation

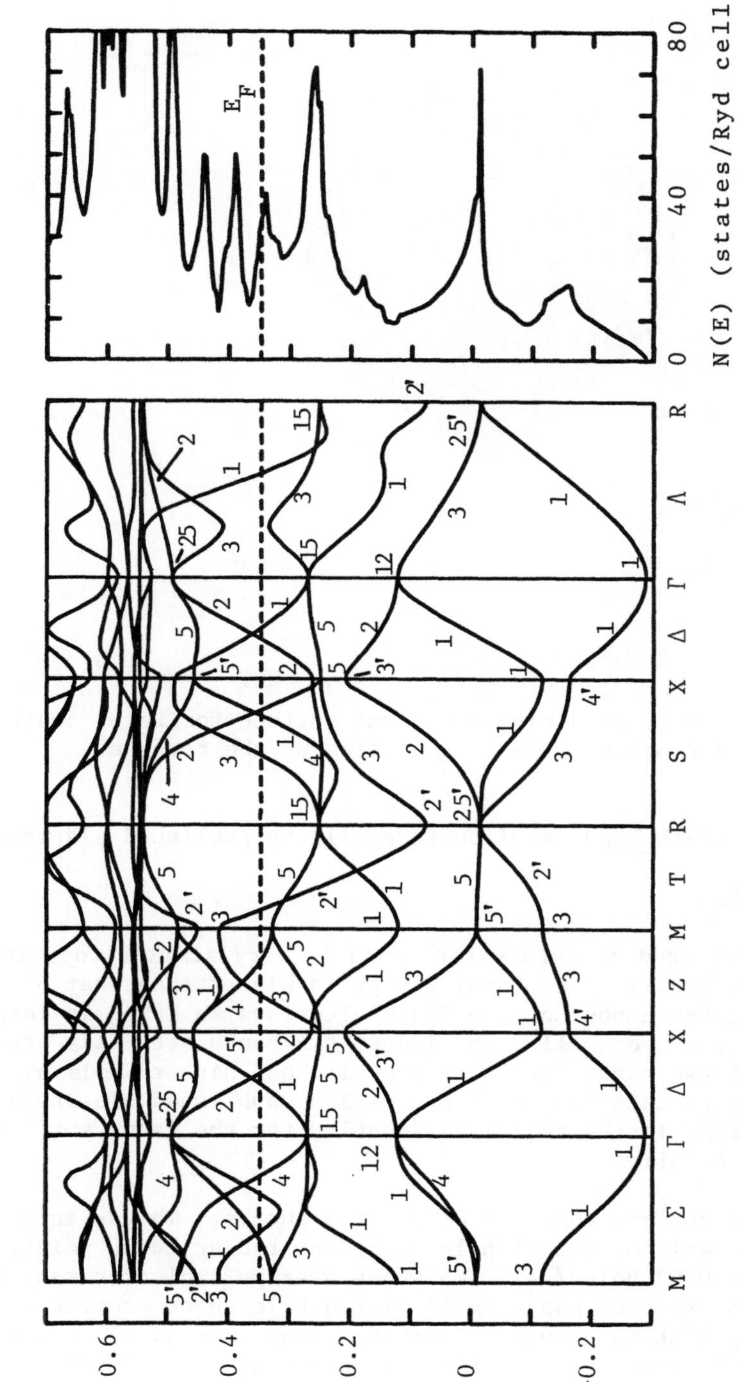

Fig. 1. Self-consistent APW energy band structure and the density of states, $N(E)$, for $LaIn_3$ with the local-spin-density approximation exchange. E_F shows the Fermi energy.

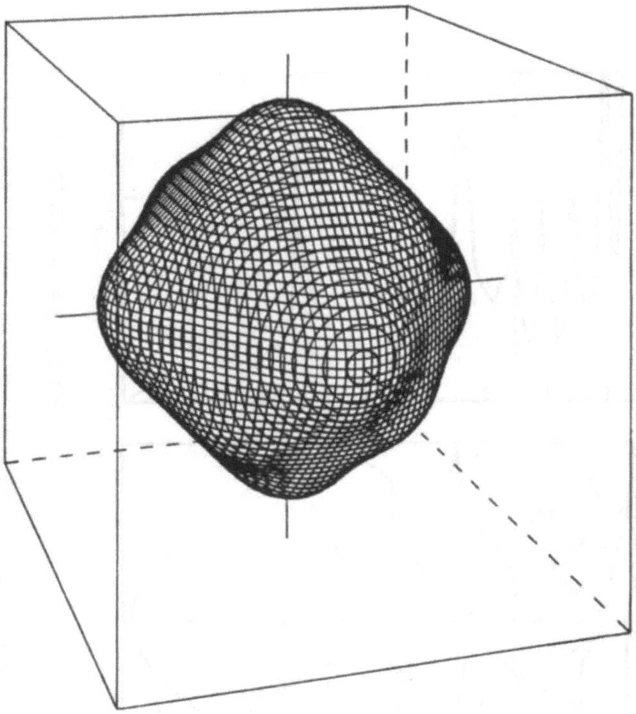

Fig. 2. The seventh-band electron Fermi surface for $LaIn_3$ centred at the R point. The Γ point lies at the corners.

including the sixth-band Fermi surface will be published elsewhere.

RESULT FOR $LaSn_3$

The energy band structure for $LaSn_3$ is very similar to that for $LaIn_3$ shown in Fig. 1. The Fermi energy is higher than that of $LaIn_3$, because the number of the Bloch electrons in $LaSn_3$ is larger by three electrons per cell. The lowest six bands are fully occupied, the next two bands partially occupied and higher bands are empty. Details of the calculation for $LaSn_3$ will be published elsewhere (Hasegawa, 1981). In the following, results for the Fermi surface are explained briefly.

The Fermi surface consists of various sheets. The seventh band has a small, nearly spherical hole sheet centred at the Γ point, which contains 0.02 holes/cell and whose average cyclotron mass is 0.18m. The eighth band has a small closed hole sheet centred at the Γ point, also, which is heavily distorted along Δ axes. In addition, it has small electron pockets which are centred at the M points and nearly spheroidal, and a large sheet around the R point, whose shape is very complex and multiply connected.

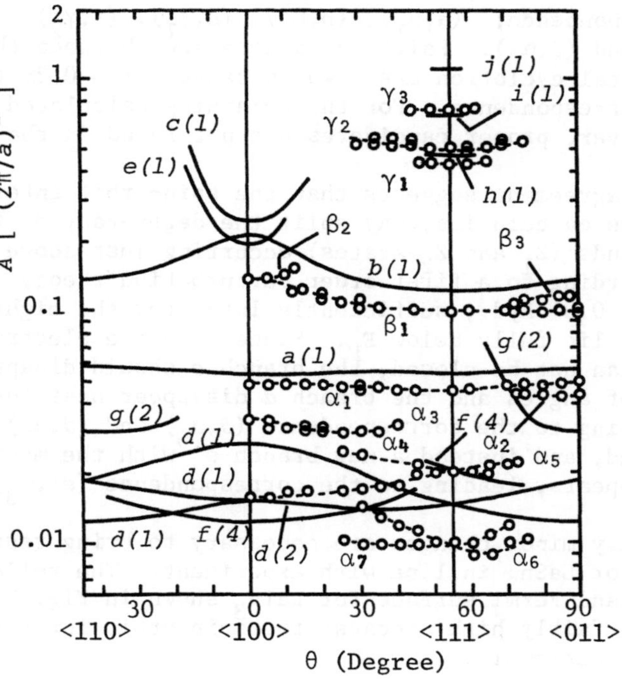

Fig. 3. Calculated and experimental extremal cross-sectional areas, A, of the Fermi surface for $LaSn_3$ as a function of the direction of magnetic field on {100} and {110} planes. The experimental results of Crabtree et al (1981) are shown by open circles with dotted lines and labelled by Greek letters.

We list here a detailed description of the origin of each branch shown in Fig. 3. The branch *a* originates from the seventh-band hole surface, and all others from the eighth-band surface;

b the isolated hole sheet centred at the Γ point,
c four-blade propeller-shaped orbit around the X point,
d the small isolated electron sheet centred at the M point,
e cross-shaped orbit around the M point,
f small orbit centred at (0.19, 0.19, 0.5)$(2\pi/a)$ on the S axis, distributed around the *c* orbit,
g small orbit centred at (0.1, 0.1, 0.5)$(2\pi/a)$ on the S axis, being perpendicular to the S axis,
h, i, j large, nearly circular orbits centred at the R point.

Each branch label is shown with a number in parentheses, which means the degeneracy of the branch, i.e. the number of branches which the branch splits into when the magnetic field direction deviates from the symmetry plane.

The comparison between theory and experiment suggests the following correspondence; (a,α_1), (b,β_1), (h,γ_1), (i,γ_2), (j,γ_3), (g,α_2), (f,α_4) and (f,α_5). Fairly good agreement between the calculated and experimental cyclotron mass, which ranges from $0.3m$ to $1.2m$, confirms this correspondence. For the remaining calculated branches, c, d and e, however, proper candidates are not found in the experiment.

This disagreement suggests that the spin-orbit interaction, neglected in this calculation, may split the degeneracy of the eighth and ninth bands (Z_1 and Z_2 states) occurring just above E_F along the Z axis. According to a first order perturbation theory, this splitting is about 0.01 Ryd., sufficiently large for the eighth band along the Z axis to lie fully below E_F. Since then the electron pockets at the M point can not be closed, the branch c should disappear in the whole range of angles and the branch d disappear near the <100> direction, leading to the correspondence (d,α_6) and (d,α_7). The branch e is destroyed, and instead a new branch e' with the magnitude of about 0.04 appears, leading to the correspondence (e',α_3).

Thus, only minor changes are necessary to bring results of this calculation for $LaSn_3$ in line with experiment. The reliability of the seventh-band Fermi surface for $LaIn_3$ shown in Fig. 2 can then be considered as fairly high, because the spin-orbit interaction does not affect it seriously.

REFERENCES

Crabtree, G.W., Johanson, W.R., Edelstein, A.S., and McMasters, O.D., International Conference on Valence Fluctuations in Solids, Santa Barbara, Ca., Jan. 27-30, 1981.

Freeman, A.J., and Koelling, D.D., J. Physique Colloq. 33, C3-57, (1972).

Gray, D.M. and Meisel, L.V., Phys. Rev. B5, 1299 (1972).

Grobman, W.D., Phys. Rev. B5, 2924 (1972).

Gunnarsson, O. and Lundqvist, B.I., Phys. Rev. B13, 4274 (1976).

Hackenbracht, D. and Kübler, J., Z. Phys. B35, 27 (1979).

Hasegawa, A., J. Phys. Soc. Japan 50, in press (1981).

Hasegawa, A. and Yanase, A., J. Phys. F: Metal Phys. 7, 1245 (1977).

Johanson, W.R., Crabtree, G.W., Edelstein, A.S. and McMasters, O.D., Phys. Rev. Lett. 46, 504 (1981).

Koelling, D.D. and Harmon, B.N., J. Phys. C: Solid St. Phys. 10, 3107 (1977).

Toxen, A.M., Gambino, R.J. and Welsh, L.B., Phys. Rev. B8, 90 (1973).

THE APPLICATION OF THE SELF-CONSISTENT MORI FORMALISM TO ANALYZE THE DYNAMICAL RESPONSE OF VAN VLECK SYSTEMS IN THE VICINITY OF THE CURIE POINT

L. Kowalewski, A. Lehmann-Szweykowska, M. Thomas and R. Wojciechowski

Solid State Theory Division, Institute of Physics
A. Mickiewicz University, Poznań, Poland

The problem of the dynamical mechanism of magnetic phase transitions in Van Vleck magnets is still open. We show this problem can be solved by the use of adequate relaxation functions in the vicinity of the Curie point. The relaxation processes in a system which are dynamically perturbed and which relax back to equilibrium are related to the normal modes of the system, including the critical central or soft modes[1-6]. We will proceed now to outline in some detail our formalism based on Mori's work. A good technique for theoretical analysis of such problems is Mori's projection formalism[7]. The key problem of the method applied lies in the choice of initial variables.

Let us consider a set of starting dynamical variables $a_o^{(\alpha)}(\vec{k})$, $(\alpha=1,\dots,n)$, which are arranged in a column:

$$a_o(\vec{k}) \equiv a_o = \mathrm{Col}(a_o^{(1)}, a_o^{(2)}, \dots, a_o^{(n)}), \tag{1}$$

where $\vec{k}$ is sufficiently small to ensure that the variables are macroscopic. In the following formulae we will omit the k-index. We choose the dynamical variables generating excitations which can be suspected as being responsible for magnetic phase transitions. For example, these variables can be the densities of the integrals of motion, the operators which determine the critical fluctuations of the order parameter, and those that create the soft mode. This set of initial variables depends on the actual system, especially on the structure of the crystal-field system such as singlet-singlet, singlet-triplet, doublet-doublet, and so on. It is also possible to include magnetoelastic interactions in the Hamiltonian and to choose a set of initial operators which also includes adequate phonon creation operators. After having chosen the initial variables, $a_o(1)$, we generate subsequently new sets of dynamical variables[1]

$$a_j=\mathrm{Col}(a_j^{(1)},a_j^{(2)},\ldots,a_j^{(n)});\quad j=1,\ldots,m \tag{2}$$

by the recurrence equation

$$a_{j+1}^{(\alpha)} = \prod_{i=0}^{j}(1-P_i)iLa_j^{(\alpha)} \tag{3}$$

where L denotes the Liouville operator. The projector is defined as follows by the matrix expression

$$P_iB=(B,a_i^+)(a_i,a_i^+)^{-1}a_i \tag{4}$$

where B is an arbitrary operator and

$$(a,b^+)=\frac{1}{\beta}\int_0^\beta d\lambda \quad \langle e^{\lambda H}ae^{-\lambda H}b^+\rangle \tag{5}$$

is the usual Mori's scalar product[7]. The operator a^+ denotes the Hermitian conjugated operator to a. Finally, one arrives at the column matrix A(t) of n·m elements:

$$A(t)=e^{iLt}A=\mathrm{Col}(a_o^{(1)}(t),\ldots,\ a_o^{(n)}(t),\ldots,a_m^{(1)}(t),\ldots,a_m^{(n)}(t)). \tag{6}$$

Then the Mori procedure is applied to A(t)[1].

Let us define the normalized matrix relaxation function

$$\theta(t)=(A(t),A^+)(A,A^+)^{-1}. \tag{7}$$

The equation of motion for $\theta(t)$ takes the following form[1,7]

$$\frac{d}{dt}\,\theta(t) - i\Omega\theta(t) + \int_0^t ds\phi(t-s)\theta(s) = 0 \tag{8}$$

where

$$i\Omega=(iLA,A^+)(A,A^+)^{-1} \tag{9}$$

is the first moment or frequency matrix,

$$\phi(t)=(F(t),F^+)(A,A^+)^{-1} \tag{10}$$

is the memory-function matrix, and $P=\sum_{i=0}^{m}P$ (11) is a random-force matrix orthogonal to A=A(0).

An adequate equation for the Laplace transform has the following matrix form

$$(z\cdot 1-i\Omega+\phi(z))\cdot\theta(z)=1 \tag{12}$$

or explicitly

$$\left[\begin{array}{c:c:c}
A_{oo} & \begin{matrix}-1 & & \\ & \ddots & \\ & & -1\end{matrix} & 0 \\ \hdashline
\Delta_1 & A_{11} & \begin{matrix}-1 & & \\ & \ddots & \\ & & -1\end{matrix} \quad 0 \\ \hdashline
0 & \begin{matrix}\Delta_2 \\ 0\end{matrix} & A_{22}
\end{array}\right]
\left[\begin{matrix}\theta_{oo}(z) & & \\ & \theta_{11}(z) & \\ & & \ddots\end{matrix}\right] = 1 \tag{13}$$

where

$$A_{oo} = \begin{bmatrix} z-i\omega_o^{(1,1)} & -i\omega_o^{(1,2)} & \cdots \\ -i\omega_o^{(2,1)} & z-i\omega_o^{(2,2)} & \cdots \\ \cdots & \cdots & \cdots \end{bmatrix}, \quad A_{11} = \begin{bmatrix} z-i\omega_1^{(1,1)} & -i\omega_1^{(1,2)} & \cdots \\ -i\omega_1^{(2,1)} & z-i\omega_1^{(2,2)} & \cdots \\ \cdots & \cdots & \cdots \end{bmatrix} \tag{14}$$

$$A_{22} = \begin{bmatrix} z-i\omega_2 & & & \\ & \ddots & & \\ & & z-i\omega_m^{(1,1)}+\phi_m^{(1,1)} & -i\omega_m^{(1,2)}+\phi_m^{(1,2)} \cdots \\ & & -i\omega_m^{(2,1)}+\phi_m^{(2,1)} & z-i\omega_m^{(2,2)}+\phi_m^{(2,2)} \cdots \\ & & \cdots & \cdots \\ & & & \ddots \end{bmatrix}$$

and

$$i\omega_i^{(\alpha,\beta)} = (iLa_i^{(\alpha)}, a_i^{(\beta)+})(a_i, a_i^+)^{-1}_{\beta\beta}, \tag{15}$$

$$\Delta_i = (a_i, a_i^+)(a_{i-1}, a_{i-1}^+)^{-1} \tag{16}$$

$$\phi_m(t) = (e^{(1-P)iLt}(1-P)iLa_m[(1-P)iLa_m]^+)(a_m, a_m^+)^{-1}. \tag{17}$$

Following the method of matrix calculations we obtain the final formula in which there occur square matrices only, taken from the matrix equation for the Laplace transforms:

$$\theta_{11}(z) = (A_{11} + \Delta_1 A_{oo}^{-1} + A_{22}^{-1}\Delta_2)^{-1}. \tag{18}$$

After very simple algebra we arrive at the following equation:

$$A_{oo}\theta_{oo}(z)A_{oo} + \theta_{11}(z) = A_{oo} \tag{19}$$

or in another form:

$$(z-i\omega_o)(a_o(z),a_o^+)(z-i\omega_o)+(a_1(z),a_1^+)=z-i\omega_o, \tag{20}$$

where

$$a_1(z)=\int_0^\infty dte^{-zt}a_1(t). \tag{21}$$

Due to Eq. (3), we have

$$a_1(t)=e^{iLt}a_1=\dot{a}_o(t)-i\omega_o a_o(t) \tag{22}$$

$$\dot{a}_o=iLa_o=i[H,a_o]. \tag{23}$$

Equation (20) links the Laplace transforms of successive relaxation matrix functions. The matrix relaxation functions $\theta_{oo}(z)$ include all the initial dynamical variables. The poles of the elements of $\theta_{oo}(z)$ determine all excitations of interest in a system, in particular the central and soft modes in the vicinity of the Curie point. The second matrix relaxation function, $\theta_{11}(z)$, includes these dynamical variables which are generated from the initial variables as the first step in the projection procedure.

This formalism could be applied to different Van Vleck systems. As an example, let us consider the transverse Ising paramagnet:

$$=-\Delta\sum_i S_i^z-\sum_{i,j}J_{ij}S_i^x S_j^x. \tag{24}$$

For simplicity we restrict ourselves to a set of initial dynamical variables which are connected to the order parameter of the system. Therefore, let us take[8]

$$a_o = \begin{pmatrix} S^+_{-\vec{q}} \\ S^-_{-\vec{q}} \end{pmatrix} \tag{25}$$

where

$$S^\pm_{\vec{q}} = \sum_j e^{\pm i\vec{q}\cdot\vec{R}_j}\, S^\pm_j. \tag{26}$$

Using Eq. (22) we generate a new set of dynamical variables, a_1. Thus we are able to take into account the formula (20) which links the successive relaxation functions.

In order to make equation (20) self-consistent, we express the higher-order relaxation function

$$(a_1(z),a_1^+)=\int_0^\infty dte^{-zt}(a_1(t),a_1^+) \tag{27}$$

in terms of the lower order one. We notice from Eq. (22) that $(a_1(t),a_1^+)$ contains the following terms: $(\dot{a}_o(t),\dot{a}_o^+)$, $(\dot{a}_o(t),a_o^+)$ and $(a_o(t),a_o^+)$. Moreover,

$$(\dot{a}_o(z),\dot{a}_o^+)=-z(\dot{a}_o(z),a_o^+)+(\dot{a}_o,a_o^+) \tag{28}$$

so we have to deal with the matrix

$$(\dot{a}_o(t),a_o^+) = \begin{bmatrix} (\dot{S}_{\vec{q}}^+(t),\ S_{\vec{q}}^-) & (\dot{S}_{\vec{q}}^+(t),\ S_{-\vec{q}}^+) \\ (\dot{S}_{\vec{q}}^-(t),\ S_{-\vec{q}}^-) & (\dot{S}_{\vec{q}}^-(t),\ S_{-\vec{q}}^+) \end{bmatrix} \tag{29}$$

and we also have to deal with the relaxation functions which describe the behavior of joined pairs of initial variables and their time derivatives.

The next step is to substitute the explicit equation of motion for $S_{\vec{q}}^{\alpha}(t)$ $(\alpha=+,-)$ and then to substitute for the operator $S_{\vec{q}}^z(t)$ appearing as the result of the first substitution, the formal solution to its equation of motion. Thus we obtain the higher order correlations:

$$(\dot{S}_{k_1}^{\alpha 1}(t),\ S_{k_2}^{\alpha 2}) \to (S_{k_3}^{\alpha 3}(t)S_{k_4}^{z}(t),S_{k_2}^{\alpha 2}) \to$$
$$\to \int_0^t d\tau(S_{k_3}^{\alpha 3}(t)S_{k_4}^{\alpha 4}(\tau)S_{k_5}^{\alpha 5}(\tau),S_{k_2}^{\alpha 2}) \tag{30}$$

which can be decoupled following the method of Kubarev[2,9]. Kubarev's method enables us to decouple higher-order relaxation functions with distinct times as follows:

$$(A(t_1)B(t_2)C(t_3),D^+)=\Sigma\langle A\rangle\langle B\rangle(C(t_3),D^+) +$$
$$+ \Sigma\langle A\rangle(B(t_2)C(t_3),D^+)+\Sigma(A(t_1)B(t_2))(C(t_3),D^+) \tag{31}$$

Using the expansion for the system under consideration (in the paramagnetic phase), we obtain:

$$\int_0^t d\tau(S_{k1}^{\alpha 1}(t)S_{k2}^{\alpha 2}(\tau)S_{k3}^{\alpha 3}(\tau),S_{k4}^{\alpha 4}) \to \int_0^t d\tau(S_{k1}^{\alpha 1}(t-\tau),S_{k2}^{\alpha 2})(S_{k3}^{\alpha 3}(\tau),S_{k4}^{\alpha 4}) \to$$
$$\to (S_{k_1}^{\alpha 1}(z),S_{k_2}^{\alpha 2})(S_{k_3}^{\alpha 3}(z),S_{k_4}^{\alpha 4}). \tag{32}$$

Finally we see that the higher order relaxation function $(a_1(z),a_1^+)$

can be approximately expressed by the matrix $(a_o(z), a_o^{+})$ as well as the terms which also include the relaxation functions of the initial variables for a definite $\vec{q}$-vector. We also consider the dynamical correlations of similar variables with other vectors.

As an example, let us take the matrix element

$$\begin{aligned}(a_1^{(1)}(\vec{q}z)a_1^{(1)+}(\vec{q})) &= iz(S_{\vec{q}}^{+}, S_{\vec{q}}^{-})\{\Delta\cdot X_{\vec{q}}(z)+\frac{i}{N}J(\vec{q})\langle S_o^z\rangle(X_{\vec{q}}(z)+Y_{\vec{q}}(z))+\\ &+\frac{i}{N^2}[(\Sigma_{\vec{\lambda}}J^2(\vec{\lambda})-J(\vec{q})\Sigma_{\vec{\lambda}}\ (\vec{\lambda}))(X_{\vec{q}}(z)Y_{\vec{\lambda}}(z)-X_{\vec{\lambda}}(z)Y_{\vec{q}}(z)-X_{\vec{\lambda}}(-z)X_{\vec{q}}(z)-Y_{\vec{\lambda}}(z)\\ &\qquad Y_{\vec{\lambda}}(z)Y_{\vec{q}}(z))\\ &+\Sigma_{\vec{\lambda}}J(\vec{\lambda})(J(\vec{\lambda})-J(\vec{q}))(X_{\vec{\lambda}}(z)X_{\vec{q}}(z)+Y_{\vec{\lambda}}(z)X_{\vec{q}}(z)-Y_{\vec{\lambda}}(z)X_{\vec{q}}(z)+X_{\vec{\lambda}}(-z)\\ &\qquad Y_{\vec{q}}(z))]\},\end{aligned} \tag{33}$$

where

$$X_q(z)=(S_q^{+}(z),\ S_q^{-}) \tag{34}$$

$$Y_q(z)=(S_q^{+}(z), S_{-q}^{+}). \tag{35}$$

Restricting our analysis to the region of the Curie point, we will focus our attention on the dynamical relaxation functions with wave vector $\vec{q}=0$. As was shown above in the equations for the dynamical spin, $\vec{q}=0$ relaxation functions, the correlations of spin Fourier transforms also appear with non-zero wave vectors. Therefore, in order to calculate the relaxation functions whose behavior is expected to be critical and which give rise to the spontaneous polarization of a system, we must postulate approximate expressions for all other dynamical correlations. For example, we can calculate the relaxation functions with $\vec{q}\neq 0$ in RPA and the static relaxation functions in the Ornstein-Zernike approximation. Moreover, in the vicinity of the Curie point, the essential terms are only those with very small $\vec{q}$.

So one can derive the $\vec{q}=0$ relaxation function expressed in terms of the system parameters (Δ, J). The poles of the initial relaxation function obtained provide information about both the position and line-width of the collective excitations of the system. In the case of the transverse Ising paramagnet, the dynamical soft mode generated by the single ion transition between its energy levels is coupled with the static (not time-dependent) fluctuation of the occupation numbers. This is expressed by $\langle S_o^z\rangle$, and this coupling gives rise to the correct critical behavior of the dynamics of the system in the vicinity of its phase transition.

The proposed approach can be made self-consistent if one 1) derives the relaxation function for $\vec{q}=0$ from Eq. (33), in which relaxation functions for $\vec{q}\neq 0$ have been calculated in RPA, and 2) derives from Eq. (33) the relaxation for an appropriate non-zero $\vec{q}$, substituting the results for the first step for $\vec{q}=0$, and so on.

REFERENCES

1. T. Karasudani, K. Nagano, H. Okamoto, H. Mori, Prog. Theor. Phys. 61, 850 (1979).
2. S.J. Kubarev, Sov. Phys. - Solid State 9, 305 (1967).
3. R. Kubo, J. Phys. Soc. Japan 12, 570 (1957).
4. T. Schneider, G. Srinivasan and C.P. Enz, Phys. Rev. A5, 1528 (1972).
5. R.J. Wojciechowski, Acta Phys. Polon. A56, 177 (1979); A57, 223 (1980).
6. A. Lehmann-Szweykowska, Acta Phys. Polon. A60, (1981) in press.
7. H. Mori, Prog. Theor. Phys. 33, 423 (1965).
8. R. Micnas, L. Kowalewski, Physica 97A, 104 (1979).
9. T. Kawasaki, Prof. Theor. Phys. 41, 322 (1969).

COMMENTS

McEWEN: Could you indicate the qualitative differences between the singlet-singlet and singlet-doublet systems which your theory predicts?

KOWALEWSKI: The most important qualitative difference between the singlet-singlet and singlet-doublet systems is as follows: In the case of the singlet-singlet system the frequency spectrum of the spectral intensity displays one peak of finite energy which becomes overdamped due to the coupling between the soft mode and the static occupation number correlation. Whereas in the case of the singlet-doublet system, from the beginning there are two peaks in the frequency spectrum: the softening mode (longitudinal magnetic excitons) and the central peak (dynamical correlation of the difference in the degenerated energy level occupation numbers). The modes intersect each other. Within the framework of our approach, we find this intersection in the necessity of having to choose two different initial sets of dynamical variables (2 dynamical variables for the singlet-singlet system and 3 for the singlet-doublet case).

KELLER: In our calculations of the linewidth for a 2 singlet level system we found a coupling between a critical mode (magnetization) and a noncritical mode (occupation number fluctuation). Does your approach give similar results?

KOWALEWSKI: In the vicinity of the Curie point our approach gives the coupling between the longitudinal magnetic exciton at $\vec{q}=0$ with both the longitudinal magnetic modes with non-zero wave vectors, and the occupation number fluctuation, the latter being static (independent of time). The coupling between the dynamical soft mode and the static fluctuation results in the overdamping of the former and the appearance of the central peak.

FULDE: Have you already obtained any numerical results for the spectral density?

KOWALEWSKI: We have not yet finished our numerical calculations. They are in progress for the singlet-singlet, singlet-triplet and two Kramers doublets systems.

THE ORIGIN OF THE CRYSTAL FIELD FOR $4f^n$ IONS IN INSULATORS

F. Anisimov and R. Dagys

The Institute of Semiconductor Physics,
237600 Vilnius, K. Pozelos 52, U.S.S.R.

INTRODUCTION

The 4f-ligand overlap for the rare earth ions in crystals is considerably smaller than the 3d-ligand one for transition metal ions. Therefore the theory of low-lying electronic levels of these ions is often based on the hypothesis that major forces acting upon 4f electrons are of a classical electrostatic nature. However, as it follows from simple calculations for the divalent ions, the ratio of the square of the 4f-ligand overlap integral to the 3d-ligand one is proportional to the ratio of the energy level splittings of the 4f and 3d electrons in the crystal field. This means that, as in the case of 3d electrons, the overlap effects must play an important role in determining both the energy level splittings of the 4f electrons and the corresponding crystal field parameters. To illustrate this, calculations of the crystal field splittings were performed for the simple systems $CaF_2:Tm^{2+}$ and $SrF_2:Tm^{2+}$, where Tm^{2+} is at a cubic site and has one 4f hole in the 4f electronic shell. The $^2F_{7/2}$ and $^2F_{5/2}$ levels of free Tm^{2+} split in the cubic field into levels Γ_7, Γ_8, Γ_6 and Γ_7', Γ_8'. Using only one spin-orbit coupling parameter ζ_{4f} for all 4f electrons in the crystal field, one can determine from the experimental data the values of ζ_{4f} and two crystal field parameters B_4 and B_6 (the experimental position of the Γ_6 level is unknown). Theoretical investigations were performed for the crystal field splittings $\Delta E_{12}=E_{t1}-E_{t2}$ and $\Delta E_{13}=E_{t1}-E_{\alpha 2}$ (E_γ is the energy of the crystal field configuration with a hole in the γ shell). These are connected with B_4 and B_6 by simple relations. However, as the experimental value of ΔE_{12} is determined less accurately than ΔE_{13} (due to the use of a single value of the spin-orbit coupling constant instead of four values) only the results for the ΔE_{13} splitting will be presented.

RESULTS AND DISCUSSION

The ab initio calculations of ΔE_{13} were performed within the framework of a cluster of $(TmF_8)^{6-}$ by the Heitler-London method, using one electron functions obtained from the Hartree-Fock calculations for the free Tm^{2+} and F^- ions. The influence of the rest of the crystal was assumed to be constant in the region of the cluster. The distances between atoms were taken from the data for the perfect lattices CaF_2 and SrF_2. To investigate the contribution of various mechanisms to $\Delta E_{13} \equiv \Delta$, at first calculations were performed for the purely ionic model $Tm^{2+}F_8^-$ of the cluster. Taking into account only the overlap of the 4f wave functions of Tm^{2+} and the 2s, 2p wave functions of F^-, the ionic part Δ_i of Δ may be written as the sum of three terms $\Delta(I)$, $\Delta(II)$ and $\Delta(III)$. Here, $\Delta(I)$ is the point charge contribution, $\Delta(II)$ is the correction due to the distribution of electronic charge at the ligand sites, and $\Delta(III)$ is the contribution from the 4f-2s, 2p non-orthogonality. $\Delta(II)$ may be divided into two terms $\Delta_c(II)$ and $\Delta_{ex}(II)$, corresponding to the Coulomb and exchange contributions. The calculated values of these terms are: $\Delta(I)=387$, $\Delta_c(II)=-358$, $\Delta_{ex}(II)=-381$, $\Delta(III)=978$ for CaF_2, and $\Delta(I)=281$, $\Delta_c(II)=-193$, $\Delta_{ex}(II)=-245$, $\Delta(III)=568$ for SrF_2. (All units are cm^{-1}.) Summing up all the contributions, we have $\Delta_i=626$ cm^{-1} for CaF_2 and $\Delta_i=411$ cm^{-1} for SrF_2.

After this, the role of the superposition of configurations (covalency) in which electrons are transferred from 2s and 2p ligands to the vacant 4f one, was investigated. For simplicity, the same single-electron atomic wave functions in the ground and excited configurations were used. The values of the covalency contributions to Δ are 222 and 132 cm^{-1}, respectively, for CaF_2 and SrF_2. Taking these contributions into account we have finally $\Delta=854$ cm^{-1} for CaF_2 and $\Delta=534$ cm^{-1} for SrF_2.

For $CaF_2{:}Tm^{2+}$ additional corrections to Δ were calculated. These are due to the non-orthogonality of the 5s and 5p wave functions of Tm^{2+} and 2s and 2p ones for F^-, as well as 2s and 2p wave functions of the neighboring ligands. The corresponding values of the corrections are -105 cm^{-1} and 288 cm^{-1}. Taking these contributions into account we obtained very good agreement of the calculated value $\Delta=1037$ cm^{-1} with the corresponding empirical one ($\Delta_{exp}=966$ cm^{-1} for CaF_2 and 830 cm^{-1} for SrF_2). However we should consider this good agreement to be accidental. There is some uncertainty in the theoretical results due to approximations made in the calculation and the small basis of functions used in the calculations. These results should be considered primarily as qualitative, illustrating the role of the various mechanisms in the crystal field splittings. The situation for 4f electrons of divalent rare earth ions in insulators is very similar to the one for 3d electrons of divalent 3d transition metal ions. As in the case of 3d electrons the pure ionic model with all ions in the ground state leads to the opposite sign for the splitting Δ compared to the experimental one. The largest contribution which finally gives the correct sign arises from 4f-ligand non-orthogonality. The covalent contribution is of experimental significance, although much smaller than the non-orthogonality terms.

It was shown that rare earth-ligand overlap effects are also quite important in determining crystal field shielding factors. In the cluster approximation, consideration of the shielding problem comes into the investigation of the corrections to E_γ, due to the superposition of the configurations in which the electron of the closed Tm^{2+} shells is excited to the vacant outer orbital. For simplicity, the only case considered was when the excitation of the 5p electron is to the empty t_{1u} orbital. This corresponds to the 5p→4f excitation in the usual theory. It was shown that in the cluster approximation the calculated value of the correction to the energy level E_{t1}, is about ten times greater than the value given by the point charge approximation.

REFERENCE

1. Z.J. Kiss, Energy Levels of Divalent Thulium in CaF_2, Phys. Rev. 127, 718 (1962).

COMMENTS

KASUYA: In the insulator magnets the point charge model usually agrees fairly well both in sign and even in magnitude. Is your calculation consistent with this fact?

DAGYS: Yes, in the sign, but not in the magnitude.

FULDE: Did you find that your results are very sensitive to the particular choice of basis sets?

DAGYS: They are sensitive, but not so much as to change the qualitative picture illustrating the role of various mechanisms in the splitting.

PIERRE: Do you expect a similar importance of 4f-ligand nonorthogonality for metallic systems?

DAGYS: It depends on the system, but as in metallic systems, the rare earth ion is usually in the trivalent state, so the 4f-ligand nonorthogonality effects must be less important.

CONDUCTION ELECTRON EFFECTS ON LOCALIZED SPIN EXCITATIONS IN THE RKKY-THEORY OF MAGNETISM

V. Christoph, A.L. Kuzemsky and Th. Frauenheim

Joint Institute for Nuclear Research, Dubna, 10100 Moscow
P.O. Box 79, USSR

INTRODUCTION

The magnetic scattering of thermal neutrons is a unique technique for establishing both the static and the dynamic properties of magnetic correlations. For interpreting neutron inelastic magnetic scattering data on heavy rare earth metals, the calculations of the magnetic susceptibility and the magnetic excitation spectrum are of particular interest[1-11]. In rare earth metals the exchange interaction between the localized 4f electrons and the extended (conduction) electrons (RKKY exchange) is basic for understanding their magnetic and electric properties[1].

In order to avoid the difficulties connected with crystal field and anisotropy effects, we restrict our consideration to rare earth metals like Gd. Recent detailed experimental and theoretical examinations[9-14] confirm the spin moment of the Gd ion to be a good quantum number. The d-band, having a width of 5-7 eV, lies well above the 4f level and is about one fourth occupied. The density of states of the d-electrons at the Fermi level is much higher than the density of states of the s(p) electrons[8-12]. The general conclusion drawn from these investigations was that the conduction electrons, which play an essential role in the mechanism of RKKY exchange in the heavy rare earth metals, cannot be considered as free s-electrons as assumed in early papers[2,3], but rather as similar to tight-binding d-electrons in transition metals[4-14]. In particular, the magnetic excitation spectrum of Gd has been calculated[11] taking into consideration the d-like character of the extended electrons. Starting with an APW calculation of the band structure and wave functions, the RKKY exchange matrix elements were obtained. Agreement of the calculated magnon spectrum with the experimental one[15] could be obtained by reducing the calculated values by a scale factor of about four. In a more recent paper[16] the generalized spin susceptibility has been calculated using the KKR method.

In the present report the generalized spin susceptibility and the magnon spectrum of Gd metal has been calculated. The tight-binding d-like character of the conduction electrons and the electron-electron and electron-phonon interaction are taken into account in a unified manner. The contributions of different interactions to the magnon damping are estimated, and the lower temperature dependence of the magnon width is calculated. As has been noted[17], the magnon lifetime investigations in Gd at low temperature should give information on different interactions and their roles and significance in the heavy rare earth metals.

RARE EARTH METAL MODEL

Neglecting crystal field and anisotropy effects, we describe the rare earth metal by localized 4f spins, interacting with d-like tight-binding conduction electrons. We take into consideration the electron-electron and electron-phonon interactions in the framework of a model given by S. Barisic et al.[18]. The Hamiltonian[18] is a generalization of the well known Hubbard Hamiltonian[19] and has been investigated in detail in Ref. 20.

The total Hamiltonian is:

$$\mathcal{H}=H_d+H_{d-f}+H_{d-ph}+H_{ph}, \tag{1}$$

where

$$H_d=\sum_{k\sigma} E(k)\alpha^+_{k\sigma}\alpha_{k\sigma}+\frac{U}{2N}\sum_{\substack{kk'\\ q\sigma}}\alpha^+_{k+q\sigma}\alpha_{k\sigma}\alpha^+_{k'-q,-\sigma}\alpha_{k',-\sigma} \tag{2}$$

is the Hubbard Hamiltonian[19]. For the tight-binding d-electrons we use $E(k) = 2\sum_\alpha t(\vec{a}_\alpha)\cos(\vec{k}\vec{a}_\alpha)$, where $t(\vec{a}_\alpha)$ is the hopping integral between next nearest neighbors[18], and $\vec{a}_\alpha$ (α=1,2,3) denotes the lattice vectors in a simple lattice with inversion center. The second term in (2) describes the Coulomb interaction of electrons with opposite spins at the same lattice site. The RKKY Hamiltonian describing the interaction of the total 4f spin $\vec{s}$ with the spin density of the conduction electrons has the form

$$H_{d-f}=-\frac{J}{\sqrt{N}}\sum_{kq}\{(\alpha^+_{k\uparrow}\alpha_{k+q\uparrow}-\alpha^+_{k\downarrow}\alpha_{k+q\downarrow})S^z_{-q}+(\alpha^+_{k\uparrow}\alpha_{k+q\downarrow}S^+_{-q}+c.c.)\} \tag{3}$$

where J is the local RKKY exchange integral. In general the exchange integral strongly depends on the wave vectors $\vec{k}$ and $\vec{q}$ having maximum values at $k=q=0$[11]. For simplicity we restrict ourselves to a local exchange. The generalization to non-local exchange is straightforward. For the electron-phonon interaction we use[18]:

$$H_{d-ph}=\sum_{kq}\sum_{\sigma\nu} V^{\nu}(k,k+q)Q_{q\nu}\alpha^{+}_{k+q\sigma}\alpha_{k\sigma} \tag{4}$$

where

$$V^{\nu}(\vec{k},\vec{k}+\vec{q})=\frac{2iq_o}{\sqrt{NM}}\sum_{\alpha} t(\vec{a}_{\alpha})e^{\alpha}_{\nu}(\vec{q})\{\sin \vec{a}_{\alpha}\vec{k}-\sin \vec{a}_{\alpha}(\vec{k}+\vec{q})\}. \tag{5}$$

In (5) q_o is the Slater coefficient originating in the exponential decrease of the d-functions[18], N is the number of unit cells in the crystal, and M is the ion mass. $\vec{e}_{\nu}(\vec{q})$ (ν=1,2,3) are the polarization vectors of the phonon modes.

For the vibrating ion system we have

$$H_{ph}=\frac{1}{2}\sum_{q\nu}(P^{+}_{q\nu}P_{q\nu}+\omega_o^2(q\nu)Q^{+}_{q\nu}Q_{q\nu}) \tag{6}$$

where $P_{q\nu}$ and $Q_{q\nu}$ are the normal coordinates and $\omega_o(q\nu)$ are the accoustical phonon frequencies. Thus, as in the Hubbard model, the d- and s-(p)bands are replaced by one "effective" band in our model[1]. However, the s-electrons give rise to screening effects and are taken into account by choosing proper values of J, U and the accoustical phonon frequences $\omega_o(q\nu)$[21].

GENERALIZED SPIN SUSCEPTIBILITY

We are interested in the Fourier transform of the generalized susceptibility of the localized f-spins $\langle\langle S^{+}_{k}|S^{-}_{-k}\rangle\rangle$ where

$$\langle\langle S^{+}_{k}(t)|S^{-}_{-k}(0)\rangle\rangle=-i\theta(t)\langle[S^{+}_{k}(t)S^{-}_{-k}(0)]\rangle \tag{7}$$

is the usual double-time commutator Green's function[22] and $\vec{S}_k$ is the Fourier transform of the f-spin $\vec{S}_i$. For the calculation of $\langle\langle S^{+}_{k}|S^{-}_{-k}\rangle\rangle$ we use the irreducible Green's function technique[23] already applied to the Hubbard model in the atomic and band limits[24].

The method of calculation is based on using the generalized matrix Green's function

$$\bar{G}=\begin{pmatrix} \langle\langle S^{+}_{k}|S^{-}_{-k}\rangle\rangle & \langle\langle S^{+}_{k}|\sigma^{-}_{-k}\rangle\rangle \\ \langle\langle \sigma^{+}_{k}|S^{-}_{-k}\rangle\rangle & \langle\langle \sigma^{+}_{k}|\sigma^{-}_{-k}\rangle\rangle \end{pmatrix}, \tag{8}$$

where the Fourier components of the conduction electron spin densities, $\sigma^{+}_{k}=\Sigma_q\alpha^{+}_{k\uparrow}\alpha_{k+q\downarrow}$ and $\sigma^{-}_{k}=\Sigma_q\alpha^{+}_{k\downarrow}\alpha_{k+q\uparrow}$, have been introduced. It may be shown that the equation of motion for G [Eq.(8)] may be exactly transformed to the Dyson equation by using the irreducible Green's functions with an explicit representation of the mass operator. The exact Dyson equation is given by

$$\bar{G}=\bar{G}_o+\bar{G}_o\bar{M}\bar{G}, \tag{9}$$

where $\bar{M}$ is the mass operator and $\bar{G}_o$ is the mean field Green's function. Hence the determination of $\bar{G}_o$ has been reduced to the determination of $\bar{G}_o$ and $\bar{M}$, when the mean field Green's function $\bar{G}_o$ coincides with the RPA-result[25].

DAMPING OF SPIN WAVES IN THE COUPLED LOCAL MOMENT-ELECTRON SYSTEM

In order to estimate the damping of the localized spin excitation spectrum due to electron-magnon and electron phonon scattering, we calculate the local spin susceptibility for small k and ω

$$<<S_k^+|S_{-k}^->>_\omega = \frac{2N^{-1/2}<\tilde{S}_o^z>}{\omega-\varepsilon_k-2N^{-1/2}<\tilde{S}_0^z>\Sigma(k,\omega)}, \tag{10}$$

which contains the matrix elements of the mass operator $\bar{M}=(M_{ij})$ in a linear approximation

$$\Sigma(k,\omega)=M_{11}+(M_{12}+M_{21})\frac{J^{1/2}\chi_o^{df}}{1-U\chi_o^{df}} + \frac{J^2N(\chi_o^{df})^2M_{22}}{(1-U\chi_o^{df})^2}. \tag{11}$$

Here we have used the notations

$$<\tilde{S}_o^z>=<S_o^z>(1+\frac{N_\uparrow-N_\downarrow}{2N^{1/2}<S_o^z>});\quad \chi_o^{df}=\chi_o^{df}(k,\omega)=\frac{1}{N}\sum_q\frac{f_{q+k\downarrow}-f_{q\uparrow}}{\omega_{q,k}}$$

$$\omega_{q,k}=\omega+E(q)-E(q+k)-\Delta;\quad N_\sigma=\sum_q f_{q\sigma}=\sum_q<n_{q\sigma}> \tag{12}$$

$$\Delta=2JN^{-1/2}<S_o^z>+UN^{-1}(N_\uparrow-N_\downarrow).$$

The spectral density of the spin wave excitations with wave vector k then reads

$$g(k,\omega)= -\frac{1}{\pi}\mathrm{Im}<<S_k^+|S_{-k}^->>=\frac{2N^{-1/2}<\tilde{S}_o^z>\Gamma(k,\omega)}{[\omega-\varepsilon_k-\Delta(k,\omega)]^2+[\Gamma(k,\omega)]^2}, \tag{13}$$

where

$$\Delta(k,\omega)=2N^{-1/2}<\tilde{S}_o^z>\mathrm{Re}\Sigma(k,\omega),$$
$$\Gamma(k,\omega)=-2N^{-1/2}<\tilde{S}_o^z>\mathrm{Im}\Sigma(k,\omega+i\varepsilon) \tag{14}$$

describes the shift and the damping of the acoustical magnons.

Finally we estimate the temperature dependence of $\Gamma(\vec{k},\omega)$ due to the mass operator terms in (11). Considering the first contribution in (11) at low temperatures we obtain

$$\mathrm{Im}M_{11} \sim T. \tag{15}$$

The other electron-magnon contributions to $\Gamma(\vec{k},\omega)$ can be treated in the same way, where M_{12}, M_{21} and the electron-magnon contribution to M_{22} are proportional to T also. For the electron-phonon contribution to M_{22} we find

$$\mathrm{Im}M_{22}^{ph} \sim T^3. \tag{16}$$

Hence the damping of the acoustical magnons at low temperatures can be written as

$$\Gamma(\vec{k},\omega)|_{k,\omega\to 0} \sim \Gamma_1 + \Gamma_2 T + \Gamma_3 T^3 \tag{17}$$

where the coefficients Γ_i (i=1,2,3) vanish for k=ω=0. For the case when J=0, the electron-electron term Γ_1 vanishes in (17).

DISCUSSION

In the present paper the spectrum of the elementary magnetic excitations and their lifetimes have been calculated for the microscopic model of heavy rare earth metals. In the mode used the electron-electron interaction of the extended (conduction) electrons and the electron-phonon interaction in the Barisic-Labbe-Friedel manner have been taken into account. Neglecting the damping terms the result obtained (10) coincides with that of L.C. Bartel[25], obtained in the RPA. Furthermore, using a generalized matrix Green's function here the expressions for the spin susceptibility of the conduction electrons and mixed terms have been derived. Therefore, starting with this formalism the contribution of the extended (conduction) electrons to the total magnetization of the rare earths[4,5] may also be considered.

The temperature dependence of the acoustical magnon damping is given mainly by the electron-magnon and electron-phonon interactions. the electron-electron contribution Γ_1 is almost temperature independent, but vanishes at J=0, also. It should be emphasized here that the real electronic structure of a rare earth metal such as Gd is much more complicated than given by our tight-binding model[1,4-14]. For a realistic first principle calculation of the stiffness constant about ten bands should be included[4,5,8,10,11]. Furthermore, the dependence of the RKKY exchange integral J(k,k+q) on the wave vector should be taken into account[11]. Nevertheless, taking into consideration all interactions playing a part in the rare earth metals, this estimate of the temperature dependence of the magnon damping should be reasonable. Unfortunately there are no neutron scattering measurements of the low temperature magnon damping in Gd[17]. Such measurements and their theoretical interpretation would greatly improve our understanding of the interactions in the heavy rare earth metals.

REFERENCES

1. A.J. Freeman, in: Magnetic Properties of Rare Earth Metals, ed. R.J. Elliott (Plenum Press, New York, 1972)Chap. 6.
2. S.H. Liu, Phys. Rev. 121, 451 (1961).
3. T.A. Kaplan, D.H. Lyons, Phys. Rev., 129, 2072 (1963).
4. A.J. Freeman, J.P. Dimmock, R.E. Watson, Phys. Rev. Lett. 16 94 (1966).
5. R.E. Watson, A.J. Freeman, J.P. Dimmock, Phys. Rev. 167, 497 (1968).
6. W.E. Evenson, S.H. Liu, Phys. Rev. Lett. 21, 432 (1968).
7. W.E. Evenson, S.H. Liu, Phys. Rev. 178, 783 (1969).
8. S.C. Keeton, T.L. Loucks, Phys. Rev. 168, 672 (1968).
9. J.F. Herbst, D.H. Lowy, R.E. Watson, Phys. Rev. B6, 1913 (1972).
10. B.N. Harmon, A.J. Freeman, Phys. Rev. B10, 1979 (1974).
11. P.A. Lindgard, B.N. Harmon, A.J. Freeman, Phys. Rev. Lett. 35, 383 (1975).
12. J.F. Herbst, R.E. Watson, J.W.E.Wilkins, Phys. Rev. B13, 1439 (1976); Phys. Rev. B17, 3089 (1978).
13. A.J. Freeman, J.P. Desclaux, Int. J. Magnetism 3, 311 (1972).
14. A.J. Freeman, Physica 91B, 103 (1977).
15. W.C. Koehler et al., Phys. Rev. Lett. 24, 16 (1970).
16. J.F. Cooke, J. Appl. Phys. 50, 1782 (1979).
17. A.R. Mackintosh, J. Magn. Magn. Mat. 15-18, 326 (1980).
18. S. Barisic, J. Labbe, J. Friedel, Phys. Rev. Lett. 25, 919 (1970).
19. J. Hubbard, Proc. Roy. Soc. A276, 238 (1963).
20. A. Holas, N.M. Plakida, A.L. Kuzemsky, Communications JINR, p.17-80-741, Dubna, 1980.
21. E.G. Brovman and Yu.M. Kagan, Uspekhi Fiz. Nauk. 112, 369 (1974).
22. S.V. Tyablikov, Methods in the Quantum Theory of Magnetism, Plenum Press, NY (1967).
23. N.M. Plakida, Phys. Lett. A43, 481 (1973).
24. A.L. Kuzemsky, Teor. Math. Phys. 36, 208 (1978).
25. L.C. Bartel, Phys. Rev. B7, 3153 (1973).

COMMENTS

GREIDANUS: What is your argument for the statement that in the neutron spectrum of $PrNi_2$ the observed linewidth cannot be explained by dispersion effects?

FRAUENHEIM: I believe that dispersion averaging is important only if you average over the whole Brillouin zone, but not if you average over the directions in a polycrystal at a fixed momentum transfer. As known from the single crystal neutron inelastic measurements in $PrAl_2$, direction averaging in a polycrystal does not give such a linewidth effect. Maybe the new theoretical results from Dr. Keller's lecture may give the explanation.

OLES: You mentioned quite a big discrepancy between the measured and the theoretical values of the exchange constant in Gd. What is your explanation of that?

FRAUENHEIM: I have no explanation for the $J(\vec{q})$ reduction in Gd. I am only sure that the electron correlation effects would lead to an enhancement of the $J(\vec{q})$ value, which is in contrast to J. F. Cooke's explanation.

MAGNETIC FIELD DEPENDENCE OF THE
CONDUCTION ELECTRON MASS IN PRASEODYMIUM

Peter Fulde* and R.M. White†

*Max-Planck-Institut für Festkörperforschung
7 Stuttgart 80, Fed. Republic of Germany
†Xerox Research Center, Palo Alto, CA 94304, USA

INTRODUCTION

Recent low temperature specific heat experiments by Forgan[1] have revealed that Pr metal has an anomalously large conduction electron effective mass m^*. Furthermore, when a magnetic field was applied, m^* was found to decrease considerably. The measurements suggest that both effects are connected with the presence of 4f electrons in Pr as they are found to be absent in La. We want to show in the following that they can be understood when the contribution of the magnetic excitons to the conduction electron mass is taken into account. While contributions of phonons or paramagnons to the conduction electron mass in paramagnetic metals are well recognized it seems to us that the experiments by Forgan demonstrate for the first time an analogous contribution of the coupled crystalline electric field (CEF) excitations.

Pr metal has a dhcp structure with two inequivalent atoms per unit cell. The local symmetry of the two sites is hexagonal and approximately cubic, respectively. The crystal fields lead to singlet ground states at both sites but the first excited state at the hexagonal sites is much lower in energy than the one at the cubic sites, i.e., 3.2 meV versus 8 meV. For that reason we will neglect the cubic sites in the following, although a more accurate theory of m^* should take them into account as well. At the hexagonal sites the ground state is $|0\rangle$ and the excited doublet is denoted by

$$|s\rangle = 2^{-1/2}(|+1\rangle + |-1\rangle)$$
$$|a\rangle = 2^{-1/2}(|+1\rangle - |-1\rangle) \; . \tag{1}$$

The interaction of the conduction electrons with the CEF levels at the hexagonal sites $\vec{R}_n$ is assumed to be of the form

$$H_{int} = - J_{sf}(g-1)\sum_n \vec{J}_n\vec{\sigma}(\vec{R}_n) \tag{2}$$

in standard notation. The only non-vanishing matrix elements of $\vec{J}$ are

$$\begin{aligned} \langle 0 | J_x | s\rangle &= \langle 0|J_y|a\rangle \\ &= (J(J+1)/2)^{1/2} = \alpha \\ \langle a|J_z|s\rangle &= 1 \end{aligned} \tag{3}$$

CALCULATION OF CONDUCTION ELECTRON EFFECTIVE MASS

The conduction electron self energy $\sum(\vec{p},i\omega_n)$ due to the interaction H_{int} in finite temperature notation[2] is

$$\sum(\vec{p},i\omega_n) = (g-1)^2 J_{sf}^2 T \sum_m \int \frac{d^3p'}{(2\pi)^3} G_o(\vec{p}',i\omega_m) \cdot \mathrm{Tr}\,[\chi_{\alpha\alpha}(\vec{p}-\vec{p}',\; i(\omega_n-\omega_m))] \; . \tag{4}$$

$G_o(\vec{p},i\omega_m)$ is the conduction-electron Green´s function in the absence of H_{int} and Tr denotes taking the trace. The susceptibility tensor $\chi_{\alpha\alpha}(\vec{q},\omega)$ of the coupled ionic system is related to the single ion susceptibility tensor $U_{\alpha\alpha}(\omega)$ through

$$\chi_{\alpha\alpha}(\vec{q},\omega) = \frac{U_{\alpha\alpha}(\omega)}{1-J_{eff}(\vec{q})U_{\alpha\alpha}(\omega)} \tag{5}$$

The effective interaction $J_{eff}(\vec{q})$ between the Pr ions is assumed to be ω independent for the present purpose. At low temperatures, when the thermal population of the doublet may be neglected, $U_{zz}(\omega) = 0$ and

$$U_{xx}(\omega) = U_{yy}(\omega) = \frac{2\alpha^2\Delta}{\Delta^2-\omega^2} \; . \tag{6}$$

The CEF splitting energy $\Delta = 3.2$ meV.

From $\sum(\vec{p}, i\omega_n)$ the effective mass is calculated as

$$\frac{m^*}{m} = 1 - \mathrm{Re}\, \frac{\partial \sum(p_F,\omega)}{\partial \omega} \Bigg|_{\omega = 0} \tag{7}$$

With the help of Eq. (4) this is rewritten as

$$\frac{m^*}{m} = 1 + \frac{N(0)\ (g-1)^2 J_{sf}^{\,2}}{2p_F^2} \int_0^{q_1} dq\ q\ \mathrm{Tr}\ [\chi_{\alpha\alpha}(\vec{q},0)]. \tag{8}$$

Here q_1 is the smaller of the Debye momentum q_o for the magnetic excitons and $2p_F$. In order to evaluate the integral one must know $J_{eff}(\vec{q})$ which can be determined from the measured dispersion curves of the magnetic excitons as obtained by inelastic neutron scattering[3].

We have replaced the optical branches, which are split by anisotropic interactions, by the simplified Debye type of model

$$\omega = a+b(q/q_o) \tag{9}$$

with a = 1 meV and b = 2.5 meV (see Fig. 1) and have neglected the acoustic branches altogether. This simplification will certainly introduce inaccuracies into the final numerical results. However, we are interested here only in approximate answers and it is clear from Eq. (8) how a more accurate calculation can be performed. When Eq. (8) is evaluated one finds

$$\frac{m^*}{m} = 1+N(0)(g-1)^2 J_{sf}^{\,2}\ \frac{\Delta^2}{b^2} U_{xx}(0)\left(\frac{q_o}{p_F}\right)^2 \Omega \tag{10}$$

with $\Omega = \ln z + 1/z - 1$ and $z = 1 + \frac{b}{a}\,(q_1/q_o)$. For $q_o < 2p_F$ it is z = 3.5. One notices that for $a \to 0$, m^* diverges logarithmically. The result resembles then that of the paramagnon theory[4]. In the present case, however, the largest contributions to m^* come from large momenta q as in the phonon case. The approximation (9) which is rather poor for small q is therefore not severe. With $\alpha^2 = 10$ and $N(0)\ (g-1)^2\ J_{sf}^{\,2} = 9.10^{-4}$ eV one finds

$$\frac{m^*}{m} = 1+4.95\ \left(\frac{q_o}{p_F}\right)^2 . \tag{11}$$

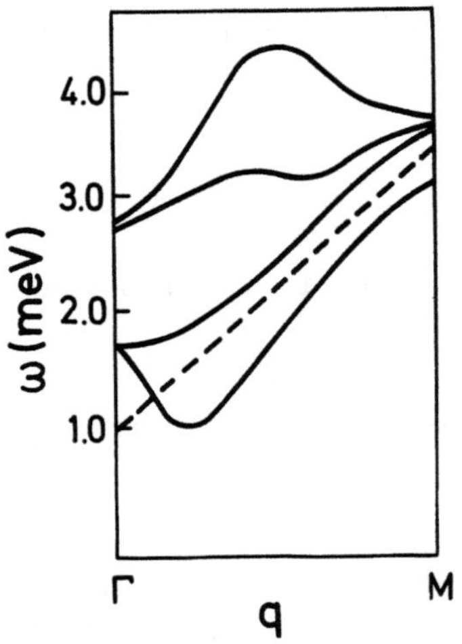

Fig. 1 Dispersion curves for the magnetic excitons at low T propagating on the hexagonal sites (from Ref. 3). In the calculations the lower two branches are replaced by the dashed line (see Eq. (9)) while the upper two branches (acoustic modes) have been neglected.

One expects q_o/p_F to be of order unity and the enhancement is then of the order as found experimentally.

APPLICATION OF A MAGNETIC FIELD

When an external magnetic field is applied one must distinguish between fields in the basal plane and perpendicular to it.

1. H parallel to the basal plane

When H is chosen along the x axis the Zeeman interaction term is $H_{Ze} = - g\mu_B J_x H$. The induced molecular field changes H into $H_{eff} = SH$, where the enhancement factor S is given by the ratio $\chi_{xx}(0,0)/U_{xx}(0)$ and can be determined experimentally. We have calculated the changes in the CEF eigenvalues and eigenstates in the presence of H_{eff} by second order perturbation theory. With them the $\chi_{\alpha\alpha}(q,0)$ are calculated anew and from Eq. (8) $m^*(H)$ is obtained. It is found to be of the same form as Eq. (10) but with Ω=0.538 replaced by

$$\Omega(H) = 0.538 - 0.334 \cdot \left(\frac{\alpha g \mu_B S H}{a}\right)^2 . \qquad (12)$$

It is seen that $m^*(H)$ decreases with H^2. However, one should be aware of the fact that the approximation (9) leads to larger errors in the H^2 term than it does in Eq. (10). For a detailed comparison between theory and experiments, particularly at large fields, one should thus 1) use the measured dispersions $\omega(q)$ instead of Eq. (9), 2) include the acoustic excitons and 3) also compute the eigenvalues and eigenstates in the field H_{eff} instead of doing perturbation theory. The decrease of $m^*(H)$ with field can be understood also from the neutron-scattering data[3]. They show that the application of a magnetic field increases the exciton energy of the magnetic excitons. This results in a decrease of m^*. Setting S = 2.8 and for H = 2T one finds [with $q_o/p_F = 1$ from Eqs. (10) and (12)] that $m^*(H)/m^*(0) = 0.66$.

2. H perpendicular to the basal plane

In this case the Zeeman term $H_{Ze} = -g\mu_B J_z H$ leads to a splitting of the states $|\pm 1\rangle$ of the doublet. Consequently one obtains $U_{+-}(0) \neq U_{-+}(0)$ and the self energy of the conduction electrons is spin dependent. Neglecting the thermal population of the states $|\pm 1\rangle$ as before, $\sum_\sigma(\vec{p},i\omega_n)$ is calculated from Eq. (4) but with $Tr[\chi_{\alpha\alpha}]$ replaced by χ_{+-} or χ_{-+}, depending on the sign of σ. It is found that

$$\chi_{+-}(q,H) = \frac{4\alpha^2\Delta}{-g\mu_B H\Delta + (a+bq/q_o)^2} \tag{13}$$

and $\chi_{-+}(q,H) = \chi_{+-}(q,-H)$. After performing the q integration one ends up with an expression of the form of Eq. (10) but with Ω replaced by a spin dependent

$$\Omega_\sigma = 0.538 + 0.133\,\sigma g\mu_B H \frac{\Delta}{a^2} + 0.048\left(g\mu_B H \frac{\Delta}{a^2}\right)^2. \tag{14}$$

For a field of H = 4.35 T this would imply

$$\frac{m^*_\sigma(H)}{m^*(0)} = 1.03 + 0.12\,\sigma \tag{15}$$

and hence a considerable difference in the effective masses of spin up and down electrons.

Again one should realize that the approximation (9) introduces a relatively large error in the H dependent terms. Thus for a quantitative comparison between theory and experiments a more accurate treatment of the magnetic exciton dispersion is required. The appearance of a spin dependent m^*_σ caused by an applied magnetic field is an interesting prediction of the above theory. There should be ways of checking this prediction experimentally.

ACKNOWLEDGEMENT

We thank Dr. E.M. Forgan for sending us his experimental results prior to publication and for drawing our attention to the problem.

REFERENCES

1. E.M. Forgan, Proceedings of the XVIth International Conference on Low Temperature Physics, 1981, to be published.
2. P. Fulde and I. Peschel, Adv. Physics 21, 1 (1972).
3. J.G. Houmann, B.D. Rainford, J. Jensen and A.R. Mackintosh, Phys. Rev. B20, 1105 (1979).
4. N.F. Berk and J.R. Schrieffer, Phys. Rev. Letters 17, 433 (1966); S. Doniach and S. Engelsberg, Phys. Rev. Letters 17, 750 (1966).

COMMENTS

McEWEN: The de Haas-van Alphen effect in Pr has been measured by R.C. Young. I recall that he found it much harder to understand his data for a field parallel to the c-direction than for fields in the basal plane. Perhaps this reflects the spin dependent m* for $\vec{H}||\vec{c}$. What do you expect to happen to m* for a helical magnetic structure, such as found in Dy?

FULDE: Thank you for pointing out that de Haas-van Alphen data which were not known to me. In the case of magnetically ordered Dy there will be a contribution of the spin waves to m*. When pressure is applied to Pr such that the energy of the magnetic excitons equals zero at finite q-value, then the effective mass m* should diverge.

KASUYA: There have been several calculations made for the mass enhancement due to electron-magnon interactions: for example, such is the case for Gd metal with a well-defined spin wave mode and in some itinerant magnets in a critical region for the formation of magnetism, where we really have very large enhancements at the lowest temperature due to the critical paramagnon mode. Your case seems to be rather similar to the latter case.

PHONON COUPLING MECHANISMS IN INTERMETALLIC RARE EARTH COMPOUNDS

B. Lüthi, M. Niksch, R. Takke, W. Assmus and W. Grill

Physikalisches Institut der Universität
Postfach 111932, Robert-Mayer-Str. 2-4
D-6000 Frankfurt a.M., F.R.G.

INTRODUCTION

In this review we shall discuss the various interaction mechanisms of phonons with magnetic ions in intermetallic rare earth compounds. One might expect a variety of effects due to the anisotropic charge distribution of the localized 4f electrons and the high density of states of 5d-band electrons. During the past few years we were able to identify 3 main coupling mechanisms of long wavelength phonons in intermetallic rare earth compounds:

a) The magnetoelastic interaction, which describes the coupling of the electronic quadrupole (or multipole) moments of the localized 4f electrons to the macroscopic strains.

b) A deformation potential coupling of the strain to 5d-band states.

c) An anharmonic phonon coupling to zone boundary phonons for cases where these zone boundary phonons exhibit a strong temperature dependence (soft mode).

THEORETICAL BACKGROUND

Rare earth atoms with the configuration $4f^n5d^16s^2$ retain their localized 4f wave functions with unquenched orbital angular momenta and anisotropic charge distributions in a compound. Strong spin orbit interactions give well separated multiplets characterized by the total angular momentum J, leading to free ion like behaviour at higher temperatures. Small crystal electric field splittings give rise to typical crystal field effects at lower temperatures. On the other hand the 5d electron forms a virtual bound state in the dilute case and 5d bands in the case of a concentrated intermetallic rare earth compound. The 6s electrons hybridize into s and p-like bands.

a) Magnetoelastic Interaction

In simple cubic materials with NaCl, CsCl or Cu_3Au type structures one can have a direct coupling of the macroscopic strain of a given symmetry Γ to the multipole moments of the 4f electrons. For general structures one has to use the cluster model to describe the appropriate coupling [1,2]. Since, with few exceptions, we treat cubic structures of the types mentioned above, we give only a treatment of the direct coupling. The magnetoelastic interaction reads $H_{me} = -\sum_i g_\Gamma^{(n)} O_{\Gamma_i}^{(n)} \varepsilon_\Gamma$ where ε_Γ denotes the symmetry strain, $g_\Gamma^{(n)}$ the magnetoelastic coupling constants and $O_\Gamma^{(n)}$ the multipole operator of symmetry Γ and order n for the ion site i. Time reversal symmetry limits n to 2, 4, 6 etc. Until now only quadrupolar, n=2, coupling effects have been investigated. We shall give clear cut evidence for higher order coupling later on in the result section.

The magnetoelastic coefficients g_Γ can be thought of as the strain derivatives of the CEF parameters. $g_\Gamma^{(2)}$ can be written as $g_\Gamma^{(2)} = G\,\alpha_J <r^2>$ where $\alpha_J <r^2>$ are the parameters depending on the rare earth ion itself due to the equivalent operator description above and G gives the effect of the surroundings. Similar to the CEF parameters themselves, G depends on a number of different contributions (ligands, Coulomb and exchange contributions of the 5d conduction electrons etc.). Although we shall present much experimental data for $g_\Gamma^{(2)}$, direct microscopic calculations are scarce [3]. For cubic symmetry and n=2 the magnetoelastic interaction reads

$$H_{me} = \sum_i g_{\Gamma_3}^{(2)} \{\varepsilon_2\, O_{2_i}^2 \sqrt{3} + \varepsilon_3\, O_{2_i}^0\} + g_{\Gamma_5}^{(2)} \{\varepsilon_{xy}\, O_{xy} + \text{cycl.}\} \quad (1)$$

with the usual definitions for the Γ_3 and Γ_5, strains ε_{ij} and O_{ij} [5]. The magnetoelastic Hamiltonian including n=4 terms reads [4]

$$H_{me} = -\sum_i \varepsilon_2 (g_{\Gamma_3}^{(2)}\, O_{2_i}^2 \sqrt{3} + g_{\Gamma_3}^{(4)}\, O_{4_i}^2) - \sum_i \varepsilon_{xy} (g_{\Gamma_5}^{(2)}\, O_{xy_i} + g_{\Gamma_5}^{(4)}\, \tilde{O}_{4_i}^2) \quad (2)$$

where $\varepsilon_2 = (\varepsilon_{xx} - \varepsilon_{yy})/\sqrt{2}$, $O_2^2 = J_x^2 - J_y^2$, $O_{xy} = J_x J_y + J_y J_x$,

$$O_4^2 = \left[(7J_z^2 - J(J+1) - 5)(J_+^2 + J_-^2) + (J_+^2 + J_-^2)(7J_z^2 - J(J+1) - 5)\right] /4$$

$$\tilde{O}_4^2 = \left[(7J_z^2 - J(J+1) - 5)(J_+^2 - J_-^2) + (J_+^2 - J_-^2)(7J_z^2 - J(J+1) - 5)\right] (-i/4).$$

In equation (2) we did not give the complete expression, because for the elastic constants in the paramagnetic phase we only need a particular component, e.g. ε_2, ε_{xy}.

In this review we shall be concerned mainly with effects on long wavelength phonons, i.e. soundwaves. The interesting physical quantities are then the elastic constants and the ultrasonic attenuation. The influence of magnetoelastic interactions of H_{me} on elastic constants has been thoroughly treated in the literature.[1,2,5,6,7] Therefore we only state the main result here:

$$c_\Gamma = c_\Gamma^o - g_\Gamma^2 N \chi_s^\Gamma (1 - K_\Gamma \chi_s)^{-1} \tag{3}$$

Here c_Γ^o is the background elastic constant, χ_s is the strain susceptibility determined from H_{me} and K_Γ denotes a quadrupole-quadrupole interaction parameter originating from $H_{int} = \Sigma'_{ij} K_\Gamma O_{\Gamma_i} O_{\Gamma_j}$. Equation (3) covers a large number of situations (cooperative Jahn Teller effect, CEF effect on elastic constants, etc.).

b) Deformation Potential Coupling

The conduction electron-phonon interaction one gets by expanding the periodic potential in terms of symmetrized strains $H_{ep} = - \Sigma \, \underset{\sim}{\varepsilon}_\Gamma \, \underline{R}_m \cdot \mathrm{grad}\, V(R_m)$. This interaction is responsible for the scattering of electrons by phonons. For long wavelength phonons it leads to local energy shifts of the energy bands $E_k = E_k^o - \varepsilon_\Gamma d_\Gamma$ with d_Γ = deformation potential coupling constant. For degenerate energy bands with a constant density of states N_o near the Fermi energy one obtains for the elastic constant [2]:

$$c = c^o + \frac{d_\Gamma^2}{2} \int_o^\infty N(E) \frac{\partial f}{\partial E} dE = c_\Gamma^o - \alpha d_\Gamma^2 N_o (1 - e^{-T_o/T}) \tag{4}$$

where $k_B T_o = E_F(T{=}0)$. Because in lowest order of deformation potential coupling (linear in the strain) one needs degenerate electron bands to get Eq. (4), this effect is often called the band Jahn-Teller effect. This approximation corresponds to the Curie type terms in the magnetoelastic case, whereas the nondiagonal Van Vleck type terms correspond to terms quadratic in the strain.

The 5d-bands in cubic intermetallic rare earth compounds are split into e_g- and t_{2g}-like bands. Therefore, e.g. Γ_3-strains ε_2, ε_3 can lift the e_g- and t_{2g}-symmetry and Γ_5-strains can lift the t_{2g}-symmetry at the Γ-point of the B.Z. From a symmetry point of view this type of coupling can lead to observable effects provided the density of states N_o is large enough. We shall give a thorough discussion of under what circumstances one can observe this effect, which in principle should always be present in intermetallic rare earth compounds.

c) Anharmonic Interactions

Anharmonic phonon-phonon interactions can be important for insulators as well as metallic compounds. We shall give evidence that an anharmonic coupling between soundwaves and zone boundary phonons is responsible for the temperature dependence of elastic constants in intermetallic rare earth compounds undergoing structural transitions. The interaction Hamiltonian for anharmonic processes reads[2,8] $H_{anh} = \sum_{ij\Gamma} g_\Gamma [Q_i\ Q_j]_\Gamma\ \varepsilon_\Gamma$ where the Q_i denote components of the local normal phonon coordinates and $[Q_i\ Q_j]_\Gamma$ denotes the symmetrized product. The actual form for a given crystal structure can be deduced using group theory. To calculate the elastic constants involves, even with simplifying molecular field type approximations, complicated k-space integrations. In the case of a soft zone boundary phonon mode one can expect either step-like discontinuities at the phase transition T_c or power law dependences of $(T-T_c)$ depending on whether order parameter fluctuations are important or not [2].

RESULTS AND DISCUSSION

In the following we give experimental results which clearly show the influence of the three coupling mechanisms on the elastic constants. Other experimental information (thermal expansion, magnetostriction, magnetoelastic birefringence, ultrasonic attenuation etc.) will also be briefly discussed.

a) Magnetoelastic Coupling

Here we discuss such topics as magnetoelastic coupling constants, higher order magnetoelastic effects and structural transitions.

1) Magnetoelastic coupling constants. The influence of the magnetoelastic interaction on elastic constants can be demonstrated best for rare earth ions with large magnetoelastic coupling constants i.e. for $g_\Gamma = \alpha_J <r^2> G$ for large α_J, G. Therefore S-state ions with $L=0$ (Gd^{3+}, Eu^{2+}) show no observable effects (GdSb [9]). Magnetoelastic coupling constants have been studied systematically for RESb [9], REZn [10], $REAl_2$ [11], and for doped crystals of Ag (Ag:RE)[12]. In some of these cases the information was gained also from magnetostriction and parastriction experiments. In Table I we give a compilation of $g_{\Gamma 3}^{(2)}$ values for the rare earth compounds listed above as adapted from the literature. Since the sign of $g_{\Gamma 3}^{(2)}$ follows the sign of α_J, according to magnetostriction measurements we give g and not $|g|$. The G values no longer vary more than a factor of 2 for a given compound series REX because the neighborhood (bandstructure, lattice constant etc.) is very similar. The many different competing contributions to G (ligands, Coulomb- and exchange contributions of the 5d, 6s, 6p electrons etc.) make general trend predictions impossible at present.[3]

2) Higher order magnetoelastic coupling. In the previous section we discussed the quadrupolar magnetoelastic coupling constants. One can ask whether higher order (l=4) coupling mechanisms can be neglected. In the very first paper on elastic constants in interme-

Table I: Magnetoelastic coefficients $g^{(2)}_{\Gamma 3}$ for RESb,REZn,REAl$_2$ in (K)

	Ce^{3+}	Pr^{3+}	Nd^{3+}	Sm^{3+}	Tb^{3+}	Dy^{3+}	Ho^{3+}	Er^{3+}	Tm^{3+}
$\alpha_J \langle r^2\rangle \cdot 10^2$ (a.u.)	-8.49	-2.85	-0.80	4.49	-0.92	-0.55	-0.18	0.2	0.76
$g^{(2)}_{\Gamma 3}$(RESb)	-71	-50	-40	147		-29	-8	11	26
$g^{(2)}_{\Gamma 3}$(REZn)					11.1	10.4	3.6	-9	-27
$\lvert g^{(2)}_{\Gamma 3}\rvert$(REAl$_2$)	71	10.5	7.0						15.2

tallic rare earth compounds [13] one has considered this question for the c_{44} mode in PrSb. Similarly as for the crystal field coefficients, where the B_6 term plays a non negligible role compared to the B_4 term, one might expect $g^{(4)}_{\Gamma 3}$, $g^{(4)}_{\Gamma 5}$ to make some contribution for the anomalous temperature dependence of the elastic constants. We shall briefly discuss here two cases where these higher order contributions have been observed: The $c_{11}-c_{12}$ modes for $PrPb_3$ [14] and the c_{44} mode for PrSb [13]. In Fig. 1 we give experimental results for the $c_{11}-c_{12}$ and c_{44} modes for these substances. For $PrPb_3$ a minimum in c_{44} at T=6K can be well explained with a quadrupolar magnetoelastic coupling. However, the minimum in $c_{11}-c_{12}$ at T=8K can only be accounted for using higher order coupling constants. It is interesting to note that the ratio $\delta=g^{(4)}/g^{(2)}= 1/135$ is close to the one calculated from a point charge model. For PrSb the unsolved

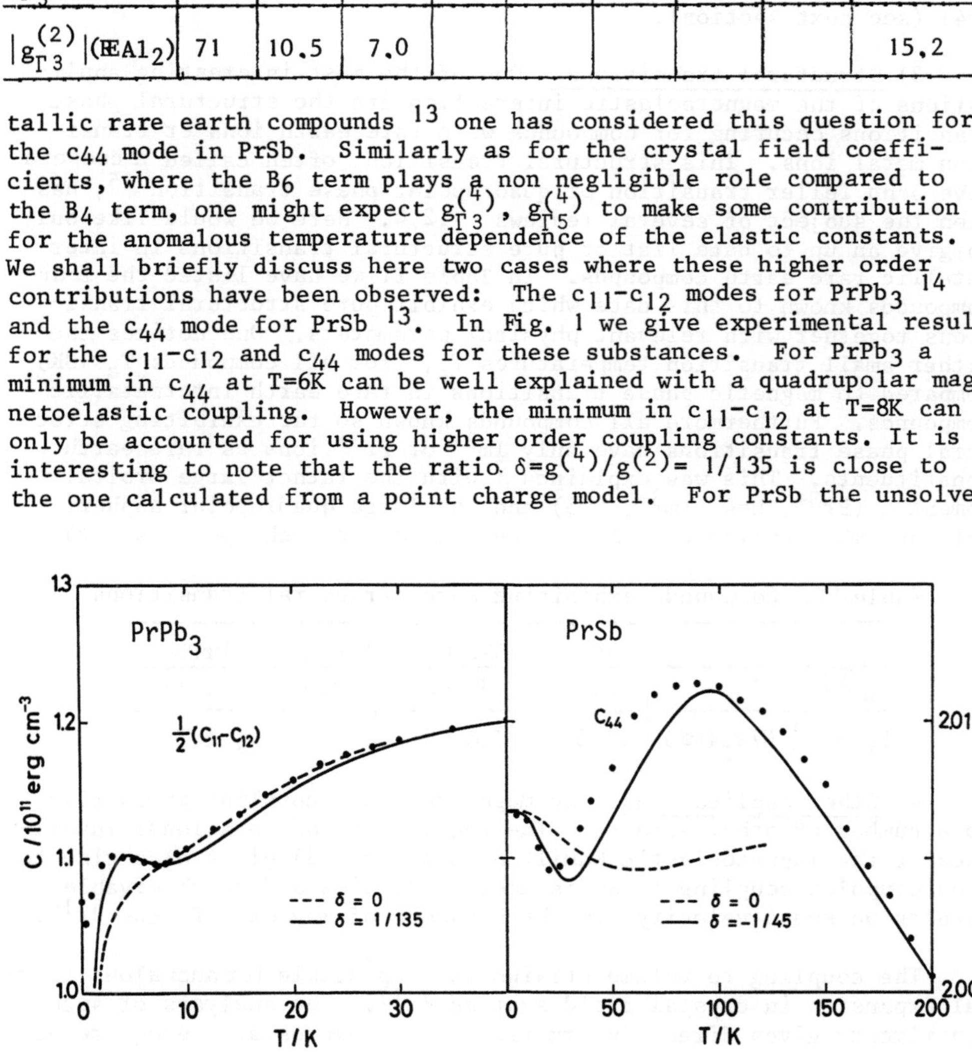

Fig. 1 Elastic constants $(c_{11}-c_{12})/2$ for $PrPb_3$ and c_{44} for PrSb. Dotted line $\chi_s(O_2^2)$, full line $\chi_s(O_2^2 + \delta O_4^2)$.

problem was the occurence of a minimum at 25K for the C_{44}-mode, whereas the quadrupolar coupling gave a minimum at 60K[13]. Again calculating the strain susceptibility including l=4 terms i.e. calculating $\chi_s(O_2^2 + \delta O_4^2)$ gives a satisfactory fit for $\delta = -1/45$.

These two cases are the first examples where higher order (l=4) magnetoelastic contributions show up in a clear-cut way in the temperature dependence of elastic constants. Usually higher order contributions do not give distinct temperature dependences for c_Γ. The reason why these contributions show up so far only for Pr-compounds is probably due to the large orbital moment for the Pr^{3+}-ion (L=5, J=4) (see next section).

3) Structural transitions. One of the most interesting applications of the magnetoelastic interaction are the structural phase transitions occuring for compounds with rare earth ions or transition metal ions. This structural transition, often called a cooperative Jahn Teller transition or quadrupolar phase transition [15], has been the subject of several reviews [1,2,6]. Here we would like only to give an up to date list of pure structural transitions in intermetallic rare earth compounds. In Table II we have listed the four compounds known to this date which exhibit pure structural transitions together with relevant physical parameters. One notices the rather small transition temperatures T_a, (for all compounds T_a<10K) compared to magnetic phase transitions in rare earth intermetallic compounds. Furthermore all compounds known so far exhibiting structural phase transitions have only Tm^{3+} or Pr^{3+} ions as rare earth constituents. This was explained [7] with the rather large orbital moment L (Pr^{3+}, L=4, Tm^{3+}, L=5) and the large quadrupolar aspherical Coulomb coupling constant (compared to the exchange constant).

Table II: Compounds exhibiting pure structural transitions

	TmCd	TmZn	$PrCu_2$	$PrPb_3$
T_a(K)	3.16	8.55	7.5	0.45
$K_\Gamma + g_\Gamma^2\ N/c_\Gamma$(mK)	11.3	25.7		6.5

4) Other applications. The magnetoelastic coupling gives rise to a number of other effects: The requirement of rotational invariance of the magnetoelastic Hamiltonian (1) or (2) gives also rise to a quadrupolar coupling to antisymmetric strains and to observable effects on sound velocity in the presence of magnetic fields [5,16].

The coupling to volume strains is responsible for anomalous thermal expansion in crystal field systems [7,17]. The analysis of such experiments gives direct information on the hydrostatic pressure dependence of CEF energy levels, the Grüneisen parameters.

Ultrasonic attenuation experiments for rare earth ions in metallic environment gives valuable dynamical information[18]. The strong ultrasonic attenuation near structural quadrupolar induced phase-transitions allows one to study the dynamics of quadrupolar excitations, in particular its relaxation. Such experiments have been performed for TmZn, TmCd.

The large magnetoelastic coupling constants observed in such systems (see Table I) also allow magnetoacoustic birefringence experiments to be performed [7,19]. The magnetoacoustic analog of the Cotton-Mouton-Voigt effect and of the Faraday effect have been measured in the paramagnetic region of $CeAl_2$. Similar magnetoelastic effects as discussed for 4f-systems have also been observed for a 3d-system ($FeCl_2$ [20]) and a 5f-system (UPd_3 [14]).

b) The Deformation Potential Coupling

From equation (4) it follows that one can observe a deformation potential effect if the deformation potential coupling and the density of states at the Fermi level $d_\Gamma^2 N_o$ is large. Indeed temperature dependent acoustic anomalies involving a deformation potential coupling have been observed in Pd and its alloys [21]. The most important case are the A-15 compounds [2]. For intermetallic rare earth compounds with rather narrow 5d-bands one would expect this effect to occur frequently. In fact it has not been seen for RE-pnictides Refs. 7,9. On the other hand it has been observed for some CsCl-type structure materials, such as LaAg, $LaAg_{1-x}In_x$ [22]. Without taking recourse to detailed band structure calculations and deformation potential calculations, electronic specific heat measurements ents clearly show the different density of states for the various groups of materials mentioned above. In Table III one notices that the density of states for RE pnictides is an order of magnitude smaller than the one for say LaAg or La_3Se_4. Therefore only these latter systems can show this effect. In Fig. 2 we give results for the temperature dependence for the $(c_{11}-c_{12})/2$-mode for LaAg, $RE_xLa_{1-x}Ag$. The curves show a fit to the experiment using Eq. (4). As seen in Fig. 2 and as discussed before [22] the T-dependence for the case of LaAg can be explained very well with Eq. (4). For the case of CeLaAg one finds a combined effect of magnetoelastic coupling and deformation potential coupling. For the other systems Pr, Tm the magnetoelastic effect is much smaller because of the small α_J. The case of CeLaAg is the only case so far where one has seen both the magnetoelastic and deformation potential effects together.

Table III: Density of states N_o

Material	LuN	LaS	LaAg	La_3Se_4
γ (mJ/K^{2}Mol)	0.6	3.28	7.7	6.3
N_o (eV atom)$^{-1}$	0.13	1.39	1.65	0.9

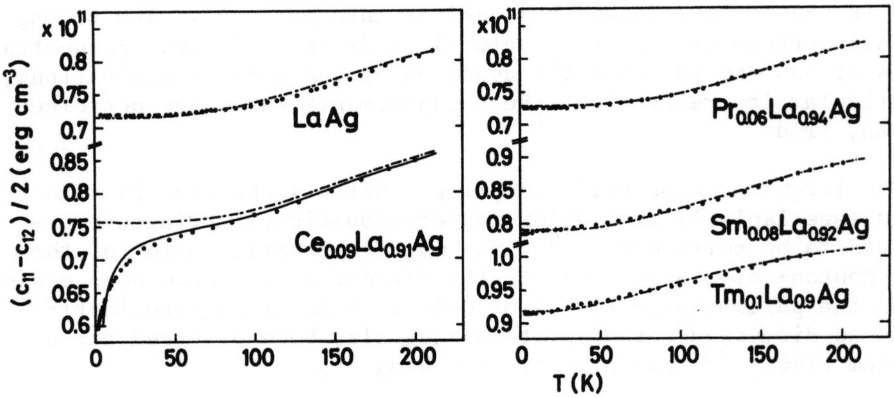

Fig. 2 $(c_{11}-c_{12})/2$ mode for $RE_xLa_{1-x}Ag$
dotted line: fit with Eq. (4), full line: fit with Eqs. (3) and (4)

The other cases LaAgIn and La_3Se_4, La_3S_4, CeAg are less clear because anharmonic effects described below can play an important role.

c) Anharmonic Interactions

Anharmonic interactions generally contribute to c_Γ^o. In some cases it can lead to a strong temperature dependence of some symmetry elastic constants. A rare earth system which was analysed in detail is $LaAg_xIn_{1-x}$ [22]. It was found that the deformation potential coupling gives an observable contribution, but that it is not strong enough to induce a structural phase transition. Phonon spectra measurements revealed a soft acoustic phonon mode at the M-point of the Brillouin zone of the CsCl structure. Anharmonic coupling of the M-point with the Γ_3-type macroscopic strain then gives rise to a pronounced temperature dependence. It was found empirically that the $(c_{11}-c_{12})/2$-mode follows a $(T-T_c)^{-1}$ T-dependence over more than 300K for $LaAg_{.78}In_{.22}$. As pointed out in the theory section it is not straightforward to calculate $c_\Gamma(T)$. For smaller In-concentrations one observed both deformation potential- and anharmonic coupling together.

Another case where strong temperature dependent anharmonic interactions have been observed [23] is CeAg which exhibits a structural transition at 15K and a magnetic transition at 7K. Again the experimental results of $(c_{11}-c_{12})$ could empirically be fitted to $c_\Gamma = c_\Gamma^o - A(T - T_c)^{-1}$. This result has to be compared to the results for CeLaAg presented above where the magnetoelastic and deformation potential contributions dominate. Phonon spectra of CeAg have not yet been measured.

Even less clear are the cases of La_3Se_4, La_3S_4 [24]. All these compounds exhibit structural transitions at rather high temperatures (60, 80K). It is not clear yet whether these transitions are induced by soft phonons (anharmonic interactions) or deformation potential coupling (band Jahn Teller effect). Inelastic neutron scattering indicate considerable softening of phonons at the B.Z.in a(110)-direction [25] for La_3Se_4. However, due to the many possible modes, a quantitative analysis has not been made.

OUTLOOK

In the previous sections we have shown how the strain modes can couple to magnetic ions and to zone boundary phonons in the case of intermetallic rare earth compounds. These different mechanisms explain the elastic constant for a wide variety of intermetallic rare earth compounds. Now we would like to mention some new recent developments.

We have dealt so far with compounds whose rare earth constituents have not unstable moments (Kondo ions, Kondo lattices and mixed valence compounds). Elastic constant measurements of a number of such compounds show that the usual magnetoelastic interaction is only operative at high temperatures (if at all). Typical examples of such compounds are $CeAl_3$, $CeSn_3$, $CePd_3$, TmSe, SmYS. In all these cases no CEF-effects were found (in $CeAl_3$ a CEF-effect at higher T was observed [26]),but an anomaly in the bulk modulus c_B was observed or deduced ($CePd_3$ [27], $CeSn_3$ [27], TmSe [28], SmYS [29]). This anomaly can be described phenomenologically with a Grüneisen parameter giving the volume dependence of the fluctuation temperature T_o $\Omega = \delta \ln T_o/\delta \ln V$. The temperature and magnetic field dependence of the various thermodynamic quantities can be described in this way [27].

We have shown how the magnetoelastic interaction gives rise to an anomalous temperature dependence of the elastic constant through a thermal population of CEF-levels. Recently the elastic constant change through laser pumping of electronic energy levels was investigated [30]. Through pumping of the R-lines in ruby using a ruby laser a sound velocity change of ca. 1 part in 10^6 was observed, which was in reasonable agreement with a calculation similar to the thermodynamic one used in deriving Eq. (3). A thermodynamic type of calculation is correct as long as the relaxation time τ between electronic energy levels is short compared to the sound wave period. Otherwise dispersive effects enter which can be accounted for: $c_\Gamma = c_\Gamma^o - g_\Gamma^2 N \chi_s^\Gamma (1 + \omega^2 \tau^2)^{-1}$. Here χ_s is the thermodynamic strain susceptibility and ω is the circular sound wave frequency. It is hoped that this technique can be applied to study a variety of problems (e.g. a study of relaxation times τ or structural transitions).

REFERENCES

1. H. Thomas, in:"Electron-Phonon Interactions and Phase Transitions", ed. by T. Riste, Plenum Press, New York (1977); G.A. Gehring, in:"Structural Phase Transitions", ed. by K.A. Müller and H. Thomas, Springer Verlag, to be published.
2. B. Lüthi and W. Rehwald, in:"Topics in Current Physics", Vol.23, ed. by K.A. Müller and H. Thomas, Springer Verlag, Berlin (1981).
3. P. Morin and D. Schmitt, Phys. Rev. B23, 2278 (1981).
4. E.R. Callen and H.B. Callen, Phys. Rev. 129, 578 (1963).
5. P. Fulde, in:"Handbook on the Physics and Chemistry of Rare Earths", ed. by K.A. Gschneidner and L. Eyring, North-Holland Publishing Comp., Amsterdam, chapter 17 (1978).
6. G.A. Gehring and K.A. Gehring, Rep. Progr. Phys. 38, 1 (1975); R.L. Melcher, in:"Physical Acoustics", Vol. XII ed. by W.P. Mason, R.N. Thurston, Academic Press, New York (1976).
7. B. Lüthi, J. Magn. Magn. Mat. 15-18, 1 (1980); B. Lüthi, in:"Dynamical Properties of Solids", Vol.3, ed. by G.K. Horton and A.A. Maradudin, North-Holland Publishing Comp., Amsterdam (1980); B. Lüthi, A.I.P. Conf. Proc. 34, 7 (1976).
8. J.C. Slonczewski and H. Thomas, Phys. Rev. B1, 3599 (1970).
9. M.E. Mullen, B. Lüthi, P.S. Wang, E. Bucher, L.D. Longinotti, J.P. Maita and H.R. Ott, Phys. Rev. B10, 186 (1974).
10. P. Morin, D. Schmitt and E. du Trémolet de Lacheisserie, in: "Crystalline Electric Field and Structural Effects in f-Electron Systems", ed. by J.E. Crow, R.P. Guertin and T.W. Mihalisin, Plenum Press, New York, p. 61 (1980).
11. C. Lingner and B. Lüthi, to be published.
12. G. Creuzet and I.A. Campbell, Phys. Rev. B23, 3375 (1981).
13. B. Lüthi, M.E. Mullen and E. Bucher, Phys. Rev. Lett. 31, 95 (1973).
14. unpublished results from our laboratory
15. P.M. Levy, P. Morin and D. Schmitt, Phys. Rev. Lett. 42, 1417 (1979).
16. V. Dohm and P. Fulde, Z. Phys. B21, 369 (1975); P.S. Wang and B. Lüthi, Phys. Rev. B15 2718 (1977); L. Bonsall and R.L. Melcher, Phys. Rev. B14, 1128 (1976).
17. H.R. Ott and B. Lüthi, Z. Phys. B28, 141 (1977).
18. K.M. Leung, D.L. Huber and B. Lüthi, J. Appl. Phys. 50, 1831 (1979).
19. P. Thalmeier and P. Fulde, Z. Phys. B29, 299 (1978); B. Lüthi and C. Lingner, Z. Phys. B34, 157 (1979).
20. G. Gorodetsky, A. Shaulov, V. Volterra and J. Makovsky, Phys. Rev. B13, 1205 (1976).
21. J.A. Rayne, Phys. Rev. 118, 1515 (1960); E. Walker, J. Ortelli and M. Peter, Phys. Lett. 31A, 24 (1970)

22. W. Assmus, R. Takke, R. Sommer and B. Lüthi, J. Phys. C11, L575, Corr. L793 (1978);
K. Knorr, B. Renker, W. Assmus, B. Lüthi and H.J. Lauter, Z. Phys. B39, 151 (198).
23. R. Takke, N. Dolezal, W. Assmus and B. Lüthi, J. Magn. Magn. Mat. 23, 247 (1981).
24. E. Bucher, J.P. Maita, G.W. Hull, L.D. Longinotti, B. Lüthi, P.S. Wang, Z. Phys. B25, 41 (1976);
P.J. Ford, W.A. Lambson, A.J. Miller, G.A. Saunders, H. Bach and S. Methfessel, J. Phys. C13, L697 (1980).
25. A. Loidl, private communication
26. M. Niksch, B. Lüthi and K. Andres, Phys. Rev. B22, 5774 (1980).
27. R. Takke, M. Niksch, W. Assmus, B. Lüthi, R. Pott, R. Schefzyk and D.K. Wohlleben, to be published.
28. B. Batlogg, H.R. Ott, E. Kaldis and P. Wachter, Phys. Rev. B19, 247 (1979).
29. T. Penney, R.L. Melcher, F. Holtzberg and G. Güntherodt, A.I.P. Conf. Proc. 29, 392 (1975).
30. W. Grill and B. Lüthi, Phys. Rev. B24, September (1981).

COMMENTS

FRANSE: What is the reason for the small thermal effect at 77 K in comparison with that at 300 K?

LÜTHI: The thermal relaxation (from specific heat and thermal conductivity) explains this difference in ruby very well.

GREIDANUS: Has a magnetic phase transition also been observed in $PrPb_3$?

LÜTHI: Down to 10 mK no magnetic transition was observed.

QUADRUPOLE INTERACTION AT ^{169}Tm IN CUBIC TmZn

G.A. Stewart

Institut für Atom- und Festkörperphysik
Freie Universität Berlin

INTRODUCTION

Cubic TmZn is of special interest because it undergoes a pure quadrupolar structural phase transition at T_Q=8.55 K closely followed by a ferromagnetic transition at T_C=8.12 K[1,2]. The tetragonal strain induced below T_Q by the ordered quadrupoles is preceeded at higher temperatures by a strong softening of the $C_{11}-C_{12}$ elastic mode. Because the quadrupole splitting of the ^{169}Tm, 8.4 keV Mössbauer resonance is an extremely sensitive probe of the local 4f quadrupole moment[3], it was decided to apply this microscopic technique to TmZn in the vicinity of T_Q.

EXPERIMENTAL DETAILS

The two polycrystalline samples were induction melted under 400 mbar argon pressure in sealed molybdenum crucibles. A temperature of 1200 C was maintained for 4 hours to ensure homogeneous reaction. One sample was quenched from 1000 C while the second was tempered for 1 week at 500 C. After their respective heat treatments, both samples were removed from their crucibles and stored in a dry argon atmosphere to avoid oxidation. X-ray powder diffraction analysis revealed only the cubic CsCl structure and excellent agreement with the value of a_0=3.515 Å given in the literature[4]. For the Mössbauer measurements, ∿10 mg/cm^2 TmZn powder absorbers were cooled in a transmission cryostat. The ^{169}Er:^{168}Er$_{1.5}$Al$_{98.5}$ source was moved sinusoidally and mounted outside at room temperature. At room temperature both samples gave a single line ^{169}Tm Mössbauer spectrum as expected for cubic symmetry. However, after the absorbers had been left exposed to air at room temperature for several weeks, additional broader lines were observed. Therefore, new absorbers were prepared and built directly into the cryostat for each series of measurements.

RESULTS

The ^{169}Tm Mössbauer spectra below 8 K showed the expected magnetic splitting although complicated by severe relaxation broadening. However, in the cubic phase above 8.55 K, the spectra appeared as typical quadrupole-split doublets (see Fig. 1) for both samples. There was no evidence for a central single line in addition to the doublet. Simple fits employing 2 Lorentzian lines were carried out to obtain a quadrupole splitting ΔE_Q, which is given in Table 1 as a function of temperature.

DISCUSSION

Although it follows from symmetry arguments that the electric field gradient (EFG) at a cubic site is zero, the ^{169}Tm Mössbauer spectra for cubic TmZn appear to be split by a quadrupole interaction. Possible explanations for this anomalous behaviour are as follows:

1. Full Hyperfine Interaction. The cubic crystal field (CF) parameters for Tm^{3+} in TmZn are well established from neutron spectroscopy[5]. The first excited 4f electron state (the Γ_3 doublet) lies 36 K above the $\Gamma_5^{(1)}$ triplet ground state. Therefore, at low temperatures, the full hyperfine (hf) interaction between the triplet ground state and the nuclear states can be considered. Ignoring the quadrupole interaction, the magnetic hf interaction can be described as $A_I^{\prime}\underline{I}\cdot\underline{\tilde{S}}$ with pseudospin $\tilde{S}=1$ (triplet) and $A_I^{\prime}=\langle\tilde{1}|J_z|\tilde{1}\rangle\simeq -2.7\ A_I$. A_I is the usual hf coupling constant for the nuclear level having spin I. Setting $F=I+\tilde{S}$, the excited (I=3/2) nuclear level couples with the triplet ground state pseudospin to give $F_e=1/2$, 3/2, and 5/2 and the ground (I=1/2) nuclear level couples to give $F_g=1/2$ and 3/2. A simple calculation places the 6 Mössbauer transitions at −253.8,

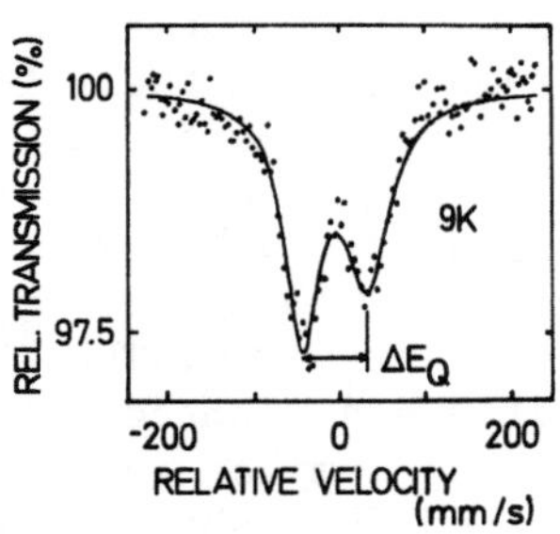

Fig. 1: ^{169}Tm Mössbauer spectrum for TmZn at 9 K.

-17.6, +37.9, +213.0, +274.1, and 449.2 mm/s. Inclusion of the full quadrupole hf interaction[7] and electronic relaxation (which would smear out the lines at larger velocities) could lead to a spectrum which appears to be quadrupole-split. However, Tm^{3+} is a non-Kramers ion so that small crystal distortions are sufficient to remove the CF level degeneracies. This will further complicate a full hyperfine interaction treatment (the analogous situation for a Γ_8 quartet has been discussed by Nowik et al.[6]).

2. Local Tetragonal Distortion. Another possibility is that sufficiently large local tetragonal distortion induces an EFG at the Tm nucleus via perturbation of the 4f shell. The local distortions may flip between the "soft" cube axes in a non-cooperative manner but at a rate slow in comparison with the Mössbauer measurement time of about 10^{-9} s. In order to obtain a feel for the distortion required, a term $B_2^oO_2^o$ can be added to the cubic CF Hamiltonian. The B_2^o values listed in Table 1 are those values necessary to generate the observed ΔE_Q values. The low temperature values correspond to a significant rearrangement of the CF level scheme but inclusion of the higher order distortion contributions to the CF Hamiltonian may remedy this.

Table 1. Quadrupole splitting ΔE_Q of the ^{169}Tm Mössbauer resonance in TmZn. The B_2^o values are those required to generate ΔE_Q if a $B_2^oO_2^o$ term is added to the cubic CF Hamiltonian.

T[K]	ΔE_Q[mm/s]	B_2^o[K]
9	75.9(1.4)	-1.04(2)
10	77.2(1.4)	-1.07(2)
17	71.4(1.1)	-1.10(2)
40	48.3(1.7)	-0.93(4)
80	28.1(1.0)	-0.85(6)
300	0	0

3. Non Cubic Phases. A final possibility is that the samples are of a single non-cubic phase which is difficult to distinguish from the cubic phase by means of X-ray powder diffraction. It has been suggested that such distortion could result from the grinding procedure which is necessary to produce a "thin" absorber[8]. In order to eliminate this possibility, measurements are planned using slices of single crystal TmZn polished to the appropriate thickness.

CONCLUSIONS

The ^{169}Tm Mössbauer investigation of TmZn suggests that a local EFG persists at the Tm site well into the cubic phase. Further measurements and theoretical calculations are necessary before definite conclusions can be made concerning the origin of this effect.

ACKNOWLEDGEMENTS

This work was supported by the Deutsche Forschungsgemeinschaft Sfb-161 and the Kernforschungszentrum Karlsruhe GmbH.

REFERENCES

1. B. Lüthi, R. Sommer, and P. Morin, J. Magn. Magn. Mater. 13,198 (1979).
2. P.M. Levy, P. Morin, and D. Schmitt, Phys. Rev. Lett. 42,1427 (1979).
3. R.G. Barnes, R.L. Mössbauer, E. Kankeleit, and J.M. Poindexter, Phys. Rev. A 136,175 (1964).
4. A. Iandelli and A. Palenzona, "Crystal Chemistry of Intermetallic Compounds", in: "Handbook on the Physics and Chemistry of Rare Earths", K.A. Gschneider Jr. and L. Eyring, eds., North-Holland (1979).
5. P. Morin, J. Rouchy, and D. Schmitt, Phys. Rev. B 17,3684 (1978).
6. I. Nowik, I. Felner, and R. Yanovsky, J. de Physique 37, C6-431 (1976).
7. A. Abragam and B. Bleaney, "Electron Paramagnetic Resonance of Transition Ions", p. 684, Clarendon Press, Oxford (1970).
8. R. Day, Private Communication.

COMMENTS

LÜTHI: One probably has a dynamic Jahn-Teller effect of the Γ_5 ground state where one has a tunneling between 3 positions. The tunneling frequency is higher than any quasi-static experiment (lattice constant, elastic constant) but smaller than the typical Mössbauer frequency.

WOHLLEBEN: Has the presence of the dynamic splittings above 8 K been detected in diffuse x-ray scattering? How large do you estimate the distortions?

STEWART: To the first question, no. To the second: I have considered the addition of B_2^0 to the cubic CF Hamiltonian and computed how large it should be to give the appropriate quadrupole splittings at various temperatures. It is about 1 K at low temperature which causes quite a rearrangement of the cubic CF levels. However, Professor Lüthi notes that a B_2^2 term is also required.

FIELD DEPENDENCE OF THE MAGNETIC ANISOTROPY OF GADOLINIUM AT 4.2 K UNDER HIGH PRESSURES

J.J.M. Franse, R. Gersdorf and E. Koops

Natuurkundig Laboratorium der Universiteit van Amsterdam
Valckenierstraat 65, 1018 XE Amsterdam
The Netherlands

INTRODUCTION

In a previous publication (Franse and Gersdorf, 1980) we reported the anomalous behaviour of the torque, due to magnetic anisotropy, of a gadolinium single crystal in the (c,a) plane at low temperatures. The anomalies were ascribed to a peak in the density of states curve that moves through the Fermi level upon rotating the magnetization vector with respect to the hexagonal axis. By studying the pressure dependence of the magnetic torque at 4.2 K some of the peak parameters could be determined. From preliminary experiments under high pressures at 25 K the width in energy of this peak was estimated and subsequently the spin orbit interaction parameter and the number of electron states contained in the peak (about 5×10^{-4} electrons/atom). In order to understand this anomaly further we studied the field dependence of the magnetic torque at 4.2 K under high pressures with magnetic field intensities of 1.8 T and 4.8 T.

EXPERIMENTAL

A new torque magnetometer was designed with a split-coil superconducting magnet of 5 T. The magnet can be rotated in the liquid helium bath around the sample holder with steps of 0.2 degree. The sample holder consists of a beryllium-copper high pressure vessel that is connected to the pressure generating unit by a capillary tube of low torsion rigidity (diameter 1 mm, length 1300 mm). The magnetic torque is determined by means of a second torsion wire of adjustable torsion rigidity connected directly to the high pressure vessel. The maximum torque corresponds in most cases with an angle of approximately 300 seconds of arc, which angle can be read within 0.5" by optical means.

At low temperatures helium gas was frozen at constant pressures in order to achieve hydrostatic conditions. Pressures were read by means of strain gauges mounted on the pressure vessel. The experiments were performed on the same single crystalline sphere of 2.8 mm diameter that was used in previous measurements.

RESULTS AND DISCUSSION

A few experimental torque curves are presented in Fig. 1, where the magnetic torque is plotted as a function of the angle θ between the magnetization vector and the c-axis. These experimental data can be represented fairly well with the model introduced in Ref. 1. In this model the anisotropy energy of gadolinium is assumed to consist of two parts: a regular part that contains only a few anisotropy constants, and an anomalous part that is connected with a small peak in the density of states curve near the Fermi level. To illustrate the proposed model, in Ref. 1 the situation was chosen in which the relevant peak belongs to the minority spin electron band with the lowest position in energy of the peak for $\theta = 0$. In the present paper we shall develop a more general description in which the field dependence of the anomalous contribution can also be discussed. In this description we suppose that both the majority (↑) and the minority (↓) subbands have a small peak in the density of states curve. With respect to the Fermi level the energy of the peaks is assumed to vary slightly as a function of the direction of magnetization θ, the pressure P and the magnetic field intensity B. The energy difference between the center of the peak and the Fermi level is denoted for the ↑ subband as

$$\varepsilon_{\uparrow} = -Q\,\{\cos 2\theta - C_{\uparrow} - \lambda P\} - \mu_B B, \tag{1a}$$

whereas for the other subband

$$\varepsilon_{\downarrow} = Q\,\{\cos 2\theta + C_{\downarrow} + \lambda P\} + \mu_B B. \tag{1b}$$

Q is a spin-orbit parameter, $C_{\uparrow}$ and $C_{\downarrow}$ determine the positions of the peak at $\theta = \pi/4$, $P = 0$ and $B = 0$; $Q\lambda$ is the shift of the peak with pressure and μ_B is the Bohr magneton. If both peaks are completely filled or completely empty no contributions from these peaks to the anisotropy energy arise. In case of one peak completely filled with n electrons and the other peak empty the anisotropy energy is enlarged with $-nQ\cos 2\theta$. In most cases this additional term can not be distinguished from other contributions to the anisotropy energy. From the anomalous behaviour of the anisotropy energy of gadolinium at low temperatures we conclude, however, that one of the peaks is very close to the Fermi level. The experiments even show that this peak crosses the Fermi level upon rotation of the magnetization vector with respect to the c-axis. In that case we write

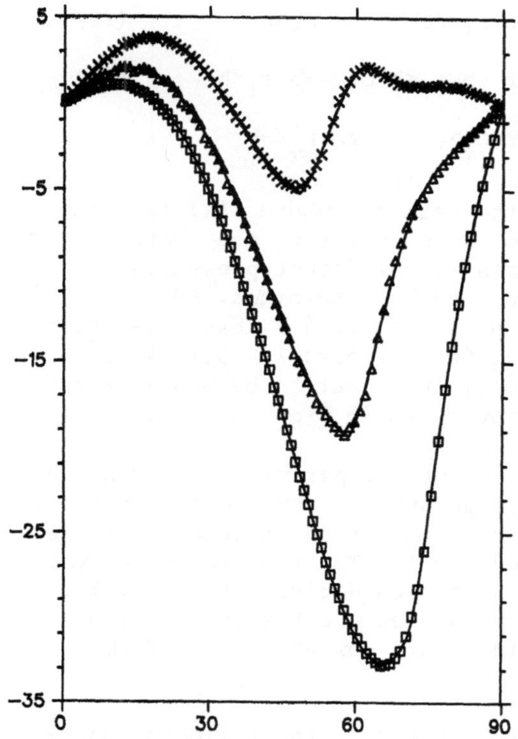

Fig. 1: The magnetic torque of a gadolinium single crystal in the (c,a) plane at 4.2 K for different pressures at an external magnetic field of 1.8 T. Squares are 1 bar, triangles are 3.5 kbar, and x's are 6.52 kbar. The full curves represent a description with the model developed in Ref. 1. The small arrows mark the transition region in which Eq. (2b) is applied. Horizontal axis is angle between magnetization vector and c-direction; vertical axis is magnetic torque (J/kg).

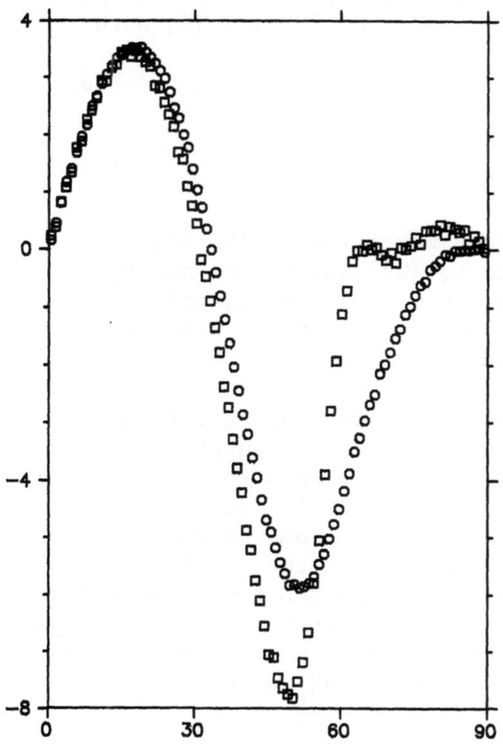

Fig. 2: The magnetic torque of a gadolinium single crystal in the (c,a) plane at 5.95 kbar for different temperatures, B=1.8 T. Squares are 4.2 K and circles are 25 K. Axes are described in Fig. 1 caption.

$$E'_A = n\varepsilon_\uparrow \qquad \text{for } \varepsilon_\uparrow < -b \quad \text{and } \varepsilon_\downarrow > b \tag{2a}$$

$$E'_A = -nbf(\varepsilon_\uparrow/b) \quad \text{for } -b<\varepsilon_\uparrow<b \text{ or } -b<\varepsilon_\downarrow<b \tag{2b}$$

$$E'_A = 0 \qquad \text{for } \varepsilon_\uparrow > b \quad \text{or } \varepsilon_\downarrow < -b \tag{2c}$$

where 2b is the width in energy of the peak and where $f(x)$ (defined for $-1 < x < 1$) is a smooth function of x representing the shape of the peak. By fitting the experimental data for different pressures at 4.2 K the parameters λ, $n|Q|$ and $b/|Q|$ can be determined. Additional experiments at 4.2 K, B = 1.8 T and pressures of 4.7 kbar, 5.95 kbar and 6.52 kbar confirm the results obtained before: $\lambda = 0.10\ \text{kbar}^{-1}$; $b/|Q| = 0.22 \pm 0.04$. The parameter $n|Q|$ turns out to be nearly pressure independent above 4 kbar and amounts to (5.1 ± 0.5) J/kg.

To determine the individual values of the parameters n, Q and b an additional extensive parameter (temperature or external magnetic field) has to be varied. The effect of a variation in temperature is shown in Fig. 2 for P = 5.95 kbar and B = 1.8 T. From similar curves approximate values for n, $|Q|$ and b have been derived (see Ref. 1). By carrying out experiments under pressure at 4.2 K with different values of the external magnetic field we are now able to confirm these estimates.

In analysing the experimental torque curves the following parameters are determined: S_2, S_4, S_6, S_8, 2nQ, $\cos 2\theta_o$ and b/Q with

$$\cos 2\theta_o = C_\uparrow + \lambda P - \mu_B B/Q \tag{3a}$$

or

$$\cos 2\theta_o = -C_\downarrow - \lambda P - \mu_B B/Q \tag{3b}$$

depending on the $\uparrow$ or $\downarrow$ peak crossing the Fermi level. The angle θ_o denotes the transition from Eq. (2a) to Eq. (2c); S_2, S_4, S_6 and S_8 represent the regular part of the anisotropy energy. Experimental results for two different values of the external field are shown in Table I.

The accuracy of the data decreases with increasing B value due to certain problems in the rotation of the magnet at higher field strengths. We note that all parameters for the two different B values are equal within the error limits with the possible exception of the parameter $\cos 2\theta_o$. By applying Eq. 3 we derive from the data of Table I a value for Q of about 12 meV, in agreement with the previous estimate of 14 meV for the absolute value of Q from the temperature variation of the magnetic torque curves. The number of electrons per atom contained in this peak at pressures above 4 kbar amounts to 7×10^{-4} as calculated with the above given numbers for nQ and Q.

Due to the large uncertainty in the experimental data at 4.8 T the error in the present Q value is relatively large. Further improvements of the experimental setup will certainly reduce these inaccuracies. Nevertheless we have shown that the model introduced in Ref. 1 is able to describe also the field dependence of the magnetic torque in gadolinium. Moreover we succeeded in determining the sign of the parameter Q which was not possible from measurements at one fixed value of B. From this sign follows that the anomalous contribution to the anisotropy is present at $\theta = 0$ (case 2a), whereas at $\theta = \pi/2$ it is absent at pressures above 3 kbar (case 2c). Values for the coefficient S_2 of Ref. 1 and Table I contain therefore a contribution from the anomalous part and should be corrected with the corresponding values for 2nQ. The S_2 value of the regular part in Table I is in fact equal to 1.3 J/kg.

Field dependent experiments on the magnetic anisotropy energy of gadolinium, however, can not answer the question which the two subbands is the peak crossing the Fermi Level.

Table I. Analysis of the magnetic torque data in the (c,a) plane of Gd at 4.2 K under a hydrostatic pressure of 4.7 kbar for different values of the external magnetic field.

	1.8 T	4.8 T
S_2 (J/kg)	-9.34 ± 0.06	-9.42 ± 0.24
S_4 (J/kg)	6.69 ± 0.09	6.74 ± 0.34
S_6 (J/kg)	1.74 ± 0.07	1.96 ± 0.26
S_8 (J/kg)	-1.00 ± 0.03	-1.16 ± 0.11
2nQ (J/kg)	10.6 ± 0.35	10.5 ± 1.4
$\cos 2\theta_o$	-0.539± 0.002	-0.553± 0.008
b/Q	0.181± 0.004	0.180± 0.02

REFERENCES

1. J.J.M. Franse and R. Gersdorf, Phys. Rev. Letters 45, 50 (1980).

COMMENTS

McEWEN: Presumably these effects could not be observed in the other heavy rare earth metals because they are much less than the 4f magnetic anisotropy.

FRANSE: These effects could equally be present in other heavy rare earth metals. They will be observed, however, by the anisotropy of the 4f moments. Nevertheless, since these anomalous effects dominate the anisotropy of Gd, there is some chance to observe them.

PIERRE: It seems that there is a close analogy with the anisotropy of Ni, where many high order terms were observed at low temperatures by Aubert and Escudier. This was also attributed to the existence of a small pocket of the Fermi surface.

FRANSE: The present model has much resemblance to the description by Gersdorf of Ni. The anomalous term in Ni, however, contributed to the anisotropy energy for only a few percent. In the case of Gd, this contribution is much larger. The nature of the anomalous term is also different for Gd. For Ni a small hole pocket near the X point in the Brillouin zone was found to be responsible for the anomalous effect. In Gd it is a peak in the density of states curve near E_F that is very sensitive to variations of θ, ρ and H.

KASUYA: Is it possible to fit the temperature dependence of the anisotropy as a function of the 4f magnetic moment? Is it possible to make an experiment on La and Y diluted systems?

FRANSE: Up to now we only studied the temperature dependence of the anomalous contribution to the anisotropy in the temperature interval $4.2\ K<<T<25\ K$. This temperature dependence is not related to the variation with temperature of the 4f moment. These experiments are indeed planned. In these diluted alloys we expect a decrease of the 5d electron polarization by the 4f moments. The anomalous contribution to the anisotropy is probably very sensitive to this decrease.

MAGNETOSTRICTION OF AN YTTRIUM MONOCRYSTAL DOPED WITH TERBIUM IMPURITIES

P. Pureur, G. Creuzet and A. Fert

Laboratoire de Physique des Solides
Université de Paris Sud
91405 Orsay France

INTRODUCTION

The interest of magnetostriction experiments on dilute rare-earth alloys to understand the microscopic magnetostriction mechanisms has been demonstrated by several recent publications[1,2,3]. The direct information on the single ion magnetoelastic couplings given by such experiments should be useful to probe the theoretical models of the ion-lattice couplings and also to separate the single and two ion effects in the magnetostriction of the concentrated rare earth systems. In this communication, we report on magnetostriction measurements on an yttrium single crystal doped with 2 at % of terbium impurities. We have measured the magnetostriction in the helium temperature range, along the a, b and c axes and for several orientations of the applied magnetic field. We estimate the magnetoelastic coefficients $\lambda_1^{\alpha,2}$, $\lambda_2^{\alpha,2}$ and $\lambda^{\gamma,2}$ from a preliminary analysis of our results.

EXPERIMENTAL

We prepared a single crystal of YTb 2at% by the recrystallization method described by McEwen and Touborg[4]. After orienting the crystal by the Laue diffraction method, a cubic sample of about 4 mm side, with faces perpendicular to the a, b, and c axes respectively, was cut by electroerosion. Magnetostriction measurements, along the a, b and c axes successively were made at helium range temperatures using a standard capacity bridge [1]. The magnetic field was given by superconductive coils and could be applied along the strain measurement direction (up to 70 kG) or rotated in the plane perpendicular to this direction (up to 35 kG).

RESULTS AND DISCUSSION

We show in Figure 1 the field dependence of the magnetostriction along several axes and for several field directions at 12.7 K and 19.9 K. The magnetostriction along the a and b axes is shown for applied fields along the a, b and c axes. The magnetostriction along the c axis is shown only for $\vec{H}$ along the c axis because our measurement with $\vec{H}$ in the basal plane gives some embarassing results. We observe a very small magnetostriction the variation of which as a function of the field orientation in the basal plane is not consistent with the crystalline symmetry. Also, in this case, small reorientations of the crystal with respect to the plane in which the field rotates change significantly the magnetostriction. A likely explanation of these results for $(\frac{\Delta\ell}{\ell})_c$ with $\vec{H}$ in the basal plane is that a small disorientation of the sample faces with respect to the crystal axes gives rise to a shear mode distorsion and to an additional spurious contribution. We have not applied the simple analysis explained below to our results with $\vec{H}$ in the basal plane.

The classical second order expression for the magnetostriction of an hexagonal dilute alloy can be written as[5] :

$$\frac{1}{c}\frac{\Delta\ell}{\ell} = \left[\lambda_1^{\alpha,2} Q_c + \frac{1}{2}\lambda^{\gamma,2}(Q_a - Q_b)\right]\beta_a^2 + \left[\lambda_1^{\alpha,2} Q_c - \frac{1}{2}\lambda^{\gamma,2}(Q_a - Q_b)\right]\beta_b^2 + \lambda_2^{\alpha,2} Q_c\, \beta_c^2 \qquad (1)$$

where β_a, β_b, β_c are the cosines of the angle between the strain measuring direction and the a, b, c axes respectively, and

$$Q_\alpha = \left[J_\alpha^2 - \frac{J(J+1)}{3}\right] / J(J+1)$$

for α = a, b or c.

In order to interpret our experimental results, we have computed the expectation values of Q_a, Q_b and Q_c at T = 12.7 K and T = 19.9 K for several applied fields along the a, b and c axes by using the crystal field parameters derived by Touborg[6] for the YTb system. When we introduce these expectation values in Eq.(1), the best fit with our experimental curves is obtained with the following values of the λ parameters:

$$\lambda_1^{\alpha,2} = 0 \quad (|\lambda_1^{\alpha,2}| < 2 \times 10^{-4})$$

$$\lambda_2^{\alpha,2} = 234 \times 10^{-3} \ (\pm\ 50 \times 10^{-3})$$

$$\lambda^{\gamma,2} = 8.4 \times 10^{-3} \ (\pm\ 2 \times 10^{-3}).$$

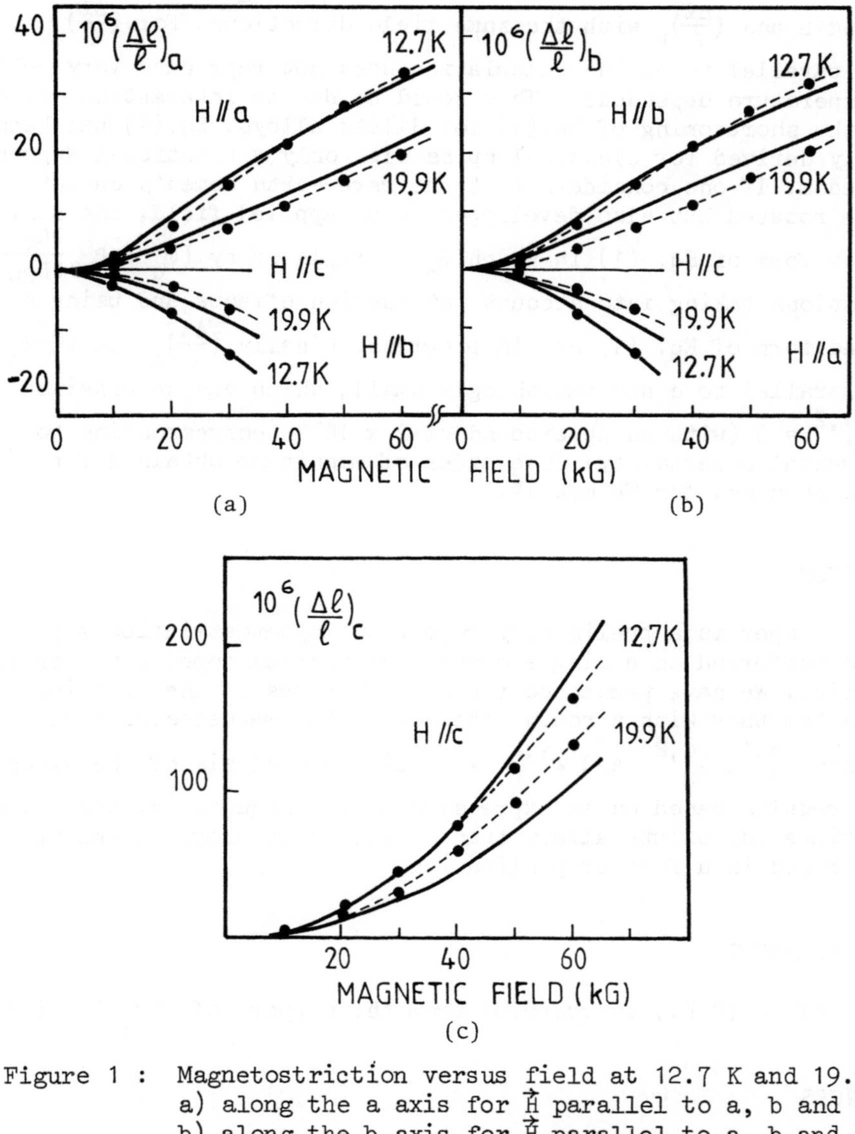

Figure 1 : Magnetostriction versus field at 12.7 K and 19.9 K :
a) along the a axis for $\vec{H}$ parallel to a, b and c,
b) along the b axis for $\vec{H}$ parallel to a, b and c,
c) along the c axis for $\vec{H}$ parallel to c.
Solid lines : experimental results.
Dashed lines : calculated curves.

The experimental and calculated curves can be compared in Figure 1. A reasonably good agreement is obtained for $(\frac{\Delta\ell}{\ell})_a$ with H parallel to a and b and $(\frac{\Delta\ell}{\ell})_b$ with the same field directions. For $(\frac{\Delta\ell}{\ell})_c$ with H parallel to c, the calculation does not reproduce very well the temperature dependence. This could be due to interaction effects or to the shortcoming of Eq.(1) for dilute alloys. Eq.(1) has been actually derived for classical spins with only a rotational degree of freedom. If one considers that the rare-earth moments cannot only be rotated but also developped by an applied field, one arrives to a new form of Eq. (1) in which Q_c is replaced by $(Q_c + B_{20}\frac{\partial Q_c}{\partial B_{20}})$. Calculations taking into account interaction effects and using a modified form of Eq. (1) are in progress. Finally $(\frac{\Delta\ell}{\ell})_a$ and $(\frac{\Delta\ell}{\ell})_b$ for $\vec{H}$ parallel to c are vanishingly small, which can be obtained with $\lambda_1^{\alpha,2} = 0$ (with an upperbound of 2×10^{-4} corresponding to the experimental uncertainty). The order of magnitude obtained for $\lambda^{\gamma,2}$ is that obtained for Tb metal[6].

CONCLUSION

This paper is a preliminary report on magnetostriction experiments performed on a single crystal of yttrium doped with terbium impurities. We have presented the main features of the experimental results together with a rough estimate of the magnetostriction constants $\lambda_1^{\alpha,1}$, $\lambda_2^{\alpha,2}$ and $\lambda^{\gamma,2}$. A detailled analysis of the experimental results based on an expression of the magnetostriction more appropriate for dilute alloys than Eq.(1) is in progress and will be described in a further publication.

ACKNOWLEDGEMENT

One of us (P.P.) is grateful from the support of CNP_q (Brazil)

REFERENCES

1. I.A. Campbell, G. Creuzet and J. Sanchez, Phys. Rev. Letters 43, 234 (1979).
   G. Creuzet and I.A. Campbell, Phys. Rev. B 23, 3375 (1981).
   I.A. Campbell and G. Creuzet, this Conference.

2. H.W. Ludwigs, H.U. Häfner, E. Holland-Moritz, W. Zell, B. Roden and D. Wohlleben, J. Magn. Magn. Mater. 15-18, 607 (1980).
H.U. Häfner, H.W. Ludwigs, K. Nicholson and D. Wohlleben, Zeitschrift für Physik B - Condensed Matter 42, 219 (1981).
3. M.H. de Jong, H. Hölscher, P. F. de Chatel, J. Bijvoet, J. Magn. Magn. Mater. 15-18, 17 (1980).
4. K.A. Mc Ewen and P. Touborg, J. Phys. F : Metal Phys. 3, 1903 (1973).
5. Eq.(1) is an extension for dilute alloys of the classical expression for ferromagnets, given for example by Rhyne [6].
6. J. Rhyne, Magnetic Properties of Rare-Earth Metals (edited by R.J. ELLIOT, Plenum 1972), 165.
7. P. Touborg, Phys. Rev. B 16, 1201 (1977).

CRYSTAL FIELD SPLITTING AND THERMAL EXPANSION IN DILUTE MAGNESIUM-RARE EARTH SINGLE CRYSTALS

M.H. de Jong, J.Bijvoet and P.F. de Châtel

Natuurkundig Laboratorium der Universiteit van Amsterdam
Valckenierstraat 65, 1018 XE Amsterdam,The Netherlands

Crystal field effects have been observed by Lüthi and Ott in the thermal expansion of both cubic rare earth compounds[1] and hexagonal rare earth compounds.[2] In the case of hexagonal compounds they were ascribed to the coupling between the strains and the quadrupolar operators of the magnetic ions only, while in the case of cubic symmetry, a coupling of the volume strain to crystal-field operators of fourth and sixth order was assumed.

We have measured the thermal expansion of single crystals of magnesium with about 0.1 at % of heavy rare earth impurities along and perpendicular to the hexagonal axis. Although the effects are small for some of the alloys, it is clear that the effects cannot be described by quadrupolar contributions only, as they do not always show the same temperature dependence along the two crystallographic axes for one particular alloy.

Crystal field parameters for dilute alloys of magnesium with the heavy rare earth metals Tb through Tm have been derived by a variety of techniques i.e. susceptibility, magnetization in high fields (up to 32 T) and inelastic neutron scattering.[3] With these data, the multipole moments of the aspherical 4f charge distribution have been calculated and their contributions to the thermal expansions have been obtained. The total energy is written as $\mathcal{H} = \mathcal{H}_{CF} + \mathcal{H}_{ME}$. In a hexagonal lattice the crytalline field energy is usually written as a sum

$$\mathcal{H}_{CF} = \sum_{\ell,m} C_{\ell m} \langle J||\theta_\ell||J\rangle\, O_\ell^m \qquad (1)$$

where only the combinations $(\ell,m) = (2,0),(4,0),(6,0)$ and $(6,6)$

contribute. For an explanation of the other symbols see Ref. 4. In the magnetoelastic part one has to consider a sum over the same operator equivalents:

$$H_{ME} = \sum_{\ell=2,4,6} \sum_{m=0,6} \sum_{\alpha} C^{\alpha}_{\ell m} \varepsilon_{\alpha} <J||\theta_{\ell}||J> O^{m}_{\ell} \qquad (2)$$

where the two ε_{α}'s stand for the symmetry strains usually denoted by ε_{v} and ε_{3}.

We have measured the dilatation along the c axis and the a axis of our samples. The measured strains are related to the symmetry strains as

$$\varepsilon_{zz} = \frac{1}{3}(\varepsilon_{v} + \varepsilon_{3})$$

and (3)

$$\varepsilon_{xx} \simeq (\varepsilon_{xx} + \varepsilon_{yy})/2 = (2\varepsilon_{v} - \varepsilon_{3})/6.$$

Lüthi and Ott have shown[2] that the equilibrium strains resulting from this magnetoelastic Hamiltonian are proportional to the expectation values of the operators involved, hence

$$\varepsilon_{\alpha}(\text{equilibrium}) = \sum_{\ell,m} K^{\alpha}_{\ell m} <J||\theta_{\ell}||J><O^{m}_{\ell}> \qquad (4)$$

from which are found the thermal expansions along the crystallographic axes

$$\frac{1}{a}\frac{da}{dT} = \text{concentration} \cdot \sum_{\ell,m} K^{a}_{\ell m} <J||\theta_{\ell}||J> \frac{d}{dT} <O^{m}_{\ell}>$$

$$\frac{1}{c}\frac{dc}{dT} = \text{concentration} \cdot \sum_{\ell,m} K^{c}_{\ell m} <J||\theta_{\ell}||J> \frac{d}{dT} <O^{m}_{\ell}>. \qquad (5)$$

The $K_{\alpha\ell m}$ depend on the magnetoelastic constants of the rare earth in $C^{\alpha\ell m}_{\ell m}$ and on the elastic constants of the matrix.

With three exceptions good fits could be obtained with four parameters, $K_{\ell m}$ free. However they varied strongly with the rare earth impurities as the contributions of the multipoles are in some cases linearly dependent. Fits with coefficients more or less constant throughout the series are shown in the figures; these coefficients $K_{\ell m}$ are given in Table 1.

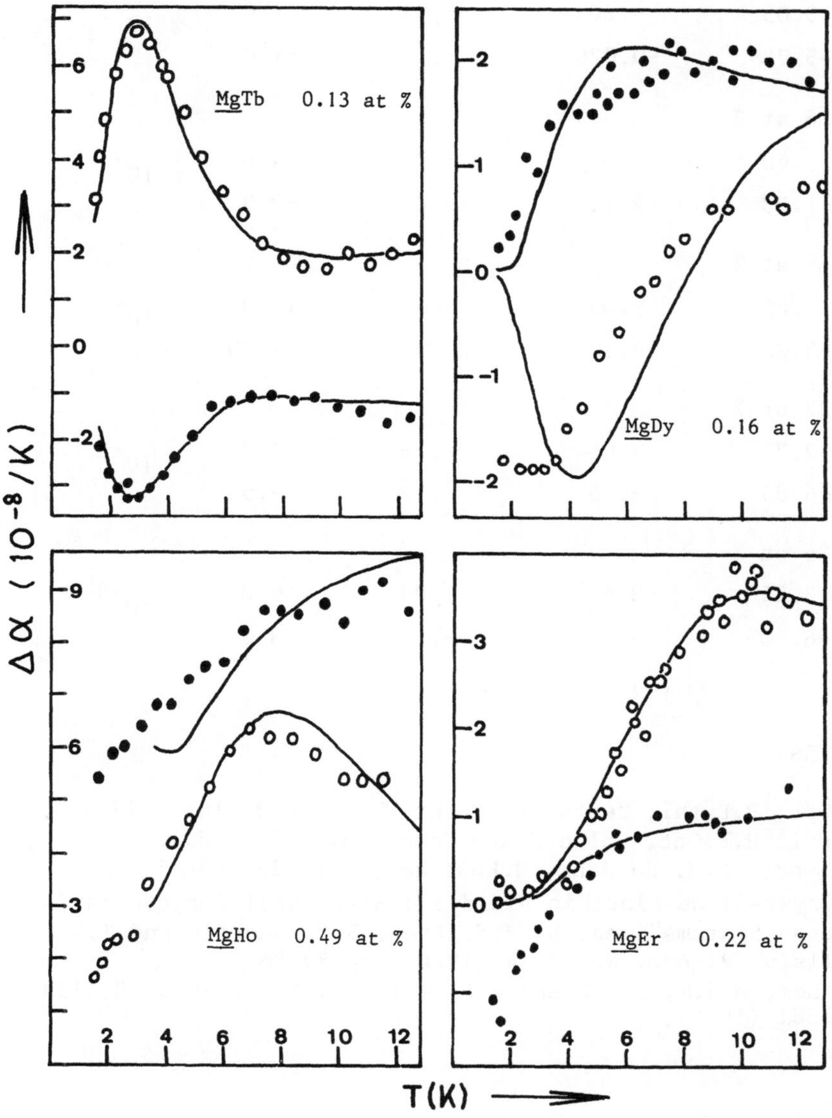

Fig. 1: Thermal expansion coefficients vs T for various Mg(R) alloys. o parallel to hexagonal axis, ● perpendicular to hexagonal axis. Coefficients of calculated curves are given in text.

Table 1

MgTb 0.13 at %				
$K^a_{\ell m}$ = 3.05	-.20	+.09	1.0	x 10^{-3}
$K^c_{\ell m}$ =-5.78	-0.026	-.10	-1.0	
MgDy 0.16 at %				
$K^a_{\ell m}$ = .60	-1.00	+.05	14.6	x 10^{-3}
$K^c_{\ell m}$ =-1.03	-2.10	+.83	-9.0	
MgHo 0.49 at %				
$K^a_{\ell m}$ = .65	-.05	+.05	10.0	x 10^{-3}
$K^c_{\ell m}$ =-3.22	-.10	-.01	-2.20	
MgEr 0.22 at %				
$K^a_{\ell m}$ = 2.77	-.01	+.18	1.0	x 10^{-3}
$K^c_{\ell m}$ =-6.85	-.14	-.10	-.5	
MgTm 0.24 at %				
$K^a_{\ell m}$ = 3.25	-2.65	+.80	6.8	x 10^{-3}
$K^c_{\ell m}$ =-6.50	-.11	-.10	-3.0	

REFERENCES

1. H.R.Ott, B.Lüthi, Zeitschrift für Physik B28, 141 (1977).
2. B.Lüthi, H.R.Ott, Solid State Comm. 33, 717 (1980).
3. J.Bijvoet, M.H. de Jong, H.Hölscher, P.F. de Châtel, in "Crystalline Electric Field and Structural Effects in f-Electron Systems", ed. by J.E. Crow, R.P. Guertin and T.W. Mihalisin, Plenum, New York (1980), pp.83-86.
4. K.R. Lea, M.J.M. Leask and W.P. Wolf, J. Chem. Phys. Solids 23, 1381 (1962).

COMMENTS

GIGNOUX: Which technique do you use to measure thermal expansion?

De JONG: The capacitance cell.

BABERSCHKE: It has been shown that approximately 1% or lower concentrations of RE in Y and Sc single crystals show interaction effects (spin glass). The solubility of RE's in Mg is obviously not better. Are your magnetostriction results influenced by interactions?

De JONG: From the susceptibility measurements we know that there is some exchange, indeed, especially for concentrations above 0.2%. Thus our results on Mg(Ho) might be affected by this. For the other alloys with lower concentrations, however, I do not expect that it will have greatly affected the results.

MAGNETOSTRICTION IN DILUTE ALLOYS OF RARE EARTHS

G. Creuzet and I.A. Campbell

Laboratoire de Physique des Solides
Université Paris-Sud, Centre d'Orsay
91405 Orsay, France

INTRODUCTION

A number of results have been obtained recently on the magnetostriction and thermal expansion of dilute alloys of rare earths in non-magnetic hosts. This data, together with EPR work on strained dilute alloy samples, can be analyzed to obtain the rare earth ion lattice strain coupling parameters and the volume dependence of the crystal fields. The results give a considerable amount of information on the interactions between an ion and its environment to supplement the standard crystal field values. Interesting comparisons can be made with magnetostriction in concentrated magnetic systems, and with data on dilute rare earths in insulators.

By the term magnetostriction, one understands the change in volume or shape of a sample under the application of a magnetic field. For systems containing localized moments, we can distinguish between two classes of effect: ion-ion interaction effects and single ion-environment effects. Here we will discuss mainly dilute systems so we will concentrate on the second class of effects, which are due to strain related crystal field terms. This is just one of the many types of problems which can be attacked using magnetostriction.

Consider a rare earth ion in a cubic environment. As the rare earth 4f shell is not spherical (except for Gd) there will be static cubic crystal field terms with parameters conventionally written C_4 and C_6.[1] A great deal of work using many different techniques has been done to determine these parameters for rare earths in various metallic hosts and in particular in noble metals. Now, if the lattice is strained in a particular direction, <001> or <111>, tetragonal or trigonal distortions will appear, giving extra crystal field terms appropriate to the new lowered symmetry[2]. The strength of these terms will be given by the amplitude of the distortion times a strain crystal field parameter. Alternatively, if the lattice is compressed uniformly, the static crystal field parameters C_4 and C_6 will change. This change is denoted by volume strain parameters.

Now, instead of straining the lattice, we can magnetize the rare earth ion by an external magnetic field. Its shape will change and the surrounding lattice will react by deforming so as to minimize the total energy. This local deformation, which leads to the macroscopic magnetostriction, is governed by the same set of strain crystal field parameters. We can write explicitly[3]

$$\left(\frac{\Delta\ell}{\ell}\right)_{<001>} = \frac{-c}{(C_{11}-C_{12})}\left[\frac{1}{2}\alpha_J V_2^3 <O_2^0> - \frac{\sqrt{5}}{16\sqrt{3}}\beta_J V_4^3 <O_4^0 - 7O_4^4> + \frac{\sqrt{14}}{64}\gamma_J V_6^3 <O_6^0 + 3O_6^4>\right] \quad (1)$$

and

$$\left(\frac{\Delta\ell}{\ell}\right)_{<111>} = \frac{c}{4C_{44}}\left[\alpha_J V_2^5 <O_2^0>_{111} + \ldots\right] \quad (2)$$

for the shape magnetostriction along the <001> and <111> direction and

$$\frac{\Delta V}{V} = \frac{-3c}{C_{11}+2C_{12}}\left[\frac{\sqrt{7}}{16\sqrt{6}}\beta_J V_4^1 \delta <O_4^0 + 5O_4^4> + \frac{1}{64}\beta_J V_6^1 \delta <O_6^0 - 21O_6^4>\right] \quad (3)$$

for the volume magnetostriction, where c is the rare earth concentration, C_{ij} are elastic coefficients, V_i^j the orbit-lattice parameters, and the $<O_\ell^m>$ are thermal averages of the operators where the quantization axis is <001>. We note $<O_2^0>_{111} = <3J_{111}^2 - J(J+1)>$ and $\delta<O_\ell^m>$ is the difference between $<O_\ell^m>$ in a given applied field and in zero field, at a given temperature.

As we go from RE to RE in a given environment the sign of the different contributions follow the signs of the Stevens factors α_J, β_J and γ_J. If we make the assumption that a point charge model (PCM) is valid in a given host, there are certain general rules, of which the most important is $V_2^5 = \frac{2}{3}V_2^3$. According to this rule, if the shape magnetostriction is dominated by the leading quadrupole term in Eqs. (1) and (2), the magnetostriction measured parallel to the magnetic field should be of opposite sign for the <001> vs the <111> orientation. To simplify, we will now concentrate on <u>Ag</u>RE and <u>Au</u>RE dilute systems and we will postpone till later the comparison with other data.

RESULTS

In the <u>Ag</u>RE and <u>Au</u>RE alloys, certain strain terms have been estimated with EPR techniques[4,5], using an ingenious sample preparation method. A thin film of the alloy is deposited on a massive insulating substrate at room temperature or above and the sample is then cooled to low temperature. Differences in thermal contraction between the metal and the substrate mean that at low temperature the metal film is under considerable tension. The EPR g-factor for resonance in a RE Γ_7 state becomes anisotropic: it depends on whether the magnetic field is applied parallel or perpendicular to the plane of the film, i.e. parallel or perpendicular to the strain direction. With suitably prepared single crystal films, the parameters V_2^3 and V_2^5 can be estimated. Unfortunately, because the films can slip somewhat relative to the substrate, the true strain is always less than that estimated, so the V_2^3 and V_2^5 values obtained this way can only be lower limits.

For the same alloys, we have prepared single crystal samples at RE concentrations of about 0.5% and we have measured the magnetostriction from 1.2 K to 15 K in parallel fields up to 70 kG or perpendicular fields up to 40 kG. The measurements were done using a capacity technique with a sensitivity in $\Delta\ell/\ell$ of about 10^{-8}. We can measure all rare earths (not only those with an EPR resonance signal) and we do not have problems analogous to the film slipping.

The behavior observed in a given host varies considerably from rare earth to rare earth, but in each case the qualitative form of the field and temperature dependence is just what one expects, knowing

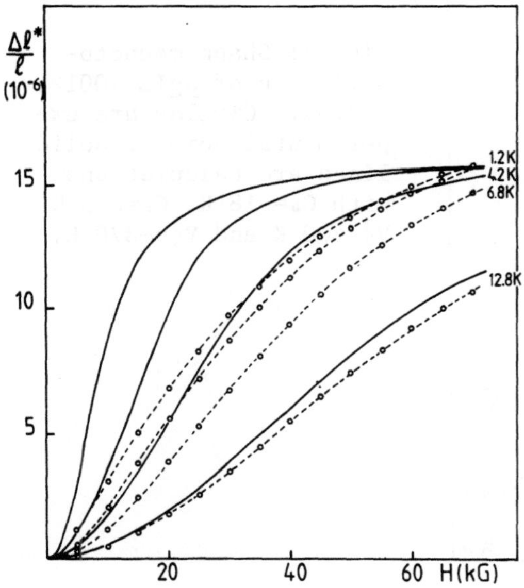

Fig. 1: Shape magnetostriction of AgTb <001>, c=0.5%. Circles are experimental points, and solid lines are calculations with C_4=-18 K, C_6=7.6 K, V_2^3=440 K and V_4^3=-520 K.

the static crystal field level schemes. In each case the shape magnetostriction is dominated by the quadrupolar term, so for a given crystal orientation relative to the applied field direction, the sign of the magnetostriction follows that of the Stevens factor α_J of the rare earth. We can give examples. For AgTb, where in the crystal field level scheme given by Lea, Leask and Wolf[1], the ground state is near a point of accidental degeneracy with the low energy states containing almost all the $|J_z=6\rangle$ character (for <001> quantization), the parallel magnetostriction for H along <001> tends to saturate rapidly with field at 1.2 K and varies considerably with temperature (Fig. 1).

For AgTm, where the ground state is a singlet isolated by about 25 K the <001> parallel magnetostriction is almost parabolic up to 70 kG and there is little temperature dependence (Fig. 2). (The sign of the magnetostriction is opposite to that in AgTb because α_J has changed sign.)

An interesting case is that of AgEr: the <001> parallel magnetostriction at high field (∿70 kG) actually grows as the temperature increases (Fig. 3). In these fields, above the ground state there is a first excited state with a higher magnetostriction than the ground state. The same behavior is observed for the AgDy <111> parallel magnetostriction, for the same reason.

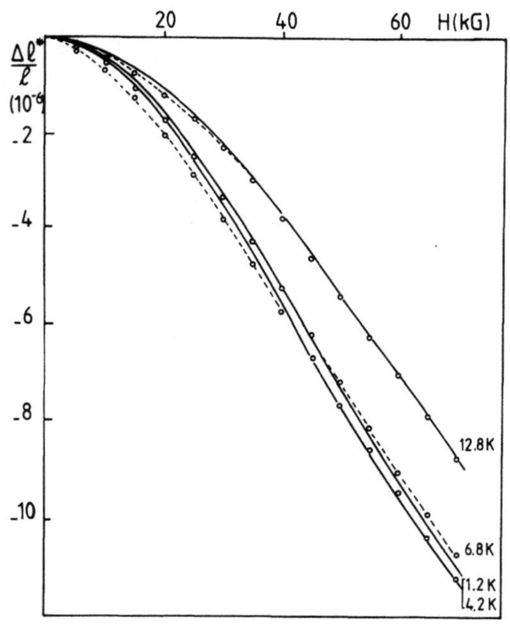

Fig. 2: Shape magnetostriction of AgTm <001>, c=0.5%. Circles are experimental points, solid lines are calculations with C_4=-18 K, C_6=6.5 K, V_2^3=400 K and V_4^3=-370 K.

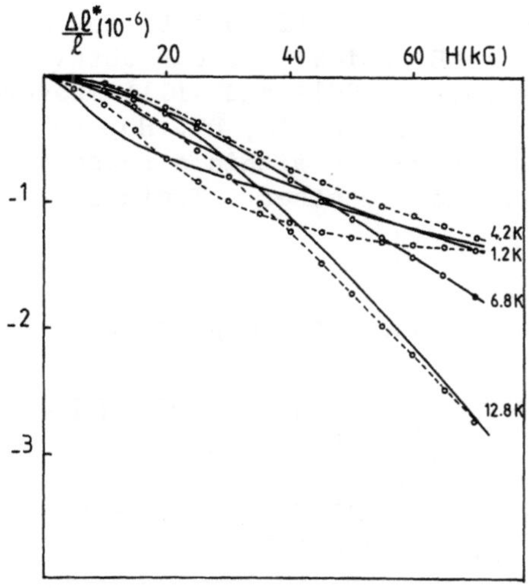

Fig. 3: Shape magnetostriction of AgEr <001>, c=0.5%. Circles are experimental points, solid lines are calculations with C_4=-18 K, C_6=7.6 K, V_2^3=440 K and V_4^3=-300 K.

We can go beyond these qualitative considerations and fit the results quantitatively, once we know C_4 and C_6 for each rare earth. To get this fit, we have used C_4 and C_6 parameters close to those of Kikkert and Niesen[6] rather than the much higher values of earlier authors[7]. Secondly we include fourth order shape magnetostriction terms as well as second order (quadrupolar) terms in the fit. If we did not introduce these fourth order terms, the fit for certain rare earths would remain almost as good but for others the agreement between experimental and calculated curves would be considerably worse. In addition to the shape magnetostriction terms, by comparing magnetostrictions in fields parallel and perpendicular to the capacity cell axis, we have found volume magnetostriction terms which are related to the volume dependence of C_4 and C_6.

The accuracy with which we can estimate the parameters is limited not by the sensitivity of the method but for a number of more basic reasons: Firstly, exact values of C_4 and C_6 are not available as input parameters, Secondly, the samples inevitably contain imperfections (dislocations, etc.). Finally, and most seriously, even at quite low concentrations there will be magnetic and electrostatic impurity interaction effects. A given impurity will not be in a truly cubic environment in zero field. This understandably produces the most notable effects in cases such as the AgTb <001> parallel magnetostriction (Fig. 1), where the effect should saturate very rapidly with field at low temperature. In these cases the higher temperature curves are more reliable as they are less susceptible to effects of small random crystal fields produced by the presence of neighboring impurities.

Taking these factors into account, it is difficult to say if the V_i^j remain strictly constant across the RE series. We can obtain a reasonably coherent fit to all the results (<001> and <111> measurements as a function of field and temperature for Tb, Dy, Ho, Er and Tm in Ag) with a set of parameters which varies within a reasonably narrow range going from one rare earth to the other. Typical results are given in Table 1.

Table 1: Crystal-field parameters and orbit-lattice parameters for AgRe and AuRE (all values in K).

	Cubic Term		Shape Terms				Volume Terms	
	C_4	C_6	V_2^3	V_2^5	V_4^3	V_4^5	V_4^1	V_6^1
AgRE	-18 (a)	7.6 (a)	+ 440	- 1000	- 400	- 1100	+ 1000	- 5000
AuRE	-30 (b)	6 (b)	+ 400	+ 1150	+ 200	+ 1000	-	-
PCM			+ 520	+ 370			+ 444	- 1135

a) Ref. 6. b) Ref. 7.

We have also made measurements on AuTb and AuYb monocrystals. Again, the results on the two rare earths from opposite ends of the series give consistent values of the V_i^j parameters. There was no sign that AuYb behaved in an anomalous manner. The AuRE results are also represented in Table 1 as are point charge estimates for certain parameters. Fig. 4 shows, for example, the shape magnetostriction of AuYb <111>. For both AgRE and AuRE the volume magnetostriction is always positive but considerably smaller than the shape magnetostriction (less than 10%), so it is impossible to obtain an estimate for V_4^1 and V_6^1 in AuRE. It is important to note that while in AgRE alloys, for each RE parallel to <001> and <111>, magnetostrictions had the same sign, but in AuRE they had opposite signs. This appears in Table 1 as a change of sign in V_2^5. EPR results[4,5] are in agreement with our data as to the sign of V_2^3 and V_2^5, but they underestimate the coefficients by factors running from 2 to 10.

At the moment, we have no clear understanding as to why the AgRE and AuRE cubic crystal fields appear to be quite similar, whereas the strain crystal field coefficients are very different in the two hosts. One can invoke d electron terms, but it is remarkable that these should be so different in Ag and Au.

COMPARISON WITH OTHER DATA

We can compare with data on other systems in which strain terms have been estimated for RE's in cubic metal environments. Nicholson et al.[8] have measured magnetostriction in dilute YPd_3(RE) alloys but only on polycrystal samples. It is difficult to estimate coupling parameters from polycrystal data, especially as <001> and

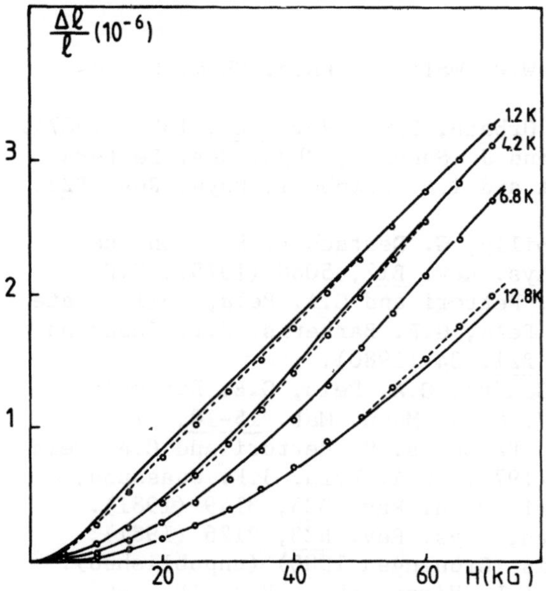

Fig. 4: Shape magnetostriction of AuYb <111>, c=0.5%. Circles are experimental points, and solid lines are calculations with C_4=-30 K, C_6=6 K, V_2^5=1320 K and V_4^5=-1000 K.

<111> contributions to the polycrystal average may have opposite signs. Morin and Schmitt[9] have measured magnetostrictions in the CsCl type ZnRE compounds. These are concentrated, but the single ion terms dominate. The coefficient V_2^3 is about 3200 K, considerably larger than 400 K for the noble metal hosts. The observed V_2^3 coefficients vary in an apparently random way from RE to RE. There is also a great deal of experimental data on the technologically important Laves phase $REFe_2$ systems[10].

Until now we have restricted discussion to RE's in cubic environments, but there is of course data on the hexagonal pure RE metals[11] and more recently on dilute RE's in the non-magnetic hexagonal metals Mg[12] and Y[13]. The number of magnetostriction coefficients is much higher than in the cubic case.

In conclusion, in every system in which there are cubic or lower order crystal fields, there will also be a whole family of strain crystal field parameters. If suitable samples can be made, these strain parameters can be very conveniently studied by magnetostriction measurements. But although we know how to estimate these parameters experimentally, we do not have a clear understanding of the interactions between the local ion and its environment which gives rise to these parameters. We can hope this situation will improve in the future.

REFERENCES

1. K.R. Lea, M.J.M. Leask and W.P. Wolf, J. Phys. Chem. Solids 23, 1381 (1962).
2. See, e.g., M. Blume and R. Orbach, Phys. Rev. 127, 1587 (1967).
3. I.A. Campbell, G. Creuzet and J. Sanchez, Phys. Rev. Letters 43, 234 (1979). G. Creuzet and I.A. Campbell, Phys. Rev. B23, 3375 (1981).
4. See, e.g., for AgRE, D. Arbilly, G. Deutscher, E. Grunbaum, R. Orbach and J.T. Suss, Phys. Rev. B12, 5068 (1975); G.E. Barberis, J.F. Suassuna, C. Rettori and C.A. Pela, Solid State Comm. 23, 603 (1977); C.A. Pela, G.E. Barberis, J.F. Suassuna and C. Rettori, Phys. Rev. B21, 34 (1980).
5. See, e.g., for AuRE, F.G. Gandra, C.A. Pela, G.E. Barberis, D.Davidov and C. Rettori, J. Magn. Magn. Mat. 15-18, 1112 (1980); J.F. Suassuna, G.E. Barberis, C. Rettori and C.A. Pela, Solid State Comm. 22, 347 (1977); C.A. Pela, J.F. Suassuna, G.E. Barberis and C. Rettori, Phys. Rev. B23, 3149 (1981).
6. P.J.W. Kikkert and B. Niesen, Phys. Rev. B23, 2126 (1981); P.J.W. Kikkert, Ph.D. Thesis (Groningen 1980) (unpublished).
7. See, e.g., G. Williams and L.L. Hirst, Phys. Rev. 185, 407 (1969).
8. H.U. Häfner, H-W. Ludwigs, K. Nicholson and D. Wohlleben, Z. Phys. B42, 219-230 (1981).
9. P. Morin and D. Schmitt, Phys. Rev. B23, 2278 (1981) and references therein.
10. A.E. Clark, in "Ferromagnetic Materials, Vol. 1", edited by E.P. Wohlfarth, North Holland (1980).
11. J. Rhyne and S. Legvold, Phys. Rev. 138, A507 (1965); A.E. Clark, B. de Savage and R. Bozorth, Phys. Rev. 138, A216 (1965).
12. M.H. de Jong, J. Bijvoet and P.F. de Châtel, this conference.
13. P. Pureur, G. Creuzet and A. Fert, this conference.

A MARTENSITIC TRANSFORMATION TRIGGERED BY MAGNETIC ORDERING: TbCu

J. Pierre[+] and B. Hennion[*]

[+]Laboratoire Louis Néel, C.N.R.S., 166X
38042 Grenoble-cédex, France
*Laboratoire Léon Brillouin, Orme-les-Merisiers
B.P. n° 2, 91190 Gif-sur-Yvette, France

INTRODUCTION

Equiatomic rare earth-copper compounds RCu with heavy rare earths have at room temperature a cubic CsCl-type structure (except with Yb), whereas light rare earth compounds have a FeB structure. Anomalous behaviours of the magnetic susceptibility (1) and Knight shift (2) have been observed for GdCu, which were attributed to a structural instability at low temperatures. Morin (3) observed by x-ray investigations a crystallographic transition when a CsCl-type sample of GdCu was cooled below room temperature; annealing at about 700 K was necessary to restore the CsCl-structure. A martensitic transformation has also been detected in YCu below 150 K by resistivity measurements (4).

TbCu orders at T_N = 115 K, its magnetic structure (5) being tetragonal with alternate ferromagnetic layers parallel to (110) planes, and a magnetic cell ($a\sqrt{2}$, $a\sqrt{2}$, $c \simeq a$). A splitting of Bragg reflections was observed in the ordered range by x-ray diffraction (3,6). This was attributed to a tetragonal magnetostriction ($c/a - 1 = 1.13 \times 10^{-2}$ at 4.2 K). However, only a few reflections were then followed, and no complete diagram was taken.

EXPERIMENTS

We have investigated the electrical resistivity and performed elastic and inelastic neutron scattering experiments on TbCu, using polycrystalline samples as well as single crystals which were prepared by the Bridgman or Chokralski methods. Some resistivity data are summarized in Figure 1. When cooling virgin CsCl-type samples the resistivity R first decreases, then shows a sharp discontinuity at T = 115 K, which corresponds to the Néel temperature. Some time-dependent effects occur until 100 K, then the resistivity decreases again. When increasing the temperature, there is a continuous increase of resistivity with a change of slope at T_N.

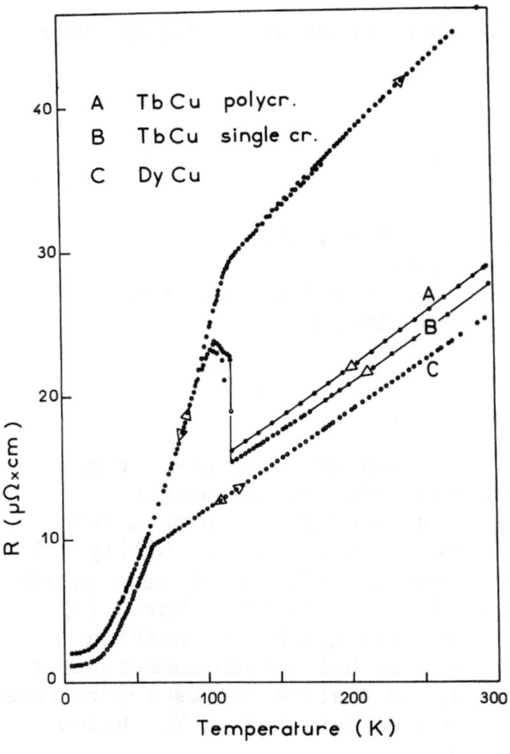

Fig. 1. Resistivity of:
A – a polycrystalline TbCu sample (second cycle after annealing at 700 K)
B – a TbCu single crystal (first cycle)
C – a DyCu polycrystal.

The hysteresis anomaly is very broad, it is necessary to warm up the sample well above room temperature in order to restore the initial resistivity. The slope dR/dT in the paramagnetic range for the untransformed samples is close to that of DyCu (Fig. 1), which has a perfectly reversible behaviour. The slope for transformed samples is higher, which was observed earlier on YCu (4) and GdCu (7). Optical micrographs were taken on the polished surfaces of single crystals. After cooling, a characteristic network of thin needles reveals the shape effects of a martensitic transformation (8). The CsCl phases of RCu compounds are easily strained. Filing the samples generally gives very poor x-ray patterns, so we performed new neutron diffraction experiments using the D1B diffractometer at the Institut Laue-Langevin. A first pattern was taken at 160 K on a virgin sample, a second one at 90 K and a third at 160 K after cooling.

The neutron pattern at 90 K shows additional lines of the type (h/2, k/2, 1) corresponding to the antiferromagnetic structure ($\mu = 4.7 \pm 0.2\ \mu_B$) compared to a value of 8.8 μ_B at 4.2 K (5). All reflections are better indexed, assuming a slight distortion $c/a - 1 \simeq 8\times10^{-3}$, close to that found by x-ray experiments at 90 K (3,6). Nuclear superstructure lines, if any, may be hidden by the intense magnetic lines. The two patterns taken at 160 K are identical, no additional lines appear after cooling. However, a broad diffuse scattering is observed on both patterns at the position of the magnetic (1/2, 1/2, 0) line, which may correspond to short range magnetic order, but it is still present well above T_N (5). The coherent martensitic phase may indeed disappear quite rapidly above T_N, leaving only stacking faults which are difficult to cure.

Inelastic neutron scattering experiments have been performed on a single crystal using the IN2 triple-axis spectrometer at the Institut Laue-Langevin. Some phonon branches have been followed at various temperatures. Figure 2 shows room temperature data, taken before or after cooling. Elastic constants do not change much with temperature down 140 K, within the precision of the experiment. Room temperature values are $(c_{11} + c_{12})/2 + c_{44} = 12.7\pm0.25$ erg/cm^3, $c_{44} = 3.7 \pm 0.2$ erg/cm^3 and $c_{11}-c_{12} = 7.1 \pm 0.2$ erg/cm^3. The anisotropy ratio $2c_{44}/(c_{11} - c_{12})$ is close to 1, as in TmCu (9). The transverse mode with the (0, 0, 1) polarization could be followed along the ΓM direction at room temperature. It has an unusual shape, bending down towards lower energies at the edge of the Brillouin zone. Similar features have been observed for $LaAg_xIn_{1-x}$ compounds (10), which also exhibit a martensitic transformation. For this latter case, a softening of the transverse phonon occurs

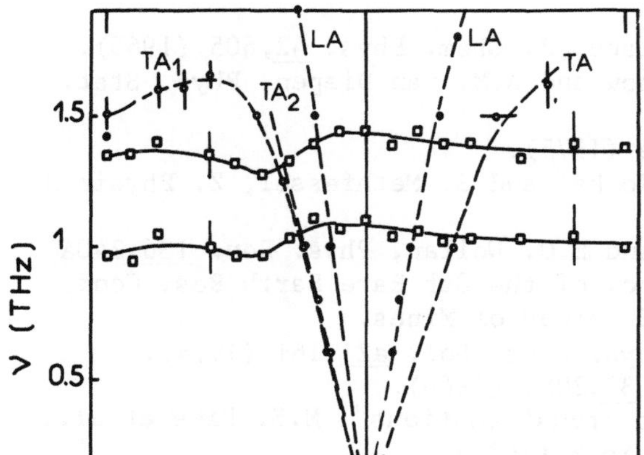

Fig. 2. Energy of excitations versus scattering vector in the Brillouin zone of the CsCl phase. Circles represent the phonon spectrum at room temperature, squares the magnetic excitations at 9 K.

at the M point when lowering the temperature. The temperature dependence of this mode should be investigated in TbCu also.

A very broad quasielastic spectrum of magnetic origin is observed in the paramagnetic range, no distinct inelastic crystal field transitions could be seen. However, well defined magnons appear in the ordered range. A complete spectrum was recorded at 9 K (figure 2). Broad neutron groups were found in the energy range from 0.95 to 1.4 THz (3.9 to 5.8 meV). Taking into account the spectrometer resolution, these groups could possibly be resolved into two peaks of similar intensity, with intrinsic widths of about 0.1 THz. Their branches have then a small q-dependence, except along the ΓM direction in a range corresponding to the crossing of phonon branches. This suggests a strong dynamical magnon-phonon coupling. Inelastic spectra taken with the IN4 spectrometer show that the next magnetic excitation is at about 17 meV. As the crystal field is unknown, and the distortion large, it is not possible to give account for these data. The presence of two low energy excitation branches may be due either to inequivalent crystallographic sites, or to two close excited levels.

It is thus necessary to get additional data on the crystal field and the structure of the low temperature phase, in order to understand how the magnetic ordering drives the crystallographic transition.

ACKNOWLEDGEMENTS

The single crystals were grown by R. Perrier de la Bathie and E. Siaud. We warmly thank J. Bouillot for his efficient help during experiments on IN2 spectrometer and J. Féron for helpful discussions.

REFERENCES

1. R.E. Walline and W.E. Wallace, J. Chem. Phys. 42,605 (1965).
2. H.W. de Wijn, K.H.J. Buschow and A.M. van Diepen, Phys. Stat. Sol. 30,759 (1968).
3. P. Morin, Thesis, Grenoble (1975).
4. H. Balster, H. Ihrig, A. Kockel and S. Methfessel, Z. Physik B 21,241 (1975).
5. J.W. Cable, W.C. Koehler and E.O. Wollan, Phys. Rev. 136,240A (1964). and J. Pierre, Proc. of the 8th Rare Earth Res. Conf. Reno, Nevada, 1970, Ed. US Bureau of Mines.
6. P. Morin and J. Pierre, Phys. Stat. Sol. a21,161 (1974).
7. C.C. Chao, J. Appl. Phys. 37,2081 (1966).
8. Z. Nishiyama, "Martensitic transformations", M.E. Fine et al., eds., Academic Press, New York (1978).
9. C. Jaussaud, P. Morin,and D. Schmitt, J.M.M.M. 22,98 (1980).
10. K. Knorr et al, Z. Physik B 39,151 (1980).

NEUTRON DIFFRACTION STUDIES OF MAGNETIC ORDERING IN RARE EARTH AND ACTINIDE INTERMETALLICS OF THE $CeAl_2Si_2$ TYPE

Janusz Leciejewicz

Institute of Nuclear Research, Swierk Research Establishment, Otwock, Poland

Among isostructural lanthanide and actinide intermetallic compounds with 3d and 4d transition metals containing silicon and germanium, those with the $CeAl_2Si_2$ type of crystal structure turn out to exhibit a variety of magnetic ordering schemes which have been recently observed in a number of neutron diffraction experiments.

The crystal structure of $CeAl_2Si_2$ is tetragonal. The space group is I4/mmm, the atomic sites are as follows:

2Ce	in	2(a)	0, 0, 0	
4Al	in	4(d)	0, ½, ¼	½, 0, ¼
4Si	in	4(e)	0, 0, z	0, 0, $\bar{z}$
			+ body-centering translation.	

The c/a ratio in all compounds is close to 2.5. Neutron diffraction data published during the last few years indicate that all magnetically ordered $CeAl_2Si_2$ type intermetallics can be classified as systems in which only lanthanide (or actinide) moments are ordered and systems in which only the transition metal orders magnetically. To the first class belong lanthanide and actinide compounds with 4d and 3d metals, except manganese. Thus URh_2Si_2[1], UNi_2Ge_2[2], $HoCo_2Ge_2$[3], $TbCo_2Si_2$ and $TbCo_2Ge_2$[4], $NdCo_2Ge_2$[5] are all simple antiferromagnets. Their magnetic structure is displayed in Fig. 1. Magnetic moments lying in the (001) planes are coupled ferromagnetically with a fairly small distance between two neighboring ions amounting to the lattice constant a, which in all compounds is close to 4 Å. Adjacent planes are coupled antiferro-

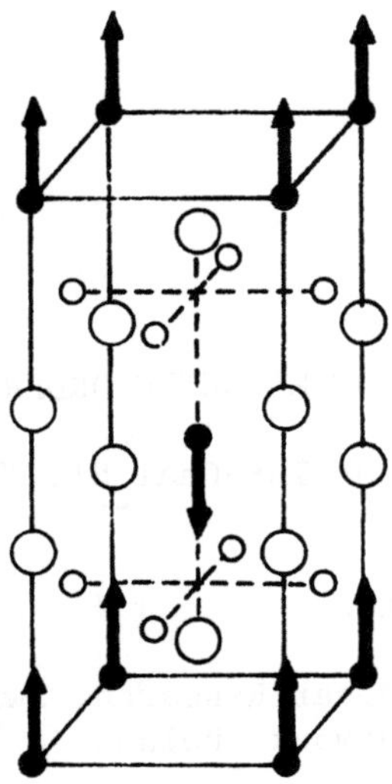

Fig. 1: Antiferromagnetic ordering of the + - + - type.

magnetically so that their sequence in the direction of the c-axis is +-+-. Two adjacent planes are, however, separated by three atomic layers composed of non-magnetic atoms i.e. Si (or Ge) and transition metal ions which, as experiment shows, do not carry a magnetic moment. Therefore direct magnetic interactions between lanthanide (or actinide) atoms belonging to adjacent planes seems to be very improbable. In $NpCo_2Si_2$ which also has the +-+- type antiferromagnetic structure Np magnetic moments deviate from the c axis by an angle of 52±15°, in contrast to the previously mentioned compounds in which the magnetic moments are aligned along the tetragonal axis. On the other hand $NpCu_2Si_2$ is a ferromagnet with a Curie point at 41±2 K with Np moments lying along the c axis. Susceptibility data obtained for $NpFe_2Si_2$ indicate the occurrence of antiferromagnetic ordering. No neutron diffraction data are available. However, a Mössbauer study shows that no localized moment is present on the Fe ion, since the hyperfine field on the ^{57}Fe nucleus was found to be very small[6].

A magnetic phase transition has been recently found in UCu_2Ge_2[7]. At 4.2 K magnetic moments on the U ion order antiferromagnetically. The magnetic structure is of the collinear +--+ type, requiring doubling of the magnetic unit cell with respect to the chemical one in the direction of the c axis (see Fig. 2). This structure transforms into a ferromagnetic one in the temperature range between 25K and 40 K. This is shown in Fig. 3 which displays the temperature dependence of peak heights characteristic of the relevant ordering schemes. The +--+ type antiferromagnetism has also been discovered in $NdFe_2Si_2$[8].

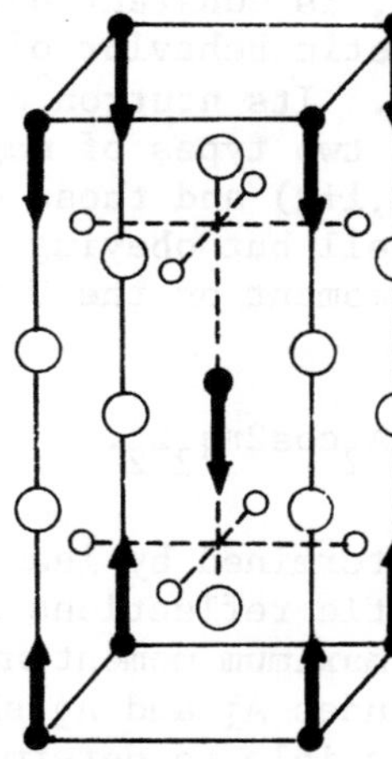

Fig. 2: Antiferromagnetic ordering of the + - - + type.

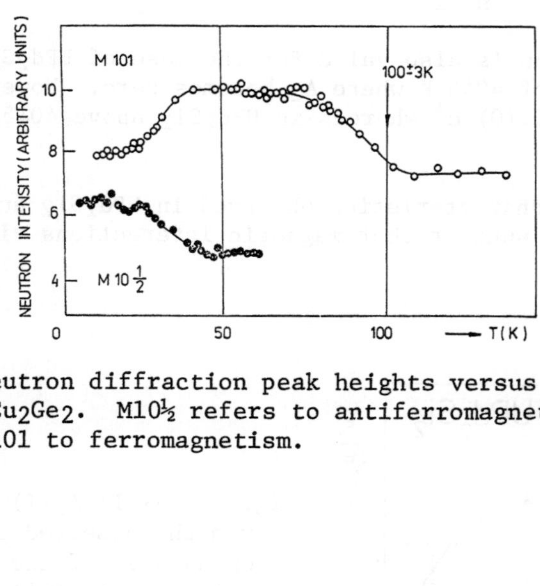

Fig. 3: Neutron diffraction peak heights versus temperature in UCu_2Ge_2. M10½ refers to antiferromagnetic ordering, M101 to ferromagnetism.

Neutron diffraction studies of magnetic properties carried out for $PrCo_2Ge_2$[3,9], UPd_2Si_2 and UPd_2Ge_2[1] have shown that their magnetic structure can be described as a consinusoidally modulated longitudinal spin wave (LSW) propagating along the c-axis. In the case of $PrCo_2Ge_2$ the magnetic propagation vector is 0.73 c^*. This remains constant up to the Néel point at 28 K. A similar situation was found in UPd_2Ge_2; the propagation vector of the magnetic structure, amounting to 0.662 c^*, is constant between 4.2 K and the Néel point at 140±4 K. The magnetic behavior of UPd_2Si_2 turned out to be particularly interesting. Its neutron diffraction pattern at 4.2 K shows the presence of two types of magnetic reflections: those with the indices (h,k,l+q) and those with (h,k,l) related to the crystallographic unit cell but obeying the rule h+k+l=2n+1. Consequently, the magnetic moment on the U ion can be expressed by two Fourier components:

$$\vec{\mu}^{(z)}=A_1\cos 2\pi(q_1 z_1+\phi_1)+A_2\cos 2\pi q_2 z_2. \quad (1)$$

The amplitudes A_1 and A_2 determined by least-squares fitting to the intensities of proper magnetic reflections at 4.2 K are A_1=2.05 μ_B and A_2=1.32 μ_B. Hence the maximum moment on a U ion is 3.37±0.25 μ_B at 4.2 K. Since the amplitudes A_1 and A_2 show a different variation with temperature, it was possible to determine the temperature at which A_2 becomes zero. This follows from Fig. 4. In UPd_2Ge_2 the magnetic moment is represented by only one Fourier component in the whole temperature range at which long range magnetic order occurs:

$$\vec{\mu}_n^{(z)}=A_1\cos 2\pi(qz_n+\phi_1). \quad (2)$$

The same expression is also valid for the case of UPd_2Si_2 above the transition point of 40±5 K where A_2 becomes zero. However, in UPd_2Ge_2 q=(0.748±0.10) c^* whereas in UPd_2Si_2 above 40±5 K it is (0.662±0.010)c^*.

Oscillatory characteristics observed in UPd_2Si_2 and UPd_2Ge_2 magnetic ordering suggest that magnetic interactions via conduction

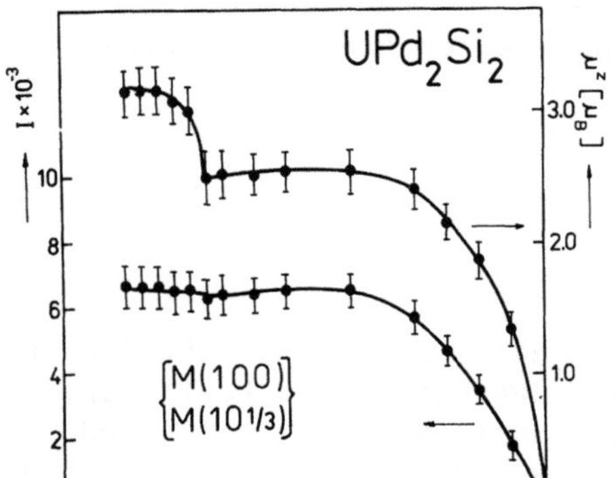

Fig. 4: $A_1(T)+A_2(T)$ for UPd_2Si_2 from the observed temperature dependences of the 102 and 101^+ magnetic reflections (upper curve). The lower curve shows the temperature dependence of the intensity of overlapping 100 and 101^- magnetic reflections.

electrons, postulated by the well-known RKKY theory should be incorporated to explain the stability of the above structures. There are other arguments which favor the above suggestion: 1. In the basal planes the U-U distances are quite large, amounting to about 4 Å. The shortest distance between two U ions belonging to neighboring planes is about 5.8 Å. Such a pair is, however, separated by three layers of Si (or Ge) and Pd (or Rh) ions. This makes any direct interactions between the U pairs highly improbable. 2. Since the isostructural alloy $ThPd_2Si_2$ is non-magnetic, the absence of a moment on a Pd ion can be safely assumed. Thus the observed magnetic order is only due to interactions among U ions. 3. All alloys studied exhibit metallic type electric conductivity, and 4. the magnetic interactions are of long range character. We also adopted the assumptions that the magnetic moment is localized on the U ion and that the Fermi surface is spherical.

For a spin system with a screw-like order, the expression for the energy in terms of the RKKY theory is given by

$$E= -NS^2 J(\vec{q}) \sim -[J(\vec{q})-J(0)] \sim -F(\vec{q}), \qquad (3)$$

where N is the total number of magnetic ions and S the magnitude of a spin. $J(\vec{q})$ represents the Fourier transform of the exchange integral $J(\vec{R}_j-\vec{R}_i)$ of two spins. Assuming that J is a constant of motion for an electron in a 5f state, S can be replaced by $(g_J-1)J$, (g_J is the Lande factor).

The stability of a magnetic structure means that the function $F(\vec{q})$ exhibits a maximum for a non zero value of $\vec{q}_o$. The function $F(\vec{q})$ for $\vec{q}=(0,0,q)$ was computed for all reciprocal lattice points up to $H=18.7(2\pi/c)$. It shows maxima in two narrow regions for z (or $2k_F$) (z=average number of free electrons per U ion). This is seen in Ref. 1, Fig. 3, which displays the value of q_o vs z where a screw-like structure is stable. The maxima appear at $\vec{q}_o$ values which depend on z. For example, the maximum for UPd_2Ge_2 is observed for $\vec{q}_o=(0.748\pm0.010)c^*$. It corresponds to z=3.86. For the same value of z, the theoretical value of $\vec{q}_o$ in the case of UPd_2Si_2 is $0.69c^*$, very close to the observed value for the magnetic structure vector of $(0.662\pm0.010)c^*$. On the other hand, computations performed for URh_2Si_2 indicate, that for z=3.86 the maximum of $F(\vec{q})$ occurs for $\vec{q}_o=1$, i.e. a collinear antiferromagnetic ordering of + - + - type should be stable, in accordance with experiment.

Since a spherical Fermi surface and only isotropic RKKY type interactions have been assumed in this model, the above quantitative relations might be quite coincidental. However in general, they indicate that the interactions via conduction electrons are dominant, at least in UPd_2Si_2, UPd_2Ge_2 and URh_2Si_2.

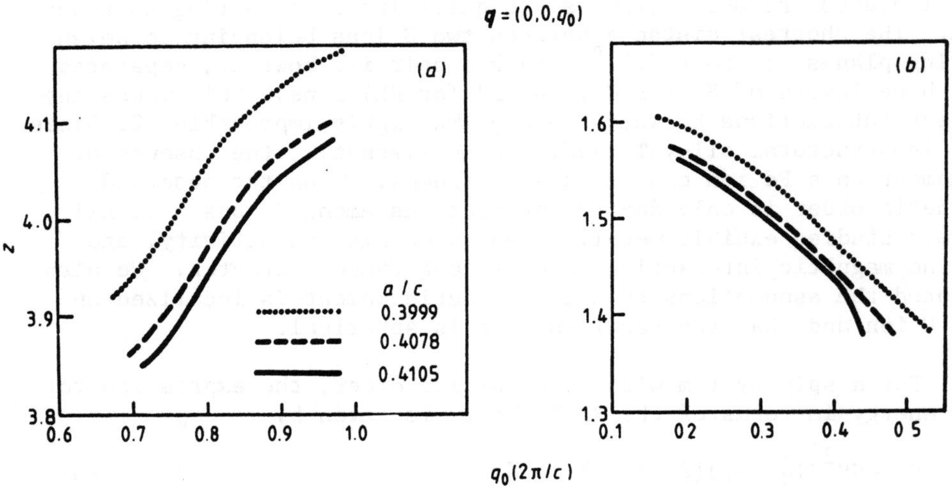

Fig. 5: The curves representing the dependence of $\vec{q}_o$ on z for UPd_2Si_2 (broken curve), UPd2Ge_2 (full curve), and URh_2Si_2 (dotted curve). The function $F(\vec{q})$ exhibits a maximum at $\vec{q}_o$ for which a screw-like structure is stable. (a) and (b) represent the region of z-values for which the screw-like magnetic structures are stable.

Recently a neutron diffraction study was carried out for $PrFe_2Ge_2$ showing that its magnetic structure is sinusoidal along the c axis with a propagation vector $\vec{q}=(0,0,q)$ with q=0.55 c*[5]. The other compound investigated, $ErMn_2Si_2$, shows ferromagnetic ordering at 4.2 K of the Er moments. The ferromagnetic axis makes an angle of roughly 45° with the tetragonal axis[5]. A series of superlattice lines was observed on the neutron diffraction pattern for $TbMn_2Si_2$ at 4.2 K. These reflections are due to Tb ordering and vanish at 55 K. The magnetic structure has a very complex character[5].

Except for $ErMn_2Si_2$ and $TbMn_2Si_2$, where both Er (or Tb) and Mn sublattices produce long range ordering, the other compounds of Mn and Si amounting to 2.54 A, an overlap of electronic shells may be assumed. Consequently, spin transfer seems very probable. This is accompanied by a decrease of the magnetic moment on the Mn ion, in accordance with experimental results.

studied until now by neutron diffraction exhibit the presence of antiferromagnetic order of the Mn moments only. This effect has been observed in $CeMn_2Si_2$[10], $PrMn_2Si_2$ and $NdMn_2Si_2$, YMn_2Si_2 and YMn_2Ge_2[11], $ThMn_2Si_2$ and $ThMn_2Ge_2$[12], and UMn_2Si_2[13]. All of them are antiferromagnetic with Néel temperatures in the range 450-500 K. At 80 K the magnitude of the magnetic moment on the Mn ion was determined to be about 1.8 μ_B for all compounds. In each case, the Mn moment is along the tetragonal axis or close to it. The magnetic structure of the Mn sublattice is shown in Fig. 6. The shortest Mn-Mn distance within the (001) plane is $a\sqrt{2}/2$. This distance is always smaller than 3 Å, suggesting that ferromagnetic couplings within the (001) plane are quite strong, in agreement with observed

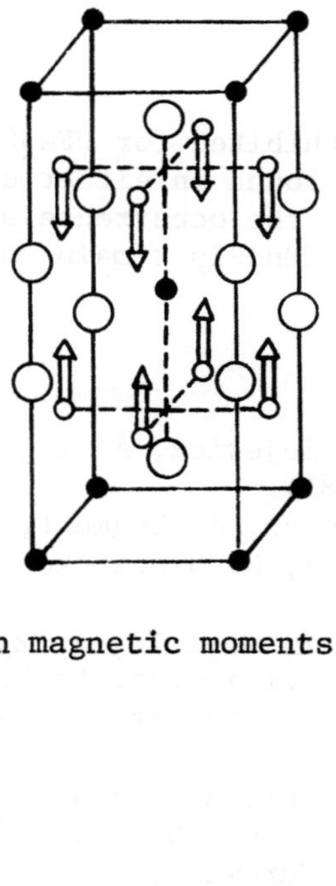

Fig. 6: Ordering of Mn magnetic moments in RMn_2Si_2 compounds.

positive paramagnetic Curie temperatures. However, fairly high Neel temperatures indicate that antiferromagnetic interactions between Mn moments belonging to neighboring planes cannot be neglected. The presence of indirect magnetic interactions via Si ions may be deduced when one compares the variation of Mn-Mn interplane distance of c/2 with the observed Neel temperatures. The latter decrease as the interplane distance rises. This follows from Fig. 7 where Neel (or Curie) temperatures are plotted against c/2. The shortest Mn-Si distance is 2.39 Å. Comparing this with the sum of the ionic radii

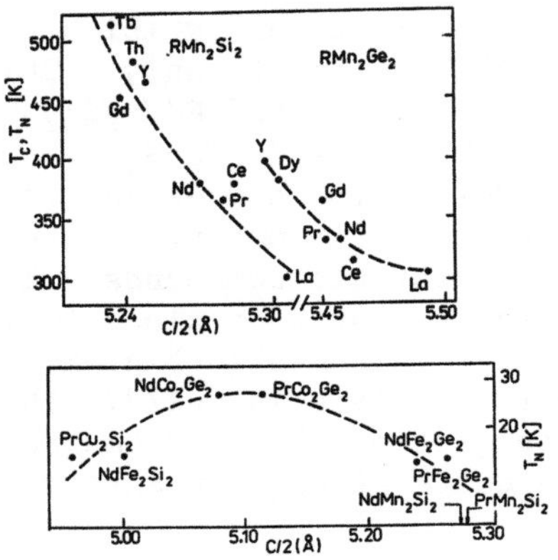

Fig. 7: Observed Néel (or Curie) temperatures vs. c/a in a number of RMn_2X_2 compounds.

A surprising result is exhibited for UFe_2Si_2 and UFe_2Ge_2. No long range magnetic order was found in either down to 4.2 K[5]. This result as well as that of the occurrence of magnetic ordering for Mn moments in RMn_2Si_2 and UMn_2Si_2 remain open for theoretical treatment.

REFERENCES

1. H. Ptasiewicz-Bak, J. Leciejewicz, A. Zygmunt, J. Phys. F: Metal Phys. 11, 1225 (1981).
2. L. Chełmicki, J. Leciejewicz, A. Zygmunt, to be published.
3. A. Szytuła, J. Leciejewicz, H. Binczycka, Phys. Stat. Solid (a) 58, 67 (1980).
4. J. Leciejewicz, S. Siek, A. Szytuła, A. Zygmunt, this conference.
5. A. Szytuła, S. Siek, J. Leciejewicz, to be published.
6. C.H. de Novion, J. Gal, J.L. Buevoz, J. Magn. Magn. Mat. 21, 85 (1980).
7. J. Leciejewicz, L. Chełmicki, A. Zygmunt, S.S. Comm. 39, 863 (1981).
8. H. Pinto, H. Shaked, Phys. Rev. B7, 3261 (1973).
9. H. Pinto, M. Melamud, E. Gurewitz, Acta Cryst. A35, 533 (1979).

10. S. Siek, A. Szytuła, J. Leciejewicz, Phys. Stat. Solid (a) 46, K101 (1978).
11. S. Siek, A. Szytuła, J. Leciejewicz, to be published.
12. Z. Ban, L. Omejec, A. Szytuła, Z. Tomkowicz, Phys. Stat. Sol. (a) 27, 333 (1975).
13. A. Szytuła, private communication.

COMMENTS

WOHLLEBEN: It seems odd to me that neither Fe nor Co seem to show indications of magnetic moments. Does one have any indication on how close they are to carrying moments? (E.g. modification of the Curie constant away from that of the RE or temperature dependence of the Co compounds with La or Lu or neutron paramagnetic scattering in high magnetic field?)

LECIEJEWICZ: The absence of long range magnetic ordering of Fe or Co is deduced from the lack of magnetic peaks on neutron diffraction patterns. Antiferromagnetic ordering would produce additional magnetic reflections which in the case of $REMn_2Si_2$ compounds, where the long range ordering of Mn moments is observed, are fairly strong. If the Fe (or Co) moment would be present, its magnitude would be less than 0.2 μ_B. This follows from the accuracy with which the magnetic peak intensity is measured. There is, however, direct evidence, at least in the particular case of $NpFe_2Si_2$: Mossbauer studies show very small hyperfine fields on the ^{57}Fe nucleus even at 4.2 K.

QUADRUPOLE EFFECTS IN THE LATTICE PARAMETERS AND MAGNETIC ORDERING TEMPERATURES OF $RECu_2Si_2$, (RE=RARE EARTH)

W. Schlabitz, J. Baumann, G. Neumann, D. Plümacher and K. Reggentin

II Physikalisches Institut der Univ. zu Köln
Zülpicherstr. 77
5000 Köln 41, FRG

INTRODUCTION

For a number of reasons the physical properties of the $RECu_2Si_2$ intermetallic compounds have been studied rather thoroughly in the last few years[1,2]. Firstly this is the only series of compounds known to contain three members (RE=Ce, Eu and Yb) with unstable 4f shells (intermediate valence)[2]. Secondly the magnetic interaction temperatures are quite low, so that single ion effects can be studied with relatively little interference from ion-ion interactions. Thirdly the $RECu_2Si_2$ structure crystallize in the tetragonal $ThCr_2Si_2$ structure, i.e. in a noncubic structure, where one expects large anisotropies in the physical properties.

Here we concentrate on the last feature, more precisely on the interaction between the (quadrupolar) charge asphericity of the 4f ion with the electric field gradient of the noncubic RE sites. This interaction leads to strong magnetoelastic coupling, causing anomalies of the temperature dependence of the ratio c/a between the c and a lattice constants by x-ray diffraction between room and helium temperature, as well as the static magnetic susceptibility over the same temperature range.

EXPERIMENTAL DETAILS

The polycrystalline samples were made using RE (4n's), Cu (5n's) and Si (6n's), which were melted in correct proportions in an induction furnace in a water cooled copper crucible under the protective atmosphere of argon (5n's).

For x-ray measurements the ingots were first homogenized at 1100 K for 80 hours. After powdering in a boron-carbide mortar the samples were then further annealed at 700 K for 20 hours. Within the resolution of the x-ray diffractometer, we did not find (even at small angles) any lines which we could not index on the basis of the I 4/mmm bodycentered tetragonal type structure. The samples for the magnetic measurements were not subjected to any metallurgical treatment. We found for $TmCu_2Si_2$ that annealing at 800 K for 80 hours does not change the magnetic susceptibility within the accuracy of the measurement. The temperature dependent x-ray diffraction measurements were done with a Siemens diffractometer with Ni filtered CuK-radiation in an evaporation cryostat. The resolution was $\Delta a/a=5.10^{-5}$ and $\Delta c/c=1.10^{-4}$ respectively. The magnetic susceptibility measurements were carried out between 4.2 K and 300 K in magnetic fields of 250 - 10^4 Oe using a pendulum Faraday balance. The accuracy was ±1%. All data were registered automatically by a computer.

RESULTS

Our x-ray measurements performed at 300 K (now with a higher resolution) principally confirm the results of Rieger and Parthé[3]. Our room temperature data are collected in Table 1. Fig. 1 shows the volume of the unit cell of the $RECu_2Si_2$ compounds plotted against the ionic radius.

Table 1: Lattice constants c and a and volume of the unit cell of $RECu_2Si_2$ at room temperature

RE	a(Å)	c(Å)	V(Å^3)
La	4.1521	9.9316	171.22
Ce	4.100	9.9244	166.83
Pr	4.0847	9.9311	165.7
Nd	4.0638	9.9322	164.03
Sm	4.027	9.9422	161.23
Eu	4.0553	9.9175	163.1
Gd	3.9987	9.9538	159.16
Tb	3.9765	9.9616	157.52
Dy	3.9654	9.9616	156.64
Ho	3.9507	9.972	155.64
Er	3.9369	9.9809	154.7
Tm	3.9247	9.9922	153.91
Yb	3.9232	9.9871	153.72
Lu	3.9073	9.9955	152.6

One recognizes the usual smooth lattice contraction due to the lanthanide contraction of the trivalent RE ions, interrupted

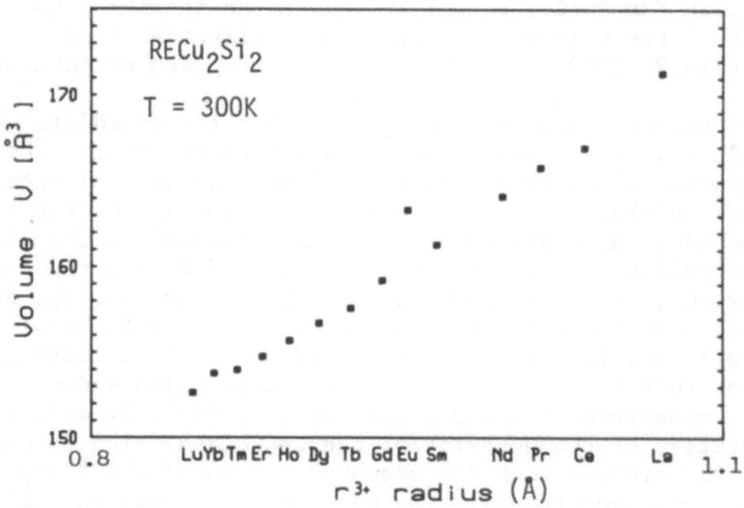

Fig. 1: Volume of the unit cell of $RECu_2Si_2$ vs ionic radius of trivalent ions at room temperature

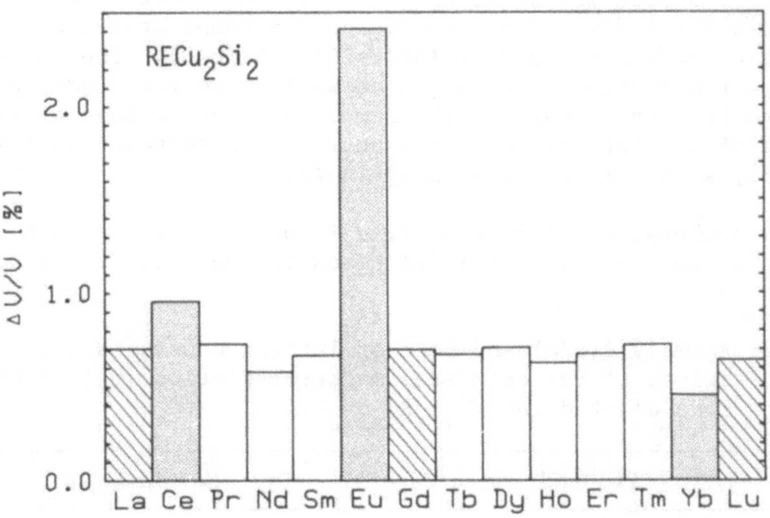

Fig. 2: Relative volume change of the unit cell of $RECu_2Si_2$ between 4 K and 300 K

by anomalies at Ce, Eu and Yb. Fig. 2 shows the relative volume expansion between helium and room temperature. Again there are clear anomalies for RE=Ce, Eu and Yb, while the thermal expansion for the rest of the compounds scatters only slightly around a common value $\Delta V/V=(7\pm0.2)\ 10^{-3}$ (with the possible exception of $NdCu_2Si_2$).

The two volume anomalies set the unstable (intermediate) valence compounds (RE=Ce, Eu and Yb) clearly apart from the rest. The detailed behavior of these compounds will be described elsewhere[4]. In this paper we shall focus on the thermal expansion of the stable valent compounds, which are apparently characterized by the absence of volume anomalies. This class can be subdivided into the compounds with L=0 (La, Gd, Lu) and those with L≠0 (Pr, Nd, Sm, Tb, Dy, Ho, Er and Tm) where L is the total angular momentum of the 4f shell. Figures 3 to 6 show the temperature dependence of both lattice constants for four members of the second class. Also shown is the temperature dependence of c and a for $GdCu_2Si_2$ (thin lines), which is representative of the behavior of the L=0 class. The Gd data are shifted to coincide with the measured values of the $RECu_2Si_2$ compounds at room temperature. The difference between the thin lines and the data points constitutes the anomaly of the temperature dependence of both lattice constants for the L≠0 class. The most prominent feature of the data is the reversal of sign of the temperature dependence of the c-axis which is visible in all figures (most dramatically in Fig. 6, $TmCu_2Si_2$).

There is a concomitant acceleration of the contraction of these compounds in the a direction with decreasing temperature, such that there remains no volume anomaly ($\Delta V/V=2\Delta a/a+\Delta c/c\sim 0$). The anomaly seems to be restricted to temperatures well below room temperature. One measure for the anomaly is the sum of absolute values of the difference of the lattice constants between the compound in question and $GdCu_2Si_2$ at 4.2 K, $\Delta(c/a)=|\Delta a/a|+|\Delta c/c|$.

These numbers, multiplied by five in order to refer them to a distortion of the RE cell only) are given in Table 2, (first line).

Table 2: Anomaly $\Delta(c/a)$ and $\Delta c/c$ of lattice constants and values of the saturated magnetostriction $|Q_s|$ of RE ions diluted into $Y_{1-x}RE_xCu_2Si_2$

	Pr	Nd	Sm	Gd	Tb	Dy	Ho	Er	Tm
$\Delta(c/a)\cdot 10^3$	3	2.8	4.5	0	4.5	8.2	5.3	5.5	10.3
$Q_s\ \ 10^3$	-4.9	-0.4	<u>1.5</u>	-0.1	-6.3	-9.4	-3.1	<u>2.5</u>	<u>1.2</u>
$Q_s/2\ \cdot 10^3$	-<u>2.5</u>	-<u>0.2</u>			-<u>3.2</u>	-<u>4.7</u>	-<u>1.6</u>		
$\Delta c/c\ \cdot 10^3$	2	1.9	3.0	0	3.0	5.5	3.5	3.7	6.9

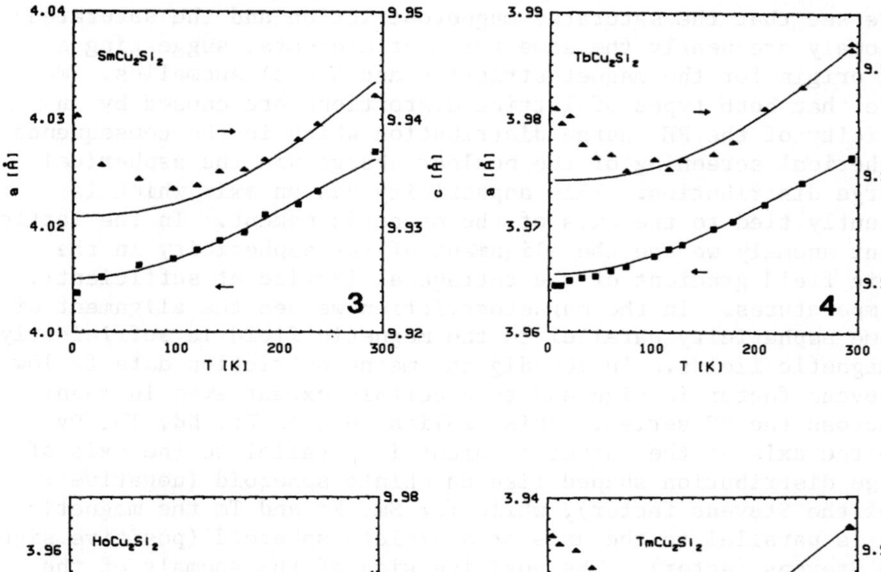

Figs. 3-6: Lattice constants c and a for $SmCu_2Si_2$, $TbCu_2Si_2$, $HoCu_2Si_2$ and $TmCu_2Si_2$ vs temperature.

We note that the $\Delta(c/a)$ values are of the order of a few times 10^{-3}. This is also the general order of magnitude of the saturated magnetostriction Q_s of the RE ions in metals[5]. Q_s is defined as the saturated value of $(\Delta\ell/\ell)_{\parallel}$, which is the relative change of the length of the sample parallel to the applied magnetic field. The relevant values of Q_s as obtained by magnetostriction measurements on dilute $(Y_{1-x}RE_x)Cu_2Si_2$ are also listed in Table 2 (second line)[5].

We see that the saturated magnetostriction and the saturated c/a anomaly are nearly the same for most elements, suggesting a common origin for the magnetostriction and $\Delta(c/a)$ anomalies. We believe that both types of lattice distortions are caused by an asphericity of the RE charge distribution which is the consequence of aspherical screening of the nuclear charge via the aspherical 4f charge distribution. This asphericity has an axis which is permanently tied to the axis of the magnetic moment. In the lattice constant anomaly we see the alignment of the asphericity in the electric field gradient of the tetragonal lattice at sufficiently low temperatures. In the magnetostriction we see the alignment of the same asphericity parallel to the magnetic field in sufficiently high magnetic fields. In YCu_2Si_2 the magnetostriction data follow the Stevens factor in sign and to a certain extent even in magnitude across the RE series. This implies that in Pr, Nd, Tb, Dy and Ho the axis of the magnetic moment is parallel to the axis of a charge distribution shaped like an oblate spheroid (negative sign of the Stevens factor), while for Sm, Er and Tm the magnetic moment is parallel to the axis of a prolate spheroid (positive sign of the Stevens factor). The positive sign of the anomaly of the c-axis (Figs. 3-6) then indicates that for Sm, Er and Tm the axis of the magnetic moment has a tendency to become aligned parallel to the c-axis at low temperatures, while in Pr, Nd, Tb, Dy and Ho it will become aligned perpendicular to the c-axis. For a positive Stevens factor therefore the (saturated) thermal expansion anomaly $\Delta c/c$ should be compared with the saturated longitudinal magnetostriction Q_s directly, while for negative Stevens factors $\Delta c/c$ should be compared with the transverse magnetostriction $(\Delta\ell/\ell)_{\perp} = Q_s/2$. Table 2 shows $Q_s/2$ in the third line and $\Delta c/c$ in the fourth. The ratio between the relevant numbers for magnetostriction and $\Delta c/c$ is quite close to one for all cases except Nd and Tm. The magnetostriction values tend to be somewhat smaller than $\Delta c/c$. The reason for the lower magnetostriction values is that for Nd and Tm the temperatures at which the values of Q_s were extrapolated were not high enough[5]. In order to see the fully saturated quadrupole moment it is necessary that either the Zeeman energy $\mu\Delta H$ or the thermal energy $k_B\Delta T$ be of the order or larger than crystal field splitting energies. This condition is much easier to fulfill in magnetostriction than in thermal expansion measurements. Therefore the $\Delta c/c$ values from thermal expansion give more reliable (higher) values for the saturated RE quadrupole than magnetostriction.

MAGNETIC MEASUREMENTS

The magnetic data for the $RECu_2Si_2$ compounds are listed in Table 3 except for the three IV systems. At high temperature the measured molar susceptibility is given by $\chi_m = \chi_o + C/(T-\theta)$, where χ_o is the temperature independent contribution extrapolated from the plot χ_m vs 1/T for $T \to \infty$. Figs. 7 and 8 show the temperature dependence of $\Delta\chi = \chi_m - \chi_o$ of Tb and Tm. Between 300 K and 150 K the magnetic susceptibility of $RECu_2Si_2$ for RE=Gd, Tb, Dy, Ho, Er and Tm shows a Curie Weiss law (CWL) with an effective moment of the trivalent RE ion for the Hund's rule ground state (Table 3). The good agreement between the experimental and theoretical p_{eff} values indicates that at higher temperatures a moment reduction by CEF effects does not occur. Below 100 K the susceptibility deviates slightly from CWL, apparently due to CEF effects. $SmCu_2Si_2$ shows the usual anomalous Van Vleck susceptibility of $4f^5$ (trivalent Sm). $PrCu_2Si_2$ and $NdCu_2Si_2$ exhibit Curie Weiss behavior down to 25 K and 10 K with a reduced magnetic moment. The compounds with RE = Pr, Sm, Gd, Tb, Dy and Ho order antiferromagnetically at low temperatures, as is evident from clear maxima in $\chi(T)$. We have

Table 3: Magnetic moments, ordering temperatures from susceptibility, magnetostriction and resistivity measurements, deviation of T_N from the spin-only de Gennes behavior and ratio of T_N deviation and $\Delta(c/a)$ anomaly for $RECu_2Si_2$

	Pr	Nd	Sm	Gd	Tb	Dy	Ho	Er	Tm
$p_{eff}(\mu_B)$ experiment	2.61	2.39	1.12	7.93	9.28	10.58	10.42	9.49	7.14
$p_{eff}(\mu_B)$ theory	3.58	3.62	0.84	7.94	9.72	10.63	10.60	9.59	7.57
T_N(K) from χ	20.8 3.3		9.0	14.2	12.5	11.8	7.7	 1.6	7.0 3.0
T_N(K) from $\Delta l/l$	22.5			12.0		10.5	7.0	2.0	7.0 3.5
T_N(K) from ρ				13.5	11.5	10.9	6.8	4.8 2.2	6.9 3.6
$T_N - T_{dG}$(K)	2.6		5.0	0	3.0	5.4	3.6		5.8
$\frac{T_N - T_{dG}}{\Delta(c/a)}$(K)	0.87		1.10	0	0.67	0.66	0.68		0.56

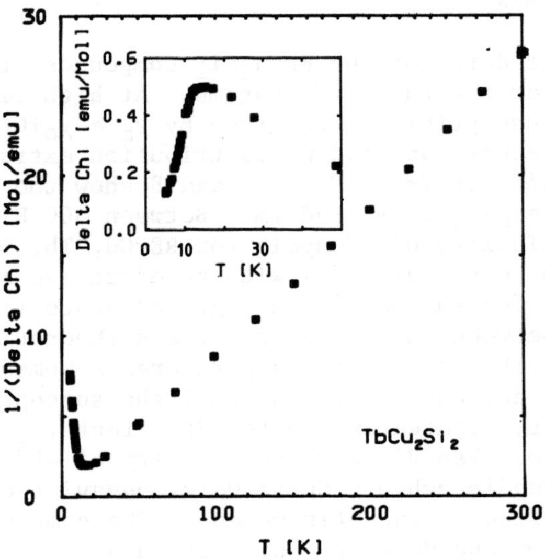

Fig. 7: Inverse molar susceptibility $(\Delta\chi)^{-1}$ vs temperature of $TbCu_2Si_2$ Insert shows $\Delta\chi$ vs temperature.

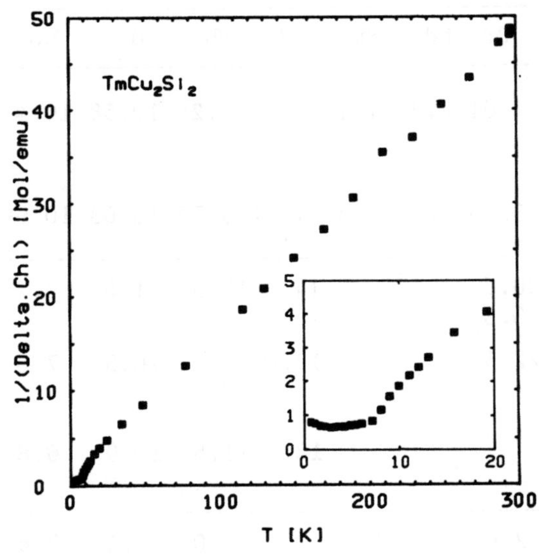

Fig. 8: Inverse molar susceptibility $(\Delta\chi)^{-1}$ vs temperature of $TmCu_2Si_2$ Insert shows low temperature part.

defined the Néel temperatures by the intercept temperatures of the slopes of $\chi(T)$ before and after the maximum. These temperatures are listed in Table 3. Also listed are the Néel temperatures as extracted by measurements of the electrical resistivity[6] and the magnetostriction[5]. The magnetic behavior exhibited by $RECu_2Si_2$, where RE=Nd, Er and Tm, is somewhat more complicated. In the case of $ErCu_2Si_2$ and $TmCu_2Si_2$ antiferromagnetic order is indicated by a very flat absolute maximum near the temperatures where other methods also detect phase transitions (at 1.6 K and 3.0 K respectively [Table 3]). No absolute maximum is seen in $NdCu_2Si_2$. However, in all three compounds there is an additional structure of $\chi(T)$ in the helium temperature region which appears to be connected with ordering effects as well. In particular, there is a clear break in the slope of $\chi(T)$ for $TmCu_2Si_2$ at 7 K. A comparison of T_N between various measurements shows reasonable agreement. There is a tendency for the T_N values from χ data to lie slightly above those from other measurements. The second to last line shows the difference of the measured Néel temperature vs the spin-only (de Gennes[7]) temperature extrapolated from $GdCu_2Si_2$. Clearly, the deviations from the de Gennes behavior are large in the $RECu_2Si_2$ series. The last line shows the ratio of the deviation and the c/a anomaly. As noted previously this ratio is quite constant for the heavy rare earths indicating some connection between the deviation from the de Gennes temperature and the quadrupoles. We note that all measured Néel temperatures are larger than the de Gennes temperatures. We speculate that the c/a anomaly corresponds to some alignment of the quadrupolar axis parallel to some common axis. This alignment helps magnetic ordering.

SUMMARY

Our temperature dependent x-ray measurements reveal the continuous development of an anomaly of the c/a ratio with decreasing temperature for all compounds which have a 4f orbital angular momentum $L \neq 0$. The anomaly can be regarded as a volume conserving distortion of the unit cell. We suggest that the origin of the distortion is the alignment of the quadrupolar moment of the RE charge distribution in the electric field gradient of the noncubic lattice sites for the tetragonal $ThCr_2Si_2$ structure. The low temperature value of the anomaly is nearly the same value as that of the saturated magnetostriction of the same RE ions diluted into YCu_2Si_2. The magnetic ordering temperatures for the $L \neq 0$ RE ions are larger than expected on the basis of $GdCu_2Si_2$ (spin-only interaction. The deviation scales with the c/a anomaly i.e. with the size of the quadrupolar alignment in the electric field gradient. Static susceptibility, inelastic neutron scattering and elastic constant measurements on single crystals will be required to understand the observed correlation between lattice distortion and magnetic ordering more completely.

The authors wish to thank Prof. D. Wohlleben for many helpful discussions and the critical reading of this manuscript.

Work supported by the Deutsche Forschungsgemeinschaft SFB 125.

REFERENCES

1. E.R. Bauminger, D. Froindlich, I. Nowik, S. Ofer, I. Felner and I. Mayer, Phys. Rev. Lett. 21, 1053 (1973); H. Oesterreicher, Phys. Stat. Sol. (a) 34, 723 (1976).
2. B.C. Sales and R. Viswanathan, J. of Low Temp. Phys. 23, 449 (1976).
3. R. Rieger and Parthé, Monatsh. Chem. 100, 444 (1969).
4. G. Neumann, Diplomarbeit, II. Physikalisches Institut der Universität zu Köln (1981); D. Plümacher, Doktorarbeit, II. Physikalisches Institute der Universität zu Köln (1980); G. Neumann, D. Plümacher, W. Schlabitz and D. Wohlleben, to be published.
5. N. Rüssman, H.U. Häfner and D. Wohlleben, this conference.
6. E. Cattaneo and D. Wohlleben, to be published.
7. P.G. de Gennes, Compt. Rend. 247, 1836 (1958).

COMMENTS

KASUYA: On the slide for resistivity I found only a sharp drop at the Néel temperature. Is there any evidence of the anomaly due to the quadrupolar scattering?

SCHLABITZ, WOHLLEBEN: Yes, we have seen a different temperature dependence of the resistivity for the RE ions with $L \neq 0$ compared with $GdCu_2Si_2$ and a weak anomaly in the first derivative of ρ at 6.9 K for $TmCu_2Si_2$. But the measurements have to be repeated on single crystals.

FULDE: The fact that T_N does not scale with the deGennes curve seems to indicate that the magnetic ordering is not due to the isotropic exchange interaction but rather to the anisotropic interactions. The corresponding curves have been calculated by Hirst and can show a maximum at Pr, for example, because they involve orbital effects.

GREIDANUS: Why is the ordering temperature of the Pr compound so high? Is this related to the ground state?

SCHLABITZ: From the susceptibility point of view it looks like the phase transition is very complicated.

GREIDANUS: Is there not a relation between $d\rho/dT$, the specific heat and the crystal-field splitting?

SCHLABITZ, WOHLLEBEN: Yes, to see crystal field effects in the resisitivity it is best to analyze $d\rho/dT$.

STEGLICH: In your $RECu_2Si_2$ compounds, can any temperature dependence be resolved in the $\rho(T)$ dependence in the paramagnetic state which is associated with CF excitation effects?

SCHLABITZ, WOHLLEBEN: Yes, there is some T-dependence for some compounds. This differs from the $\rho(T)$ behavior of $GdCu_2Si_2$ and should be correlated with the CF excitation effects. Unfortunately, the CF level scheme is not known by neutron scattering so that a quantitative check has not been possible so far.

INFLUENCE OF THE CRYSTAL FIELD ON Dy^{3+} IONS IN DyM_2Si_2 COMPOUNDS AS REVEALED BY INVESTIGATIONS OF THEIR MAGNETIC PROPERTIES AND NUCLEAR HYPERFINE INTERACTIONS

E.A. Görlich*, R. Kmiec*, B. Janus† and A. Szytuła*

*Institute of Physics, Jagiellonian University
Krakow, Poland
†Institute for Low Temperature and Structural Research
Wrocław, Poland

The crystal field interaction exhibits a pronounced effect on the nuclear hyperfine coupling in the tetragonal dysprosium compounds DyM_2Si_2 (M=Mn, Re, Co, Ni, Cu)[1]. The arrangement of atoms surrounding the dysprosium ion is shown in Fig. 1. The crystalline electric field influence is also visible in the results of the bulk magnetic measurements. Fig. 2 shows the magnetization and the inverse susceptibility for $DyFe_2Si_2$ and $DyCo_2Si_2$. The latter compound undergoes an antiferromagnetic transition at T_N = 19 K while the former remains paramagnetic down to helium temperatures. The influence of

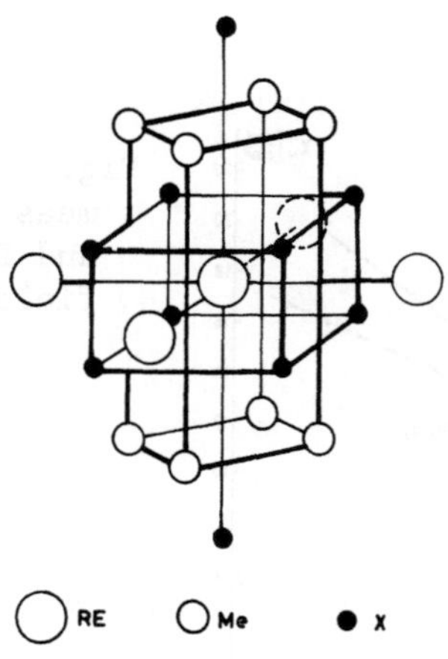

Fig. 1: The surrounding of the dysprosium ion in DyM_2Si_2 compounds (point symmetry D_{4h}).

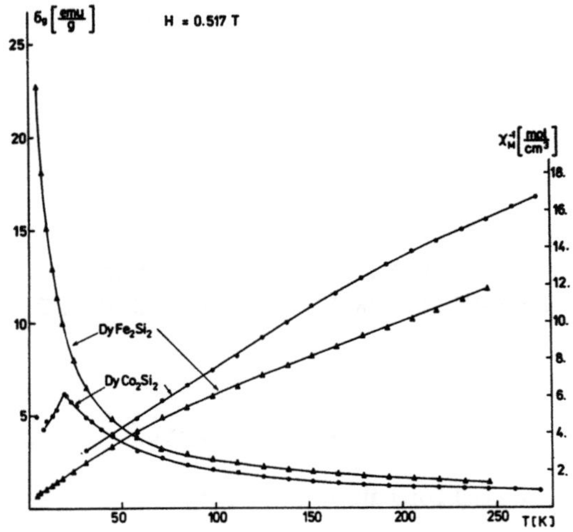

Fig. 2: Measurements of the bulk magnetic properties of $DyFe_2Si_2$ and $DyCo_2Si_2$ taken on the Foner type magnetometer in a 0.517 T field.

the crystal field is clearly reflected in the temperature dependence of the inverse molar susceptibility of $DyNi_2Si_2$ (Fig. 3), since this compound stays paramagnetic at least down to 4.2 K. The susceptibility of $DyCu_2Si_2$ obeys a Curie-Weiss law with an effective moment of 10.3 μ_B (Fig. 3) and the compound orders antiferromagnetically at T_N = 11 K. In Fig. 4 we see the dependence of the magnetization of $DyCu_2Si_2$ at 4.2 K upon the application of a magnetic field. This indicates a spin reorientation process being forced by the external field. The effect vanishes close to the Néel temperature (see the 11 K curve in Fig. 4). The less smeared spin reorientation behavior for another polycrystalline sample of $DyCo_2Si_2$ can also be seen.

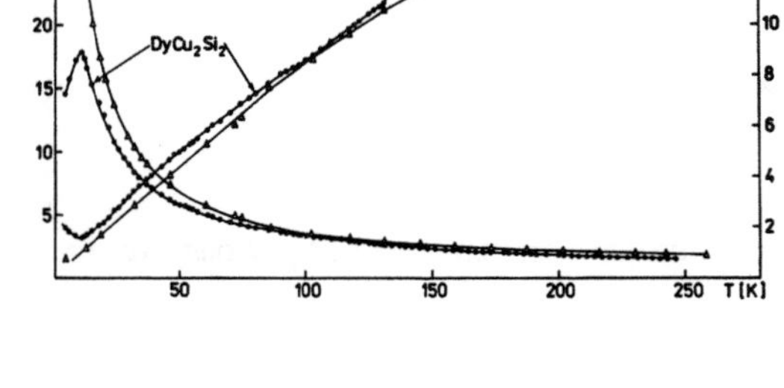

Fig. 3: The magnetic measurements for $DyNi_2Si_2$ and $DyCu_2Si_2$ in the applied field of 1 T.

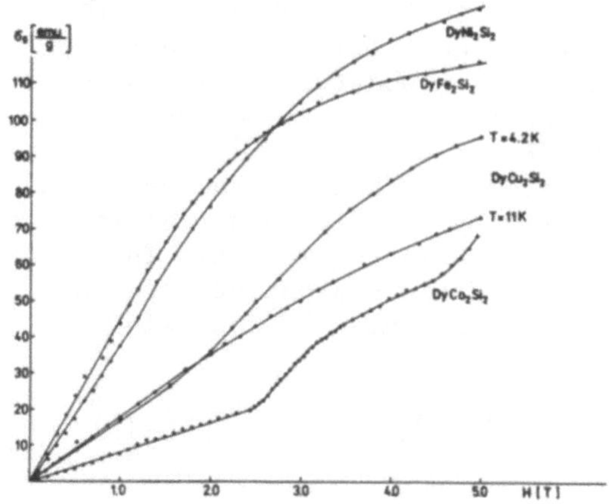

Fig. 4: The magnetization of DyM_2Si_2 polycrystalline samples as a function of applied field. For $DyCu_2Si_2$ a curve taken at 11 K is included.

As far as the hyperfine (hf) interactions of the dysprosium nuclei are concerned we have observed the Mössbauer effect for the 25.6 keV transition in ^{161}Dy. The trivalent dysprosium Kramers ion ($4f^9$) has the Hund's rule ground state of $^6H_{15/2}$, and therefore the coupling of the nucleus to the 4f shell is most important in its hf interactions. All dynamical effects involving the 4f angular momentum find reflection in the observed hf pattern of the resonantly absorbed gamma radiation if their rate is comparable with the frequency of the Larmour precession of the nuclear spin in an internal field. As in the non-S state rare earth ions, this field is unusually high (Dy^{3+} free ion value H_{hf}^{4f} = 580 T). On the other hand, intermetallic compounds exhibit strong magnetocrystalline anisotropy, and it is not uncommon to observe the static hf structure in the macroscopic paramagnetic state. This is also the case with the compounds investigated here. The shape of the hf Mössbauer spectrum (Fig. 5) of $DyCo_2Si_2$ does not lose its character up to about 40 K above the Néel temperature.

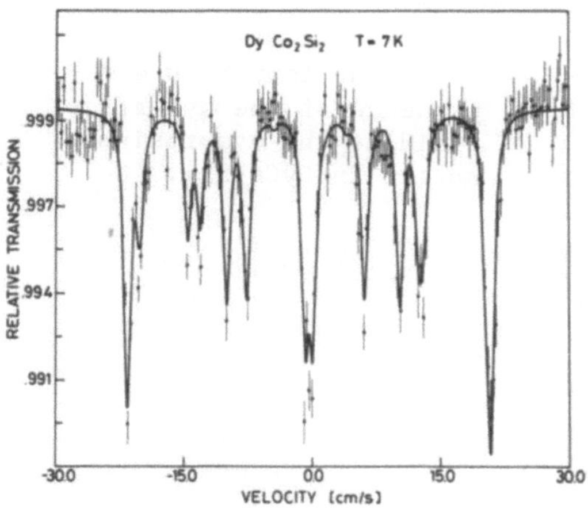

Fig. 5: The Mössbauer spectrum for the 25.6 keV transition in ^{161}Dy taken at 7 K for $DyCo_2Si_2$.

Fig. 6 shows the hf field vs T of Dy nuclei in $DyCo_2Si_2$. The striking feature is an increase of its value below T_N. Hereafter, we will show semi-qualitatively that such an observation may be expected when one considers the impact of the effective molecular field on the ground state within the $^6H_{15/2}$ term of Dy^{3+}. The dysprosium site has tetragonal (D_{4h}) symmetry (Fig. 1) and we assume that CF interactions may be approximated by a cubic potential in addition to the leading quadrupole term, while the magnetic interaction is accounted for by the molecular field approximation. The Hamiltonian with respect to the fourfold axis has the following form:

$$H = g_J\mu_B\vec{J}\cdot\vec{H}_{max}+B_2^0O_2^0+B_4^0(O_4^0+5O_4^4)+B_6^0(O_6^0-21O_6^4). \qquad (1)$$

In order to investigate the properties of the solutions of the eigenproblem we have used the parametrization analogous to that of Lea, Leask and Wolf for a cubic potential:

$$H = W\{2zg_J\mu_B\vec{J}\cdot\vec{H}_{max}+(1-|y|)\frac{O_2^0}{F_2^0}+y[x\frac{O_4^0+5O_4^4}{F_4^0}+(1-|x|)\frac{O_6^0-21O_6^4}{F_6^0}] \ , \qquad (2)$$

where

$$B_2^0 = \frac{W(1-|y|)}{F_2^0}, \quad B_4^0 = \frac{Wxy}{F_4^0}, \quad B_6^0 = \frac{Wy(1-|x|)}{F_6^0} .$$

The effective hf field and the axial electric field gradient at the nucleus, both originating from the 4f shell, are proportional to $\langle\Gamma|J_z|\Gamma\rangle$ and $\langle\Gamma|3J_z^2-J(J+1)|\Gamma\rangle$, respectively. Energies and wave functions have been obtained by numerical diagonalization of the Hamiltonian. The exchange field dependence of the expectation value of the z-component of the total angular momentum is plotted in Fig. 7. Magnetostrictive effects, considered as being of secondary

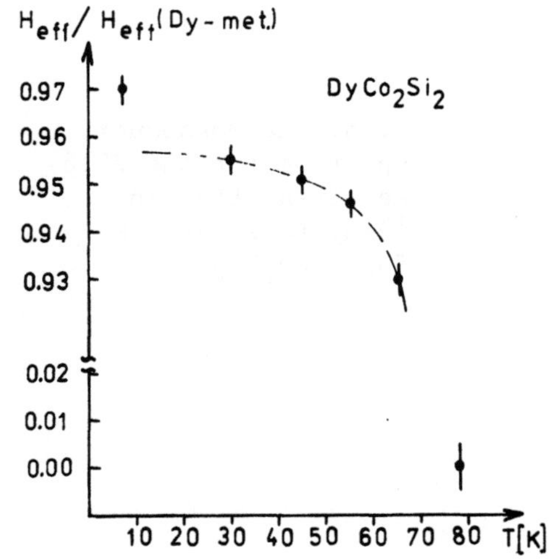

Fig. 6: The temperature dependence of the hyperfine field at the dysprosium nuclei in $DyCo_2Si_2$.

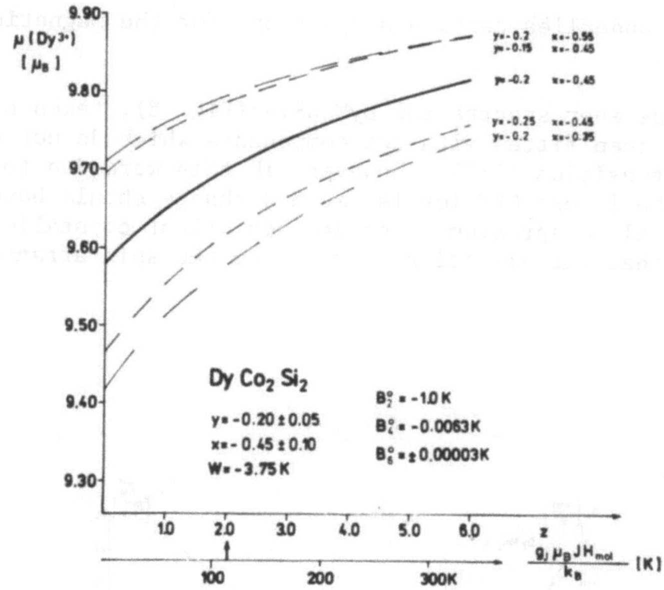

Fig. 7: The expected modification of the ground state magnetic moment of Dy^{3+} by the molecular field (H_{mol} along the c-axis).

importance, have not been taken into account. The main curve was obtained for the crystal field parameters as indicated in Fig. 7. They were estimated from the nuclear hyperfine parameters. An arrow on the bottom scale points to the value of the molecular field which results from using the given T_N, and the increase of the hf field is predicted correctly. The dashed lines are intended to show the sensitivity to the CF coefficients. Essentially the same kind

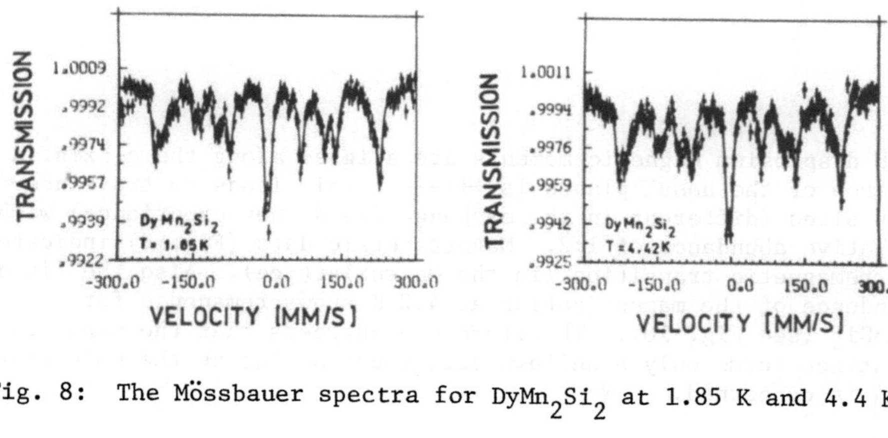

Fig. 8: The Mössbauer spectra for $DyMn_2Si_2$ at 1.85 K and 4.4 K.

of argument underlies certain suggestions for the magnetic structure of $DyMn_2Si_2$.

The Mössbauer spectra for $DyMn_2Si_2$ (Fig. 8), taken at 1.8 K and 4.4 K, have been fitted with two components which do not change their relative intensities (1:2). However, if this were due to the population of the lowest CEF levels, such a change should have been expected. As all dysprosium sites are equivalent crystallographically, we believe that the situation arises from the spin arrangement in

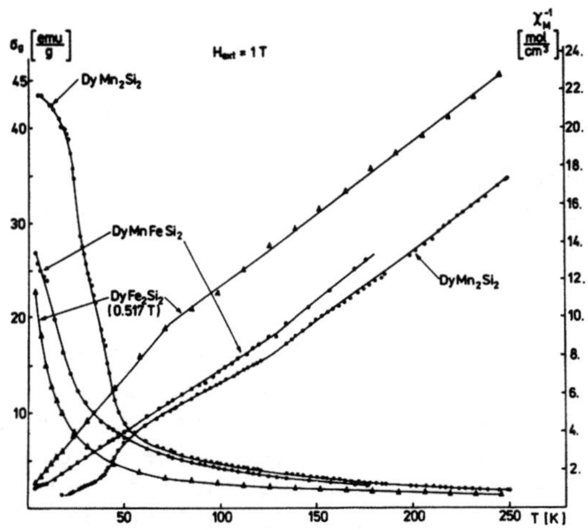

Fig. 9: Magnetization vs. T of $DyMn_2Si_2$ in a field of 1 T.

which dysprosium magnetic moments are aligned along the c-axis. A sequence of the basal planes is +++---. This leads to two kinds of Dy sites (different in the exchange field they experience) with a relative abundance of 1:2. Magnetometric data (Fig. 9) indicate a ferromagnetic transition (in the Mn sublattice). Also the field dependence of the magnetization at 4.2 K shows remanence for $DyMn_2Si_2$ (see Fig. 10). Therefore one suggests that the manganese sublattice forms only a uniform background as far as the molecular field is concerned.

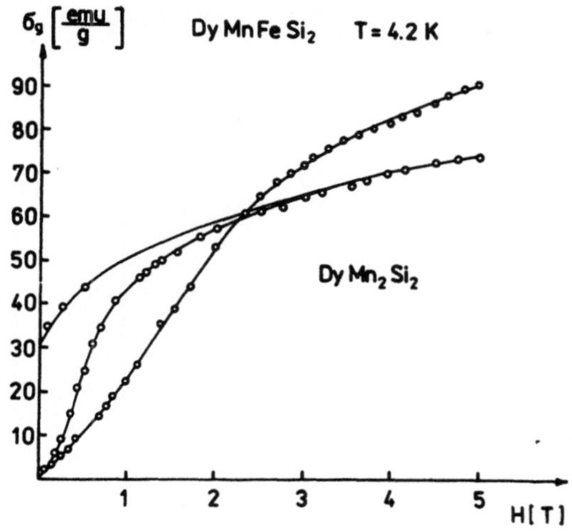

Fig. 10: The field dependence of the magnetization of $DyMn_2Si_2$ at 4.2 K.

REFERENCE

1. E.A. Görlich, A.Z. Hrynkiewicz, R. Kmiec, K. Tomala (to be published).

COMMENTS

WALTER: The CF Hamiltonian for tetragonal symmetry has 5 parameters, in general. What is the reason that you used a Hamiltonian for cubic symmetry plus a second-order term?

GORLICH: Certainly, the correct Hamiltonian for tetragonal symmetry should introduce 5 parameters, but as the quadrupole term is the leading one, we have accepted the above approximation. Moreover, we have shown that the discussed tendency of the changes of $\mathcal{H}_{4f}$ and U^{latt} with B_2 value is hardly affected by the higher order terms (at least in the region where the second order term is dominant).

MAGNETIC PROPERTIES OF RCo_2Si_2 and RCo_2Ge_2 COMPOUNDS

M. Kolenda*, A. Szytuła* and A. Zygmunt†

*Institute of Physics, Jagellonian University
30-059 Krakow, Poland
†Institute of Low Temperature and Structure Research
Polish Academy of Sciences, Wrocław, Poland

INTRODUCTION

The ternary compounds described by the formula RM_2X_2, where R is a rare earth, M is transition metal and X is Si or Ge, have been extensively studied[1-6]. These compounds crystallize in the $ThCr_2Si_2$ structure of the space group I4/mmm[1]. Previous work reports that the compounds were ferromagnets or antiferromagnets depending on the transition metal[2-4]. The present study is a systematic investigation of the magnetic properties of RCo_2Si_2 and RCo_2Ge_2 compounds and a comparison of their magnetic properties.

EXPERIMENTAL RESULTS

The samples were prepared by melting together rare earth metal (4n's), cobalt (4n's) and silicon or germanium (5n's purity) in an arc furnace under a purified atmosphere of argon gas. The samples were then annealed in a quartz tube at 800 C for 100 hours and cooled to room temperature. Powder samples were examined by x-ray diffraction using Fe K_α radiation. The powder patterns of all the samples could be indexed on the basis of the I4/mmm body-centered tetragonal type structure.

Lattice parameters of the compounds determined by the x-ray examination were found to be in good agreement with those obtained by Rossi et al.[5] and McCall et al.[7]. According to Rossi et al.[5] and Szytuła et al.[6] the R atoms occupy the 2(a) sites, the Co occupy the 4(d) sites and Si or Ge the 4(e) sites. The structure consists of atomic layers perpendicular to the c-axis, stacked with the sequence of R-X-M-R. The observed values of the average atomic volumes as a function of the atomic number of the rare earth vary linearly except for the typical deviations of the Ce compounds (especially for the Si compounds) and for the Yb compounds (especially for the Ge compounds).

Magnetometric measurements were carried out in the temperature range from 4.2 to 300 K using a vibrating sample magnetometer (Foner type) and in the temperature range 120 to 500 K using an electronic balance of the Sucksmith type. The magnetic data are listed in Table 1 and in Figs 1(a,b) and 2(a,b). The results of the magnetometric measurements show:

1. The RCo_2Si_2 compounds with R=La and Lu are Pauli paramagnets: their susceptibilities are independent of temperature.

2. The temperature dependence of the susceptibility of $CeCo_2Si_2$ has a complicated character.

3. The reciprocal magnetic susceptibility of the RCo_2Si_2 compounds (R=Sm, Er and Tm) obeys the Curie-Weiss law down to 4.2 K. This shows the lack of ordering of magnetic moments above 4.2 K.

4. The temperature dependence of the magnetic susceptibility for the compounds RCo_2Si_2 (R=Pr, Gd, Tb, Dy and Ho) and RCo_2Ge_2 (R=Pr, Nd, Tb and Ho) show characteristic maxima typical of the Néel temperatures. Above the Néel temperatures, the reciprocal magnetic susceptibility obeys the Curie-Weiss law.

5. For the samples $PrCo_2,X_2$ (X=Si, Ge) and $NdCo_2Ge_2$ below the Neel temperature there is a jump in the magnetic susceptibility at 12, 18 and 10 K respectively. These results show the existence of an additional magnetic phase.

For the antiferromagnetic compounds additional measurements of the magnetic field dependence on the magnetization were made up to H=50 kOe at a temperature of 4.2 K. The results are given in Figs. 1(a,b) and 2(a,b) and show:

1. The field dependence of the magnetization of $GdCo_2Si_2$ is linear up to 50 kOe.

2. For the remaining compounds the variation of magnetization as a function of the external field has a complex character. In low external field the magnetization vs magnetic field is linear. In high external field deviations from linearity with one or two kinks are observed. The values of the critical field are given in the figures.

3. For all compounds the values of the magnetic moments at T=4.2 K and H=50 kOe are smaller than those expected for the R^{3+} free-ion (gJ). (See Table 1).

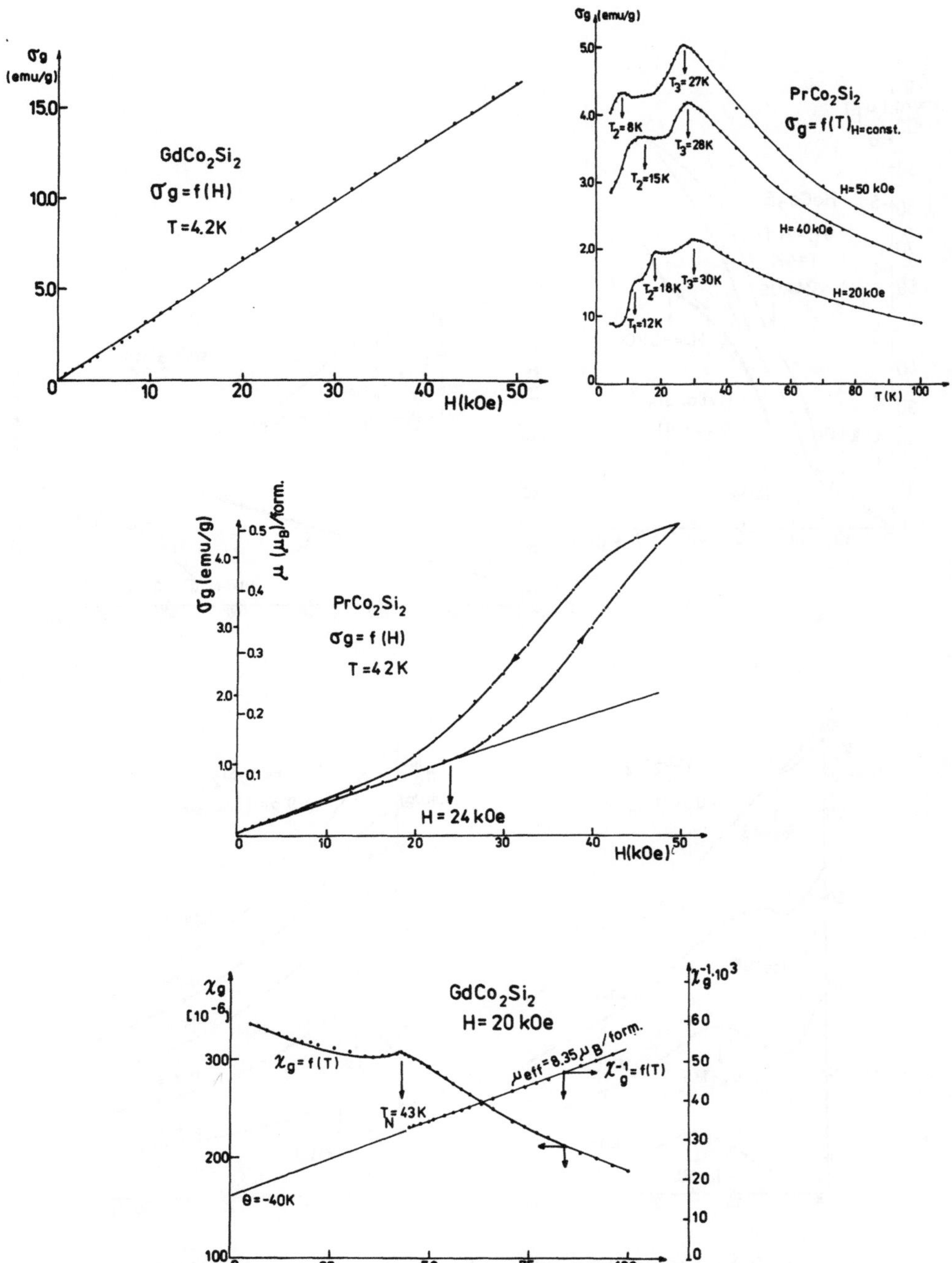

Fig. 1 (a): Magnetic properties of RCo_2Si_2 compounds.

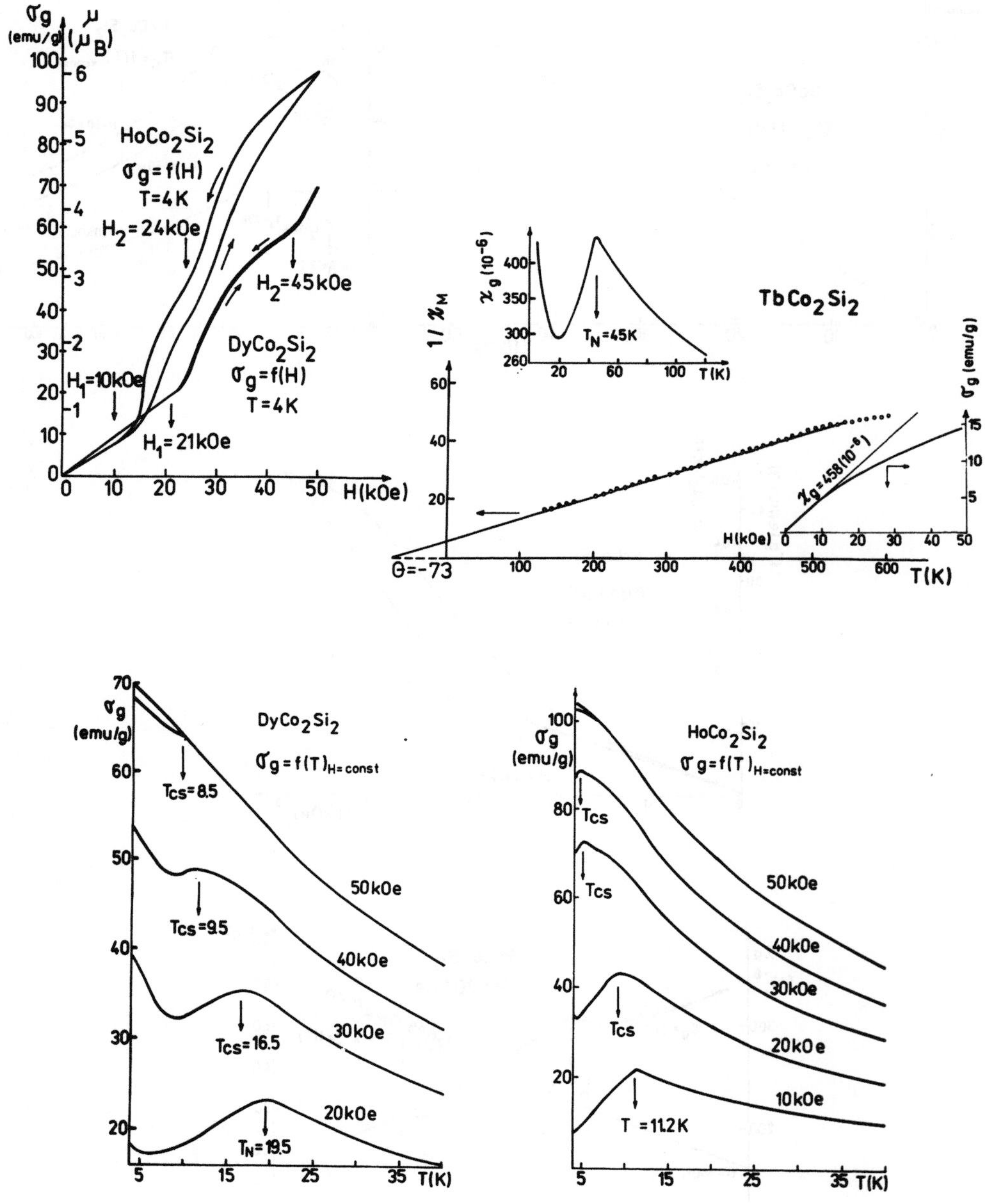

Fig. 1 (b): Magnetic properties of RCo_2Si_2 compounds.

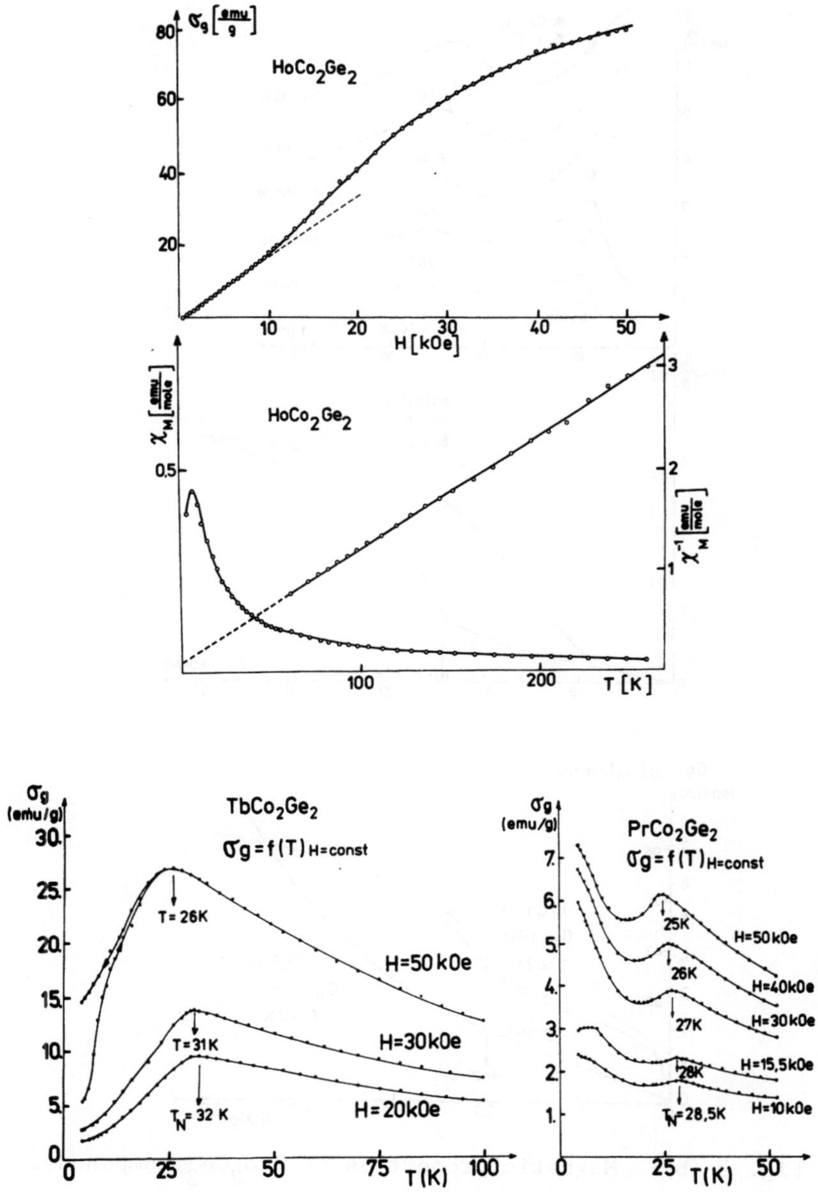

Fig. 2 (a): Magnetic properties of RCo_2Ge_2 compounds.

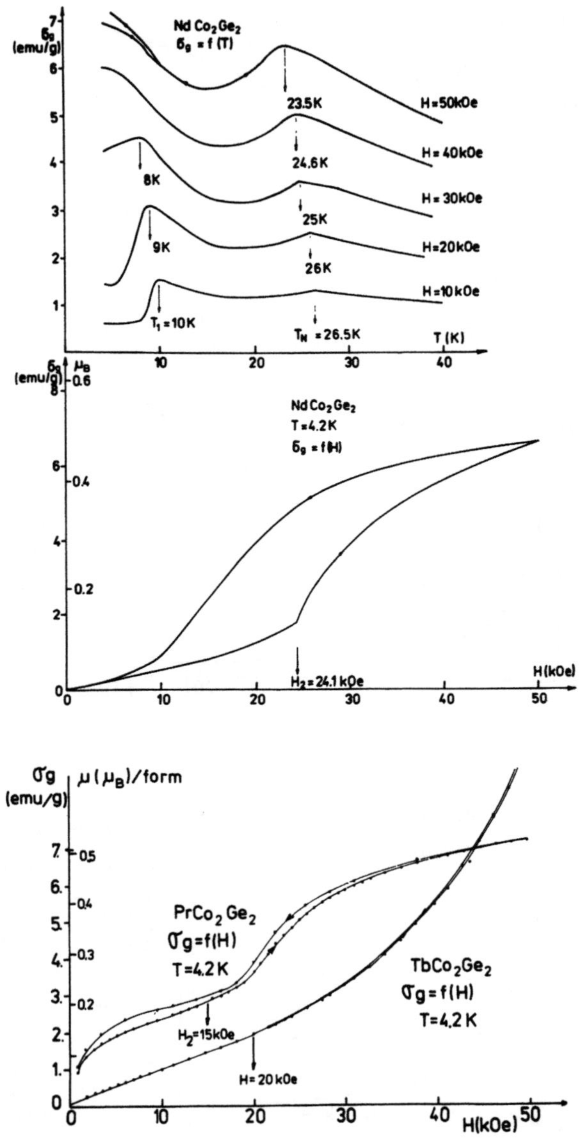

Fig. 2 (b): Magnetic properties of RCo_2Ge_2 compounds.

Table 1. Magnetic data for RCo_2Si_2 and RCo_2Ge_2 compounds

R	RCo_2Si_2				RCo_2Ge_2				Theoretical Values for R^{3+} ions:	
	T_N(K)	θ_p(K)	$\mu_{eff}(\mu_B)$	$\mu_n(\mu_B)$	T_N(K)	θ_p(K)	$\mu_{eff}(\mu_B)$	$\mu_n(\mu_B)$	gJ	$g\sqrt{J(J+1)}$
Ce						-152†	2.74		2.14	2.54
Pr	30	-15	3.7	0.5	28.5 28†	-9 -9	2.9 2.9	0.55	3.20	3.58
Nd					26.5 28†	-5.5 -32.5	4.09 3.34	0.4	3.28	3.62
Sm					†	-43.8	1.12		0.72	0.85
Gd	43 44*	-40 -55	8.3 9.9	1.0	40†	-39.5	6.96		7.0	7.94
Tb	45 8*	-73 -18	10.2 9.4	0.85	32 30†	-22 -44	10.3 9.27	0.7	9.0	9.72
Dy	19.5 21*	-27 -35	10.5 10.2	4.5	14†	-4.7	10.17		10.0	10.64
Ho	11.2 15*	-10 -8	10.5 10.7	6.1	8 8†	-10 6	10.0 10.38	6.2	10.0	10.61
Er	6*	-9	10.4		†	8.7	7.6		9.0	9.58
Tm		-8	6.9	3.2	†	-0.8	5.89		7.0	7.56

T_N = Neel temperature; θ = paramagnetic temperature; μ_{eff} = effective moment; μ_n = value of magnetic moment at T=4.2 K and H=50 kOe.

(*) Obtained by Yakinthos et al.[4].

(†) Obtained by McCall et al.[2].

DISCUSSION

The results of the magnetometric measurements indicate that the compounds RCo_2Si_2 (R=Pr, Gd, Tb, Dy, and Ho) and RCo_2Ge_2 (R=Pr, Nd, Tb and Ho) exhibit antiferromangetic properties. This confirms the results of neutron diffraction experiments for RCo_2Ge_2 (R=Pr, Nd, Tb and Ho) and $TbCo_2Si_2$[6,7]. In all cases the Néel temperatures for germanides are smaller than those for silicides. This suggests that the Néel temperature depends on interlayer distances. The paramagnetic temperatures are negative, giving evidence of the predominant antiferromagnetic interaction. The values of the effective moment for all the compounds are in good agreement with the theoretical values for free rare earth ions. This result indicates that the magnetic moment is due to the R atoms only and that the cobalt in these compounds does not order magnetically. This is in good agreement with the neutron diffraction data[6,7].

Low values for the temperatures of magnetic phase transitions and large R-R atom distances suggest that the magnetic exchange interaction occurs most probably via conduction electrons. The fact that similar compounds RFe_2Si_2 (R=Nd, Eu and Dy)[9] and $GdMn_2Ge_2$[10] are good electrical conductors additionally supports this idea. According to the RKKY model, the Néel temperatures follow the deGennes function. However, most of the RCo_2X_2 (X=Si and Ge) compounds are antiferromagnetically ordered and their Néel temperatures do not follow the deGennes function. These results indicate that in these compounds another magnetic interaction exists. It is possible that this is a super-exchange via two X atoms. The magnetic structures of all RM_2X_2 compounds[6,7,11] contain ferromagnetic layers perpendicular to the c axis, which are antiferromagnetically coupled to adjacent layers.

The measurements of magnetization vs magnetic field at 4.2 K for all antiferromagnetic compounds (except $GdCo_2Si_2$) show one or two phase transitions. The magnetization curves for different compounds have different characters (see Figs. 1(a,b) and 2(a,b)). For all compounds (except $PrCo_2Ge_2$) in low external fields the mangetization vs magnetic field is a linear function up to a critical value H_c. Neutron diffraction measurements for some compounds[6,7] indicate a colinear antiferromagnetic structure. In high external fields ($H>H_c$) deviations from linearity are observed. This suggests metamagnetic or spin flop phase transitions. Magnetic moments measured at 50 kOe and T=4.2 K are much smaller than gJ, indicating that saturation was not reached. This indicates the existence of strong magnetocrystalline anisotropy. The values of crystal field parameter B_2^0 determined for the samples $DyCo_2Si_2$[12] and $GdCo_2Ge_2$ confirm the above results. For these samples, $B_2^0 \simeq 1.0$ K is unexpectedly large, and suggests a strong one-ion magnetocrystalline anisotropy.

REFERENCES

1. W. Rieger and E. Parthé, Monatshefte für Chemie 100, 444 (1969).
2. W.M. McCall, K.S.V.L. Narasimhan and R.A. Butera, J. Appl. Phys. 44, 4724 (1973).
3. I. Felner, J. Phys. Chem. Solids 36, 1063 (1975).
4. J.K. Yakinthos, Ch. Routsi and P.F. Ikonomou, J. of Less-Common Metals 72, 205 (1980).
5. D. Rossi, R. Marazza and R. Ferro, J. of Less-Common Metals 58, 203 (1978).
6. A. Szytuła, J. Leciejewicz and H. Binczycka, Phys. Status Solidi (a) 58, 67 (1980).
7. W.M. McCall, K.S.V.L. Narasimhan and R.A. Butera, J. Appl. Cryst. 6, 301 (1973).
8. J. Leciejewicz, S. Siek, A. Szytuła and A. Zygmunt, These Proceedings.
9. I. Felner and I. Mayer, Mater. Res. Bull. 8, 1317 (1973).
10. K.S.V.L. Narasimhan, V.U.S. Rao, R.L. Bergner and W.E. Wallace, J. Appl. Phys. 46, 4957 (1975).
11. H. Pinto and H. Shakec, Phys. Rev. B7, 3261 (1973).
12. K. Łatka (private communication).

COMMENTS

WOHLLEBEN: What is the magnetic state of cobalt in the $RECo_2Si_2$ compounds? What are the high temperature susceptibility measurements telling you about this when the RE is magnetic? What about when the RE is La or Lu?

SZYTUŁA: The magnetic moment is localized only on the RE^{3+} ions. The observed paramagnetic moment is in good agreement with the theoretical values of the free RE^{3+} ions. This result is in good agreement with the neutron diffraction data for these compounds (presented by Prof. J. Leciejewicz during this Conference). There is no magnetic contribution to the neutron peaks coming from the Co ions. For the compounds containing La or Lu the susceptibility is temperature independent, thus indicating the existence of Pauli paramagnetism.

^{169}Tm MÖSSBAUER STUDY OF $TmCu_2Si_2$

Glen Alan Stewart and Jan Zukrowski[+]

Institut für Atom- und Festkörperphysik
Freie Universität Berlin
[+]On leave from Solid State Physics Dept., AGH, Cracow
Poland

INTRODUCTION

Until now, the only detailed crystal electric field (CEF) information available for the tetragonal RCu_2Si_2 (R=rare earth) series has been that obtained recently for $CeCu_2Si_2$ by means of neutron diffraction measurements[1]. In order to obtain the CEF parameters for a heavy rare earth compound we have measured the quadrupole splitting, ΔE_Q, of the ^{169}Tm, 8.4 keV Mössbauer resonance in $TmCu_2Si_2$ over the temperature range 1.5-400 K.

EXPERIMENTAL DETAILS

The $TmCu_2Si_2$ sample was prepared in an argon arc furnace and then vacuum-annealed for 2 weeks at 800 C. X-ray diffraction analysis revealed a single tetragonal phase passing to the RCu_2Si_2 systematics[2]. For the Mössbauer measurements a 12 mg/cm^2 $TmCu_2Si_2$ absorber was cooled (or heated) in a transmission cryostat. The ^{169}Er:$^{168}Er_{1.5}Al_{98.5}$ source was moved sinusoidally and mounted outside at room temperature.

THEORETICAL BACKGROUND

The CEF Hamiltonian for a rare earth ion in a site of tetragonal symmetry is

$$\mathcal{H}_{CEF} = B_2^0 O_2^0 + B_4^0 O_4^0 + B_4^4 O_4^4 + B_6^0 O_6^0 + B_6^4 O_6^4 \quad (1)$$

with the c-axis taken as quantization axis for the O_k^m operator equivalents. For this symmetry the quadrupole splitting of the ^{169}Tm, I=3/2, excited state can be expressed as

$$\Delta E_Q = \left|\frac{1}{2} eV_{zz}Q\right| \tag{2}$$

where V_{zz} is the axial component of the electric field gradient (EFG). A value of Q=-1.2(1)b will be taken for the nuclear quadrupole moment[3].

There are two main contributions to V_{zz}. The 4f-electron contribution

$$V_{zz}^{4f} = -e<J||\alpha||J>(1-R_Q)<r^{-3}>_{4f}<3J_z^2-J(J+1)>_T \tag{3}$$

is attenuated by the atomic Sternheimer effect (R_Q). The last term in Eq. (3) is the average of the expectation values from each of the (thermally populated) CEF levels. Consequently, V_{zz}^{4f} has a strong temperature dependence, the nature of which is determined by the full set of B_k^m values. The second contribution is the direct lattice EFG which is enhanced by the Sternheimer antishielding effect (γ_∞). It is practically temperature independent and can be expressed in terms of B_2^o alone.

$$V_{zz}^{latt} = -4\left[(1-\gamma_\infty)/\{e<J||\alpha||J>\ (1-\sigma_2)<r^2>_{4f}\}\right]B_2^o \tag{4}$$

Because many of the atomic parameters depend on the chemical environment of the Tm^{3+} ion it is convenient to define the "shielding parameters"

$$\rho_1 = Q(1-R_Q)<r^{-3}>_{4f},$$
$$\rho_2 = Q(1-\gamma_\infty)/\{(1-\sigma_2)<r^2>_{4f}\}. \tag{5}$$

The magnetic hyperfine interaction is

$$\mathcal{H}_m = A_I I_z J_z = \frac{\mu(I)}{I} I_z B_{eff} \tag{6}$$

where B_{eff} is the effective hyperfine field. The moments for the nuclear ground and excited states will be taken as $\mu(1/2)=-0.231(2)$nm and $\mu(3/2)= -2.223(13)\mu(1/2)$ respectively[4,5].

RESULTS AND ANALYSIS

The 3 K and 300 K spectra shown in Fig. 1 are representative quadrupole split ^{169}Tm Mössbauer spectra for $TmCu_2Si_2$. Whereas the splitting is large at 3 K it is insufficient above 250 K to visibly resolve the two lines. However, the ΔE_Q values at these higher temperatures could be obtained by fitting two Lorentzian profiles with a fixed intensity ratio. A feature of the spectra is their asymmetry

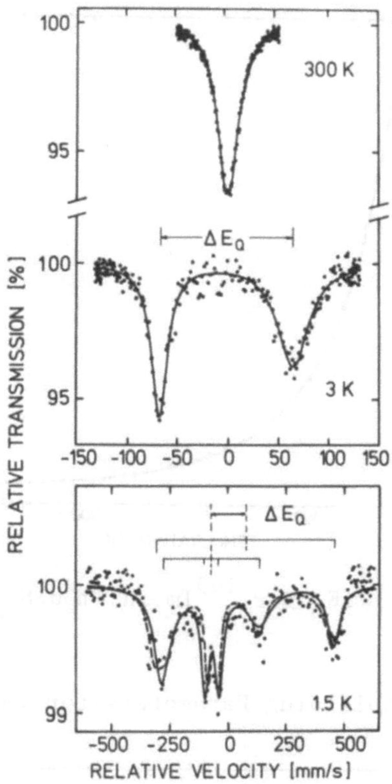

Fig. 1. Representative ^{169}Tm Mössbauer spectra for $TmCu_2Si_2$.

which persists to high temperatures and is due to the relaxing magnetic hyperfine interaction. This enables the sign of V_{zz} to be determined as negative over the full temperature range. ΔE_Q is shown in Fig. 2 as a function of temperature and is observed to pass through a maximum at 12 K. Above 300 K it has fallen to a constant value so that the lattice contribution can be estimated at V_{zz}^{latt}=-2.5(7)x10^{17} V/cm^2 using Eq. (2).

If it is assumed that the CEF's are approximately the same at rare earth ions in different RCu_2Si_2 compounds then B_k^m estimates for Tm^{3+} in $TmCu_2Si_2$ can be converted from those B_k^m values already derived for Ce^{3+} in $CeCu_2Si_2$[1]. The three values thus obtained are given in the last column of Table 1 (J=5/2 for Ce^{3+} so that the k=6 terms play no role in the CEF splitting). First attempts at fitting the $\Delta E_Q(T)$ data used shielding parameters measured elsewhere[6] for Tm^{3+} in thulium sulphate (TmES) and endeavoured to retain those B_k^m values converted from $CeCu_2Si_2$. The "restricted" fit described in Table 1 and shown as a solid curve in Fig. 2 came closest to achieving this. An additional fit was performed with

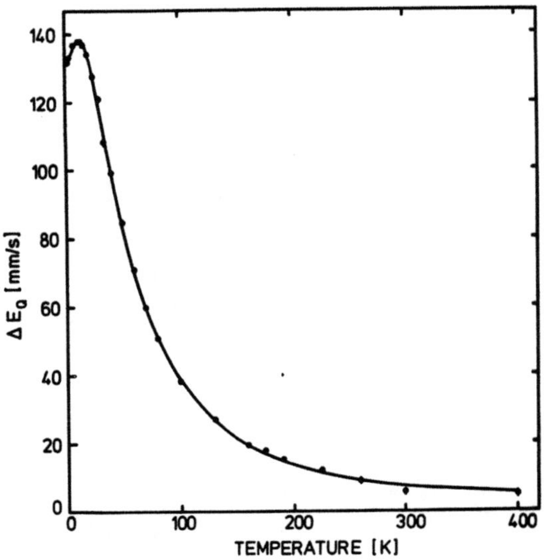

Fig. 2. ΔE_Q for ^{169}Tm in $TmCu_2Si_2$

Table 1. CEF- and Shielding Parameters for Tm^{3+} in $TmCu_2Si_2$

	"Restricted" Fit	"Full" Fit	Converted from $CeCu_2Si_2$
B_2^0 [K]	+0.4[a]	+0.12(2)	+0.4(1)
B_4^0 [mK]	-39.7(4)	-31.2(3)	-3.2(8)
B_4^4 [mK]	-19.8(5.0)[b]	-49.0(3.0)[b]	-49.6(4)
B_6^0 [mK]	-0.108(4)	-0.136(5)	
B_6^4 [mK]	+1.78(2)[b]	+1.26(1)[b]	
ρ_1 [ba_o^{-3}]	-15.0[a,c]	-13.5(1)	
ρ_2 [ba_o^{-2}]	-550[a,c]	-1560(150)	
$\rho_2 B_2^0$ [Kba_o^{-2}]	-222	-180(42)	
Δ [K]	14.4	8.3	
α	5.5	5.6	
Δ' [K]	58	78	

[a]Treated as constant.
[b]Simultaneous sign change of B_4^4 and B_6^4 leaves V_{zz} unaltered.
[c]Thulium ethylsulphate values[6].

the two shielding parameters ρ_1 and ρ_2 and all five B_k^m parameters free to vary. The results of this "full" fit are also given in Table 1. Both fits correspond to a CEF level scheme having 2 low-lying singlets well isolated from the three other singlets and four non-Kramers doublets. A 2-singlet system consisting of a ground state $|G\rangle$ and excited state $|E\rangle$ is characterized[7,9] by the energy separation Δ and the electronic matrix element $\alpha = \langle G|J_z|E\rangle$. These values are included in Table 1 together with Δ', the separation of the next excited CEF level from $|E\rangle$.

The 1.5 K Mössbauer spectrum in Fig. 1 is a magnetically split six-line spectrum. It was fitted (solid curve) with B_{eff}=4400(80) kG and a quadrupole interaction equivalent to ΔE_Q=150(4) mm/s. The magnetic splitting is too large to be attributed to slow paramagnetic relaxation because at 1.5 K only the two singlet states are populated and they possess no permanent magnetic moment. Therefore $TmCu_2Si_2$ must be magnetically ordered at 1.5 K. The central region of the spectrum was monitored as a function of temperature. Whereas the quadrupole doublet is centered near zero velocity, the central well-defined doublet of the magnetic spectrum is shifted to a negative velocity. The change-over is sudden so that the ordering temperature for $TmCu_2Si_2$ could be located at T_O=2.8(2) K.

DISCUSSION

Unfortunately, because the B_2^o term plays a relatively minor role in the CEF splitting, little information can be learned about the conduction electron contribution to V_{zz}^{latt} or the Sternheimer antishielding effect. As seen from Table 1, changes in B_2^o are compensated by changes in ρ_2 which maintain $\rho_2 B_2^o$ at a value determined by the high temperature ΔE_Q data (note that $V_{zz}^{latt} \sim \rho_2 B_2^o$).

In support of the fitted CEF scheme it should be commented that (for tetragonal symmetry) only singlet states include the $J_z = \pm 6$ components which are necessary to generate the large negative V_{zz}^{4f} values measured in $TmCu_2Si_2$ at low temperatures. The maximum for ΔE_Q is thus expected to be due to two singlets contributing slightly different EFG'S at the nucleus. However, in order for ΔE_Q to plunge as rapidly as it does after the maximum value, it is necessary that the next excited state contributes a much smaller (or positive) V_{zz}^{4f}. These requirements can be fulfilled only by a dominant B_4^o term. It is not unusual that the fourth order CEF term for Tm^{3+} in $TmCu_2Si_2$ is an order of magnitude larger than that found for Ce^{3+} in $CeCu_2Si_2$. It is difficult to allow for the conduction electron contributions to the CEF and Tm and Ce are well removed from one another in the 4f series. Moreover, Kondo-like correlations have been postulated[1] for $CeCu_2Si_2$ suggesting a mixing of f-like character into the conduction electron wave functions.

Further experimental evidence for a 2-singlet system is obtained from the Mössbauer spectra used to measure ΔE_Q. Although the low-lying Tm^{3+} singlets have no permanent magnetic moments, the asymmetric broadening of the low-temperature spectra is still magnetic in origin and still yields the same negative sign for V_{zz}. It is due to the so-called "pseudoquadrupole" interaction[7,8] which is a weak second-order effect of the magnetic hyperfine interaction. Another consequence of this interaction is a small pseudo-isomer shift, S_P, which is dependent on the relative population of the two singlet levels so that

$$S_p = \frac{\alpha^2}{4\Delta}\left[A_{1/2}^2 - 5\cdot A_{3/2}^2\right] \tanh\left(\frac{\Delta}{2T}\right). \tag{7}$$

The shift was experimentally observed for $TmCu_2Si_2$ and in Table 2 it is compared with theoretical values at several temperatures.

Table 2. Pseudo Isomer Shift, S_p, for ^{169}Tm in $TmCu_2Si_2$

Temperature [K]	3.0	4.2	12.0
S_P [mm/s]:			
Experimental	-1.6(4)	-1.3(1)	-0.6(1)
Theoretical (Δ=8.3 K, α=5.6)	-1.6	-1.4	-0.6

For magnetically ordered rare earth intermetallics the exchange interaction is generally large in comparison with the CEF interaction. However, $TmCu_2Si_2$ most likely retains its CEF level scheme and is a magnetically-ordered 2-singlet system; a problem which has been given considerable theoretical attention. Evidence for this is the B_{eff} value at 1.5 K which is about 2/3 of the "free ion" value expected for a strong exchange interaction. A molecular field treatment of the 2-singlet system[9] shows that the induced local moment is already close to saturation at $T \leq 1/2\ T_0$ and also that the saturation values can be considerably less than the free ion value. The sharply resolved central lines and broad, shallow outer lines of the 1.5 K spectrum are typical relaxation effects which usually demand an involved numerical fitting procedure. However, simple analytical line-shape expressions exist for the case of relaxation between just two electronic states[10]. Using these expressions, a good fit was obtained for the 1.5K spectrum (dashed curve in Fig. 1) employing fewer parameters than for the original 6-line fit (solid curve). A relaxation time of $\tau \simeq 10^{-10}$ s and a splitting of $\Delta \simeq 2$ K were obtained. Although the Δ value is somewhat smaller than that given by the CEF analysis of the ΔE_Q data it is believed that the physics is correct. This work is being continued.

CONCLUSIONS

The CEF level scheme of the Tm^{3+} ion in tetragonal $TmCu_2Si_2$ is determined primarily by the B_4^0 term and is characterized by two low lying singlets well isolated from the higher levels. It is believed that the ordered state below T_o=2.8(2) K is a magnetically ordered 2-singlet system.

ACKNOWLEDGEMENTS

We wish to thank Prof. Israel Nowik for his helpful comments. This work was supported by the Deutsche Forschungsgemeinschaft Sfb 161 and the Kernforschungszentrum Karlsruhe GmbH.

REFERENCES

1. S. Horn, E. Holland-Moritz, M. Loewenhaupt, H. Scheuer, A. Benoit and J. Flouquet, Phys. Rev. B23, 3171 (1981).
2. D. Rossi, R. Marazza, and R. Ferro, J. Less-Common Met. 66, 17 (1979).
3. M. Kalvius, P. Kienle, H. Eicher, W. Wiedemann, and C. Schüler, Z. Phys. 172, 231 (1963). R.L. Cohen, Phys. Rev. A134, 94 (1964).
4. D. Giglberger and S. Penselin, Z. Phys. 199, 244 (1967).
5. H.P. Wit and L. Niesen, Hyperfine Interactions 1, 501 (1976).
6. R.G. Barnes, R.L. Mössbauer, E. Kankeleit and J.M. Poindexter, Phys. Rev. A136, 175 (1964).
7. B. Bleaney, Physica 69, 317 (1973).
8. M.J. Clauser and R.L. Mössbauer, Phys. Rev. 178, 559 (1969).
9. B.R. Cooper, J. Appl. Phys. 40, 1344 (1969).
10. I. Nowik and H.H. Wickman, Phys. Rev. Lett. 17, 949 (1966).

COMMENTS

WOHLLEBEN: Did you detect the second transition (around 7 K) by your method above the one at 2.8 K?

STEWART: No discontinuity has been observed but we took measurements only at about 2 K intervals in this region.

GREIDANUS: Have specific heat measurements been performed on $TmCu_2Si_2$ between 1 K and 10 K?

WOHLLEBEN: I believe he said that such measurements have not been done.

MAGNETIC PROPERTIES OF $TbCo_2Si_2$ AND $TbCo_2Ge_2$

J. Leciejewicz[*], S. Siek[†], A. Szytuła[§], and A. Zygmunt[¶]

[*]Institute of Nuclear Research, Swierk Research Establishment, 05-400 Otwock, Poland
[†]Institute of Physics, Technical University of Kracow 30-084 Kracow, Poland
[§]Institute of Physics, Jagellonian University, 30-039 Kracow, Poland
[¶]Institute of Low Temperature and Structure Research Polish Academy of Sciences, Wrocław, Poland

INTRODUCTION

Intermetallic compounds of the general formula RM_2X_2 (where R is rare earth or actinide metal, M is a transition 3d or 4d metal, and X is Ge or Si) have become a subject of intensive investigations[1-6] in recent years. They crystallize in the body-centered tetragonal structure (space group I4/mmm) with R, M and X atoms occupying the 2(a), 4(d) and 4(e) positions. The structure consists of atomic layers perpendicular to the c-axis, stacked with the sequence R-X-M-X-R.

The magnetic properties of RCo_2X_2 compounds (where R is a rare earth element) have been reported by McCall et al.[1] and Yakinthos et al.[2]. $TbCo_2Si_2$ is a ferromagnet with Curie temperature 8K, while $TbCo_2Si_2$ is an antiferromagnet with a Neel temperature of 30 K[1].

In this paper we report the results of x-ray, neutron diffraction and magnetometric measurements on powder samples of $TbCo_2Si_2$ and $TbCo_2Ge_2$ undertaken in order to determine the crystal and magnetic structures.

EXPERIMENTAL DETAILS

The samples were prepared by melting rare earth metal (4n's), cobalt (4n's) and germanium or silicon (5n's) in an induction furnace. The samples were then annealed in a quartz tube at 800 C for 100 hours and then cooled to room temperature.

The single-phase nature of the compounds was established by x-ray analysis using Fe K_{α} radiation. All lines observed could be indexed assuming a tetragonal $ThCr_2Si_2$-type structure[3]. Lattice parameters of the compounds determined from the x-ray diffractograms were found to be in complete agreement with those obtained previously[4,5] (see Table 1).

Magnetometric measurements were carried out in the temperature range from 4.2 to 300 K, using a vibrating sample magnetometer, and in the temperature range 120 to 600 K using an electronic balance.

RESULTS

The thermal dependence of the susceptibility exhibits a maximum (see Fig. 1) which is characteristic of antiferromagnetic behavior below the Neel temperature, which is 45 K for $TbCo_2Si_2$ and 32 K for $TbCo_2Ge_2$. At high temperatures the reciprocal susceptibility obeys the Curie-Weiss law. The paramagnetic moment is in good agreement with the theoretical values for a free ion. The paramagnetic Curie-Weiss temperatures are negative, giving evidence for predominant antiferromagnetic interactions.

Figure 1 also shows magnetization curves for both compounds taken at 4.2 K in magnetic fields up to 50 kOe. The magnetization increases linearly up to 7.8 kOe for $TbCo_2Si_2$ and 20 kOe for $TbCo_2Ge_2$. In higher fields (above H_c) the magnetization has different behaviors for the two curves. In the case of $TbCo_2Ge_2$ a spin-flop phase transition is observed at the critical field.

Table 1: Structural and magnetic parameters of $TbCo_2Si_2$ and $TbCo_2Ge_2$

	$TbCo_2Si_2$		$TbCo_2Ge_2$	
a (10^{-1}nm)	3.901(1)	3.896*	3.979(1)	3.984(2)†
c (10^{-1}nm)	9.797(2)	9.766*	10.055(3)	10.059(2)†
c/a	2.511	2.507*	2.527	2.525†
z	0.3715(6)		0.3708(4)	
e	0.055(18)		0.016(7)	
R_n (%)	6.3		5.55	
T_N (K)	45	8§	32	30†
θ_p (K)	-73	-18§	-22	-44†
μ_{eff} (μ_B)	10.2	9.4§	10.3	9.27†
μ_n (μ_B)	0.85		0.7	
H_c (kOe)	7.8		20.0	

(*) Ref. 4; (†) Refs. 1 and 5; (§) Ref. 2.

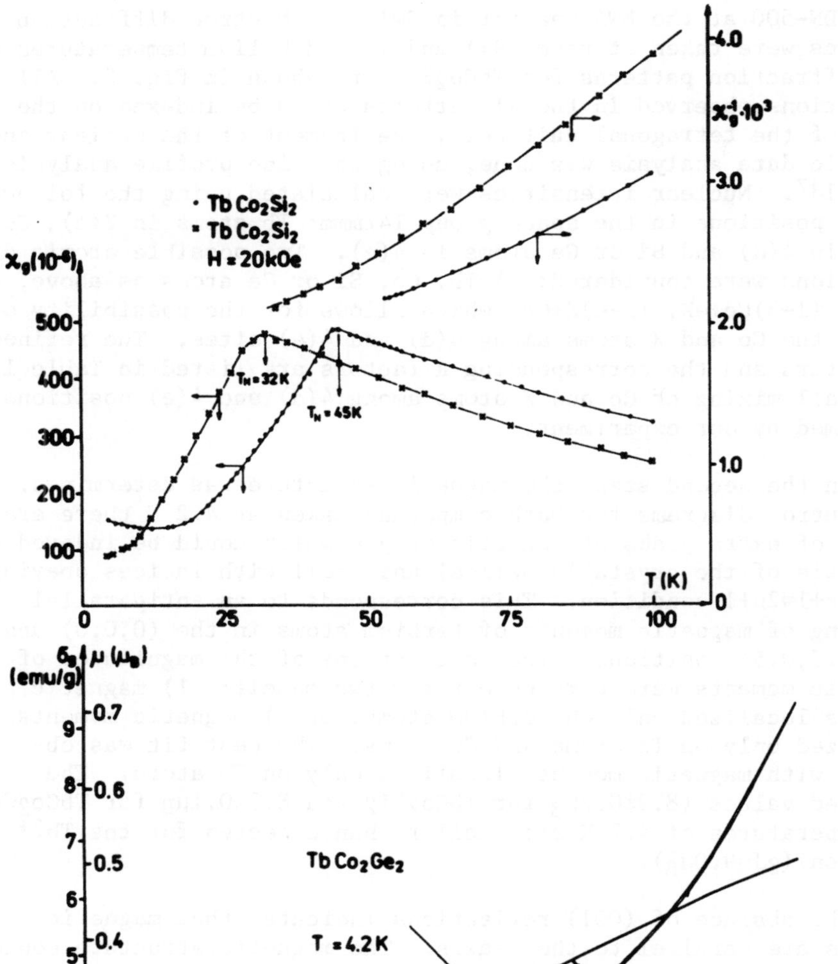

Fig. 1: Temperature dependence of the magnetic susceptibility and reciprocal susceptibility, and field dependence of the magnetization at 4.2 K for $TbCo_2Si_2$ and $TbCo_2Ge_2$.

Neutron data (λ=1.324 10^{-1}nm) were obtained on diffractometer DN-500 at the EWA reactor in Swierk. Neutron diffraction patterns were taken at room (RT) and liquid helium temperatures (LHT). The diffraction patterns for $TbCo_2Si_2$ are shown in Fig. 2. All the reflections observed in the RT patterns could be indexed on the basis of the tetragonal unit cell. Refinement of the nuclear and magnetic data analysis was done, using the line profile analysis of Rietveld[7]. Nuclear intensities were calculated using the following atomic positions in the space group I4/mmm: Tb atoms in 2(a), Co atoms in 4(d) and Si or Ge atoms in 4(e). Two possible atomic distributions were considered: 1) Tb, Co, Si or Ge atoms as above, or 2) Tb, (1-3)Co+eX, (1-e)X+Co, which allows for the possibility of mixing the Co and X atoms among 4(d) and 4(e) sites. The refined parameters and the corresponding R factors are listed in Table 1. The small mixing of Co and X atoms among 4(d) and 4(e) positions was confirmed by our experiment.

In the second step, the magnetic structure was determined. In the neutron diagrams for both compounds taken at 4.2 K there are a series of extra peaks of magnetic origin which could be indexed on the basis of the crystallochemical unit cell with indices obeying the h+k+l=2n+1 condition. This corresponds to an antiparallel coupling of magnetic moments of terbium atoms in the (0,0,0) and (0.5,0.5,0.5) positions. The calculations of the magnitudes of magnetic moments were carried out for two models: 1) magnetic moments localized only on terbium atoms, or 2) magnetic moments localized only on Tb atoms and Co atoms. The best fit was obtained with magnetic moments localized only on Tb atoms. The measured values ($8.9\pm0.2\mu_B$ for $TbCo_2Si_2$ and $8.5\pm0.1\mu_B$ for $TbCo_2Ge_2$) at temperatures of 4.2 K are smaller than expected for the Tb^{3+} free-ion ($gJ=9.0\mu_B$).

The absence of (001) reflections indicates that magnetic moments are parallel to the c-axis. The magnetic structure could be visualized as a pile of ferromagnetic sheets along the c-axis with the sequence +-+-, etc. The magnetic structure determined for $TbCo_2X_2$ compounds is similar to that for $HoCo_2Ge_2$[6].

Finally, the temperature dependence of the intensity of the (100) reflection has been measured (see Fig. 3). The temperature variation of the (100) reflection intensity revealed that the transition to an antiferromagnetic state occurs at 45 K for $TbCo_2Si_2$ and 30 K for $TbCo_2Ge_2$.

CONCLUSIONS

The compounds investigated have the same crystal and magnetic structures. The magnetic structure consists of ferromagnetic layers perpendicular to the c-axis, and they are antiferromagnetically

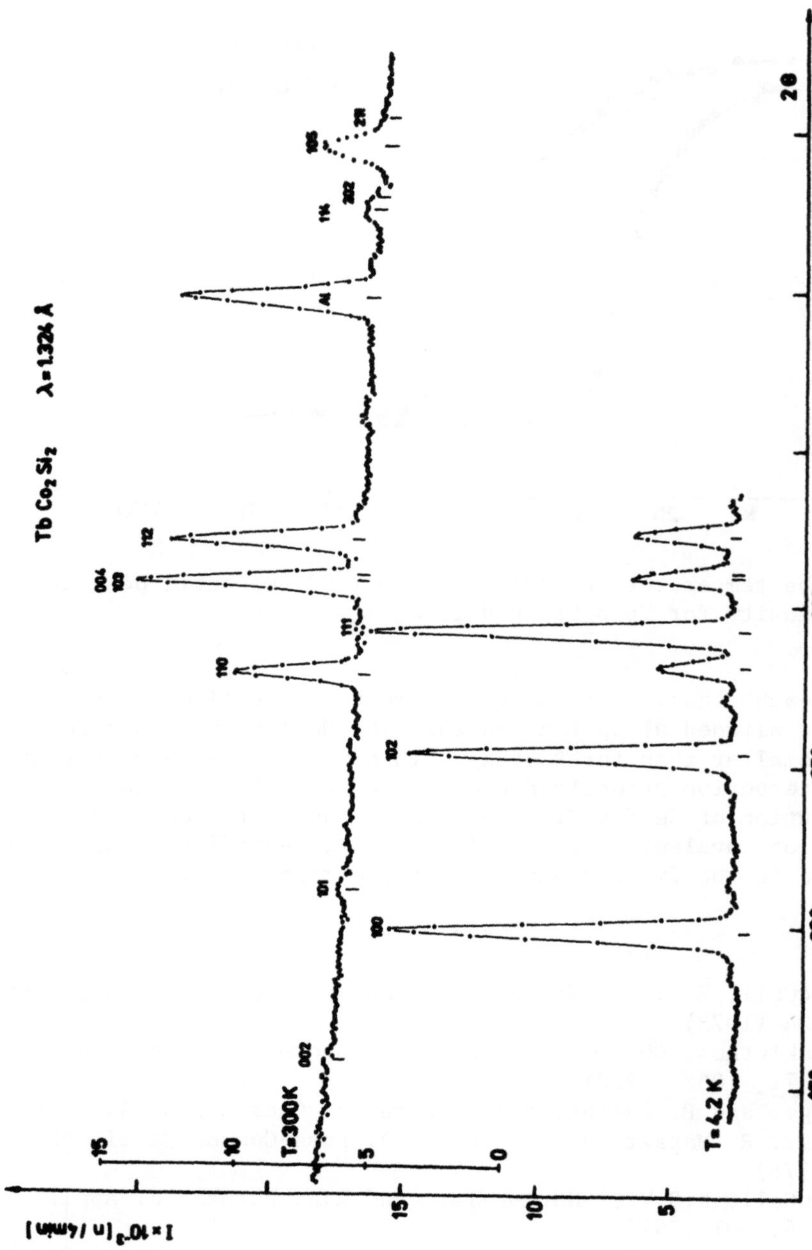

Fig. 2: Neutron diffraction patterns of $TbCo_2Si_2$ at 4.2 K and 300 K.

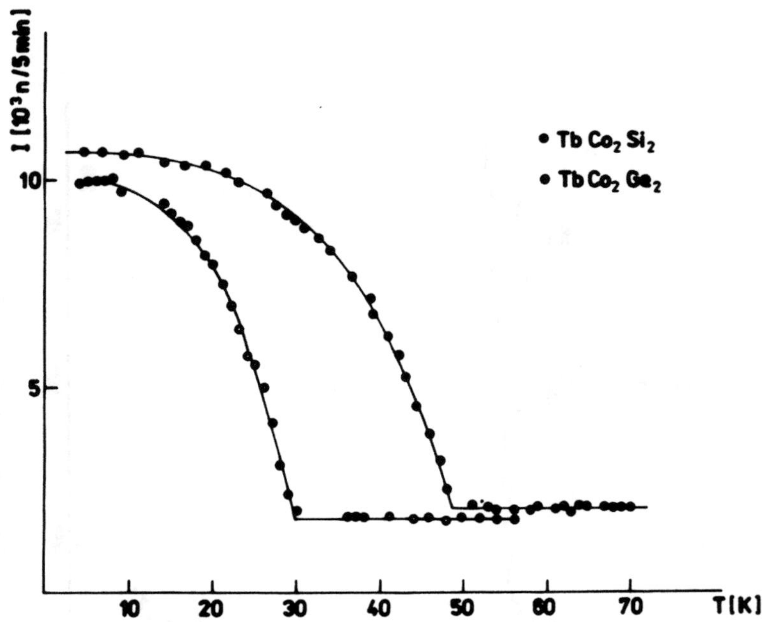

Fig. 3: The temperature behavior of the (100) magnetic peak intensity for $TbCo_2Si_2$ and $TbCo_2Ge_2$.

coupled to each other. The magnetic moment is localized only on Tb atoms and is aligned along the c-axis. The Neel temperature for RCo_2Ge_2 is smaller than for RCo_2Si_2. This result suggests that the magnetic interaction strongly depends upon inter-layers distances. The substitution of Ge for Si, i.e. the exchange of atoms with the same number of covalent electrons (Si-$3s^23p^2$, Ge-$4s^24p^2$) causes only a difference in the field dependence of the magnetization.

REFERENCES

1. W.M. McCall, K.S.V.L. Narasimhan and R.A. Butera, J. Appl. Phys. 44, 4724 (1973).
2. J.K. Yakinthos, Ch. Routsi and P.F. Ikonomou, J. Less Common Metals 72, 1063 (1980).
3. W. Rieger and E. Parthé, Monatshefte für Chemie 100, 444 (1969).
4. D. Rossi, R. Marazza and R. Ferro, J. Less-Common Metals 58, 203 (1978).
5. W.M. McCall, K.S.V.L. Narasimhan and R.A. Butera, J. Appl. Cryst. 6, 301 (1973).
6. A. Szytuła, J. Leciejewicz and H. Binczycka, Phys. Stat. Sol. (a), 58, 67 (1980).
7. H.M. Rietveld, J. Appl. Cryst. 2, 65 (1969).

MAGNETOSTRICTION OF RARE EARTH IMPURITIES IN YCu_2Si_2

N. Rüßmann, H.U. Häfner, and D. Wohlleben
II. Physikalisches Institut
Universität zu Köln
D-5000 Köln 41, Fed. Rep. Germany

INTRODUCTION

Measurements of the magnetostriction (MS) of dilute Rare Earth (RE) ions in various metallic matrices at low temperatures and high fields [1-7] have established that a nearly volume conserving lattice distortion exists around each individual RE ion in the magnetically saturated state which amounts to 10^{-3} to 10^{-2} per unit cell. The distortion depends very little on concentration but very strongly on the RE species in a given matrix. Although the distortion behaves very much like a single ion effect, it cannot be regarded as a universal property of a given RE species or compared to the magnetic moment. This follows from its dependence on the matrix for a given RE ion and also from the fact that the correlation between the observed distortion and the so called Stevens factor across the series of RE is sometimes weak or non existent. In order to better understand the origin of the distortion it seems therefore necessary to study it in many different matrices. In this paper we report the first systematic study of the RE single ion MS in a non cubic metallic environment, namely in YCu_2Si_2 which has the tetragonal $ThCr_2Si_2$ structure.

EXPERIMENTAL DETAILS

Polycrystalline samples were prepared in an induction furnace under an argon atmosphere on a watercooled copper hearth by melting together first CuSi, then the compounds $RECu_2Si_2$ and then YCu_2Si_2 and $RECu_2Si_2$ in a Y:RE ratio of 20:1. The dilute alloys in their final concentration were then obtained by adding YCu_2Si_2 to these master alloys. Powder x-ray diffraction showed only the $ThCr_2Si_2$ phase. Tests showed no influence of annealing at 300 C and 600 C for 24 hours on the MS results. On the other hand, annealing made the samples so brittle that handling became exceedingly difficult. Therefore the results reported in the following were all obtained from unannealed samples. Samples of 2x2x15 mm^3 were cut from the ingots. The MS was measured in a capacitance cell after Brändli and Grießen [8] in fields up to 5 Tesla. A new cryostat allowed measurements between 1.6 and 40 K, a considerable extension of the temperature range compared with measurements reported earlier [1,4,5,7].

RESULTS AND DISCUSSION

Figs.1 to 9 show the isothermal longitudinal MS curves of the dilute alloys $RE_xY_{1-x}Cu_2Si_2$ with x=0.05 for Pr, Nd, Sm, Gd and x=0.01 for Tb,...,Tm. The left hand ordinate gives the measured relative length change and the right hand ordinate the same value multiplied by 5 and divided by the concentration x. The right hand ordinate is then a measure of the relative length change of the individual RE cell parallel to the field.

Qualitatively the MS of the dilute alloys studied here behaves very much like that of systems investigated earlier: the impurity MS can be positive as well as negative. While the light RE impurities only show the initial nearly parabolic field dependence, a clear tendency to saturation is observed for the heavy RE impurities; the order of magnitude of the saturated MS is again a few times 10^{-3} for the heavy RE impurities. The effect is also again volume conserving at least at low fields. This was tested by measurements of the transverse MS at 4.2 K and 2 K in fields up to 7.8 kG for RE=Nd, Gd and Dy. It was found that the volume effect $\Delta V(H)/V_o = [2\Delta l_\perp(H)+\Delta l_\parallel(H)]/l_o$ is less than 2% of the longitudinal effect $\Delta l_\parallel/l_o$, our overall experimental error. The saturable MS per RE cell is again quite independent of concentration. This is best demonstrated by comparison of the behaviour of the dilute alloys $RE_xY_{1-x}Cu_2Si_2$ with that of the compound $RECu_2Si_2$ for the RE with the lowest magnetic ordering temperature amongst the compounds, namely RE=Er ($T_N \simeq 2$ K). Comparison of Fig. 10 with Fig. 9 shows that the MS per cell at 4.2 K and 5 Tesla, i.e. in the paramagnetic region, differs by only 5% between compound and alloy, i.e. the two are coincident within our experimental error.

Quantitative analysis of the measured MS curves requires knowledge of crystal field (CEF) spectra for the RE ions at high dilution in zero and finite applied magnetic field for YCu_2Si_2, as well as knowledge of elastic constants. Unfortunately very little is known about either. Moreover even with those data quantitative evaluation must await considerable progress of the theory. Nevertheless we may apply a semiquantitative phenomenological analysis which allows us to extract some important physical information. (A similar analysis was discussed in detail and applied to the RE in YPd_3 in Ref. 5.) The analysis assumes that there is an anisotropic charge distribution tied axially to the 4f magnetic moment of the RE ion. This charge distribution distorts the lattice when the magnetic moment becomes aligned

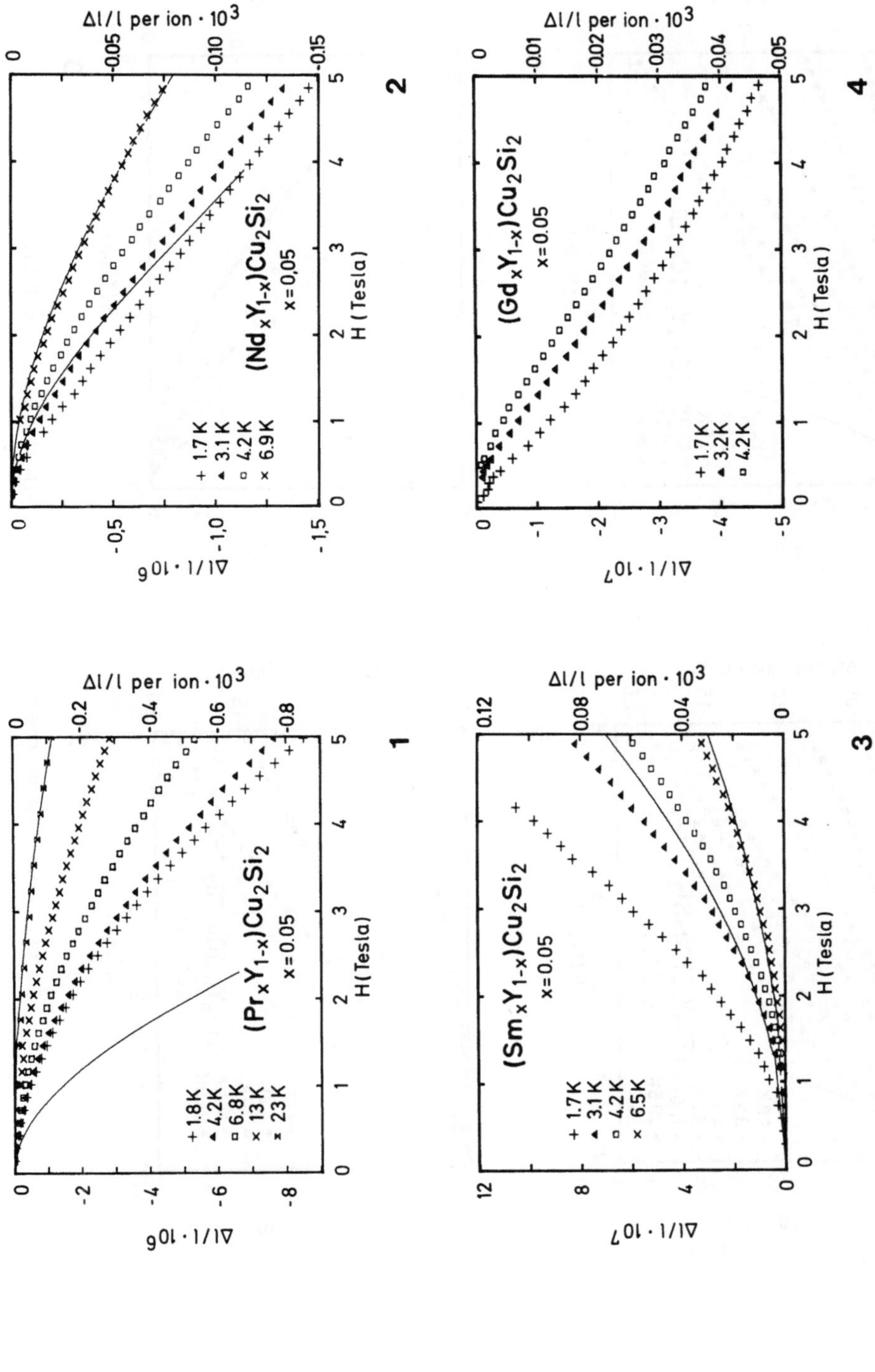

Figures 1-4: Magnetostriction of RE impurities on YCu_2Si_2.

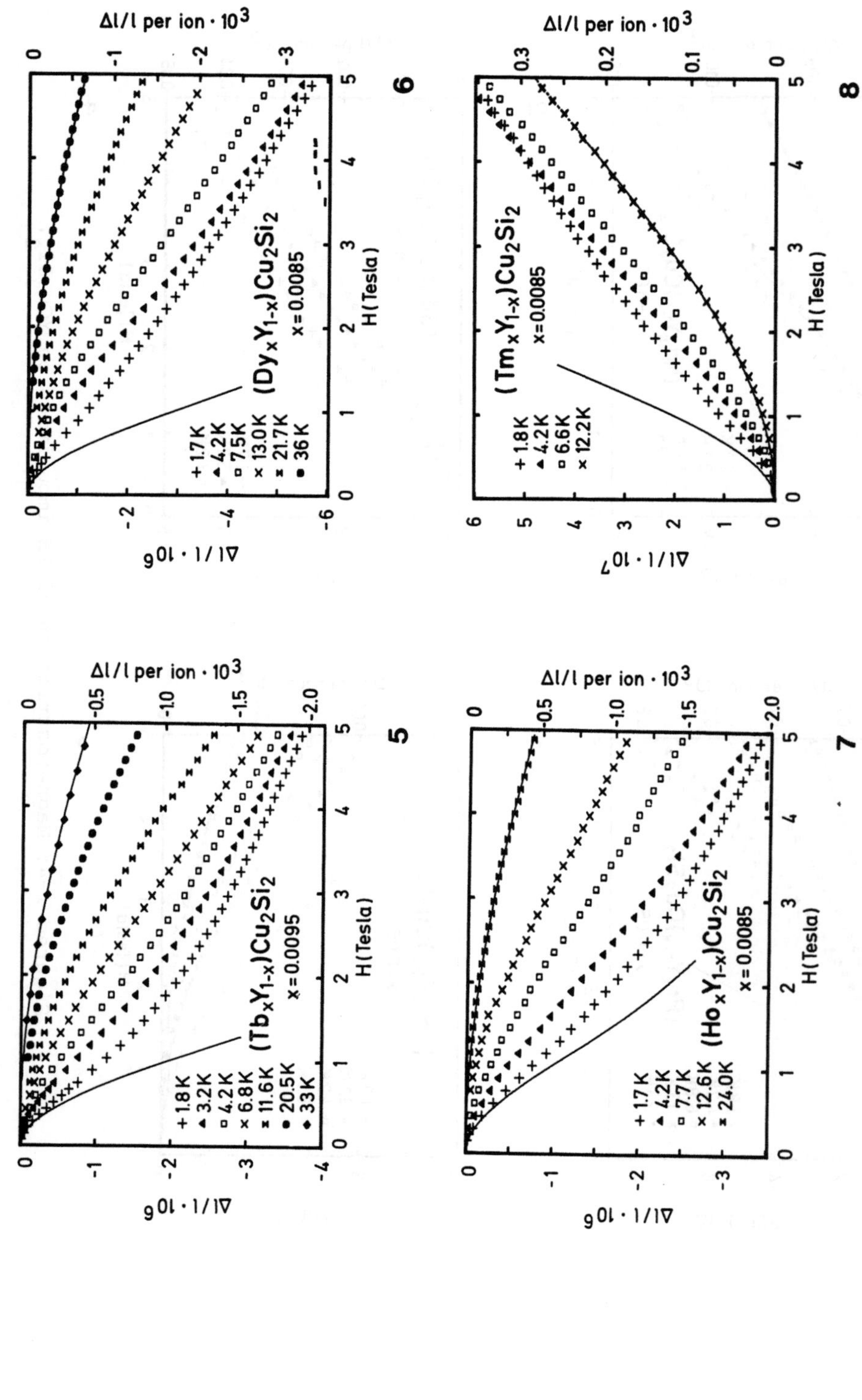

Figs. 5-8: Magnetostriction of RE-impurities in YCu_2Si_2

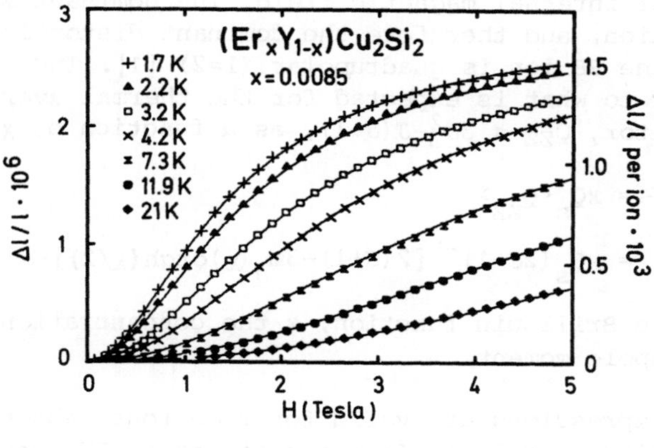

Figure 9.

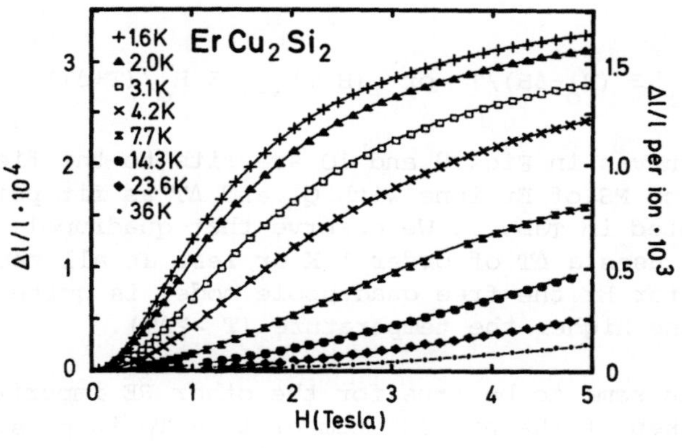

Figure 10.

Table 1a: Fitting parameters for $(Y_{1-x}Er_x)Cu_2Si_2$; x=0.0085

T(K)	ΔT^*(K)	$Q_s \cdot 10^3$
1.7	0.8	1.7
2.2	0.7	1.7
3.2	0.7	1.7
4.2	0.9	1.8
7.3	0	1.7
11.9	0	1.8
21.0	0	2.5

* $\Delta T \pm 2\%$ of T

Table 1b: Fitting parameters for $ErCu_2Si_2$

T (K)	ΔT^*(K)	$Q_s \cdot 10^3$
1.6	0.8	1.8
2.0	0.7	1.8
3.2	0.3	1.7
4.2	0.5	1.7
7.7	0.2	1.5
14.3	0	1.8
23.6	0	2.1
36.0	0	2.2

by an external or internal magnetic field. The dominant moment of the charge distribution, and therefore the dominant distortion of the lattice around the RE ion is quadrupolar (l=2) [1]. Indeed all MS curves look similar to what is expected for the thermal average of the quadrupole operator, $Q_{ZZ} = 3J_Z^2-J(J+1)$, as a function of $\chi=(g_J\mu_B/k_B)(H/T)$.

$$\frac{\Delta l}{l} = xQ_s\langle Q_{ZZ}\rangle = xQ_s(2J-1)^{-1}[2(J+1)-3m_J(\chi)\mathrm{ctgh}(\chi/2)]\cdot \tag{1}$$

Here $m_J(\chi)$ is the Brillouin function, x the concentration and Q_s the saturated quadrupole moment.

The above expressions are valid for free ions (absence of mutual ion-ion interactions and CEF). If interactions of RE ions with the crystal or with other ions resist alignment of the RE magnetic moment with the magnetic field only weakly, one may try fits by replacing H/T by

$$(H/T)_{eff} \equiv (H_o-\Delta H)/T \quad \text{or} \quad (H/T)_{eff} \equiv H_o/(T+\Delta T). \tag{2}$$

The solid curves in Figs.9 and 10 are fits to the field and temperature dependent MS of Er ions with Q_s and ΔT as fit parameters which are collected in Tab. 1. We observe that quadrupole fits are excellent if one uses a ΔT of order 1 K or less at all temperatures. It follows that for Er the free quadrupole model is quite appropriate and the better the higher the temperature ($T \gg \Delta T$).

Assuming the same to be true for the other RE impurities we have used the data taken at the highest temperature T_F in Figs. 1 to 8 for each case to fit a value for Q_s, setting $(H/T)_{eff}=H/T_F$. The second sketched line in these figures is the calculated free quadrupole MS for T=4.2 K $(H/T)_{eff}=(H_o/4.2K)$ with the Q_s obtained above. Clearly these curves lie above the measurements everywhere; in some cases by several times the measured values. This indicates that ΔT as defined above is of the order of 4 K or more in most of the RE ions studied here. Such large ΔT's cannot be explained by spin-spin interactions. The spin-spin interactions expected in our alloys on the basis of the Neel temperature of Gd (T_N=12K), the de Gennes factor, and the linearity of magnetic interaction with concentration are at most a few 10th's of 1 K. Therefore the observed ΔT's must arise from a supression of the magnetization and MS due to CEF. Indeed, as long as the thermal energy is comparable to or larger than the CEF splitting, the thermal expectation values of dipole (magnetization) and quadrupole (MS) are very similar to the free ion case. If however, $k_BT \ll E_{CEF}$, both magnetization and MS will be suppressed with respect to the free ion case. Therefore quite generally we expect the Q_s obtained from the above fits to be more representative of the free ion value, the higher the measuring temperature, and to increase with increasing temperature in accordance with experience.

Tab. 2: Saturation values Q_S, obtained by fit at temperature T_F

	Pr	Nd	Sm	Gd	Tb	Dy	Ho	Er	Tm
Q_s 10^3/ion	-4.9	-.38	1.5	-.1+	-6.3	-9.3	-3.1	2.5	1.2
T_F [K]	23	6.9	6.5		33	36	24	21	12.2
$1/2\cdot\alpha O_2^0(J)$	-.294	-.116	.206	0	-.333	-.333	-.133	.133	.333
$2Q_s/\alpha\, O_2^0(J)$	.88	.17	.4		1.01	1.50	1.24	1	.19

+estimated value

In Table 2 we have listed the Q_S values extracted from quadrupolar fits to MS curves at highest temperatures T_F at which a fit could be made with sufficient accuracy. The temperature T_F is listed in the second row; the third lists the so-called Stevens factor and the fourth the ratio of Q_S to the Stevens factor, with that ratio set arbitrarily equal to 1 for Er. We note that all signs of the MS correlate with the Stevens factor. There also exists a rather good correlation in magnitude, except for Tm, Nd and Sm, where the measured Q_S seems too small. Note that in these cases Q_S was extracted at rather low temperatures. If for Tm, Nd and Sm $\Delta T \gtrsim T_F$, one cannot expect the extracted Q_S to represent the free quadrupole. According to the above arguments, it should then come out too small.

CONCLUSION

In tetragonal YCu_2Si_2 the MS of dilute RE impurities, when referred to the individual RE cell, is of the same order of magnitude as in cubic metallic matrices studied earlier. All qualitative features are comparable to earlier results (independence from concentration, nearly quadrupolar dependence of H/T, conservation of volume, order of magnitude of Q_S). However, contrary to previous findings on cubic systems, we here have for the first time a rather good correlation with Stevens factor, which is perfect with respect to sign and even rather satisfactory with respect to magnitude.

REFERENCES

1. Nicholson, K., Häfner, H.U., Müller-Hartmann, E., Wohlleben, D., Phys. Rev. Lett. 41, 1325 (1978).
2. Campbell, I.A., Creuzet, G., Sanchez, J., Phys. Rev. Lett. 43, 234 (1979).
3. Jong, M.H. de, Hölscher, H., Châtel, P.F. de, Bijvoet, J., Mag. Mag. Mater. 15-18, 17 (1980).
4. Häfner, H.U., Meul, H.W., Nicholson, K., Wohlleben, D., in: "Crystalline Electric Field and Structural Effects in f-Electron Systems", Crow, J.E., Mihalisin, T.W., Guertin, R.P. Editors, p.95, Plenum Press London/New York (1980).

5. Häfner, H.U., Ludwigs, H.-W., Nicholson, K., Wohlleben, D., Z. Phys. B42, 219 (1981).
6. Creuzet, G., Campbell, I.A., Phys. Rev. B 23, 3375 (1980).
7. Meul, H.W., Diplomarbeit, Universität Köln (1980) to be published.
8. Brändli, G., Grießen, R., Cryogenics 13, 299 (1973).

This work was supported by Deutsche Forschungsgemeinschaft, SFB 125.

COMMENTS

LÜTHI: Were the measurements in YPd_3 and YAl_2 on polycrystalline samples? (The scaling with Stevens α_y-parameters should be checked with single crystals.)

HÄFNER: If we say that the magnetostriction across the RE series does not scale with the Stevens factor, we mean this in the sense that the observed distortion of our polycrystalline samples, for instance the longitudinal one parallel to the field near saturation, does not scale with the Stevens factor in either sign or magnitude. We believe that once the measurements on single crystals are properly averaged, these data will not differ much from the observed polycrystalline results, i.e. not much different in magnitude and much less in sign.

CAMPBELL: (Remark) There are experimental examples where in cubic crystals the <001> and <111> effects are of the opposite sign. This makes polycrystalline averaging dangerous.

SZYTUŁA: Does the additional phase transition of $TmCu_2Si_2$ compound have magnetic or structural character?

HAFNER: The high temperature phase transition of $TmCu_2Si_2$ at 7 K is seen very clearly among the various methods used in magnetostriction (see Table 3 in the article of Schlabitz et al.). However, this only means that the transition strongly affects the strain susceptibility. We cannot tell from this finding alone whether there is actually a structural phase transition. We have no thermal expansion data and no adequate x-ray results.

ELECTRONIC TRANSPORT PROPERTIES OF METALLIC Ce SYSTEMS

F. Steglich, K.H. Wienand, W. Klämke, S. Horn and W. Lieke

Insitut für Festkorperphysik, Technische Hochschule Darmstadt, D-6100 Darmstadt, FRG

INTRODUCTION

Ce ions in a metallic environment give rise to a variety of fascinating many body phenomena, for example intermediate valence (IV), valence transitions and the Kondo effect[1]. Despite a controversial debate among spectroscopists[2], there is widespread confidence that the above phenomena arise from the close proximity of the local 4f level of Ce and the Fermi level. Based on the Anderson model, in the limit of large Coulomb energy, an IV phenomenon should occur when the Fermi level E_F intersects the 4f level. In a "high-temperature picture" this can be attributed to the action of charge fluctuations ("valence fluctuations"), which occur at a rate of the order of the 4f level width Δ. In the Born approximation $\Delta \simeq \pi N(E_F) \langle v_{kf}^2 \rangle$, where $N(E_F)$ is the conduction band density of states at E_F per spin and v_{kf} the matrix element for covalent 4f-conduction electron mixing. On the other hand, a Kondo effect is expected when the 4f level, at E_{4f}, is reasonably well separated from E_F, i.e., $\Delta < \varepsilon = E_F - E_{4f}$. Therefore, to a good approximation, the Ce ions in Kondo systems can be considered to be trivalent, i.e. valence fluctuations can be considered to be unimportant. According to Schrieffer and Wolff[4] the coupling between Ce^{3+} ions and conduction electrons can then be treated as an effective antiferromagnetic exchange interaction. We note that in the limit of negligible mixing, i.e. $\Delta \ll \varepsilon$, Ce^{3+} would behave similarly to "normal" rare-earth ions like Gd^{3+}, in that its coupling to the conduction electrons would be rather weak and, eventually, ferromagnetic, i.e. governed by the intra-atomic Heisenberg exchange integral. These "normal" rare earths are sometimes referred to as exhibiting a "stable magnetic moment".

The above classification of rare earth ions with respect to their different couplings to the conduction electrons is supported by recent results of the dynamic magnetic response function as obtained from neutron scattering: 1) Crystal field (CF) split rare earths with "stable moments" show a magnetic relaxation rate, determined by the half width $\Gamma/2$ of the quasielastic magnetic neutron line, whose temperature dependence is in close agreement[5] with the theory of Becker, Fulde and Keller[6]. This theory is based on the Born approximation. Therefore, $\Gamma/2$, which is proportional to T, vanishes as $T \to 0$, in accordance with the well-known Korringa law. 2) Kondo systems containing Ce show a distinct temperature dependence $\Gamma/2(T)$, with a residual value $\Gamma/2$ (as $T \to 0$) of the order of a few K.[7] This effect is nearly independent of Ce concentration. 3) In Ce IV systems, like $CePd_3$ and $CeSn_3$, $\Gamma/2$ is almost temperature independent and of the order of several hundred K.[8] This is in agreement with a recent "high temperature" theory of Kuramoto and Müller-Hartmann[9]. We note that the residual half width $\Gamma/2$ (as $T \to 0$) has the meaning of a characteristic "spin fluctuation temperature", T^*. Whereas for Kondo systems T^* should be comparable with the "Kondo temperature" T_K, extracted from bulk experiments, in IV systems it should be of the order of the characteristic "valence-fluctuation temperature", or 4f level width Δ.

In principle, electronic transport properties are an ideal tool to study the interaction between 4f and conduction electrons, in addition to neutron scattering. Unfortunately, several difficulties are encountered in analyzing and interpreting transport data. For example, incremental (4f-derived) contributions frequently cannot be extracted from the measured data or, if this were possible in some fortunate cases, a theory is often lacking with which transport results can be compared with results from, say, susceptibility or specific heat measurements.

As far as dilute alloys are concerned, two situations are presently quite well understood: 1) For "normal" rare-earth impurities with CF-split Hund's rule ground states, experimental results on both resistivity (e.g. of (Y,Tb)Sb[10]) and thermopower (e.g. (La,Tb)Al_2[11]) are in agreement with the adequate CF theory[10,12]. To our knowledge, however, no alloy system has so far been found in which dilute Ce^{3+} ions show such a "stable moment" behavior. 2) The low-temperature transport data of certain dilute Kondo systems, like (La,Ce)Al_2[13,14] can be satisfactorily explained within the "S=1/2 exchange model" based on the Nagaoka-Suhl theory[15]. In these systems, which are cubic, Ce^{3+} has a Γ_7 doublet CF ground state, well separated (by $\Delta_{CF} \simeq 100$ K) from the excited Γ_8 quartet. The combined action of the Kondo effect and CF splitting on the electronic transport properties of such materials has been treated by Coqblin and coworkers[16]. This work is based on the Anderson model[3].

In the following section we present the first experimental

evidence that dilute Ce^{3+} ions can show "stable-moment behavior", i.e. when dissolved in certain monopnictides. However, they behave "Kondo-like" in some other hosts. Next we will study situations which are clearly beyond those of the well-established dilute Kondo system, i.e. both resistivity and thermopower results will be discussed for substitutional Ce dialuminides. In these systems, according to neutron scattering[17], there is a transition from the trivalent ("Kondo") state to an IV state which can be achieved by varying the chemical composition. Finally the resistivity of a concentrated Kondo system shall be discussed. For this purpose we have chosen the "Kondo lattice" $CeCu_2Si_2$, which has recently become interesting because of its exotic low temperature phase[18]. Some conclusions are discussed in the final section.

DILUTE Ce MONOPNICTIDES

As was mentioned in the introduction, there is no alloy system known in which Ce^{3+} impurities behave "stable-moment like". The necessary condition for this would be $\Delta/\varepsilon \ll 1$, in terms of the Anderson parameters. Although we do not know how to experimentally manipulate this ratio, we would at least expect that the metallic host should exhibit a small $N(E_F)$. Therefore, in order to search for Ce^{3+} impurities with "stable moments" we have chosen as host materials the monopnictides MZ (M=La,Y; Z=Sb,Bi). These compounds, which crystallize in the rocksalt structure, are semimetals with a concentration of free carriers $n_o \sim 10^{20}\ cm^{-3}$.[19] From the good metallic conductivity observed in each case[20], we infer the effective mass of the carriers to be quite small. Ce was substituted for La or Y at concentrations ≤ 0.1 at %. In the octahedral CF produced by the six pnictogen nearest neighbors the J=5/2 state of Ce^{3+} splits into a Γ_7 doublet and a Γ_8 quartet.

In the upper and middle part of Fig. 1 we show for several of these dilute $M_{1-x}Ce_xZ$ systems the temperature dependence of the incremental resistivity per at % Ce, defined by

$$\rho_{Ce}(T)=\Delta\rho_{4.2}(T)=x^{-1}\{[\rho(T)-\rho(4.2K)]_{alloy}-[\rho(T)-\rho(4.2K)]_{host}\}. \quad (1a)$$

For some of the samples because of many pores and microcracks, the shape factor was not known[20], and this resulted in a considerable scatter between corresponding $\Delta\rho(T)$ curves. In some other cases, samples had a well-defined shape. For these latter cases $\Delta\rho(T)$ is clearly "step-like", very similar to what was found by Hessel Anderson[10] for (Y,Tb)Sb where, as already mentioned, Tb acts as a CF-split ion with a "stable moment". More specifically, our results for both $Y_{0.97}Ce_{0.03}Sb$ and $Y_{0.9}Ce_{0.1}Bi$ can be nicely fitted by the theory of Fulde and Peschel[12], modified for degenerate CF levels[21]. In this theory, which is based on the Born approximation, only isotropic scattering processes are taken into account. The best fits

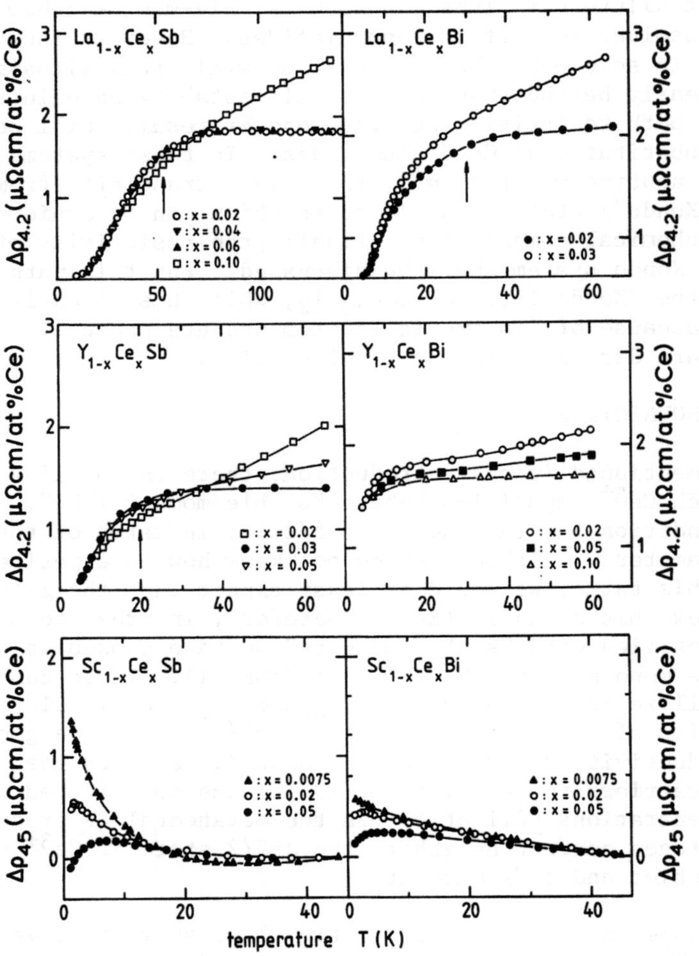

Fig. 1: Incremental resistivities [as defined by Eqs. (1a) and (1b) as a function of temperature for several dilute Ce monopnictides. Solid lines are intended as guides for the eye. Arrows indicate crystal field splitting as obtained from Ce^{3+} ESR (Ref. 20).

were obtained by choosing Γ_7 as the ground state with Γ_8 lying above Γ_7 by Δ_{CF}=20.6 K and 13.1 K, respectively, as will be shown in a separate paper[20]. In most of these alloy systems (the exception being (Y,Ce)Bi), a Ce^{3+}-ESR line could be observed. Since the samples were polycrystalline, a Γ_7 ground state must be concluded from these experiments also. In addition, the CF splitting Δ_{CF} can be extracted from the thermal broadening of the ESR line (see arrows in Fig. 1). The results for Δ_{CF} from both ESR and $\Delta\rho(T)$ are in excellent agreement for (Y,Ce)Sb. For the two La-based systems, for which the fit of the Fulde-Peschel theory to the $\Delta\rho(T)$ data is less satisfactory, the ESR results indicate that, like in the Y-based alloys, the pronounced resistivity drops near 50 K and 30 K must also be caused by thermal depopulation of excited Γ_8 states.

Fig. 1 also shows that both dilute (Sc,Ce)Sb and (Sc,Ce)Bi exhibit a temperature dependence of the incremental resistivity. In this case it is defined by

$$\rho_{Ce}(T)=\Delta\rho_{45}(T)=x^{-1}\{[\rho(T)-\rho(45K)]_{alloy}-[\rho(T)-\rho(45K)]_{host}\}, \quad (1b)$$

which is qualitatively different from that of the La- and Y-based systems. For the two most dilute samples, we find $\Delta\rho\sim(-\ln T)$ in the range $1.5\leq T\leq 20$ K, 1.5 K being the lowest temperature accessible. In addition, higher concentrated samples show "spin-glass like" maxima[22] in $\Delta\rho(T)$. These results indicate a single-ion Kondo effect of Ce^{3+} in the low concentration-high temperature regime and the existence of magnetic interactions, probably of the RKKY type, between Ce's. If free conduction electrons can be assumed, the characteristic energy scales of these phenomena are determined by

$$k_BT_K\sim(N(E_F))^{-1}\exp\{(N(E_F)J)^{-1}\} \quad (2)$$

and

$$k_BT_{RKKY}\sim xN(E_F)J^2. \quad (3)$$

Because of the negative temperature coefficient of $\Delta\rho(T)$, J must be negative. It can be decomposed into

$$J=J_o+J_1 \quad (4)$$

where $J_o>0$ is the intra-atomic exchange integral and

$$J_1=-\frac{2}{\pi}\frac{1}{N(E_F)}\frac{\Delta}{\varepsilon}<0. \quad (5)$$

J_1 is the negative exchange integral due to covalent 4f conduction electron mixing[4]. (Here it is assumed that $k_BT\ll\varepsilon$ and $U\gg\varepsilon,\Delta$, where U is the intra-atomic Coulomb energy).

Obviously, for the Sc-based systems J_1 dominates over J_o and the resulting J appears to be of considerable size. This is inferred from the magnetic interaction effects observed for $x\geq 0.02$. On the other hand, no significant interaction effects can be resolved, within the scatter of the data, for the La- and Y-based systems, at least on the concentration level of a few at.% Ce. This suggests a comparatively small value of J. Unfortunately, the "step-like" $\Delta\rho(T)$ curves do not provide the sign of J (within the Born approximation, $\Delta\rho\sim J^2$): either $|J_1|$ is smaller than J_o, i.e.

$J>0$, or $|J_1|$ exceeds J_o, i.e. $J<0$. If the latter holds, a rather small value of $|J|$ would also have to be inferred from the fact that for all these alloys $\Delta\rho$ lacks a "Kondo-type" temperature dependence above 1.5 K.

In summary, we have found from our resistivity results that Ce^{3+} impurities in LaSb, LaBi, YSb and YBi seem to exhibit a "stable moment" behavior, which was not observed hitherto. On the other hand, in both ScSb and ScBi dilute Ce^{3+} ions give rise to a Kondo effect. Using the Anderson parameters, we conclude that Δ/ε must be much larger for the Sc-based systems than for the La- and Y-based systems. At present, we are attempting to understand the origin of this qualitative difference in the Ce-conduction electron coupling. It is perhaps interesting to note that this observation seems to correlate with a systematic dependence of Δ_{CF} of the isolated Ce^{3+} impurities in the different hosts. For M=La or Y, Δ_{CF} is finite (Fig. 1), while it is almost zero for M=Sc[20]. Remarkably, the Δ_{CF} values found in the dilute monopnictides are as small as in the corresponding Ce intermetallics (for CeSb, $\Delta_{CF}\simeq 25$ K, for CeBi, $\Delta_{CF}\simeq 10$ K)[10]. Takegahara et al.[23] have explained this in terms of covalent mixing between "p-valence states" originating from the six pnictogen nearest neighbors and the "4f-core states" of Ce^{3+}. This will be discussed in more detail elsewhere[20]. This explanation seems to be supported by the aforementioned variation of Δ_{CF} with the different hosts in the dilute ($\underline{M}$,Ce)Z alloys: Δ_{CF} is found to decrease systematically on decreasing the difference between the electronegativities of the Z and M atoms (or the increasing covalent nature of the M-Z bonds). From Ref. 23, the strength of the "p-f mixing" is seen to track the covalency of the host metal. Then it appears not unlikely that "p-f mixing" also dominates the Schreiffer-Wolff exchange integral J_1 [Eq. (5)]. This would mean that the apparent increase in Δ/ε, concluded for the ($\underline{M}$,Ce)Z systems on going from M=La or Y to M=Sc, was caused mainly by a decreasing difference in the electronegativities of the pnictogen Z and the transition metal M. More theoretical work is necessary to clarify this question.

SUBSTITUTIONAL Ce DIALUMINIDES

There is an increasing interest in studying valence transitions in metallic Ce systems by as many techniques as possible. The aim of these efforts is to establish phenomenologically possible similarities and differences between the trivalent ("Kondo") state and the IV state. As was noted in the introduction, neutron scattering linewidth measurements reveal a distinctly different magnetic relaxation behavior in either state, caused presumably by the different natures of the magnetic relaxation processes. If the Ce ions are coupled to charge relaxation in an IV system , they are very likely not in a (nearly integral-valent) Kondo system. A major difficulty in comparing, e.g., the resistivity of IV and Kondo systems lies in the fact that the IV phenomenon is usually observed in concentrated rare-earth materials, notably intermetallic compounds, like $CePd_3$ or $CeSn_3$, while the Kondo effect has been studied primarily with dilute alloys, like (La,Ce)Al_2[13,14] or (La,Ce)B_6[24,25]. As will be discussed in this section, there is a fundamental difference between the low temperature resistivities of an intermetallic compound and a dilute alloy. A reliable comparison of the respective $\rho(T)$ dependences can be expected if, for example, a trivalent Ce material showing "Kondo anomalies" at ambient pressure is transformed into an IV state by application of pressure. Alternatively, such a valence transition can also be induced by suitable chemical alloying.

The substitutional alloys $Ce_{1-x}Sc_xAl_2$, which are prepared by diluting the trivalent concentrated Kondo system $CeAl_2$ (lattice parameter a=8.06 Å), appear highly suited for an investigation of transport properties on "both sides" of a valence transition. This is because a rather sharp change from trivalent ($x \leq 0.4$) to IV ($x \geq 0.5$) has been discovered by neutron scattering[17]. This means that without using a pressure cell, we will be able to compare a Kondo system (x=0.4) and an IV system (x=0.5) of almost the same chemical composition.

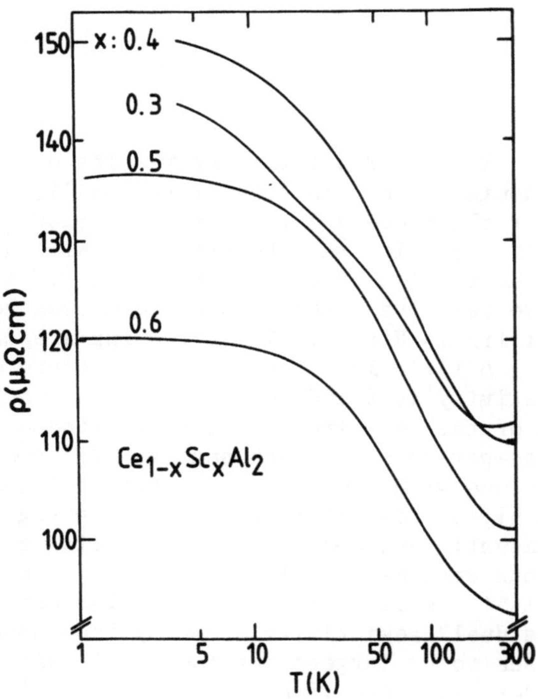

Fig. 2: Resistivity of three $Ce_{1-x}Sc_xAl_2$ alloys vs. temperature (on logarithmic scale).

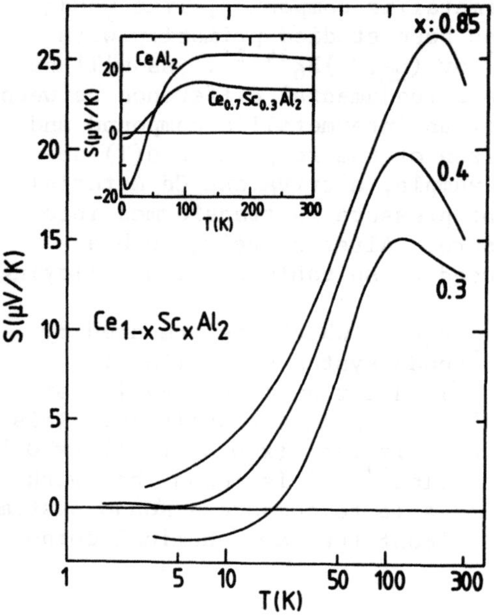

Fig. 3: Thermopower of three $Ce_{1-x}Sc_xAl_2$ alloys vs temperature (on logarithmic scale). Inset shows (on linear scale) data of $Ce_{0.7}Sc_{0.3}Al_2$ compared to data of $CeAl_2$ (Ref. 29).

In Figs. 2 and 3 we show preliminary results of the resistivity and thermopower (dominated by electron diffusion[14]), respectively, vs temperature over the range $4.2 \leq T \leq 300$ K for $Ce_{1-x}Sc_xAl_2$ for x=0.3, 0.4 and 0.5. The resistivity results display a similar temperature dependence to recent results of Levine and Croft[26]. Although these latter authors gave relative numbers, we have chosen to show the absolute resistivities in Fig. 2. The following are the salient features concerning $\rho(T)$: 1) A negative temperature coefficient (NTC) of the resistivity is found for all samples, regardless of the valence state of the Ce. Around 100 K, we find ρ to vary logarithmically with temperature for each sample. 2) Among the two trivalent alloys, the one with lower Ce concentration (1-x=0.6) shows (for $T \leq 150$ K) the higher resistivity, thus suggesting an appreciable interference of magnetic and nonmagnetic scattering processes. Similar observations were made earlier on the related ($\underline{La}$,Ce)Al_2[14] and ($\underline{La_{1-z}Y_z}$,Ce)Al_2[27] systems. In the particular case of $Ce_{1-x}Sc_xAl_2$, we expect the "residual" resistivity stemming from nonmagnetic scattering processes to be largest for the sample with maximum substitutional disorder, i.e. $Ce_{0.5}Sc_{0.5}Al_2$. If the Ce's were to remain trivalent, the incremental resistivity per single Ce ion, ρ_{CE}, should be largest for this sample also. Our preliminary results even suggest that, then, the total incremental resistivity, $\Delta\rho=(1-x)\rho_{CE}$, should be largest for x=0.5. 3) There exist several qualitative differences between the $\rho(T)$ curve of the two samples and those of the trivalent samples. Contrary to the expectation that a trivalent x=0.5 sample would perhaps show the largest low-temperature resistivity, the observed value is distinctly reduced compared to that of the (trivalent) x=0.4 sample. In addition, the $\rho(T)$ curves for $x \geq 0.5$ start to saturate at about 10 K, while no tendency toward such a saturation is visible for the two other samples down to 4.2 K. This hints at a large increase in the "Kondo temperature" on going from x=0.4 to 0.5 and might well be correlated to the factor-of-thirty increase of the "spin fluctuation temperature" T* as found by neutron scattering.[17]. Finally, while the $\rho(T)$ curves of the trivalent samples show some signatures of CF splitting effects, e.g., a flat shoulder near 70 K for x=0.3, such features are absent in the $\rho(T)$ curves of the alloys with $x \geq 0.5$. This observation is corroborated by the neutron scattering results, according to which CF effects exist for $x \leq 0.4$ and disappear in the IV state ($x \geq 0.5$)[17].

Qualitative differences between the IV alloy and the trivalent systems are also seen in Fig. 3. So far, only one IV sample (x=0.85) has been studied by TEP and was found to exhibit a pronounced peak at $T_{max} \simeq 200$ K. This coincides within experimental error with the "spin fluctuation temperature" T* obtained from the quasielastic magnetic neutron linewidth[17]. As temperature is decreased, S(T) decreases steadily. The trivalent alloys $Ce_{0.6}Sc_{0.4}Al_2$ and $Ce_{0.7}Sc_{0.3}Al_2$ show pronounced maxima and minima. Here T_{min} is in good agreement with T* from neutron scattering[17]. The origin of the positive peak at 100 K could be due either to local environmental effects or to CF effects. In the former case, one might assume that a small fraction of the Ce ions becomes transformed into an IV state precursive to the valence transition, i.e. for $x \leq 0.4$. TEP is sensitive to single ion contributions, so very few IV Ce ions could reflect themselves in the observed peak at 100 K. Further support for local environment effects in $Ce_{1-x}Sc_xAl_2$ comes from the observation of a very flat maximum in the low temperature resistivity of the IV sample with x=0.5 (see Fig. 2). This peak is presumably caused by residual magnetic interactions between "Ce^{3+} impurities" in this sample. Another possible reason for the TEP peaks of the x=0.3 and 0.4 samples near 100 K is the CF splitting of the Ce^{3+} ionic state. In fact, the S(T) curves of the trivalent $Ce_{1-x}Sc_xAl_2$ alloys resemble those previously found for the dilute Kondo system (La,Ce)Al_2[28]. A theory treating this exists for Ce^{3+} impurities[16]. Comparisons with the results of previous valence transition experiments[28-30] to the appropriate theories[31] will be important.

One important finding of these preliminary experiments is that we find a "Kondo-like" logarithmic temperature dependence of the resistivity to be clearly visible also on the IV side of the valence transition, e.g. for $x \geq 0.5$ in $Ce_{1-x}Sc_xAl_2$, thus confirming the results on IV compounds, like $CePd_3$[32], which show NTC in ρ(T) for sufficiently high temperatures (T≳150 K in the case of $CePd_3$). On the other hand, it is very interesting that, when dilute intermediate-valent Eu ions are dissolved in $ScAl_2$, the incremental (Eu-derived) resistivity exhibits a positive rather than a negative temperature coefficient[33]. Further experimental and, even more importantly, theoretical efforts on the transport properties of disordered and dilute IV systems are clearly needed to understand these phenomena.

INTERMETALLIC COMPOUND $CeCu_2Si_2$

Because most of the existing Kondo theories treat the case of a single magnetic impurity dissolved in a nonmagnetic host[15], experimentalists have generally concentrated their investigations on dilute magnetic alloys. It was only recently that it became clear that a "Kondo-type" mechanism may survive the increase of magnetic "impurity" concentration, even up to 100 at.%. Good examples of such systems, sometimes labeled "Kondo lattices"[34], are the cubic intermetallics $CeAl_2$[35] and CeB_6[36,37], which show complex antiferromagnetic order at low temperature[38,39,40]. According to neutron scattering results, both the hexagonal compound $CeAl_3$[41] and the tetragonal $CeCu_2Si_2$[42] appear to belong to the "Kondo lattices" class as well[7]. These two compounds do not show magnetic order. Instead, $CeAl_3$ remains in a "Pauli" paramagnetic state down to the low mK region[43], whereas a novel superconducting low temperature phase has recently been discovered for $CeCu_2Si_2$[18].

The electronic transport properties of $CeCu_2Si_2$ between 1.5 and 300 K have already been reported[44]. Although for the previous experiments[44,18] the only samples available were those which contained a finite amount of spurious phases, it has recently become possible, by means of a proper heat treatment, to prepare single-phase $CeCu_2Si_2$[45]. To date, all available samples are polycrystalline.

In Fig. 4 the resistivity vs temperature is shown for a very pure $CeCu_2Si_2$ sample and compared to that of an $LaCu_2Si_2$ sample. The ρ(T) dependence of the latter, which was unannealed, resulted in a rather large residual resistivity[44], and ρ(T) was similar to that of many other d-band metals. As was found for less pure $CeCu_2Si_2$ samples[44], the ρ(T) curve of our very pure sample shows several anomalies, which fit into a qualitative scheme as proposed by Mott[46]. 1) At high temperature (T>20 K), the scattering of the conduction electrons may be considered, to a good approximation, to be incoherent. In fact, the single-ion contribution to the incremental resistivity is similar to that of a dilute alloy containing CF-split Ce^{3+} ions[16]. As temperature is lowered, A "Kondo-like" NTC above 100 K is followed by a maximum near 80 K and a minimum at 60 K. These features can be understood using the CF level scheme of Ce^{3+} (appropriate to the tetragonal symmetry of $CeCu_2Si_2$) of three Kramers doublets with energy separations of 140 K and 364 K.[42] As will be

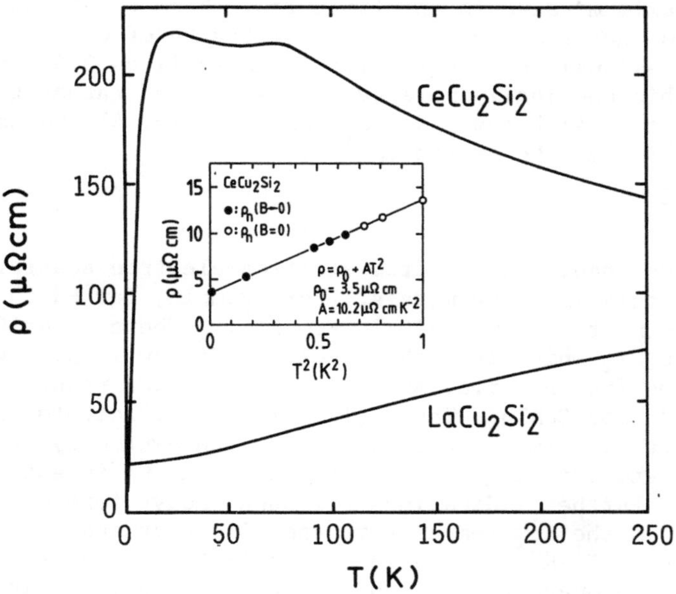

Fig. 4: Temperature dependence of the resistivity of $CeCu_2Si_2$ and $LaCu_2Si_2$ (Ref. 44). Inset shows normal-state resistivity ρ_n of $CeCu_2Si_2$ vs T^2. ρ_n was measured either at B=0 above T_c or below T_c, by extrapolating magnetoresistivity data to B=0 (Ref. 47).

shown in more detail elsewhere[47], the maximum in ρ(T) near 80 K reflects the thermal depopulation of the first excited CF doublet, and the minimum at 60 K is caused by a "Kondo-type phenomenon" associated with the ground state doublet, the corresponding "Kondo temperature" being of order $T_K \simeq T^* \simeq 10$ K.[42] 2) The second maximum in ρ(T) at $T_{max} \simeq 20$ K can by no means be explained with CF effects[47]. Rather, it seems to indicate the onset of spatial coherence among conduction electrons when scattered from the (almost) regular array of Ce^{3+} scattering centers. For $CeCu_2Si_2$, both predictions by Mott[46] concerning this kind of resistivity peak appear to be qualitatively met: (i) T_{max} is of the order of the T^* and (ii) $\rho(T_{max})$ is close to the "Mott limit", which corresponds to the minimum size for the electronic mean free path in a metal, i.e. $\ell \simeq a$, the separation of the Ce scatterers (see Ref. 44). 3) The steep depression of the resistivity below 20 K nicely illustrates the famous Bloch theorem, according to which a normal state metal with a regular lattice should show zero resistivity in the limit T→0. Note that this constitutes a fundamental difference between the transport behaviour of an intermetallic compound and a dilute alloy containing "Kondo-type" Ce^{3+} impurities, since the resistivity of the latter assumes its maximum value in the limit $T \to 0$. From the inset of Fig. 4 it can be seen that the normal state resistivity of $CeCu_2Si_2$ extrapolates to a residual value ρ_o, which is as small as 3.5μΩcm and proves that this chemically pure sample is also of rather good metallurgical quality. 4) below 1 K (see inset of Fig. 4) the normal state resistivity can be expressed[47] by

$$\rho_n = \rho_o + AT^2. \tag{6}$$

Such a T^2 dependence, arising from electron-electron scattering, has been observed only in a few metals. For example, for Pd the coefficient A was found to be $A = 3.1 \times 10^{-5}$μΩ-cm-K^{-2}. Whereas, A=10.2μΩ-cm-K^{-2} of $CeCu_2Si_2$ which is much larger. A T^2 law in $\rho(T)$ with an even larger coefficient, i.e. A=35.4 μΩ-cm-K^{-2}, was reported by Andres et al.[43] for $CeAl_3$. These gigantic A-coefficients provide strong evidence of fermion quasiparticles with extremely heavy effective masses, existing in both compounds at sufficiently low temperatures. Further indications for such "heavy fermions" are a giant term γT in the low temperature specific heat and the susceptibility ($\gamma \simeq 1$ J-mole^{-1}-K^{-2} and $\gamma \simeq 1.6$ J-mole^{-1}-K^{-2}, and a strongly-enhanced Pauli paramagnetic susceptibility, $\chi_o \simeq 6.5 \times 10^{-3}$ emu-mole^{-1} and $\chi_o \simeq 3.6 \times 10^{-2}$ emu-mole^{-1}, for $CeCu_2Si_2$[18,45] and $CeAl_3$[43], respectively).

However, while these "heavy-fermion features" can be studied in $CeAl_3$ down to ultra-low temperatures[43], they disappear in $CeCu_2Si_2$ at $T_c \simeq 0.5$ K, where this material becomes superconducting[18,45] as mentioned above. From the size of the specific heat anomaly at T_c, which was found to be comparable to the giant normal state specific heat near T_c, namely γT_c, it was concluded that in $CeCu_2Si_2$ the Cooper pairs must be formed by those "heavy-mass fermions". It is unknown why this intermetallic compound, which at sufficiently high temperatures (T>10 K) contains a dense array of magnetic moments, passes into a superconducting state at one half a degree Kelvin.

CONCLUSIONS

We have found that electronic transport properties can substantially help to distinguish phenomenologically between different physical situations realized in metallic Ce systems. For dilute Ce monopnictides, a transition taking Ce from the "stable-moment type" to the "Kondo type" was established when the host compounds were changed. For substitutional Ce dialuminides, a change from a trivalent to an intermediate-valent (IV) state occurred upon varying the Ce concentration, and this was well reflected both in the resistivity and the thermopower. An interesting difference was found between substitutional (Sc,Ce)Al_2 and dilute (Sc,Eu)Al_2, the rare earth ions being in an IV state for both systems: The incremental resisitivity showed a negative ("Kondo like") temperature coefficient for IV-Ce, but a positive one for IV-Eu. For the (nearly) trivalent

"Kondo lattice" $CeCu_2Si_2$, an extremely large T^2 coefficient in the low temperature, normal-state resistivity was observed. This supports the existence of "heavy fermions" in $CeCu_2Si_2$ below T=1 K, concluded before from specific heat and susceptibility measurements. $CeCu_2Si_2$ is unique because these previous experiments also indicated that the "heavy fermions" system undergoes a phase transition into a superconducting state at $T_c \simeq 0.5$ K. It can be expected that knowledge of further superconducting properties will help to gain quantitative information about this "heavy-fermion system".

ACKNOWLEDGMENTS

We would like to acknowledge enlightening discussions with Prof. B. Elschner and Prof. M. Croft regarding the physics of the monopnictides and dialuminides. We are also grateful to M. Schildwächter for his help in the resistivity measurements and H. Biesenkamp for her assistance in the data analysis. Special thanks are due to Prof. R.P. Guertin for critical reading of the manuscript and for several helpful comments and to Ms. C. Mees for her skillful typing. This work was supported by SFB 65, Frankfurt-Darmstadt.

REFERENCES

1. For a recent review, see: J.M. Lawrence, P.S. Riseborough and R.D. Parks, Rep. Prog. Phys. 44, 1 (1981).
2. For a discussion of recent results, see: S. Hüfner and P. Steiner, submitted to Z. Phys. B.
3. P.W. Anderson, Phys. Rev. 124, 41 (1961).
4. J.R. Schrieffer and P.A. Wolff, Phys. Rev. 149, 941 (1966).
5. M. Loewenhaupt, S. Horn and B. Frick, this conference.
6. K. Becker, P. Fulde and J. Keller, Z. Phys. B28, 9 (1977).
7. S. Horn, F. Steglich, M. Loewenhaupt and E. Holland-Moritz, Physica 107B, 103 (1981).
8. M. Loewenhaupt and E. Holland-Moritz, J. Appl. Phys. 50, 7456 (1979).
9. Y. Kuramoto and E. Müller-Hartmann in: "Valence Fluctuations in Solids", p.139, L.M. Falicov, W. Hanke and M.B. Maple, eds., North Holland, Amsterdam (1981).
10. N. Hessel Andersen in: "Crystalline Electric Fields and Structural Effects in f-Electron Systems", J.E. Crow, R.P. Guertin and T.W. Mihalisin, eds., Plenum, New York (1980), p.373.
11. E. Umlauf, G. Pepperl and A. Meyer, Phys. Rev. Lett. 30, 1173 (1973).
12. P. Fulde and I. Peschel, Adv. Phys. 21, 1 (1972).
13. K. Winzer, Z. Phys. 265, 139 (1973).
14. F. Steglich, Z. Phys. B23, 331 (1976).
15. See, e.g., the following review articles: K. Fischer, Springer Tracts Mod. Phys. 54, 1 (1970); G. Grüner and A. Zawadowski, Rep. Prog. Phys. 37, 1497 (1974).

16. Resistivity: B. Cornut and B. Coqblin, Phys. Rev. B5, 4541 (1972); Thermopower: A.K. Bhattacharjee and B. Coqblin, Phys. Rev. B13, 3441 (1976).
17. M. Loewenhaupt, S. Horn and F. Steglich, Solid State Commun. 39, 295 (1981).
18. F. Steglich, J. Aarts, C.D. Bredl, W. Lieke, D. Meschede, W. Franz and H. Schäfer, Phys. Rev. Lett. 43, 1892 (1979).
19. A. Hasegawa, J. Phys. C: Solid State Phys. 13, 6147 (1980).
20. K.H. Wienand, B. Elschner and F. Steglich, to be published.
21. P. Fulde, in: "Handbook on the Physics and Chemistry of Rare Earths, Vol. 2", Ed. by K.A. Gschneidner Jr. and L. Eyring, North Holland, Amsterdam (1978), p.295.
22. See, e.g., E.D. Ramos, J. Low Temp. Phys. 20, 547 (1975).
23. K. Takegahara, H. Takahashi, A. Yanase and T. Kasuya, Solid State Commun. 39, 857 (1981).
24. K. Winzer, Solid State Commun. 16, 521 (1975).
25. W. Felsch, Z. Phys. B29, 211 (1978).
26. M. Croft and A. Jayaraman, Solid State Commun. 29, 9 (1979).
27. F. Steglich, W. Franz, W. Seuken and M. Loewenhaupt, Physica 86-88B+C, 503 (1977).
28. F. Steglich, in: "Festkörperprobleme (Adv. Solid State Physics)" Vol. XVII, p.319, J. Treusch, ed., Vieweg, Braunschweig (1977).
29. W. Franz, private communication and to be published.
20. J. Aarts, F.R. deBoer, S. Horn, F. Steglich and D. Meschede, Ref. 9, p.301.
31. H. Lustfeld, submitted to Physica B.
32. P. Scoboria, J.E. Crow and T.W. Mihalisin, J. Appl. Phys. 50, 1895 (1979).
33. W. Franz, F. Steglich, W. Zell, D. Wohlleben and F. Pobell, Phys. Rev. Lett. 45, 64 (1980).
34. S. Doniach, Physica 91B, 231 (1977).
35. See: F. Steglich, C.D. Bredl, M. Loewenhaupt and K.D. Schotte, J. Phys. (Paris) 40-C5, 301 (1979) and references cited therein.
36. K. Winzer and W. Felsch, J. Phys. (Paris) 39-C6, 832 (1978).
37. T. Komatsubara, T. Suzuki, M. Kawakami, S. Kunii, T. Fujita, Y. Isikawa, A. Takase, K. Kojima, M. Suzuki, Y. Aoki, K. Takegahara and T. Kasuya, J. Magn. Magn. Mat. 15-18, 963 (1980).
38. B. Barbara, J. Boucherle, J.L. Buevoz, M.F. Rossignol and J. Schweizer, Solid State Commun. 24, 481 (1977).
39. S. Horn, F. Steglich, M. Loewenhaupt, H. Scheuer, W. Felsch, and K. Winzer, Z. Phys. B42, 125 (1981).
40. P. Burlet, J. Rossat-Mignod, J.M. Effantin, T. Kasuya, S. Kunii and T. Komatsubara, Proc. 27th MMM Conference, Atlanta, 1981, J. Appl. Phys. (1982).
41. A.P. Murani, K. Knorr, K.H.J. Buschow, A. Benoit and J. Floquet, Solid State Commun. 36, 523 (1980).
42. S. Horn, E. Holland-Moritz, M. Loewenhaupt, F. Steglich, H. Scheuer, A. Benoit and J. Floquet, Phys. Rev. B23, 3771 (1981).
43. K. Andres, J.E. Graebner and H.R. Ott, Phys. Rev. Lett 27, 1779 (1975).

44. W. Franz, A. Griessel, F. Steglich, and D. Wohlleben, Z. Phys. B31, 7 (1978).
45. W. Lieke, N. Rauchschwalbe, C.D. Bredl, F. Steglich, J. Aarts and F.R. deBoer, see Ref. 40.
46. N.F. Mott, Phil. Mag. 30, 403 (1974).
47. W. Klämke, H. Biesenkamp, S. Horn, M. Schildwächter, W. Lieke and F. Steglich, to be published.
48. J.T. Schriempf, Phys. Rev. Lett. 20, 1034 (1968).

COMMENTS

BABERSCHKE: Where would you place $CeRu_2$ in your universal curve of IV systems versus the covalency parameter? If $CeRu_2$ shows IV properties (see J. Allen et al. for recent resonant photoemission), it would be interesting to compare $CeRu_2$ with $CeCu_2Si_2$ as a superconductor.

STEGLICH: There are two questions: As to the first, what you call "universal curve" only concerns the dependence of the CEF-split energy of dilute Ce^{3+} impurities on the bonding nature of non-magnetic, semimetallic monopnictides. I do not see how I could include a compound like $CeRu_2$ in this scheme. As to the second, while $CeRu_2$ behaves nonmagnetically below room temperature, there is considerable evidence for CEF-split trivalent Ce ions in $CeCu_2Si_2$ above, say, 10 K. From the lattice parameter data one is tempted to classify Ce in $CeRu_2$ close to tetravalent, and in $CeCu_2Si_2$ close to trivalent. Therefore, I feel that the superconductivity should be of different origin in these compounds.

KASUYA: In CeN the valence of Ce changes continously from trivalent to mixed valent with decreasing temperature due to the entropy effect. Then the resistivity behavior also changes from Kondo-like to a rather normal behavior. In your alloy system near the critical concentration is a similar effect observed?

STEGLICH: You are asking for the actual phase boundary between the trivalent and IV states in the temperature-concentration diagram of the Ce dialuminide systems. I have to admit that our preliminary transport data, shown in my talk, cannot answer this question.

COOPER: You mentioned a small effective mass for CeSb and CeBi in the transport properties. Could you give a number for that mass? Also, can you say more about the "continuous" transformation from Kondo to IV behavior in Ce intermetallics, i.e., what role does lattice coefficient play in this? What is the simplest picture for the absence of the Kondo effect in Eu IV materials?

STEGLICH: To answer the first question, several tenths of the free-electron mass. To answer the second, the lattice parameter does not

FERMI SURFACE AND p-f MIXING MECHANISM IN CeSb

Takashi Suzuki*, Hideaki Kitazawa*, Masafumi Sera*, Isamu Oguro†, Hiroshi Shida*, Akira Yanase* and Tadao Kasuya*

*Department of Physics, Tohoku University, Sendai, Japan
†Institute of Solid State Physics, Tokyo University
Minatoku Tokyo, Japan

INTRODUCTION

Among the rare earth monopnictides which crystallize in the rocksalt type structure, Ce-monopnictides CeX (X=N,P,As,Sb and Bi) have a special position due to their anomalous properties. In particular, CeSb and CeBi have attracted much attention from both theoreticians and experimentalists due to their very unusual properties. They have very small crystal field splittings in the paramagnetic region (Birgeneau et al., 1973) but show very strong Ising type antisotropy (Bartholin et al., 1974, 1975) in the ordered states, which are separated by many phases (Rossat-Mignod et al., 1977; Fisher et al., 1978). To explain these anomalies, our theoretical group proposed the anisotropic p-f mixing model (Takegahara et al., 1980). In the paramagnetic region, according to the theory, the Γ_8 quartet states of 4f levels in the cubic field mix with the J=3/2 quartet states at Γ point, while there is no mixing for the 4f Γ_7 doublet state. Therefore, the Γ_8 states are pushed down and cancel the splitting due to the point charge model in which Γ_7 is the lowest state.

The essential point for this mechanism to work is that there should be some number of valence holes around the Γ point. Detailed band calculations have been performed by our theoretical group on both La and Y-monopnictides (Hasegawa, 1980). They showed that the bottom of the conduction band is at each X point of the Brillouin zone and the top of the valence band is at the Γ point. For example, the number of holes, and thus that of electrons, was evaluated to be 0.013/Sb in LaSb. A main difference between La and Y compounds is the position of the 4f levels. In the Y compound, they are situated so high that they have no effect. In La compounds however, they are situated several eV above the top of the valence band and thus push down the top of the valence band causing smaller overlap between the valence and the conduction bands. Actually, in YSb, the number of holes was evaluated to be 0.028/Sb. In CeSb, because Ce^{+++} is stable, one 4f electron per Ce is situated below the Fermi energy, while the $4f^2$ vacant states seem to be similar to those in LaSb. We are now considering the mixing effect on the occupied 4f state below the Fermi energy. The band structure without this mixing effect should be similar to that of LaSb, or in between LaSb and YSb because of a smaller p-f mixing matrix in CeSb compared to that of LaSb. In this sense it is important to know the band structure, that is, the Fermi surfaces, of La and Y antimonides and bismuthides experimentally. There has been no experiment thus far to check the above point. We succeeded in making very good single crystals of LaSb and LaBi. Experimental results of magnetoresistance and de Haas van Alphen (dHvA) effects on these crystals are reported in the following. Note that in CeSb and CeBi the number of holes in the paramagnetic region should be larger than that in LaSb or in LaBi, because the top of the valence band is pushed up by the p-f mixing. This effect may be seen by magnetoresistance and Hall constant in the paramagnetic region.

The above mechanism may be described by the interaction picture as follows: when a particular 4f quartet state Q_r is occupied at a site R_o, the population of the valence hole with the same Q_r symmetry increases around R_o and then this hole polarization interacts with the near 4f state with the same Q_r symmetry through the p-f mixing interaction. Therefore, this is the interaction within the same symmetry state and thus substantially different from the usual exchange interaction described by the spin operator. When the present type of p-f mixing interaction is considered, with the holes around the Γ point only, there are no interactions within Γ_7 nor between Γ_8 and Γ_7. Both in CeSb and in CeBi, Γ_7 is thought to be lower than Γ_8 in the paramagnetic state; see an accompanying paper for the interpretation of the polarized neutron scattering experiment (Takegahara et al., 1981). Then, with decreasing temperature, the population of Γ_8 decreases. But when the ordering energy among Γ_8 exceeds the crystal field splitting energy, the ground state should be the ordered quartet with a particular symmetry. Our theory shows that, because of the competition between the interactions among the quartet and the point charge crystal field energy, the actual ground state is the nearly completely polarized $J_z=5/2$ state in agreement with experiment (it is necessary to be limited to the real point charge model, but it should include all effects except the present p-f mixing effect). In this ordered state, the 4f wave function extends out from the x-y plane and thus mixes strongly with the x,y orbits of the p-functions on the nearest neighbor anions sitting on the same x-y plane. This causes the strong in-plane interaction with the same symmetry (thus it now resembles the usual anisotropic exchange interaction) and pushes up the valence band holes with the x,y character and with wave vectors along the Δ axis in the z direction. Theory predicts that, as a result, an open orbit of valence holes appears along the Δ_z axis with a neck in the middle of the Δ_z axis (Takegahara et al., 1981). Therefore, if it is possible to observe this Fermi surface directly by the dHvA effect, it should be one of the most direct proofs of the present mechanism. Fortunately, we also succeeded in growing good quality single crystals of CeSb and CeBi. Some magnetic and transport properties have been reported (Suzuki et al., 1981). Now we show the experimental results of magnetoresistance and the dHvA effect in the following. These single crystals were synthesized by direct reaction and melting of the constituent elements in vacuum fused Mo crucibles by a high frequency induction furnace at about 2250 C. Crystals were grown by the slow-cooling Bridgman method.

EXPERIMENTAL RESULTS ON LaSb

Transport properties have been measured on several samples made from different batches. All of them showed the same characteristics. The resistivity at room temperature is about 67 μΩcm and that at 4.2 K is about 0.67 μΩcm. The transverse magnetoresistance at 4.2 K is very large, increasing nearly proportionally to $H^{1.8}$. The ratio $\rho(H)/\rho(0) = 420$ in the best sample at H=85 kOe. This shows that the numbers of electrons and holes are nearly equal, indicating the good stoichiometry of the samples. Because of the near balance of electrons and holes, the Hall voltage depends on both temperature T and the applied magnetic field H, and thus we cannot determine carrier numbers from Hall voltage. In LaSb, even the longitudinal magnetoresistance increases substantially. For example, at 4.2 K and 85 kOe, the ratio $\rho(H)/\rho(0)$ becomes 70. The origin of this phenomenon is not yet clear.

We observed the dHvA effect by a field modulation method in fields up to 140 kOe and temperatures down to 0.5 K. The oscillatory component of the magnetic susceptibility was observed by the Fourier transform analysis. The sample shape was a parallelepiped with dimensions of about 1.5 x 1.0 x 2.0 mm and with a cleavage surface. The dHvA signal for LaSb is easily detected at 4.2 K. A typical example of the signal and its Fourier transform for the applied field parallel to the [100] axis are shown in Fig. 1 and Fig. 2, respectively. To avoid complexity due to a monotonic increase of magnetic susceptibility, the second harmonics signal of the modulation frequency was used in the actual analysis. Four peaks, labeled (b) to (e) look like the higher harmonics of peak (a). But the second peak is highest and shows different behavior from that of (a) as the field is rotated from the [100] axis. Therefore, (b) is independent of (a) but accidentally degenerates with the second harmonic of (a) in the particular [100] direction of the field. On

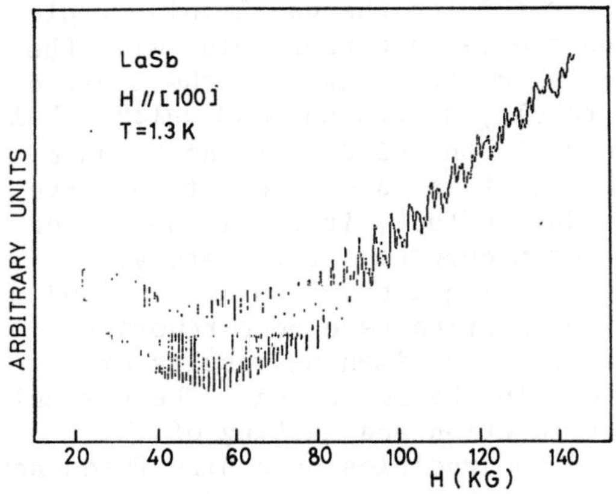

Fig. 1: The first harmonic signal of dHvA oscillation for magnetic field along the [100] axis at 1.5 K for LaSb.

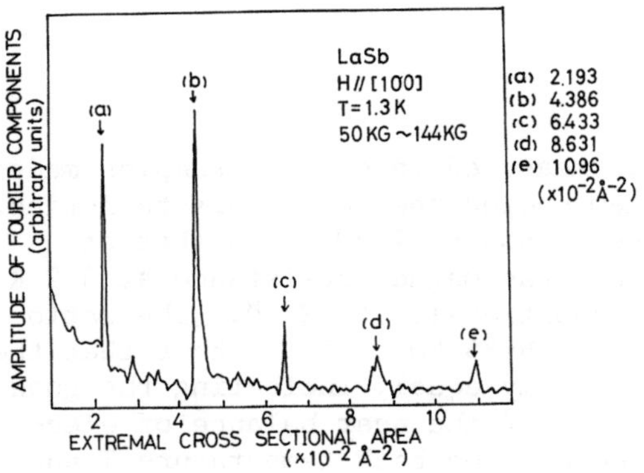

Fig. 2: The Fourier transform of data in Fig. 1.

the other hand, the peaks labeled (c), (d) and (e) are the higher harmonics of the signals (a) and (b).

When the direction of the applied field rotates from [100] to [110] on the (001) plane, another independent signal appears. We call these three independent signals α_1, β and α_2 respectively. α_1 and α_2 coincide at the [100] axis. The results are summarized in Fig. 3. With the aforementioned information of the band calculation, the following interpretation is unique and straightforward. At first, α_1 and α_2 should correspond to the electron surfaces centered at three X points. We fit the data by an ellipsoidal model elongated to the Γ point. The best fitted result is shown in Fig. 3 (solid lines). The shape of the Fermi surface is shown in Fig. 4 on some particular planes in the Brillouin zone (solid line). The fitted ellipsoidal is also shown by a solid line. In Figs. 3 and 4 we also show the results of the band calculation (dashed lines). It is surprising that, even though the calculation was done a few years ago without using any adjustable parameters, agreement is nearly perfect. The band calculation indicates that the Fermi surface is more sharply elongated than the ellipsoid along the Δ axis. However we assumed the Fermi surface of the band calculation

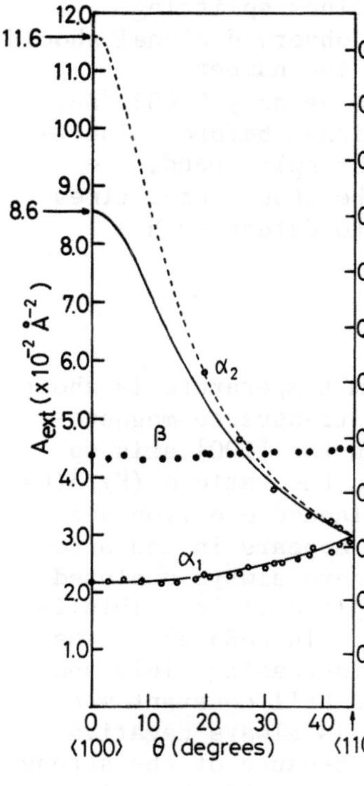

Fig. 3: Extremal cross sectional area of the Fermi surface for LaSb as a function of the field direction on the [001] plane. The experimental results are shown by the circles.

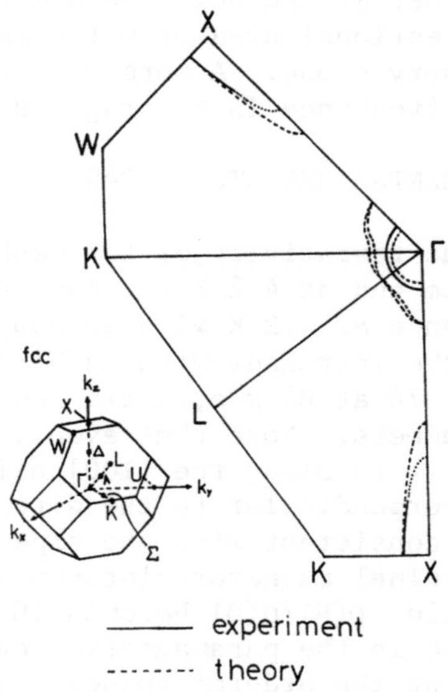

Fig. 4: The Fermi surfaces of LaSb. Experimental results are shown by the solid lines and the dashed lines are band calculation results.

to be ellipsoid as shown in Figs. 3 and 4 by dashed lines. This fact was not detected by our experiment because we could not detect a clear α_2 signal around the [100] axis. More careful measurements are now in progress. Note that in a general direction three signals should be observed corresponding to the three electron Fermi signals around X points. Here, we detected only the lowest two signals. When the ellipsoid model is applicable, the number of conduction electrons is about 0.02/La from band calculations and 0.015/La from experiment. It is also clear that the β branch comes from the hole surface, because it is nearly flat and thus corresponds to the nearly spherical Fermi surface. The Fermi surface is also shown in Fig. 4 by the solid line around the Γ point. The hole surfaces from band calculations are also shown by dashed lines. It should be noted, however, that the band calculations do not include the spin-orbit interaction. It is expected that, due to the strong spin-orbit interaction in the atomic 5p state of Sb, the quartet J=3/2 state of the valence band at the Γ point may be higher by about 0.5 eV than the doublet J=1/2 state. Therefore, in real LaSb, it is expected that the smallest hole surface should vanish, the larger dashed lines splitting further into two well separated surfaces. The observed signal should correspond to the upper split hole band. Then the number of holes on the observed branch is estimated to be only 0.0035/Sb, much smaller than the number of electrons mentioned before. Therefore, most of the holes should be in the highest split band, the cross-sectional area of which is estimated to be about three times the observed one. A more detailed experiment to detect such a higher frequency is in progress.

EXPERIMENTAL RESULTS ON CeSb

The resistivity of the CeSb sample at room temperature is about 177 μΩcm and at 4.2 K is about 0.78 μΩcm. The transverse magnetoresistance at 4.2 K with an applied field along the [100] axis up to 85 kOe increases with an $H^{1.7}$ dependence and the ratio $\rho(H)/\rho(0)$ becomes 70 at 85 kOe, again indicating well balanced electron and hole numbers. Note that even if an open orbit appears in the ordered state, it is along the [100] axis and thus we have always a closed orbit perpendicular to the direction of the applied field. This is really consistent with the experimental result. In CeSb also, the longitudinal magnetoresistance increases with increasing field and the ratio $\rho(H)/\rho(0)$ becomes 10 at 85 kOe. The Hall constant was obtained in the paramagnetic region. The sign is always negative including the ordered phases. This means that because of the strong p-f mixing interaction, the valence holes are scattered strongly by the disordered 4f states. The temperature dependence of the Hall constant is weak. When we neglect the contribution from the holes due to their smaller mobility, the number of conduction electrons is evaluated from the simple one band model to be 0.032/Ce. This is a factor of 2 larger than that in LaSb, in good agreement with the p-f mixing model. In the ordered phase, the Hall voltage has a complicated dependence on the successive phase transitions. In the perfect ferromagnetic region, the Hall constant is defined and gives the value 0.067 electrons/Ce at 4.2 K, based on the simple one band model. A large increase of the electron number is again consistent with the p-f mixing model. More details will be published in a separate paper (Sera et al.).

We observed the dHvA effect for CeSb. The sample was shaped into a sphere, since the inhomogeneity of the internal field in these magnetic substances seriously obstructs the detection of the dHvA effect. They were cemented tightly in the rotating holder so as to resist the strong torque on the sample due to extremely high magnetic anisotropy. The sample diameter was about 1.5 mm for both CeSb and CeBi. The signal amplitude of the dHvA effect is one order of magnitude smaller than that of LaSb. A typical example for the Fourier transform of the signal for the [100] direction of applied field is shown in Fig. 5. Here the first harmonic of the

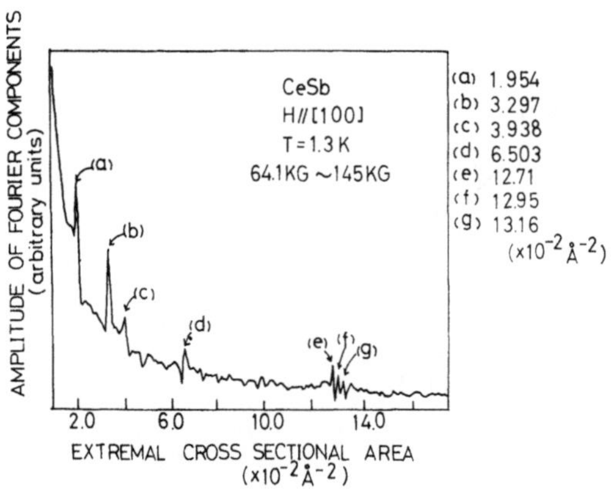

Fig. 5: The Fourier transform of the first harmonic of dHvA oscillation for magnetic field along the [100] axis at 1.5 K on CeSb.

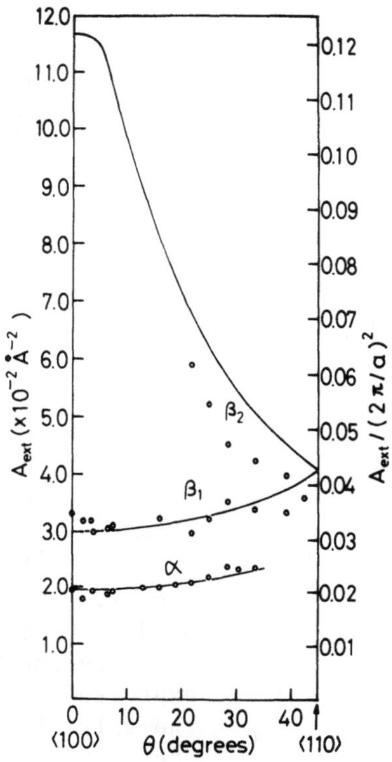

Fig. 6: The extremal cross sectional area of the Fermi surfaces as a function of the field direction on the (001) plane for CeSb.

modulation frequency is used. The extremal cross-sectional area of the lowest branch is nearly the same as that of the LaSb. It is clear that the signals labeled by (a) and (b) are independent of each other but those of (c) and (d) seem to be their higher harmonics. There are some signals labeled by (e), (f) and (g) in higher frequency regions. We neglect them in the present analysis because their signals are not so clear. These will be studied in a more careful measurement in the future. Again the applied field is rotated in the plane including the [100] and [110] axes, and only the independent signals are picked up. These results are summarized in Fig. 6. The lowest frequency signal, labeled α, is the strongest signal. However, it suddenly disappears when the direction of the applied field rotates more than 33 degrees from the [100] direction. This fact clearly indicates that this signal comes from a neck part of the Fermi surface along the [100] direction and thus should correspond to the open orbit predicted by the p-f mixing model. This part of the Fermi surface is plotted by a heavy solid line in Fig. 7, where the Fermi surface calculated by the p-f mixing model is shown by a solid line on the particular planes in the Brillouin zone. Agreement is very good. Note that the solid line drawn in Fig. 6 for the α branch corresponds to the parallelpiped shape. Therefore, the deviation from the nearly uniform neck part should occur fairly sharply. The signals labeled by β_1 and β_2 coincided in the direction

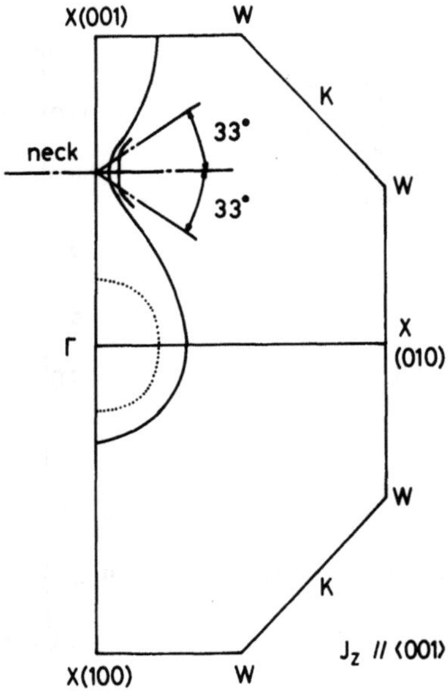

Fig. 7. The Fermi surfaces of the valence band of CeSb. Experimental results are shown by the short solid line and the well expressed cylinder-like shape as shown by the solid line on α branch in Fig. 6. The continuing solid line is the theoretical result with p-f mixing and the dotted line is without p-f mixing.

of [110] and are identifed as due to the electron part in the same way as in LaSb. However, here the data are more scattered due to weaker signals and thus we see more ambiguity. Therefore, in Fig. 6 we draw the same shaped ellipsoid as in LaSb but with a 40% larger cross-sectional area. The fitting is reasonable. It is seen that the experimental cross section for β_2 increases faster at smaller θ values, as mentioned in LaSb. If we assume that all electron Fermi surfaces are the same, the number of conduction electrons is evaluated to be 1.66 times that in LaSb, or 0.03-0.02/Ce. This number is substantially smaller than that of our theory (Takegahara et al., 1981, this conference). This discrepancy comes from the following fact: because of the d-f Coulomb exchange interaction, which is the most dominant interaction of the conduction electrons, our theory predicted that for the perfect ferromagnetic alignment of 4f spins the conduction band should split by as much as 0.1 eV. Then, what we observed experimentally should correspond to the upper bands. If the total number of conduction electrons is 0.067/Ce as the Hall constant indicates, we should expect the number of electrons in the lower split band to be about 0.05/Ce while that in the upper split band should be 0.015-0.1/Ce. In any case, more precise measurements and analysis for higher frequencies are necessary.

EXPERIMENTAL RESULTS ON LaBi AND CeBi AND CONCLUSIONS

The resistivity of LaBi at room temperature is about 76 μΩcm and that at 4.2 K is 0.20 μΩcm. The ratio ρ (H)/ρ(0) can be 1850 at 4.2 K and at 80 kOe indicating good cancellation of electron and hole numbers. The longitudinal magnetoresistance ρ(H) does not increase as much as for LaSb. So far we have measured only along the [100] axis of the applied field for the dHvA effect. We detected two signals at 0.024 $Å^{-2}$ and 0.054 $Å^{-2}$ of the extremal cross sectional areas. These signals seem to correspond to α and β in LaSb and they indicate that the number of conduction electrons increases a little bit more compared to that for LaSb. The resistivity of CeBi at room temperature is about 138 μΩcm and that at 4.2 K is 6.7 μΩcm. The ratio ρ (H)/ρ(0) at 4.2 K and 80 kOe is 2.1 and Δρ (H)/ρ(H) for H is negative. For the dHvA effect we measured only for applied fields along the [100] axis and obtained the main signal corresponding to the cross sectional area of 0.0224 $Å^{-2}$. This should correspond to the α branch in CeSb and again the cross sectional area increased about 12% in agreement with theory. More detail will be reported in a forthcoming paper.

To conclude, magnetoresistance and dHvA effects have been measured on good single crystals of LaSb, CeSb, LaBi and CeBi. In LaSb, a well defined conduction band Fermi surface was obtained, in excellent agreement with the band calculation. In CeSb, a well-defined neck part of the Fermi surface was observed, in good agreement with our theory. These results strongly support the validity of the p-f mixing model.

ACKNOWLEDGEMENTS

We would like to express our thanks to Prof. S. Tanuma for his valuable discussions, to Prof. A. Hasegawa for his continuous interest and helpful discussion, and to Dr. K. Takegahara and Miss H. Takashi for their stimulating discussions.

REFERENCES

Bartholin, H., Florence, D., Tcheng-Si, W. and Vogt, O., Phys. Stat. Sol. (a) 24, 631 (1974); 29, 275 (1975).

Birgeneau, R.J., Bucher, E., Maita, J.P., Passel, L. Turberfield, K.C., Phys. Rev. B8, 5345 (1973).

Fischer, P. Lebech, B., Meier, G., Rainford, B.D. and Vogt, O., J. Phys. C: Sol. Stat. Phys. 11, 345 (1978); 11, 1173 (1978).

Hasegawa, A., J. Phys. C: Sol. Stat. Phys. 13, 6145 (1980).

Rossat-Mignod, J., Burlet, P., Villain, J., Bartholin, H., Tcheng-Si, W., Florence, D. and Vogt, O., Phys. Rev. B16, 440 (1977).

Sera, M., Suzuki, T. and Kasuya, T., to be published.

Suzuki, T., Sera, M., Shida, H., Takegahara, K., Takahashi, H.,

Yanase, A. and Kasuya, T., in "Valence Fluctuations in Solids", Falicov, L.M., Hanke, W. and Maple, M.B., eds., North-Holland, Amsterdam,pp255-258 (1981).
Takegahara, K., Yanase, A., and Kasuya, T., J. de Phys. 41, C5:327, (1980); Erratum J. de Phys. 41, C5:1231 (1980).
Takegahara, K., Takahashi, H., Yanase, A. and Kasuya, T., J. Phys. C: Sol. Stat. Phys. 14, 737 (1981); Sol. Stat. Commun. 39, 857 (1981).
Takegahara, K., Takahashi, H., Yanase, A., and Kasuya, T., this conference.

COMMENTS

LUTHI: How did you grow the crystal and how did you get small residual resistance?

SUZUKI: The usual Bridgman method was used. The crystal was grown in a vacuum-fused Mo crucible. The stoichiometry of this sample is most important to make a good quality sample. An excess 3-5% of Sb or Bi gives a good result.

PIERRE: My question is rather general and refers also to the $CeSn_3$ compound. How is it possible to reconcile the existence of the de Haas-van Alphen effect, which implies a perfect coherence and a long mean free path, with the rapid spin flucturations inferred from the linewidths observed by neutron scattering?

WOHLLEBEN: Since nobody else seems to want to answer it, let me try. According to my experience many people are troubled by this question. Actually you should have worried about this for the first time when you saw that the resistivity goes to zero with decreasing temperature in these compounds, or for that matter, in any metal, considering the zero point motion of the nuclei. You are dealing in such cases with a random force on a Bloch electron. However, it is random only in the sense of a deviation from a potential with translational symmetry. The temporal average of the force has a perfect translational symmetry (no resistivity). How much of the random force the electron (or neutron) sees depends on the energy difference to the Fermi energy, $\Delta\varepsilon=\hbar\omega$ for the electron. If $\hbar\omega_f$ where ω_f is some measure of the width of the Fourier transform of the random force, then the effect of the random force is averaged out to nearly zero. Since $\langle\Delta\varepsilon\rangle\simeq kT$, $\rho\to 0$ for $T\to 0$. If, however, you inject electrons near T=0 at $\Delta\varepsilon >> kT$, you see very short mean free paths. This is seen in point contact measurements. In de Haas-van Alphen, both $\Delta\varepsilon\simeq\mu H$ and $\Delta\varepsilon\simeq kT$ are small compared to $\hbar\omega_f$ or $\hbar\omega_D$ in these compounds, so the random force from the fluctuations averages out to nearly zero, allowing for sufficiently long mean free paths.

MECHANISM OF UNUSUAL MAGNETIC ANISOTROPY IN THE CERIUM MONOPNICTIDES

K. Takegahara, H. Takahashi, A. Yanase and T. Kasuya

Department of Physics
Tohoku University
Sendai 980, Japan

INTRODUCTION

The inelastic neutron scattering done on various rare earth monopnictides with the NaCl crystal structure indicates that except for Ce monopnictides the point charge model of crystal field theory is quite applicable.[1] In the Ce monopnictides except CeN, each Ce is in the trivalent configuration, that is S=1/2, L=3 and J=5/2.[2] Because of a large spin-orbit interaction, the excited J=7/2 states lie at 0.28eV above the J=5/2 states and may be ignored in the following discussion. In the cubic crystal field, the J=5/2 multiplet splits into a Γ_7 doublet and a Γ_8 quartet. The point charge model predicts the Γ_7 state to be lower. Among these compounds, CeP and CeAs behave as normal trivalent magnetic materials with the Γ_7 ground state in the paramagnetic region and its antiferromagnetic order is of the type I[3]. However the crystal field splitting is about a half of that expected from the extrapolation from other rare earth monopnictides[1].

On the other hand, CeSb and CeBi show the following anomalous phenomena. 1) The observed values of the crystal field splitting in the paramagnetic state are unusually small, 37K in CeSb and 8K in CeBi, compared to those expected from the extrapolation from the other are earth monopnictides; 264K in CeSb and 247K in CeBi[1]. 2) Contrary to such small crystal field splitting values in the paramagnetic state, the magnetic anisotropies in the ordered phases are extraordinarily large with the easy axis in the <001> direction.[4] This also contradicts the usual result for the Γ_7 ground state, because the <001> direction is the hardest axis. The absolute values of the anisotropy field have not yet been obtained accurately because they are too large to be measured. In CeSb, even a magnetic field up to 15T in the <111> direction cannot induce any visible change in the direction of the moments.[5] From this fact the anisotropy field is estimated to be substantially larger than 100T. 3) Both CeSb and CeBi show very complicated magnetic ordering structures in the ordered states.[6,7] The spin arrangements in CeBi are described by various different stackings of the ferromagnetic (001) planes with two kinds of spin directions perpendicular to the plane. The moment on each site has nearly the saturated value for J=5/2 with g=6/7, that is 2.14μ_B. In CeSb, zero moment planes mix further among the ferromagnetic planes above. The zero moment sites is not the singlet state but the doublet state with very small mutual interaction.[8] 4) The above feature may suggest a large anisotropic exchange interaction. However, contrary to this, the observed Néel temperatures of CeSb and CeBi are not so high, 16K and 26K, respectively, and the former is a first order transition but the latter a second order transition. 5) Under a sufficiently large external magnetic field, about 8T, nearly the same amount of the anisotropy as that in the ordered phase was observed even at temperatures as high as 50K both in CeSb and in CeBi.[9] The induced moment under the above mentioned conditions is about 1/5 of the saturation moment. 6) In the paramagnetic state without external field, the Γ_7 state is thought to be the lower state from susceptibility data[10] but the results of polarized neutron diffraction under about 5T of external magnetic field reveal that the main character of the induced moment is that of the Γ_8 state.[11]

To explain these unusual magnetic properties, we have previously proposed the anisotropic p-f mixing model.[12] In this paper, we review the mechanism of the p-f mixing.

The trivalent rare earth monopnictides are known as typical indirect narrow gap semiconductors or weakly overlapping semimetals.[13] The APW band calculations on La and Y monopnictides show that the top of the valence bands, derived mostly from the p states on pnictogens, is at the Γ point and the bottoms of the conduction bands, derived mostly from the 5d t_{2g} states on rare earth atoms, are at the three X points.[14] The band gap becomes narrower for heavier Ce pnictides and is expected to disappear for the arsenide. Therefore CeSb and CeBi are semimetals in the paramagnetic region with an equal number of holes and electrons, about a few percents per Ce atom. The 4f level is expected to be a little below the Fermi level.[15] The mixing of the 4f state with the valence p states is much larger than that with the conduction electrons because of the shorter distance with pnictogens. Therefore the p-f mixing is expected to be most important for the magnetic properties of 4f electron systems. Note that there is no intra-atomic mixing between the 4f and 5d states. Recently, Siemann and Cooper[16] tried to explain these unusual magnetic properties by the covalent mixing of 4f and conduction electrons, but their theory can not explain the above mentioned anomalous properties self-consistently.

The wave function of $J_z=J$ is definitely oblate in shape and thus has a large mixing matrix with the p states on pnictogens in the (001) plane. This is the origin of the strong anisotropy favouring the spin direction perpendicular to the (001) plane. But the d-f mixing matrix has an opposite character and thus should give opposite results to our p-f mixing model. By the p-f mixing, the 4f level is pulled down and the valence bands are pushed up by the bonding-antibonding effect. When the pushed up valence bands are still below the bottom of the conduction band and thus still filled, no real energy gain is obtained. In CeP and CeAs, the p-f mixing matrix is smaller and furthermore the original energy separation of the valence and conduction bands is larger and thus the induced number of holes should be smaller. However, in CeSb and CeBi, the induced number of holes are larger and thus the p-f mixing effect should play an important role.

THE p-f MIXING MODEL

We consider the simplified system which contains the valence bands, the conduction bands and the 4f states. In various interactions among them, we consider the mixing term between the 4f states and the valence bands, the d-f exchange interaction in the conduction bands and the Zeeman energy when the external magnetic field is applied. The valence bands are treated by the simple tight binding model. Only the p-p and p-d transfer matrices are considered for the nearest neighbour anion-anion and anion-cation pairs, respectively. The d-d transfer matrix elements between different Ce sites are neglected because the detailed character of the conduction bands is not important in our calculation; only an appropriate density of states is introduced. The lowest part of the conduction bands is treated by the nearly free electron model with the effective mass approximation. Here we use 0.25m as the effective mass according to the APW band calculations[14] which give 0.3m for LaSb and 0.25m for YSb and GdSb, m being the free electron mass. Among the interactions between the 4f states and the conduction bands, we choose here only the d-f exchange interaction which causes the splitting of the conduction bands. We treat this d-f exchange interaction by the usual molecular field approximation. The value of the d-f exchange constant is assumed to be 0.15eV corresponding to the atomic optical data. Because of the mixed 5d states in the valence bands mentioned above, the d-f exchange interaction acts also in the valence bands. This effect is, however, smaller than that in the conduction bands and has no important effect on the present problem, so that this term is neglected. The effects of external magnetic fields both on the conduction and on the valence bands, are small compared with other interactions under consideration and thus they are neglected here. Here we consider the Zeeman effect only on the 4f electron in the usual form.

In CeSb and CeBi, the 4f state is expected to be about 1eV below the Fermi level.[15] Therefore, in the first approximation, the 4f state at each Ce site is treated as an atomic localized state under the cubic crystal field V_c and the effective field H_{eff} which is constituted by the external magnetic field and the exchange field. As the exchange field we consider here only the d-f exchange field mentioned above. Then the Hamiltonian for the 4f states may be written as

$$H_{4f}=\sum_{iMM'}|JM;i\rangle\langle JM;i|V_c+H_{eff}|JM';i\rangle\langle JM';i| \tag{1}$$

where M means the M-th eigenstate of any perfect set to describe the 2J+1 multiplet and i the Ce site. In the paramagnetic region without the external magnetic field, H_{4f} represents the usual crystal field Hamiltonian and then it is better to use the eigenstate of the latter, that is, M means one of Γ_7 or Γ_8 states.

The p-f mixing Hamiltonian may be written as

$$H_{vf}=\sum_{M_i iqk}|JM_i;i\rangle\langle JM_i;i|V|qk\rangle\langle qk|+h.c. \tag{2}$$

where q means the band suffix for the valence bands and k the wave vector. The explicit expression of the mixing matrix is given in Ref.12. To evaluate the mixing matrix elements, we considered only the nearest neighbour p-f mixing matrices which are given by the two parameters, (pfσ) and (pfπ). These values are evaluated from the band calculation of LaSb to be (pfσ)=0.5eV and (pfπ)=-0.25eV.[17] The p-p and p-d transfer matrices of the valence bands are also determined so as to fit to the valence bands of LaSb obtained by the APW method. In this process, the effects of the mixing between the valence bands and the unoccupied 4f states are subtracted.

Finally, we should check other types of interactions which were not considered in the present model. First of all, the mixing interaction between the 4f state and the conduction bands should be checked because this was considered to be important by Siemann and Cooper.[16] The d-f mixing matrix element is considered to be one order of magnitude smaller than the p-f mixing. However, the plane wave parts and the p states on pnictogens in the conduction bands have larger mixing matrices and make the effective mixing of the conduction bands larger, particularly in higher energy regions. Nevertheless, the mixing of the conduction electron with the 4f state is much smaller than the valence band. The most important term due to this d-f mixing interaction is the anisotropic self-energy of the second order perturbation, which should be written as the form of the one-ion cubic crystal field and thus can be included in V_c. This means that the point charge model part is really very complicated including various terms, among which the present term is included. Evaluation of the absolute value of the present term is not easy because the 5d and the plane wave parts are expected to enhance the point charge model but the p states mixed into the conduction bands cancel it. Anyway, the total contribution should be a fractional part of the observed so called point charge model. The second important term is the interaction term due to the fourth order perturbation. This term was claimed by Siemann and Cooper to be important, but this should be even smaller than the second the fact that nearly all the valence bands can now contribute to crystal field term in the paramagnetic region and for the origin of the large magnetic anisotropy in the ordered states.

The other important term which was not treated here explicitly is the mixing of the valence bands with the unoccupied 4f states making the $4f^2$ configurations. The mixing matrix elements have the same values as those in H_{vf} but the energy difference from the valence bands to the unoccupied 4f levels is larger by a factor three or four due to the well known correlation energy which is expected to be several eV. This factor is, however, compensated by the fact that nearly all the valence bands can now contribute to o the process. This kind of mixing process has two important effects. One is the one-ion crystal field and thus is included in the extrapolated experimental value. The other has the same form as that which is the main subject of this paper, but is smaller by a factor about five because of the larger excitation energy mentioned above. Therefore, this is included in the H_{vf} term. More detailed descriptions will be published in a separate paper.

MECHANISM OF MAGNETIC ANISOTROPY IN ORDERED STATES

At first we consider the ferromagnetic ground state, which is observed in real crystals under a large external magnetic field. In the ferromagnetic ground state, all M_i are equal to M_0. Then, by introducing the Fourier transformed 4f state, H_{vf} is rewritten as

$$H_{vf}=\sum_{qk}|JM_0;k\rangle\langle JM_0;k|V|qk\rangle\langle qk|+h.c. \tag{3}$$

Then for a given k we can solve $H_v+H_{4f}+H_{vf}$ under a given effective field on the 4f electron, where H_v is the Hamiltonian of the valence bands. The effective field is then solved consistently with the conduction band splitting due to the d-f exchange field as well as the position of the Fermi level. To do so, it may be most convenient to assume a variational state for M_0 and to make the minimum energy postulate. However, we use the following simple forms for M_0 derived from the following physical considerations. From the experimental results, both from the neutron scattering[6,7] and from the measured ferromagnetic moment,[18] the 4f moment in the ferromagnetic state is nearly completely saturated, which means that $J_z=J=5/2$. Therefore, at first, the 4f state is assumed to be in the complete ferromagnetic state. To evaluate the anisotropy field, the direction of the moment is taken along one of the three principal axes, <001>, <110> and <111>. This is the correct choice to define the anisotropy field because it is defined under the high magnetic field. Hereafter the above choices are called as the cases 1, 2 and 3. As another choice, one of the Γ_8 quartet states is chosen as M_0. This is based on the fact that the p-f mixing is largest at the Γ point where the valence band holes are concentrated when the number of holes is small. At the Γ point, only the Γ_8 state has a p-f mixing matrix. In the ferromagnetic state, the number of holes increases and thus the holes are no longer localized at the Γ point. However, the Γ_8 state should still be a fairly good choice for M_0. We call hereafter the cases 4 and 5 as the Γ_8 states of $\langle J_z \rangle=11/6$ and 1/2, respectively. The total energy is calculated for each M_0 by adjusting the position of the Fermi level so as to give equal numbers of holes and electrons. For the actual calculations we assumed that, without the p-f mixing interaction, the 4f state is situated 1.54eV below the top of the valence bands and the number of the valence band holes is 0.05/Sb.

Next we diagonalize the 7x7 matrix of $H_v+H_{4f}+H_{vf}$ for a given wave vector k. The upper parts of the valence bands are shown in Fig. 1 for wave vectors along the three Δ axes. The dispersion of the 4f state is also shown for the case 1. In all cases, a particular valence band couples strongly with a given 4f state M_0 showing a large shift, while the other bands show only small shifts or no shift. In the cases 1 and 4, the highest band becomes vacant along the Δ(00k) axis thus making the open orbit. The Fermi surfaces of these highest valence bands are given in Fig. 1. The total energy shift due to the p-f mixing and the d-f exchange interaction is evaluated. When there is no p-f mixing, the ground state is the Γ_7 doublet state whose energy is now chosen to be zero. Then the total energy shift is divided into three terms; $U=U_p+U_d+V_c$, where U_p is the contribution from the valence bands including the p-f mixing energy, U_d from the conduction bands including the d-f exchange energy and V_c the energy of the 4f state measured from the Γ_7 level. The numerical results are shown in Table 1.

First, we compare the energy shift U in the cases 1,4 and 5. It is seen that concerning the mixing energy, U_p, the cases 1, 4 and 5 have nearly equal energy gain but the case 1 can save the crystal field energy much more than the cases 4 and 5. Thus the $J_z=5/2$ ferromagnetic state, case 1, is the most stable state, in good agreement with experiments. Second, we estimate the anisotropy energy from the results of the cases 1, 2 and 3. The anisotropy energy, $\Delta U=U(\langle 110\rangle)-U(\langle 001\rangle)$, and the corresponding anisotropy magnetic field, $\Delta H=\Delta U/g\mu_B J(1-\cos\pi/4)$, are 3.83meV and 105T, respectively, in agreement with experiments. As the point charge model predicts, the V_c term makes the <111> direction most favourable. However, the strong anisotropic p-f mixing interaction

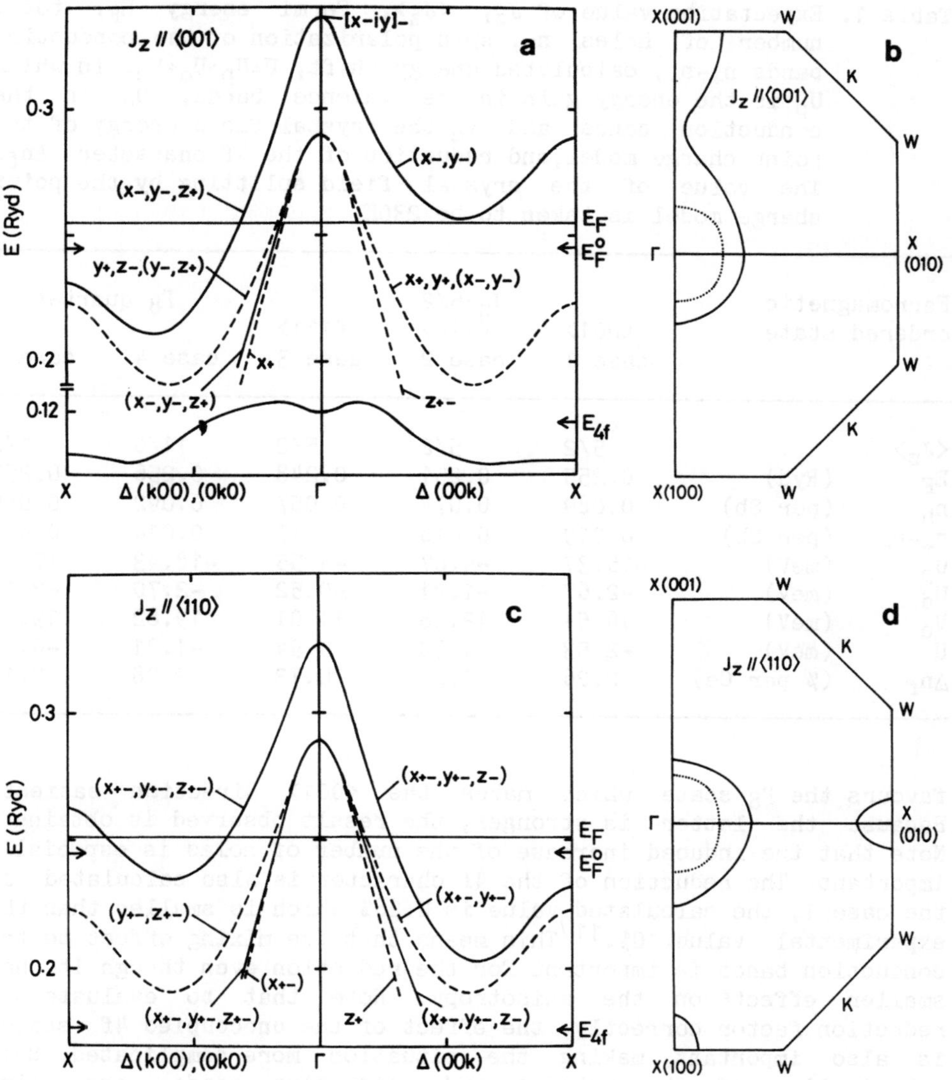

Fig. 1. Dispersion curves of the perturbed valence bands along three Δ axes (a), (c) and the calculated Fermi surfaces for the highest valence band (b), (d). In (a) and (c), shifted bands are shown by solid lines and unshifted bands by broken lines. Each state is constructed by a linear combination of states shown in the parentheses. Positions of the unperturbed Fermi energy E_F^o and the 4f level E_{4f} are indicated by arrows. Dispersion of the perturbed 4f state is also shown in the lower part of (a). In (b) and (d), solid lines show the perturbed Fermi surface and dotted lines the unperturbed Fermi surface.

Table 1. Expectation value of J_z, $\langle J_z \rangle$, Fermi energy E_F, total number of holes n_h, spin polarization of the conduction bands $n_+ - n_-$, calculated energy shift, $U=U_p+U_d+V_c$, in which U_p is the energy gain in the valence bands, U_d in the conduction bands and V_c the crystal field energy of the point charge model, and reduction of the 4f character Δn_f. The value of the crystal field splitting by the point charge model is taken to be 230K.

Ferromagnetic ordered state		$J_z=5/2$ <001> case 1	<110> case 2	<111> case 3	Γ_8 quartet case 4	case 5
$\langle J_z \rangle$		5/2	5/2	5/2	11/6	1/2
E_F	(Ryd)	0.255	0.251	0.248	0.256	0.255
n_h	(per Sb)	0.089	0.072	0.057	0.092	0.089
$n_+ - n_-$	(per Sb)	0.019	0.018	0.017	0.014	0.004
U_p	(meV)	-16.37	-9.87	-4.55	-18.43	-17.80
U_d	(meV)	-2.67	-1.21	-0.52	-2.70	-2.18
V_c	(meV)	16.51	12.38	11.01	19.82	19.82
U	(meV)	-2.53	1.30	5.94	-1.31	-0.16
Δn_f	(% per Ce)	2.26	1.24	0.73	2.38	2.17

favours the Γ_8 state which makes the <001> direction easiest. Because the latter is stronger, the result observed is obtained. Note that the induced increase of the number of holes is especially important. The reduction of the 4f character is also calculated. In the case 1, the calculated value is 2.26% which is smaller than the experimental value, 10%.[11] This means that the mixing effect to the conduction bands is important for the reduction even though it has smaller effects on the anisotropy. Note that to evaluate the reduction factor correctly, the effect of the unoccupied 4f states is also important, making the situation more complicated. Spin polarization of the valence and conduction bands are also estimated. In the case 1, considering the 2.26% reduction of the 4f character and the spin polarizations, the total magnetization is estimated to be $2.04\mu_B$ in reasonable agreement with the experimental value, $2.06\mu_B$.[11]

To do more detailed checks of the ground state, we introduce the variational function $\sqrt{(1-b^2)}|5/2,5/2\rangle + b|5/2,-3/2\rangle$ where a parameter b must be determined to minimize the total energy. The obtained value is very small, b=0.076, and this tendency agrees reasonably with the polarized neutron experiment.[11]

Calculations with different conditions such as (i) a slight

change in the number of the valence band holes as well as an unbalance in the numbers of the conduction electrons and the valence band holes, (ii) change in the d-f exchange constant, (iii) change of the effective mass of the conduction bands and (iv) change in the crystal field splitting, show no essential change in the results mentioned above.

Concluding this section, we offer some comments. 1) When the energy gap between the valence and the conduction bands is narrow, or they overlap slightly making a semimetallic state, anisotropic p-f mixing pushes up a special type of valence band and thus creates or increases the number of this special type of valence holes. This causes a strong non-linear type of anisotropic interaction among the given type of the 4f states on different sites as well as the anisotropic self-energy, substantially different from the usual bi-linear type interaction. In the usual treatment, the interaction term is derived from the fourth order perturbation. In the present system, because the p-f mixing matrix is larger than the Fermi energy of the valence band holes, the usual perturbation treatment is not applicable. Rigorous self-consistent determination for the change of the number of valence holes is especially important. 2) It is shown that the stable state in the ferromagnetic region is given well by the nearly full alignment of $J_z=5/2$ in the <001> direction, in good agreement with experiments. This is due to the balence of the p-f mixing and the crystal field energies. Because of $J_z=5/2$, the strong mixing appears within the (001) plane and thus the strong anisotropy energy is caused due to the strong anisotropic p-f mixing. It should be noted that the energy difference of this ferromagnetic state and the doublet ordered state is small because of approximate cancellation of the p-f mixing energy gain and the crystal field energy loss in the former state, in good coincidence with the observed low Néel temperature. Note that the excitations to other quartet dominated states are very large because the induced valence holes have very small p-f mixing matrices with the other type of 4f quartet states. Therefore, the first excited state observed by the neutron diffraction measurement[2,19] should be the doublet dominated state. In the ordered phases of CeSb, zero moment planes with an entropy Rln2 per site coexist with the $J_z=5/2$ planes.[8] On the other hand no such doublet state is observed in the ordered phases in CeBi. This is because V_c becomes smaller and the p-f mixing larger in CeBi leading the relatively higher Γ_7 level. The opposite may be expected in CeP and CeAs where V_c effects are dominant and thus the ground state is the usual Γ_7 antiferromagnetism as the point charge model predicted. As is seen in Table 1, this reversed condition occurs very easily with small changes in parameters. The interaction considered above is not not important now because of the small p-f mixing in the doublet. Detailed calculations of the elementary excitation is now in progress. 3) In the case of $J_z=5/2$ // <001>, the 4f wave function is extended along the x-y plane making a strong mixing with the p states on the same plane. This explains the observed strong ferromagnetic plane and the much weaker interlayer interaction corresponding to the observed various types of ordering. Similar calculations based on the various observed orderings are now in progress.

MECHANISM OF SMALL CRYSTAL FIELD SPLITTING IN PARAMAGNETIC REGION

In the paramagnetic state, the 4f state is denoted by the eigenstate of the usual crystal field Hamiltonian of the point charge model, M=Qi ($i=\kappa$, λ, μ, ν) for the Γ_8 quartet and M=Di' ($i'=\alpha''$, β'') for the Γ_7 doublet.[20] Because the number of the valence band holes is small, we use the simplified model to replace all the existing hole states by the six states at the Γ point. The spin-orbit coupling of the valence bands splits the hole states into the j=3/2 quartet, q=Pi ($i=\kappa$, λ, μ, ν), and the j=1/2 doublet, q=P'i' ($i'=\alpha'$, β').[20] It is easily shown that the only remaining p-f mixing matrices with a common value connect states Qi and Pi, i.e., $\langle JQi|V|Pi'\Gamma\rangle=\delta_{ii'}\sqrt{18/7}[(pf\sigma)-\sqrt{3}/2(pf\pi)]$. The energy shift of the $Pi\Gamma$ state is given by the second order perturbation as $E(Pi)=x(Qi)|\langle JQi|V|Pi\Gamma\rangle|^2/\Delta E_f$ where $\Delta E_f=E_F-E_{4f}$ is the energy difference of the 4f level E_{4f} and the Fermi level E_F. x(Qi) is the Qi state population ratio, 1/6 for high enough temperatures. The corresponding energy shift of the Γ_8 quartet states is given by $E(Qi)=-n(Pi)|\langle JQi|V|Pi\Gamma\rangle|^2/\Delta E_f$ where n(Pi) means the number of Pi holes per pnictogen and divides into two terms, $n_0(Pi)$ and $\Delta n(Pi)$. In the simple second order perturbation, only the $n_0(Pi)$ term is used. The Pi state energy shift, E(Pi), induces the increase in the number of Pi holes which corresponds to the $\Delta n(Pi)$ term. Thus E(Qi) comes up to the large value which nearly completely cancels out the splitting of the point charge model, in good agreement with experiments.[12] Because x(Qi) varies with temperature, the effective crystal field splitting is also temperature dependent. A big increase of the effective crystal field splitting Δ_{eff} is expected when the temperature becomes lower than Δ_{eff}. Then x(Qi) and n(Pi) have to be determined self-consistently. Calculated results show that, at high temperatures, the four x(Qi) have the same value but, with decreasing temperature, the equivalence of x(Qi) begins to break out at a critical temperature which we call the Curie temperature T_c. This critical temperature may be given by

$$kT_c=x(Qi)\rho(E_F)[|\langle JQi|V|Pi\Gamma\rangle|^2/\Delta E_f]^2 \quad (4)$$

in the second order transition among the quartet, and where $\rho(E_F)$ means the valence band density of states at the Fermi level. Here we have a new parameter $\rho(E_F)$ which changes fairly sensitively with the position of the unoccupied $4f^2$ level. When the value for the LaSb band calculation is used, $\rho(E_F)$ is large and thus T_c is too large. When, however, the value for YSb is used, in which the 4f level is situated very high, we have reasonable value of T_c for CeBi around 30K, in good agreement with experiments. Because T_c is larger than Δ_{eff}, the change of Δ_{eff} in the paramagnetic region is very small. For CeSb, we could not find T_c, which should correspond to the first order transition in CeSb. When the point charge crystal field splitting becomes a little bit larger, no ordering appears in the quartet and the doublet becomes the ground state corresponding to CeAs.

Finally the effect of the external magnetic field is considered very briefly. The polarization of the magnetic moment in the quartet under the external magnetic field is enhanced by the anisotropic interaction mentioned above, compared with nearly no enhancement in the doublet. This is especially so near T_c, in which the main character of the spin polarization is quartet. This is the situation observed by the neutron scattering at T=35.5K.[11] Large induced anisotropy in the paramagnetic region comes also from the same mechanism[9]. The details will be published in separate papers.

ACKNOWLEDGMENTS

Various useful discussions and private communications with Prof. A. Hasegawa on the band structure and Prof. T. Suzuki on experimental results are gratefully acknowledged. One of the authors (K.T.) would like to acknowledge the Yamada Science Foundation for financial support.

REFERENCES

1. R. J. Birgeneau, E. Bucher, J. P. Maita, L. Passell, and K. C. Turberfield, Phys. Rev., B8,5345 (1973).
2. H. Heer, A. Furrer, W. Hälg, and O. Vogt, J. Phys. C:Solid State Phys., 12,5207 (1979).
3. F. Hulliger and H. R. Ott, Z. Physik, B29,47 (1978).
4. G. Busch and O. Vogt, Phys. Letters, 25A,449 (1967).
5. P. Burlet, J. Rossat-Mignod, H. Bartholin, and O. Vogt, J. de Phys., 40,47 (1979).
6. J. Rossat-Mignod, P. Burlet, J. Villain, H. Bartholin, W. Tcheng-Si, D. Florence, and O. Vogt, Phys. Rev., B16,440 (1977).
7. H. Bartholin, P. Burlet, S. Quezel, J. Rossat-Mignod and O. Vogt, J. de Phys., 40,C5-130 (1979).
8. J. Rossat-Mignod, P. Burlet, H. Bartholin, O. Vogt, and R. Lagnier, J. Phys. C:Solid State Phys., 13,6381 (1980).
9. T. Suzuki, M. Sera, H. Shida, K. Takegahara, H. Takahashi, A. Yanase, and T. Kasuya, in:"Valence Fluctuations in Solids," L. M. Falicov, W. Hanke, and M. B. Maple, eds., North-Holland, Amsterdam, (1981).

10. Y.-L. Wang and B. R. Cooper, Phys. Rev., B2,2607 (1970).
11. J. X. Boucherle, A. Delapalme, C. J. Howard, J. Rossat-Mignod, and O. Vogt, Physica, 102B,253 (1980).
12. K. Takegahara, A. Yanase, and T. Kasuya, J. de Phys., 41,C5-327 (1980), Erratum J. de Phys., 41,1231 (1980).
K. Takegahara, H. Takahashi, A. Yanase, and T. Kasuya, J. Phys. C:Solid State Phys., 14,737 (1981).
T. Kasuya, K. Takegahara, M. Kasaya, Y. Isikawa, H. Takahashi, T. Sakakibara, and M. Date, in:"Physics in High Magnetic Fields," S. Chikazumi and N. Miura, eds., Springer-Verlag, Berlin, (1981).
T. Kasuya, in:"Electron Correlation and Magnetism in Narrow-Band Systems," T. Moriya, ed., Springer-Verlag, Berlin, (1981).
K. Takegahara, H. Takahashi, A. Yanase and T. Kasuya, Solid State Commun., 39,857 (1981).
13. G. Güntherodt, E. Kaldis, and P. Wachter, Solid State Commun., 15,1435 (1974).
14. A. Hasegawa and A. Yanase, J. Phys. Soc. Japan, 42,492 (1977).
A. Hasegawa, J. Phys. C:Solid State Phys., 13,6147 (1980).
15. Y. Baer, R. Hauger, Ch. Zürcher, M. Campagna, and G. K. Wertheim, Phys. Rev., B18,4433 (1978).
16. R. Siemann and B. R. Cooper, Phys. Rev. Letters, 44,1015 (1980).
17. A. Yanase and A. Hasegawa, in:"Electron Correlation and Magnetism in Narrow-Band Systems," T. Moriya, ed., Springer-Verlag, Berlin, (1981).
18. H. Bartholin, D. Florence, W. Tcheng-Si, and O. Vogt, Phys. Stat. Sol., (a)24,631 (1974).
H. Bartholin, D. Florence, W. Tcheng-Si, and O. Vogt, Phys. Stat. Sol., (a)29,275 (1975).
19. A. Furrer, W. Hälg, H. Heer, and O. Vogt, J. Appl. Phys. 50,2040 (1979).
20. J. S. Griffith, "The Theory of Transition-Metal Ions," Cambridge, London (1961).

COMMENTS

COQBLIN: In the Friedel-Anderson model that we have used to derive our Hamiltonian with J. R. Schrieffer, only the $\ell=3$ (f) part of the conduction electrons is "mixed" (or "hybridized") with the 4f electrons. In the "p-f mixing" mechanism that you consider in your paper, one mixes a p-wave function centered on a given Sb site with an f wave function centered on the nearest Ce site. I think that your "p-f mixing" mechanism is in principle similar to the classical Friedel-Anderson k-f mixing for the following reasons: In the Anderson Hamiltonian itself, one deals with a complete set of $|k\rangle$ states and an extra wave function $|f\rangle$ which is orthogonal to the $|k\rangle$ states, i.e. $\langle k|f\rangle = 0$ for all the $|k\rangle$ states, while one takes $V_{kf}=\langle k|V|f\rangle \neq 0$, which is in clear contradiction with the orthogonality relation. It has been shown, however, by several authors (see, for example, Haldane at the Rochester Conference, 1976) that the V_{kf} hybridization term originates from a non-zero mixing between f wave functions (which could be of s, p or d character) centered on the nearest site. In this sense, your "p-f mixing" mechanism is in the same line as the Friedel-Anderson mixing, but indeed you have made more sophisticated calculations to describe the conduction bands of cerium monopnictides.

EQUILIBRIUM AND DYNAMIC BEHAVIORS OF CUBIC Ce^{3+} SYSTEMS WITH ANISOTROPIC COQBLIN-SCHRIEFFER AND CRYSTAL FIELD INTERACTIONS

David Yang and Bernard R. Cooper

Department of Physics
West Virginia University
Morgantown, WV 26506, U.S.A.

INTRODUCTION

The cerium and light actinide monopnictides and monochalcogenides show many similar and striking ordering and phase transition properties.[1] The occurrence of the series of modulated magnetic structures in CeBi[2] and CeSb[3], in UAs[4,5] and possibly in USb[5], including the noncolinear structures observed[5] in uranium monopnictides, points to the existence of a strongly anisotropic exchange interaction which dominates the behavior of these systems.

One mechanism which yields an anisotropic exchange interaction is that of Coqblin and Schrieffer.[6] They showed that the mixing of Ce^{3+} ions with conduction electrons can be represented as the combined spin and orbital exchange scattering of conduction electrons by the localized ions. This leads to a Rudermann-Kittel type interaction between neighboring Ce^{3+} ions:

$$H_{ij} = \sum_{MM'} f(R_{ij})\, G(MM')(L^{i}_{M'M} - \tfrac{1}{6}\,\delta_{MM'}1_i)(L^{j}_{MM'} - \tfrac{1}{6}\,\delta_{MM'}1_j), \quad (1)$$

i,j label sites, M, M' label the quantum states of the cerium ions, with quantum axis taken along the bonding axis. The operator $L^{i}_{M'M} = |M'\rangle\langle M|$ projects an ion at site i from state M to M', and 1_i is the identity operator at site i. As shown by Siemann and Cooper[7], the only nonvanishing G(M,M') are the spin-only interaction, i.e. M, M'=±1/2, with G(M,M') the same for M, M'=±1/2. In other words, the coupling comes dominantly from the f orbital states with m=0 which pile up the charge along the bonding axis, giving the highly anisotropic Coqblin-Schrieffer (CS) interaction. This bonding axis anisotropy directly leads to the otherwise anomalous magnetic ordering properties of cerium and light actinide monopnictides and monochalcogenides.

Upon transforming to the fcc crystalline coordinate system, this anisotropic CS exchange takes the form:

$$H_{ij}=-E_{ij}\sum_{\mu\nu\varepsilon\sigma} e^{i(\mu-\nu+\varepsilon-\sigma)\psi} f^{\varepsilon\sigma}_{\mu\nu}(\theta)\, L^{i}_{\mu\nu}\, L^{j}_{\varepsilon\sigma} \tag{2}$$

$$f^{\varepsilon\sigma}_{\mu\nu}(\theta) = \sum_{MM'} (d_{\mu M'} d_{\nu M} - \frac{1}{6}\delta_{MM'}\delta_{\mu\nu})(d_{\varepsilon M} d_{\sigma M'} - \frac{1}{6}\delta_{MM'}\delta_{\varepsilon\sigma}) \tag{3}$$

where θ and ψ are the angular coordinates of the bonding axis, $\vec{R}_{ij}$, with respect to the crystal axes system. The Greek letters label the states with quantum axis taken along the <001> axis, $d_{\mu\nu}(\theta)$ is the rotational transformation matrix as conventionally defined[8] (J=5/2). As pointed out by Siemann and Cooper[7,9], as a simple consequence of the interaction being strongest when charge is piled up along the bonding axis, the CS interaction, for ions pairs with saturation moments, is strong in the (001) plane and weak between neighboring (001) planes. This effect, for the fcc lattice of Ce^{3+} ions with nearly equal nearest neighbor (nn) and next-nearest neighbor (nnn) CS interactions, is sufficient to stabilize the colinear orderings observed in the cerium monopnictides.

We have extended the investigation further to include the occurrence of noncolinear structures, and the dynamical behavior of the CS systems. The selection between colinear and noncolinear structures depends critically on the range of the CS interaction. In the dominately nn-only CS interaction system, the nn interaction can no longer stabilize the near saturation colinear ordering. The geometrical arrangement (of spins) which gives the best angular distribution of moments of each nn pair and optimizes the free energy, can be noncolinear. In the system with nn-only CS interaction the ground state is a ferromagnet with moments along <110>. In the presence of a small antiferromagnetic Heisenberg exchange, we find the noncolinear canted ferromagnetic or antiferromagnetic 2k structures. With the addition of a cubic crystal field term, the Kouvel-Kasper noncolinear 3k antiferromagnetic ordering, with moments pointing along the cubic diagonal directions, can be stabilized. The excitation spectra observed in systems with CS interaction reflect the exchange interaction as seen by the ions in their ground states. In the systems with <001> or <110> ordering we observe anisotropic dispersion, reflecting the anisotropic CS interaction, while for systems with <111> alignment, the anisotropy is greatly diminished, leaving rather isotropic spectra.

HAMILTONIAN OF SYSTEMS WITH COQBLIN-SCHRIEFFER INTERACTION

We consider an fcc lattice of Ce^{3+} ions interacting with the

CS interaction. The Hamiltonian, in Fourier transformed form, is:

$$\mathcal{H}_{CS} = - \sum_{q} \sum_{\mu\nu\varepsilon\sigma} J^{\varepsilon\sigma}_{\mu\nu} (\vec{q}) L^{\vec{q}}_{\mu\nu} L^{-\vec{q}}_{\varepsilon\sigma} . \quad (4)$$

For $\vec{q}$ along <001>, the nn exchange gives

$$J^{\varepsilon\sigma}_{\mu\nu} (\vec{q}) = \frac{4E_1}{N} [2f^{\varepsilon\sigma}_{\mu\nu} (\tfrac{\pi}{4}) \cos\frac{qa}{2} + (-1)^{\xi} f^{\varepsilon\sigma}_{\mu\nu} (\tfrac{\pi}{2})] \, \delta_{\mu-\nu+\varepsilon-\sigma,4\xi} \quad (5)$$

where the $\delta_{\mu-\nu+\varepsilon-\sigma,4\xi}$ factor is associated with the four-fold symmetry of the <001> axis. The anisotropy of the CS interaction is evident upon comparison with the exchange for $\vec{q}_{\perp}$<001>:

$$J^{\varepsilon\sigma}_{\mu\nu} (\vec{q}) = \frac{4E_1}{N} [2f^{\varepsilon\sigma}_{\mu\nu} (\tfrac{\pi}{4}) (\cos \frac{q_x a}{2} + (-1)^{\xi} \cos \frac{q_y a}{2}) \, \delta_{\mu-\nu+\varepsilon-\sigma,2\xi} + f^{\varepsilon\sigma}_{\mu\nu} (\tfrac{\pi}{2}) (-1)^{\xi} \cos \frac{(q_x + q_y)a}{2} \, \delta_{\mu-\nu+\varepsilon-\sigma,4\xi}] . \quad (6)$$

The difference is due to the angular dependence of the CS exchange.

Other terms relevant to the behavior of the cerium and light actinide monopnictides include the cubic crystal field, the tetragonal distortion, the Zeeman term, and the Heisenberg exchange. These give the total Hamiltonian of the system.

EQUILIBRIUM PROPERTIES

We examine the equilibrium behavior of the system dominated by the CS interaction within the molecular field (MF) approximation. Particular attention is paid to the determination of the ground states of two systems, one with nn-only CS interaction, and one with equal nn and nnn CS interactions, at T=0. The wide variety of magnetic orderings to be taken into consideration, and the formal complexity of the CS interaction, make the determination of the ground state rather complicated. Here, we limit ourselves to a rather selective set of magnetic structures. This includes the ferromagnets, the two and four sublattice antiferromagnets (colinear and noncolinear) and canted ferromagnets. The limited set of magnetic structures means we have omitted considering some interesting structures, such as the complex modulated orderings in CeSb. The procedure taken is to first find the lowest energy state for each of the orderings mentioned above. The ordering with the lowest eigenenergy is then taken as the ground state of the system.

The orderings of the three lowest energy states for the nn only CS interaction system are shown in Fig. 1. The ferromagnetic state, shown in (a), lies at $-1.7558\ E_1$. The moments point along <110> with components along <100> and <010> being (1.01,1.01), giving a total moment of 1.43, compared with the saturation value of 2.5. The state

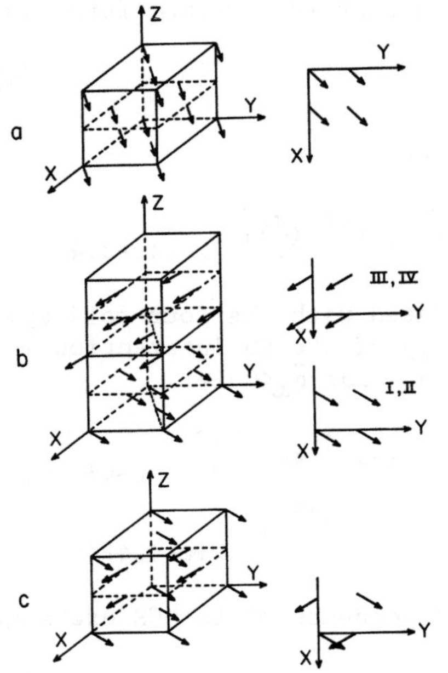

Fig. 1. Lowest energy magnetic structures in fcc lattice of Ce^{3+} ions interacting with nearest-neighbor only Coqblin-Schrieffer interaction.

which lies immediately higher in energy, at $-1.7237\ E_1$, is a noncolinear four sublattice canted ferromagnet, shown in (b). The (001) planes are ferromagnetically ordered, with moments of (1.03,0.54) in two adjacent planes, and (-1.03,0.54) in the next two. The two sublattice noncolinear canted ferromagnet shown in (c) lies at $-1.7197\ E_1$. The (001) planes are ferromagnetically ordered, with moment (1.06,0.56) on one plane and (-1.06,0.56) on adjacent planes. The small energy differences between the three states follow from the small exchange energy between ions on adjacent (001) planes. The interaction of ions on the same (001) plane is large. This anisotropy characterizes these structures. In the presence of a small antiferromagnetic Heisenberg term, the noncolinear canted ferromagnetic structure is favored. With increasing antiferromagnetic Heisenberg exchange, we obtain first a 2k antiferromagnetic structure, in which moments on (001) planes order antiferromagnetically pointing along the face diagonal directions, with moments directions on adjacent planes vertical to each other. With a strong Heisenberg term, the colinear Type I structure with moments along <001> is favored. We should also point out that more complicated structures, such as the Type IA 2k structure[5] observed in the low temperature phase of UAs, may also occur but have not been included in this calculation.

In the system with equal nn and nnn CS interactions we recover the result of Siemann and Cooper[7,9]. The three low lying states are the colinear ferromagnetic, Type IA (++--), and Type I (+-+-) antiferromagnetic states, at energies $-2.2371\ E_1$, $-2.1453\ E_1$, and $-2.0575\ E_1$, respectively. The moments are aligned along the <001>

axis, close to saturation. The small energy differences between the three states point to the small exchange interaction between ions on different (001) planes. The Type I or IA ordering can be stabilized with a small antiferromagnetic Heisenberg term. The magnetic orderings in CeBi can be explained with this model.

The effect due to the crystal field term is also interesting. In the nn-only CS interacting system, we obtain a ferromagnetic state with <111> alignment for $60B_4 < -.23\ E_1$. With the addition of a small antiferromagnetic nn Heisenberg term, the Kouvel-Kasper antiferromagnetic 3k ordering is found to be stable. This may provide a model for USb, for which the 3k structure has been suggested[5]. In the system with equal nn and nnn CS interaction, it takes a large crystal field term to yield the Kouvel-Kasper ordering. This may bear on the behavior[10] of CeP and CeAs.

In summary, the ordering depends critically on the range of the CS interaction. The model with equal nn and nnn CS interactions yields colinear structures, giving behavior as observed in cerium compounds. The nn-only CS model leads to noncolinear structures, giving behavior like that in the uranium compounds. These results, qualitative as they are, point to the dominant role of the CS interaction in the cerium and light actinide monopnictides and monochalcogenides. There is evidently a wealth of phase transition behaviors in such systems. An interesting example is the case with small but significant nnn CS interaction ($E_2=0.4E_1$). The free energies for the colinear Type I and noncolinear 2k structures are nearly equal at T=0. This may lead to a transition between the colinear and noncolinear structures at finite temperature[5].

EXCITATION BEHAVIOR

Here, we examine the excitation spectra and the intensities of the excitation modes, within the random phase approximation. For systems discussed above with <001> or <110> alignment, the small interaction between ions of different (001) planes leads to two main features, the occurrence of a low lying mode along <001> direction, and the anisotropic dispersion. The excitation spectra reflect the exchange interaction as seen by the ions in their ground state; these features thus exemplify the anisotropy of the CS interaction and the nature of the ground state. In systems which order with <111> alignment, either ferromagnetically, or in the Kouvel-Kasper (3k) antiferromagnetic ordering, the interaction as seen by the ions in the ground state is rather isotropic. As a result, the spectra becomes almost isotropic.

We start by projecting the Hamiltonian into the MF manifold:

$$H = \sum_i \sum_m E_m L^i_{mm} - \sum_q \sum_{mnm'n'} J^{m'n'}_{mn}(q)(L^q_{mn} - \langle L^q_{mn} \rangle)(L^{-q}_{m'n'} - \langle L^{-q}_{m'n'} \rangle). \tag{7}$$

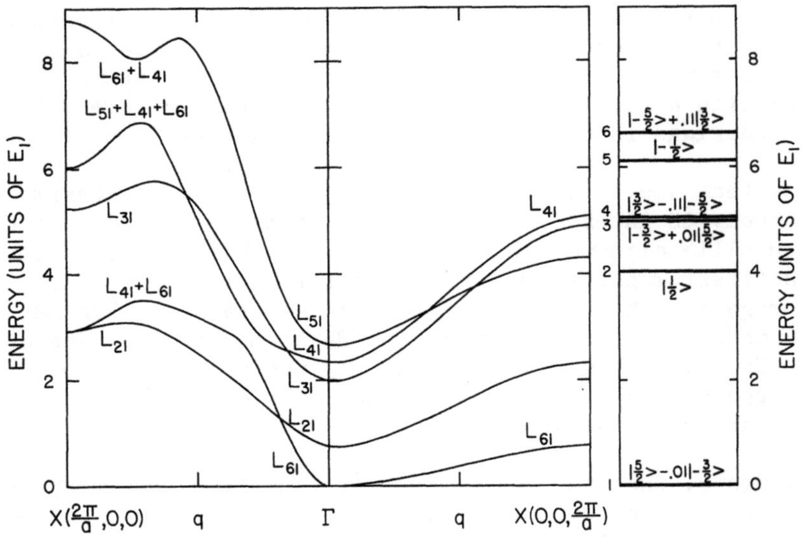

Fig. 2. Excitations at T=0 for Ce^{3+} ions on fcc lattice with equal nn and nnn CS interactions (ferromagnetic state).

Here, m, n refer to the MF states, and the projection operator L_{mn}', and the exchange matrix are defined within the MF manifold. At T=0 we need only consider the set of ten operators which take the ions from or to the ground state. The 10x10 frequency matrix:

$$\Omega_{mn}^{m'n'}(q) = (E_m - E_n)\,\delta_{mm'}\delta_{nn'} - 2\,J_{mn}^{m'n'}(q)\,\langle L_{nn} - L_{mm}\rangle \qquad (8)$$

gives ten excitation modes (five at energy gain, and five at energy losses), and the corresponding normal mode operators A_i, i=1,10. The intensity of these modes in experiment depends on the configuration of the particular measurement. For example, if the transverse response is measured, the linear response is given by Kubo's inner product[11], (J^+,J^+), which can be decomposed into:

$$(J^+,J^+) = \sum_{i,j} n_i n_j^* (A_i,A_j) = \sum_i n_i n_i^* (A_i,A_i) \qquad (9)$$

where the matrix element n_i is defined by $J^+ = \Sigma\, n_i A_i + \ldots$, and we have used the condition $(A_i,A_j) = \delta_{ij}(A_i,A_i)$. The intensity of the mode A_i is thus the product of the static susceptibility (A_i,A_i), and the square of the matrix element n_i.

The spectra for the nn and nnn CS interaction system, which orders ferromagnetically with near saturation moment along the <001> axis, is shown in Fig. 2. These results, which have been briefly discussed previously[12], give the typical behavior of the planar ordering systems. The two main features, as shown in the figure, are the strong anisotropy of the dispersion, and the existence of the low lying mode. For $\vec{q}$ along <001>, the moment direction, the dispersion is small, and the frequency is low. This behavior is particularly evident for the low lying mode labeled L_{61}, which gives the soft mode (vanishing excitation energy) at the zone centre, and remains low even at the zone boundary. The small dispersion results from the small interaction between ions on different (001) planes. For $\vec{q}$ along the <100> direction, this mode shows strong mixing, and becomes the highest frequency mode at the zone boundary. This anisotropy reflects directly the anisotropy of the CS interaction.

Of the five modes shown in Fig. 2, the mode L_{41} takes the ion from the ground state, which is essentially $|5/2\rangle$, to the third excited state which is essentially $|3/2\rangle$, and gives the dominant transverse mode. The soft mode L_{61} has small transverse matrix element [Eq.(9)]. However, due to the large static susceptibility associated with the soft mode, it gives large intensity in the immediate vicinity of q=0. All other modes have small matrix elements, and thus small intensities.

As mentioned, the anisotropic CS interaction, as seen by the ions in the ground state, is reflected in the anisotropic dispersion of the excitation spectra. This behavior is clearly present in the system with nn-only CS interaction which orders ferromagnetically with <110> alignment. The spectra are similar to that for the equal nn and nnn CS interaction system. The low lying mode intensity, which is proportional to the matrix element defined in Eq. 9 and thus depends on the ground state of the system, is large in the nn-only CS interaction system. The large anisotropy in the dispersion should thus be easily observable in the transverse response of the system.

The nn only CS system with crystal field ($60B_4=-0.5E_1$) orders ferromagnetically with <111> alignment. The excitation spectra of this system are shown in Fig. 3. This dispersion is surprisingly isotropic. Apparently this reflects the fact that in this case the CS interaction, as seen by the ions in the ground state, is rather isotropic. This is verified by examining the exchange matrix defined in the MF manifold, which indeed is almost isotropic. The intensity of the modes, as discussed before, depends on the ground state of the system through the matrix element as defined in Eq.9. In the present case, the longitudinal mode (defined with respect to the moment), L_{61}, gives the same order of intensity as that of the transverse modes, L_{21} and L_{41}. At the immediate vicinity of the zone centre, the soft mode dominates the picture.

For equal nn and nnn CS interactions, the results obtained for the Type I and Type IA antiferromagnetic orderings are similar to that obtained for ferromagnetic ordering in the absence of the small antiferromagnetic Heisenberg term. An example is that of Type I ordering for the system with a small antiferromagnetic nn Heisenberg term, as shown in Fig. 4. We note the low lying mode and the large anisotropy in dispersion. The intensity of the excitation modes

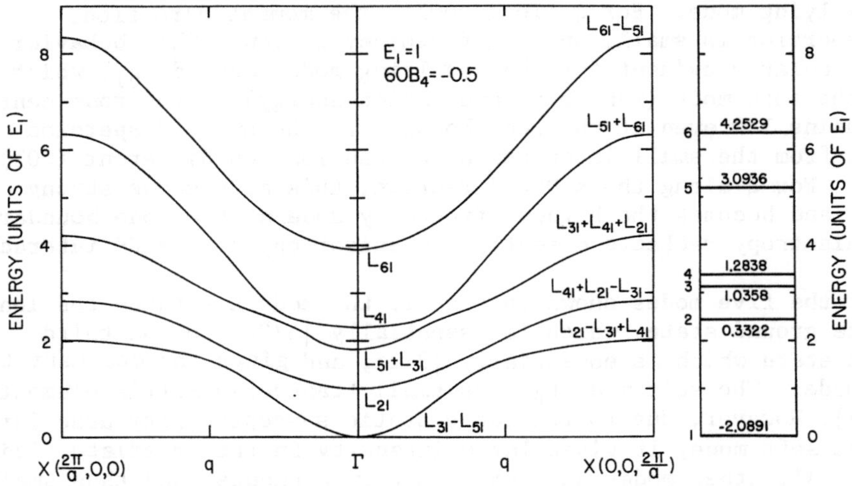

Fig. 3. Excitations at T=0 for fcc Ce^{3+} system with nn-only CS interaction and cubic crystal field (ferromagnetic state).

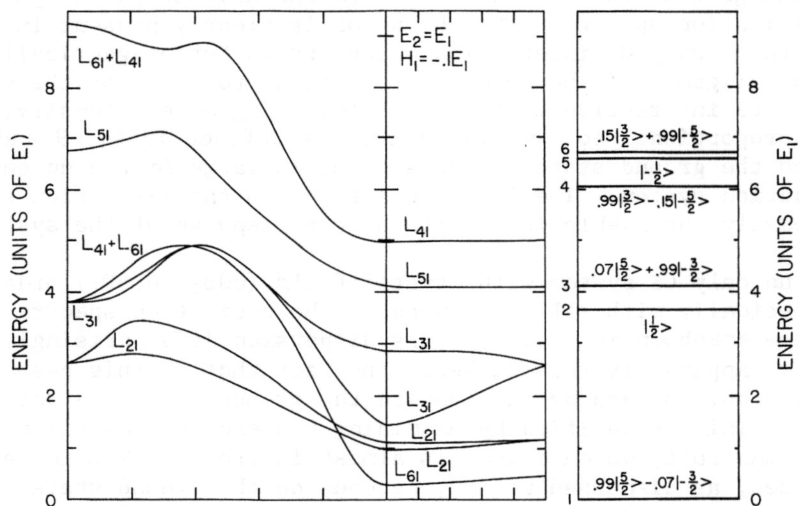

Fig. 4. Excitations at T=0 for fcc Ce^{3+} system with equal nn and nnn CS interactions plus small nn antiferromagnetic Heisenberg interaction($H_1 \Sigma \vec{J}_i \cdot \vec{J}_j$) (antiferromagnetic Type I ordering).

are dominated by the L_{41} and the L_{61} modes. The small Heisenberg exchange and the correlated motion of the two sublattices further decrease the dispersion along the <001> direction, compared to the

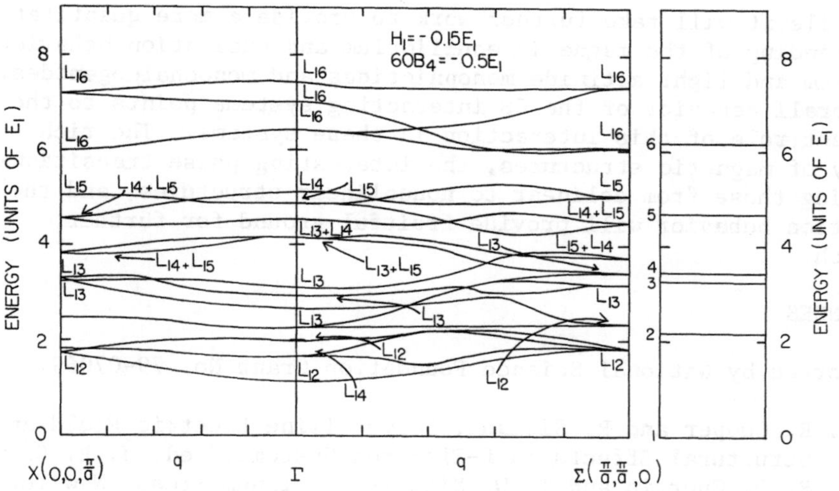

Fig. 5. Excitations at T=0 for fcc Ce^{3+} systems with noncolinear Kouvel-Kasper 3k antiferromangetic ordering.

ferromagnetic case of Fig. 2. The L_{41} mode which shows substantial dispersion in the ferromagnetic case, is now nearly flat. With increasing Heisenberg term, the low lying L_{61} mode at q=0 is lifted to higher energy and loses its intensity. The excitation is then dominated by the L_{41} mode which shows small dispersion. This agrees qualitatively with the excitation observed in CeSb[13] and CeBi[14].

For the nn-only CS interaction system, the excitation spectra obtained for the 2k structures (canted ferromagnetic or anitferromagnetic), in the presence of a small Heisenberg term, resemble those of the ferromagnetic system with <110> alignment, and those for 3k ordering resemble those of the ferromagnet with <111> alignment. Most importantly, the low lying mode and large anisotropy in dispersion are present in the spectra of the 2k ordered system, while in the 3k system, the spectra are nearly isotropic. Fig. 5 gives the spectra of a system ordered in the Kouvel-Kasper (3k) ordering with <111> alignment. The intensity is dominated by the low lying L_{21} transverse (with respect to the cubic diagonal direction) mode, and the L_{61} longitudinal mode which lies high in frequency. The spectra are notably lacking in dispersion, showing dominant crystal field effect. This lack of dispersion differs strongly with spectra for the 3k structure calculated by Jensen and Bak[15], and the experimentally observed behavior of USb[16]. We should point out, however, that here we assume a ferromagnetic CS interaction, while the dominant interaction in USb is clearly antiferromagnetic. The extension of the CS interaction to the uranium system, and detailed examination of the ordering and excitation behavior for USb is currently underway.

While it will take further work to provide a more quantitative understanding of the magnetic equilibrium and excitation behavior in cerium and light actinide monopnictides and monochalcogenides, the overall behavior of the CS interacting systems points to the important role of this interaction in these systems. The rich variety of magnetic structures, the interesting phase transitions including those from colinear to noncolinear structures, and the excitation behavior will provide fruitful ground for further research.

REFERENCES

* Supported by National Science Foundation Grant No. 79-07069.

1. B. R. Cooper and R. Siemann, "Crystalline Electric Field and Structural Effects in f-Electron Systems," ed. J. E. Crow, R. P. Guertin and T. W. Mihalisin, Plenum Press, New York, (1980), pp. 241-255.
2. J. W. Cable and W. C. Koehler, AIP Conference Proc. No. 5, "Magnetism and Magnetic Materials-1971," 1381 (1972).
3. J. Rossat-Mignod, P. Burlett, J. Villain, H. Bartholin, Wang T.-S., D. Florence, and O. Vogt, Phys. Rev. B16, 440 (1977); P. Fischer, B. Lebech, G. Meier, B. D. Rainford, and O. Vogt, J. Phys. C 11, 345 (1978).
4. G. H. Lander, M. H. Muller and J. F. Reddy, Phys. Rev. B6, 1880 (1972); S. K. Sinha, G. H. Lander, S. M. Shapiro, and O. Vogt, Phys. Rev. B23, 4556 (1981).
5. J. Rossat-Mignod, P. Burlett, H. Bartholin, R. Tchapontian, O. Vogt, C. Vettier, and R. Lagnier, Physica 102B, 177 (1980); J. Rossat-Mignod, P. Burlett, S. Quezel, and O. Vogt, Physica 102B, 237 (1980).
6. B. Coqblin and J. R. Schrieffer, Phys. Rev. 185, 847 (1969).
7. R. Siemann and B. R. Cooper, Phys. Rev. Lett. 44, 1015 (1980).
8. M. Tinkham, "Group Theory and Quantum Mechanics," McGraw-Hill, New York (1964).
9. R. Siemann and B. R. Cooper, to be published.
10. F. Hullinger, B. Natterer, and H. R. Ott, JMMM 8, 87 (1978).
11. D. L. Huber, Phys. Rev. B18, 429 (1978).
12. D. Yang and B. R. Cooper, J. Appl. Phys. 52, 2234 (1981).
13. A. Furrer, W. Haelg, H. Heer, and O. Vogt, J. Appl. Phys. 50, 2040 (1979).
14. J. Rossat-Mignod and D. Delacotte, private communication (1981).
15. J. Jensen and P. Bak, to be published.
16. G. H. Lander, W. G. Stirling, and O. Vogt, Phys. Rev. Lett. 42, 260 (1979).

COMMENTS

COQBLIN: As I have said in the discussion of the paper by K. Takegahara, I think that the "p-f mixing" described by them is a sophisticated version (including a better description of the conduction band) of the resonant scattering mechanism which was at the origin of our Hamiltonian with J. R. Schrieffer. In any case both mechanisms yield an anisotropic exchange interaction which appears to be essential to explain the properties of Ce monopnictides.

COOPER: I think that the work by Takegehara et al is complementary to our work on the Coqblin-Schrieffer coupling. It is not completely clear to me whether the "p-f mixing" theory is indeed essentially the same resonant scattering mechanism as the CS coupling. The work of Takegahara et al does not exhibit an explicit anisotropic exchange interaction, so this makes it a little more difficult to compare with our work.

KASUYA: Firstly, our mechanism is essentially a non-linear effect and thus not described by your perturbational formula. The p-f mixing should be treated more rigorously. Secondly, when d-f mixing is concerned, there is a more important exchange mechanism. Then the situation is the same as in Eu-chalcogenides, and we showed in 1969 at the IBM conference and in 1976 in the US-Japan seminar that once the d-electron is created virtually by the d-f mixing effect, the most important mechanism is the intra-atomic d-f exchange interaction with the nearest neighbor Eu site which gives a ferromagnetic exchange interaction.

COOPER: It seems to me that the work of you and your associates may be complementary to our work on the Coqblin-Schrieffer coupling. Concerning understanding the fundamental origin of anisotropic exchange via the CS mechanism, it is important to understand that this comes from covalent effects (mixing of Ce^{3+}-f states and the f component of band electrons, including, of course, that from other sites). A p state for an anion has an f wave component with respect to a Ce site. The relationship of the CS coupling to the Anderson Hamiltonian should also be kept in mind.

ROSSAT-MIGNOD: I do not understand how you explain the double $\vec{k}$ to single $\vec{k}$ transition observed in UAs and UP when the temperature changes.

COOPER: For a moderately short range of CS exchange ($E_2/E_3 \simeq 0.4$) the ground state has a noncolinear structure, but the state with a colinear structure lies only slightly higher in energy; and the nature of excited states of the two structures is such that entropy considerations favor the transition to the colinear structure with increasing temperature.

STEGLICH: Assume a cubic metallic system of Ce^{3+} ions with a Γ_7 CEF ground state, say about 100 K below Γ_8. Also, assume the temperature to be sufficiently small so that all ions are in the Γ_7 state. Will the Coqblin-Schrieffer type exchange interaction between these doublets be anisotropic?

COOPER: Yes, the crystal field splitting is small compared to the matrix element mixing the f and band electrons that gives rise to the CS interaction.

KNORR: Do you think that it is reasonable to apply the CS formalism to the isomorphic compounds PrSb and TbP which show a similar type of anisotropic exchange?

COOPER: In principle, "yes", but you must remember that we expect the effect to be significantly smaller in PrSb than in CeBi or CeSb and very much smaller for TbP.

LARGE PRESSURE EFFECTS ON THE MAGNETIC PHASE DIAGRAMS OF CeSb AND $Ce_x(La_{0.76}Y_{0.24})_{1-x}Sb$ COMPOUNDS

H. Bartholin*, J.M. Effantin†, P. Burlet†, J. Rossat-Mignod† and O. Vogt§

*Université de Toulon, 83130 La Garde, France
†Laboratoire de Diffraction Neutronique, Centre d'Etude Nucléaire, Grenoble, France
§Laboratorium für Festkörperphysik, Eidgenössische Technische Hochschule, Zürich, Switzerland

INTRODUCTION

Cerium and uranium monopnictides of rocksalt structure, exhibit unusual and very anisotropic magnetic properties compared to the other rare earth or actinide compounds[1]. This behavior is attributed to the proximity of the 4f (or 5f) level and the Fermi level. To explain this trend, Cooper et al.[2] have proposed an interaction between the f electrons via the d electrons of the rare earth, whereas Takegahara, Yanase and Kasuya[3] have suggested a strong hybridization between the f electrons of the rare earth and the p electrons of the valence band. This f-p hybridization, normally smaller than the f-d one, is enhanced by the proximity of pnictide and cerium atoms and by the existence of p-holes in the valence band due to the semimetallic character of these compounds[3-6]. CeN, which is metallic, is considered as non-magnetic. The complexity of the magnetic properties increases with the semimetallic character in the cerium monopnictides series from CeN to CeBi. A strong exchange anisotropy due to multiphase behavior arises for CeSb and CeBi and in contrast with the other rare earth monopnictides, the splitting Δ between the ground state and the excited state is unusually small[7,8].

In the paramagnetic state, with a cubic crystal field, the 4f level (J=5/2) splits into a doublet Γ_7 and a quartet Γ_8, Γ_7 being the ground state level for CeP and CeAs. When the semimetallic character increases, the value of the splitting Δ decreases very quickly and a change of the ground state level is observed from Γ_7 to Γ_8 (CeSb, CeBi)[7]. Takegahara et al.[4] and Kasuya et al.[5] have attributed this effect to the f-p hybridization which leads to a decrease of the Γ_8 energy and consequently of the splitting due to the crystal field alone.

In the ordered state, the f-p hybridization leads to strong, non-linear, very anisotropic interactions and is considered the origin of the <001> easy axis of magnetization[4-6].

Among the cerium monopnictides, CeSb is certainly the most typical, and CeAs is a relatively simple antiferromagnet. Below 16 K and with an applied field, CeSb exhibits 14 magnetic phases[8]. The number of p-holes in the valence band has been estimated to be 0.06 per cerium atom for CeSb[4-6]. Large changes are observed with only a small number of holes and consequently, experiments which can cause variation in the hole concentration are very interesting. To this end, the effects of pressure and dilution of cerium ions have been investigated and the principal results are presented in this paper.

EXPERIMENTAL RESULTS AND DISCUSSION

The magnetic measurements have been performed in high magnetic fields for several temperatures and pressures at the Service National des Champs Intenses de Grenoble[9]. The magnetic field was always applied along a <001> axis of the crystal. Compounds with the composition $Ce_x(La_{0.76}Y_{0.24})_{1-x}Sb$ were studied and in particular those with x=1, 0.95, 0.9 and 0.85. The dilution of cerium ions was controlled in such a way as to keep the lattice parameter constant.

The 14 magnetic phases of CeSb can be described by a periodic stacking of zero magnetized and ferromagnetic planes with a magnetization parallel or antiparallel to the <001> easy magnetization direction[8]. Each structure is characterized by a wave vector which is commensurate with the crystallographic cell. The low temperature phases (ℓ,t) are antiferromagnetic (AF), antiferro-ferromagnetic (AFF) or ferromagnetic (F) and contain only magnetic planes. For the high temperature phases (h,t) such as ferro-paramagnetic (FP) or antiferro-paramagnetic (AFP), non-magnetic planes exist in the structure[8].

Upon substitution for the Ce, the stability of the (h,t) phase decreases and the multiphase behavior disappears[12]. For x=0.8, only the Type I antiferromagnetic phase was seen in zero field below T_N=6.3 K.[13] A change of the easy magnetization axis from <001> to <111> also occurs with dilution together with a drastic decrease of the Néel temperature[10-12]. At least for high cerium concentrations ($x \geq 0.9$), the dilution effect on the magnetization (Fig.1) is similar to an increase of temperature (Fig. 2)[12].

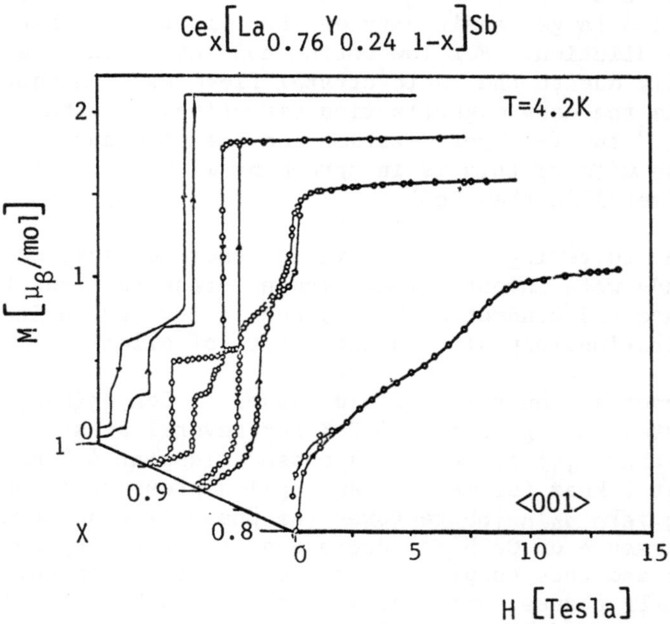

Fig. 1: Magnetization vs field and x for $Ce_x(La_{0.76}Y_{0.24})_{1-x}Sb$ at 4.2 K.

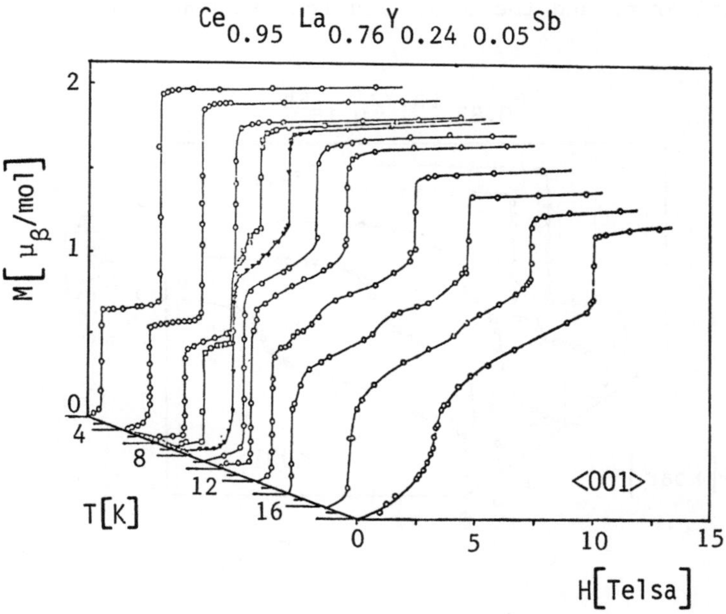

Fig. 2: Magnetization vs field and temperature for $Ce_{0.95}(La_{0.76}Y_{0.24})_{0.05}Sb$.

The sudden decrease of T_N and the change of easy magnetization axis indicate a large sensitivity of the anisotropic exchange interactions with dilution. For low cerium concentration, the anisotropy is essentially due to the cubic crystal field and consequently the <111> axis is the easy magnetization direction. In the analysis of Kasuya et al.[5] the f-p hybridization and then the number of p-holes must decrease with x; this is in agreement with the fact that LaSb is less semimetallic than CeSb.

In order to verify our analysis, pressure effects on diluted CeSb compounds were investigated. Indeed pressure leads to a widening of the valence and conduction bands, and it induces an increase of the hole concentration, opposite to the effect of dilution.

The magnetization versus field curves of $Ce_{0.85}(La_{0.76}Y_{0.24})_{0.15}$ Sb are reported in Fig. 3 at T=4.2 K for several pressures. A change from the (h,t) to the (ℓ,t) phase arises at 4.5 kbar for x=0.85 and at 2 kbar for x=0.9. An applied pressure of only 4 kbar on $Ce_{0.9}(La_{0.76}Y_{0.24})_{0.1}Sb$ restores the phase diagram of CeSb. The effect of pressure on CeSb[9,14] decreases the stability range of the (h,t) phases and thus suppresses the non-magnetic planes. For example, the FP phases vanish at 12 kbar and 33 K with 5.1 T (Fig. 4) and at 1 kbar and 34 K with 17 T.

The most striking feature of these results is the similarity between the effects of pressure, dilution and temperature as far as the phase diagrams and the magnetization are concerned. At least

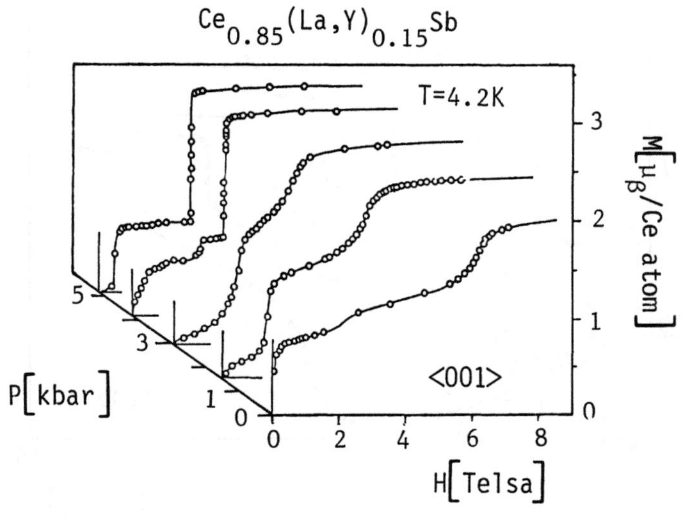

Fig. 3: Magnetization vs field and pressure for $Ce_{0.85}(La_{0.76}Y_{0.24})_{0.15}Sb$ at 4.2 K.

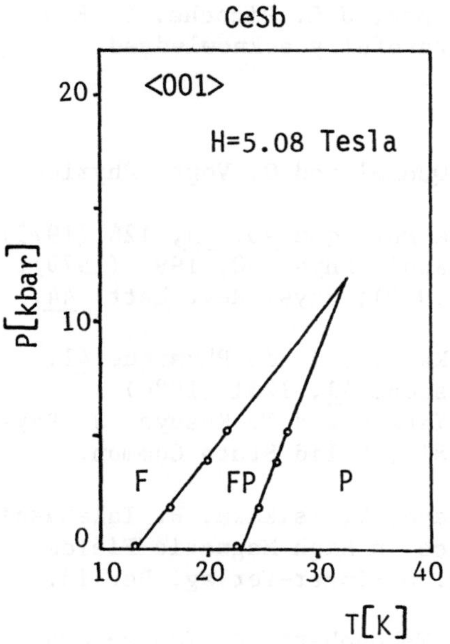

Fig. 4: Depicting the disappearance with pressure of the ferro-paramagnetic phase of CeSb.

for $x \geq 0.9$, a pressure increase of 400 kbar is equivalent to a temperature decrease of 0.4 K and to an increase in concentration of cerium ions of 0.01. With this analogy in mind, a variation of +0.75 K/kbar for the Néel temperature with pressure can be evaluated from its change with dilution. This variation is in good agreement with that obtained by direct measurements (+0.75 K/kbar)[9,14]. If the Néel temperature is assumed to represent the exchange energy, the anisotropic exchange energy depends strongly on pressure, dilution and temperature.

CONCLUSION

The experiments reported in this paper have revealed a striking analogy between the effects of pressure, dilution and temperature on the magnetic properties of the $Ce_x(La_{0.76}Y_{0.24})_{1-x}Sb$ compounds, at least for high concentration of cerium ions. The non-magnetic planes observed in the magnetic structures of CeSb disappear both with an applied pressure and dilution. Therefore, a large variation of the anisotropic exchange interactions with pressure, dilution and temperature occurs which can be considered as the origin of the multiphase behavior of CeSb. These results strongly support the idea that the high anisotropic exchange interactions are due to an f-p hybridization which is very sensitive to the p-hole concentration in the valence band.

The authors are indebted to Professor G. Aubert and R. Pauthenet for permission to perform the measurements at the SNCI in Grenoble. The assistance of G. Dampne, J.C. Picoche, P. Rub and J. Voiron in the experiments is gratefully acknowledged.

REFERENCES

1. J. Rossat-Mignod, P. Burlet, S. Quezel and O. Vogt, Physica 102 B, 237 (1980).
2. B.R. Cooper and R. Siemann, J. de Physique 40, C5, 126 (1979); Phys. Rev. B19, 2645 (1979); J. Appl. Phys. 50, 1991 (1979); J. of Magn. Magn. Mat. 15, 575 (1980); Phys. Rev. Lett. 44, 1015 (1980).
3. K. Takegahara, A. Yanase and T. Kasuya, J. de Physique 41, C5, 327 (1980). Erratum J. de Physique 41, 1231 (1980).
4. K. Takegahara, M. Takahashi, A. Yanase and T. Kasuya, J. Phys. C, Solid State Phys. 14, 737 (1981); Solid State Commun. (to appear).
5. T. Kasuya, K. Takegahara, M. Kasaya, Y. Isikawa, H. Takahashi, T. Sakakibara and M. Date, Physics in High Magnetic Fields, S. Chikazumi and N. Miura (eds.), Springer-Verlag, Berlin, (1980), p.150.
6. T. Suzuki, M. Sera, H. Shida, K. Takegahara, A. Yanase and T. Kasuya, Proc. Int. Conf. on Valence Fluctuations in Solids, L.M. Falicov, W. Hanke and M.B. Maple (eds.), Santa Barbara, (1981).
7. J.X. Boucherle, A. Delapalme, C.J. Howard, J. Rossat-Mignod and O. Vogt, Physica 102 B, 253 (1980).
8. J. Rossat-Mignod, P. Burlet, J. Villain, H. Bartholin, W. Tcheng-si, D. Florence and O. Vogt, Physica 86-88B, 129 (1977); Phys. Rev. B16, 440 (1977).
9. H. Bartholin, D. Florence and O. Vogt, J. Phys. Chem. Solids 39, 89 (1978).
10. B.R. Cooper and O. Vogt, J. Physique 32, C1, 1026 (1971).
11. B.R. Cooper, M. Landolt and O. Vogt, Proc. Int. Conf. Magnetism 1973, Moscow, V,354 (1974).
12. H. Bartholin, D. Florence and O. Vogt, Phys. Stat. Sol. (a) 52, 647 (1979).
13. G. Meier, Ph.D. Thesis ETH, Zürich (1977).
14. H. Bartholin, O. Vogt and J.P. Senateur, J. Magn. Magn. Mat. 15-18, 1247 (1980).

COMMENTS

COOPER: The effects you discuss point to the fact that the observed properties of CeSb and other such materials depend on the balance between anisotropic exchange effects and crystal field effects. They do not point to any particular model for the electronic structure giving anisotropic exchange. I suspect that if a particular microscopic model is used as the basis for interpretation, the experiments will indeed confirm that interpretation.

BARTHOLIN: We have shown that the anisotropic exchange interactions in CeSb depend strongly on dilution, temperature and pressure and our results agree with the f-p mixing model proposed by the Kasuya group. We recently performed experiments on $CeSb_xTe_{1-x}$ compounds and a drastic decrease of the Neel temperature is observed, and the multiphase behavior of CeSb disappears with only a small amount of Te. With pressure we restore the multiphase behavior. Indeed, Te increases the number of holes in CeSb. This result corroborates our analysis. I agree that these macroscopic experiments are not sufficient to decide without ambiguity what microscopic model must be chosen.

FRANSE: I did not understand the increase of the number of holes caused only by increasing the bandwidth as you do by applying pressure. For that one needs a transfer of electrons to other states.

BARTHOLIN: As shown by the Kasuya group, overlapping of the valence band and conduction band exists in those compounds and the number of holes and conduction electrons is similar. In other words the effect of pressure is to increase the overlap between the conduction and the valence bands and to transfer electrons from the valence band to the conduction band. The experiments performed with $CeSb_xTe_{1-x}$ are a confirmation of this point.

KASUYA: Your experiment agrees perfectly well with the prediction based on the p-f anisotropic mixing mechanism which will be presented. I have one minor comment about the ground state level in CeSb and CeBi. Our theory is rather consistent with a doublet ground state. The quartet dominated result in the polarized neutron experiment is understandable by a strongly enhanced interaction among the quartets, which will also be mentioned.

BARTHOLIN: In my talk I have reported the ground state expected from the polarized neutron experiment performed by Bouchele et al., in the laboratory of Rossat-Mignod.

ROSSAT-MIGNOD: The polarized neutron experiments on CeSb have been compared to the magnetic cross section calculated by using the same molecular field for the quartet Γ_8 and the doublet Γ_7. Within this assumption in the paramagnetic state the anisotropy of the form factor (of the 4f cloud) can be explained only by considering the Γ_8 quartet as the ground state.

CRYSTAL FIELD EXCITATIONS IN $CeAg_{1-x}In_x$ COMPOUNDS

H. Wehr*, K. Knorr*, A.P. Murani† and W. Assmus§

*Inst. f. Physik, Universität Mainz, Mainz, FRG
†Inst. Laue-Langevin, Grenoble, France
§Physikal. Inst., Universität Frankfurt, Frankfurt, FRG

INTRODUCTION

The substitution of the monovalent Ag for the trivalent In in $LaAg_{1-x}In_x$ and $CeAg_{1-x}In_x$ offers the possibility of influencing the conduction electron system in these compounds. The strong dependence of the spin susceptibility of La-systems on the In-concentration[1] was interpreted by band calculations which suggest that the Fermi level is pushed into a region of a high 5d-e_g-band density of states[2]. Furthermore, $CeAg_{1-x}In_x$ is one of the rare 4f systems where a structural phase transition (from cubic to tetragonal) occurs prior to magnetic ordering. The ferromagnetic Curie temperature T_C is roughly independent of x whereas the structural transition temperature T_s rises from 15 K for CeAg to about 120 K for $CeAg_{0.8}In_{0.2}$[3,4]. The simple fact that the dependence of T_s on x is almost identical for the Ce- and La-compounds shows that the early view put forward by Ray and Sivardiere[5] that the structural transition is driven by the interacting electric quadrupoles of the 4f states is incorrect. Nevertheless, the reduced lattice symmetry should affect the magnetic properties below T_s via the crystal field (CEF).

In the present article we investigate the CEF splitting of $CeAg_{1-x}In_x$ by inelastic neutron scattering. Herewith we complement and extend the work of Pierre and Murani on CeAg[6].

EXPERIMENTAL DETAILS

The polycrystalline samples (x=0,0.05,0.1,0.2) were prepared from the elements in a high frequency levitation furnace. The neutron experiments were carried out on the time of flight spectrometer IN4 of the Institute Laue Langevin. Fifty detector boxes covered scattering angles from 4.5 to 70 degrees. Since the excitation energies observed turned out to be independent of the scattering angle, the counting rates of individual detectors were summed up for better statistics in the ultimate step of data reduction. Two incident neutron energies, 12.5 and 50.4 meV, were used. With this choice energy transfers from 0.5 to 45 meV could be probed. The instrumental resolution at zero energy transfer was 0.7 and 3.3 meV (FWHM) respectively for the two incident energies. The background scattering from the sample holder and the cryostat was subtracted from each spectrum, taking into account the transmission of the sample. The calibration was performed with a vanadium standard. The time of flight data were converted into counting rates $S(\omega)$ proportional to the neutron cross section as a function of energy transfer. Thus the experimental conditions were identical to those of Ref. 6.

CEF IN THE CUBIC PHASE

According to Pierre and Murani, the J=5/2 ground multiplet of CeAg is split into a Γ_8 ground state followed by the Γ_7 state at 22.5 meV. In the pseudobinary compounds $CeAg_{1-x}In_x$ several changes of the CEF spectrum could be possible: 1) If Ag and In - as seen from the 4f-state - carry different charges or have different sizes, the CEF at a given Ce-site would depend on the local environment which is in general non-cubic. 2) The CEF-parameters - whether the CEF be cubic or a superposition of various symmetries - depend on the In-concentration via the population of the 5d-virtual bound state, which gives a signficant contribution to the CEF[7]. 3) The lifetime of the 4f-state should decrease - hence the line width of the transitions should increase with the In concentration - due to a higher density of states at the Fermi level $n(E_F)$. This in turn gives rise to an enhanced k-f-scattering rate.

Experimentally we could verify the CEF splitting of CeAg with a Γ_8-Γ_7 transition at 22.5±0.5 meV. In the samples containing In no additional transitions were observed; they were rather similar to that of CeAg with a single, admittedly very broad, inelastic transition somewhat above 20 meV and a central quasielastic distribution. Both features become broader the higher the In-concentration and the higher the temperature (Fig. 1). (Residual scattering around 30 meV did not depend in any systematic way on x and T and will be ignored in the following.) These results suggest that possibilities 1) and 2) are not realized. This in turn suggests that a) the ionic excess charge of In relative to Ag is very effectively screened - already within the Ce-In distance - presumably by the indium's own s and p electrons, and that b) the 5d-population is practically unchanged.

For the line width effects the results were analyzed in terms of the theory of Becker, Fulde and Keller[8] which considers the

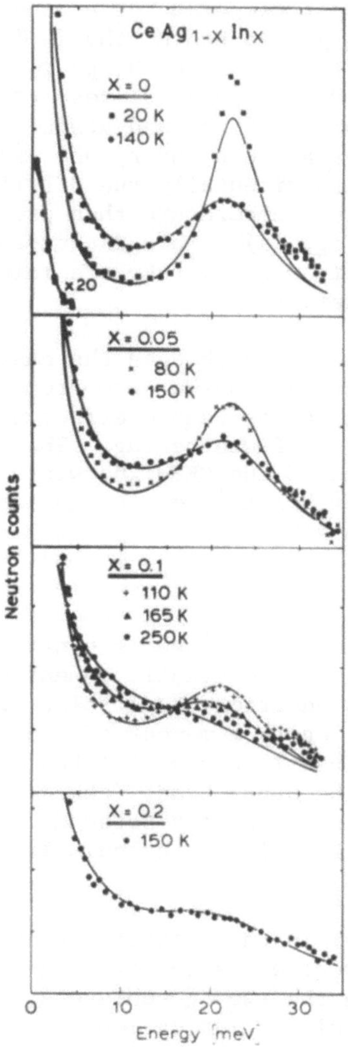

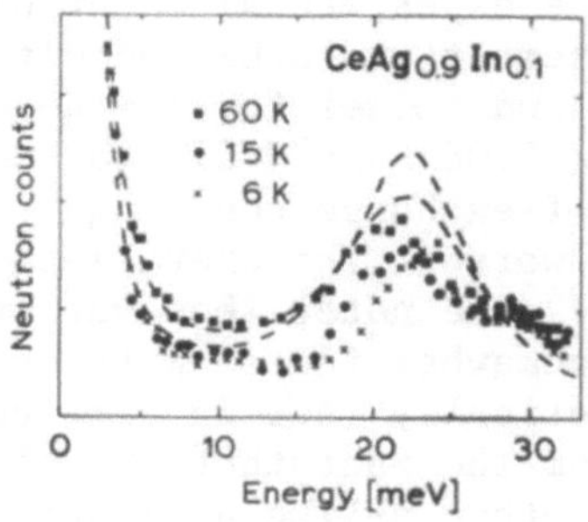

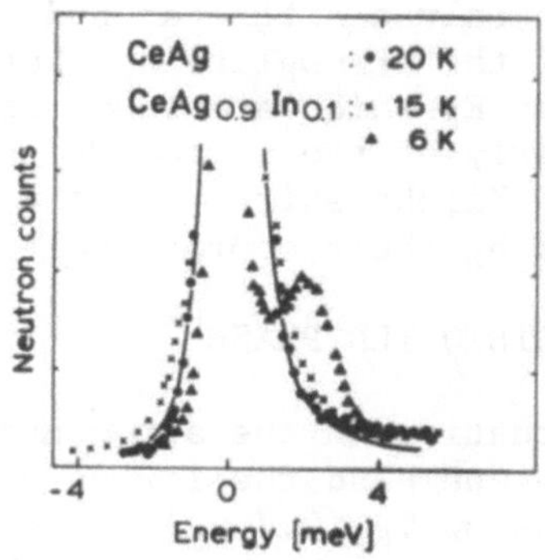

Fig. 1: (left) The inelastic neutron spectra of $CeAg_{1-x}In_x$ in the cubic phase. The solid lines are fits with the model of Ref.8.

Fig. 2: (upper right) The spectrum of $CeAg_{0.9}In_{0.1}$ in the tetragonal phase. The dashed lines are the result of the model for the cubic phase calculated for temperatures of 60 K and 16 K.

Fig. 3: (lower right) High resolution spectra for cubic CeAg at 20 K with the corresponding theoretical curve (solid line) - for $CeAg_{0.9}In_{0.1}$ between T_s and T_C (15 K) and below T_C (6 K).

damping of CEF transitions due to electron-hole excitations in an s-type conduction band. The imaginary part of the dynamic CEF susceptibility was calculated using a computer program, kindly supplied by J. Keller, then folded with the experimental resolution and compared with the experimental spectrum $S(\omega)$ at temperatures around 150 K. The best agreement was achieved with a Γ_8 ground state, the Γ_7 level at 22.5 meV, and values for the k-f-coupling parameter $N(E_F)J_{ex}$ of 0.057, 0.060, 0.065 and 0.072 for x=0, 0.05, 0.1 and 0.2 respectively. J_{ex} is the k-f-exchange coupling. Fits with a Γ_7 ground state were considerably worse. The overall agreement between theory and experiment is good. One notes, however, that experimentally the CEF-line width drops somewhat faster with decreasing temperature than predicted theoretically (see in particular Fig. 1a). This discrepancy may arise from the fact that the electronic states at E_F are mainly d-like rather than s-like as assumed in Ref. 8.

In summary, we note that the addition of In changes the charge distribution only a little around a Ce site as can be seen from the constancy of the CEF-splitting, but it has a noticeable effect on the states at E_F which are responsible for k-f-scattering. The great similarly of the x- and T-dependence of the CEF-line width and the muon Knight shift[9] and the susceptibility[1] of the La samples is explained by their common scaling with $N(E_F)$.

CEF IN THE DISTORTED PHASE

The situation in the structurally distorted phase is controversial. On one hand the low value of the ferromagnetic moment of Ce in CeAg below T_C (0.7 μ_B[10]) demands an additional splitting of the Γ_8 ground state which would otherwise carry a moment of about 1.8 μ_B. In fact, a Γ_8-splitting higher than T_s was postulated recently by Eiling and Schilling[11]. On the other hand, Takke et al.[12] have shown that the paramagnetic susceptibility between T_s and T_C still follows the CEF susceptibility well, as calculated with an unchanged cubic level scheme.

The results of the present experiment in the distorted phase are shown in Figs. 2 and 3. The most direct evidence for a Γ_8-splitting, a clear CEF transition between the two Γ_8 sublevels, could not be found. But there are two indirect indications: first, Fig. 2 shows that for $CeAg_{0.9}In_{0.1}$ at 15 K and 6 K the integrated intensity of the transition around 23 meV is only about half of that expected from a continuation of the fit which is based on the cubic scheme. One also notes a small shift to higher energies. Secondly, the quasielastic width of $CeAg_{0.9}In_{0.1}$ at 15 K in the distorted phase is significantly larger than that of cubic CeAg at 20 K. (Without a Γ_8 splitting the width of the In-sample should be slightly smaller according to an extrapolation of the fit.) Presumably this extra width is due to an overdamped and possibly dispersive transition between the Γ_8 sublevels. These findings suggest that the splitting is of the order of 1 to 3 meV. It is plausible that the sublevels have some intrinsic width and an overdamped transition connecting them will lead to a susceptibility which is hardly different from one calculated with an unsplit Γ_8.

COMMENTS ON THE MAGNETIC PHASE

In the magnetic phase of $CeAg_{0.9}In_{0.1}$ we observed a well-defined excitation at 2 meV. It can be regarded as the strongest magnetic dipole transition between the four lowest molecular field states. With this information and a reasonable guess for the Γ_8 splitting which should be appreciably lower for CeAg than for $CeAg_{0.9}In_{0.1}$ due to the lower T_s, complemented by the knowledge of the spontaneous moment, we tried to understand the ordered phase by a mean field calculation. The simplest model considers a cubic CEF plus a tetragonal perturbation $B_2\hat{O}_2^o$ split selfconsistently by a magnetic field along [100]. Positive B_2 yields a $|\pm 1/2>$ ground state which carries a moment of 0.36 μ_B. Unfortunately, one cannot find a stable solution which mixes in enough of the other Γ_8 sublevel in order to reproduce the desired moment of 0.7 μ_B. In a second step two sublattices of Ce-ions with different parameters B_2 were considered. This choice is a natural consequence of the cell doubling below T_s[3]. CeAg would then be a ferromagnet with an antiferromagnetic component with a propagation vector (1/2, 1/2, 0). Even this model fails. A better description of the experimental data is obtained by returning to the simple model and allowing for an additional CEF term of the type $B_2^1\hat{O}_2^2$ which requires an orthorhombic cell. Hoeck has indeed shown (private communication) that the soft mode behavior observed in $LaAg_{1-x}In_x$[13] can lead to either a tetragonal or an orthorhombic phase. In any case, these efforts are highly speculative and a precise determination of the magnetic and crystallographic structures of the low temperature phases is highly desirable.

This work was supported by the BMFT of the Federal Republic of Germany.

REFERENCES

1. H. Ihrig, D.T. Vigren, J. Kuebler and S. Methfessel, Phys. Rev. B8, 4525 (1973).
2. A. Hasegawa, B. Bremicker and J. Kuebler, Z. Physik B22, 231 (1975).
3. H. Ihrig and S. Methfessel, Z. Physik B24, 385 (1976).
4. W. Assmus, R. Takke, R. Sommer and B. Lüthi, J. Phys. C11, L575 (1978).
5. D.K. Ray and J. Sivardiere, Solid State Comm. 19, 1053 (1976).
6. J. Pierre, A.P. Murani, in: Crystalline Electric Field and Structural Effects in f-Electron Systems, J.E. Crow, R.P. Guertin and T.W. Mihalisin, eds., Plenum Press, New York (1980).
7. G. Williams and L. Hirst, Phys. Rev. 185, 407 (1969).

8. K.W. Becker, P. Fulde and J. Keller, Z. Physik B28, 9 (1977).
9. H. Wehr et al., (this conference).
10. J. Pierre, P. Morin and D. Schmitt, J. Magn. Magn. Mat. 8, 249 (1978).
11. A. Eiling and J.S. Schilling, Phys. Rev. Lett. 46, 364 (1981).
12. R. Takke, N. Dolezal, W. Assmus and B. Lüthi, J. Magn. Magn. Mat. 23, 247 (1981).
13. K. Knorr, B. Renker, W. Assmus, R. Takke, B. Lüthi and H.J. Lauter, Z. Physik B39, 151 (1980).

COMMENTS

WOHLLEBEN: In the analysis of the neutron spectra, how do you separate the dynamic line broadening from the static broadening due to the environmental effects in these concentrated alloys? If you cannot see static broadening, how do you explain its absence?

KNORR: Of course, there might be a residual static contribution to the total linewidth due to the effects of the local environment. The systematic variation of the linewidth with x and T suggests, however, that these contributions are small. I explain the absence of local environmental effects by a very effective screening of the excess In charge.

LOEWENHAUPT: Your interpretation of the CF level scheme in $CeAg_{0.90}In_{0.1}$ (splitting of Γ_8) agrees with our investigation of the low energy excitation in CeAg (up to 3 meV), i.e. there is only a very small matrix element between the two lowest states.

MAGNETORESISTIVITY OF CERIUM COMPOUNDS

Y. Lassailly[+], A.K. Bhattacharjee[++] and B. Coqblin[++]
[+]CRTBT, 38042 Grenoble Cedex, France
[++] Physique des Solides, 91405, Orsay, France

The low temperature resistivity and magnetoresistivity of Cerium compounds, which order at low temperatures, such as $CeAl_2$, are compared in the model which describes both the crystalline field and the Kondo effects. A good agreement with experiment is found for $CeAl_2$.

The transport properties of Cerium "Kondo compounds" have been extensively studied both experimentally and theoretically. It is well known that the model [1] describing the influence of the crystalline field on the Kondo effect can account for the main features of the resistivity of many Kondo compounds at sufficiently high temperatures (i.e. T $\gg$ 10 to 20 K). Compounds such as $CeAl_2$, $CeAl_3$ or TmS [2] have at high temperatures a resistivity decreasing as log T and passing at lower temperatures through a maximum characteristic of the crystalline field. However, the low temperature resistivity cannot be accounted for by the preceding model of combined crystalline field and Kondo effects [1] and moreover two different behaviours have been observed at low temperatures: either the resistivity is monotonically decreasing with decreasing temperature and the Cerium atoms are finally demagnetized [3] as in $CeAl_3$, or the resistivity passes through a maximum with a strong field dependence and the Cerium compound is finally magnetically ordered as in $CeAl_2$ [4] and CeB_6 [5].

$CeAl_2$ [6] and CeB_6 [7] present a modulated magnetic ordering below 3.85K and 2.8K respectively. The low temperature resistivity decreases rapidly with increasing applied field around the Neel temperature and the magnetoresistivity disappears at about 15 K. When temperature increases above the Neel temperature, there is, at zero magnetic field, a resistivity decrease which persists in CeB_6 up to at least 85 kOe [5] and disappears in $CeAl_2$ at high fields giving rise to a continuous increase at 50 kOe [4] .

The purpose of the present paper is to extend at low temperatures the resistivity calculation of Ref. 1 to the case of Cerium Kondo compounds such as $CeAl_2$ which order at low temperatures, and moreover to compute the magnetoresistivity. The resistivity and magnetoresistivity of transition alloys have been extensively computed [8] , but here we calculate them for the Hamiltonian described in Ref. 1 and appropriate for the case of Cerium impurities.

The resistivity is described in the formalism of Ref. 1. We discuss the case of a cubic crystalline field for Cerium, giving rise to a Γ_7 doublet as ground state and a Γ_8 quartet. The presence of an applied magnetic field and magnetic ordering splits the two states of the Γ_7 doublet ; we neglect here the splittings of the Γ_8 quartet since they have almost no influence on the resistivity in the typical case of $CeAl_2$ where the crystalline field splitting is sufficiently large. We call 1 the new ground state, 2 the second level of Γ_7 at the energy δ above 1, and 3 the four levels of Γ_8 at the energy Δ above 1.

Thus, we restrict our calculations to the low temperature region only, where the results will be different from those presented in Ref. 1, and we discuss here only the typical case of $CeAl_2$ where Δ is much larger than δ and the Neel temperature T_N.

There are two contributions to δ: one due to the applied magnetic field, which is given (in K) by 0.095 H (where H is expressed in kG) for the Γ_7 doublet of Cerium and the other one due to the magnetic ordering. The second term is treated within the molecular field approximation in the classical limit for a ferromagnetic ordering. The value $\langle S\rangle$ of the ordered magnetic moment is thus given by

$$\langle S\rangle = S\,\mathcal{L}\left(\frac{\delta}{kT}\right) = S\,\mathcal{L}\left(\frac{0.095\ H}{kT} + \frac{3T_N}{T}\,\frac{\langle S\rangle}{S}\right) \qquad (1)$$

where we neglect the contribution of the Γ_8 quartet. In Eq.(1) H is expressed in kG and $\mathcal{L}(x) = \coth x - 1/x$ is the Langevin function.

For the $4f^1$ configuration of Cerium, the occupation numbers $<n_i>$ of the levels 1, 2 and 3 are given by

$$<n_1> = \frac{1}{1+ e^{-\beta\delta} +4\, e^{-\beta\Delta}}$$

$$<n_2> = <n_1>\, e^{-\beta\delta} \qquad (2)$$

$$<n_3> = <n_1>\, e^{-\beta\Delta}$$

with $\beta = 1/kT$.

We take also all the $\mathcal{U}_{MM}$ and $J_{MM'}$ values of Ref. 1 respectively, equal to $\mathcal{U}$ and J. We use for the resistivity the expression written within the so-called "$f_k = 1/2$ approximation" (equation (117) - (120) of Ref.1), since we have checked that it yields results very close to the exact results [1] . Indeed, a calculation which does not use the approximation would be easy to perform.

Thus, the resistivity ρ in presence of an applied magnetic field and in both the paramagnetic and magnetically ordered regions is given by

$$\rho = \frac{3\, m^2\, \pi\, v_o\, c}{e^2\, \hbar^3\, k_F^2\, (2j+1)} \left[(\sigma^{(2)})^{-1} + \frac{\sigma^{(3)}}{(\sigma^{(2)})^2} \right] \qquad (3)$$

with the usual notations (see equation (92 b) of Ref. 1).

The calculation of $\sigma^{(2)}$ and $\sigma^{(3)}$ gives

$$(\sigma^{(2)})^{-1} = 6\, \mathcal{U}^2 + 2\, J^2 \Big[<n_1><n_2> + 10 <n_3> (1 - <n_3>)$$

$$+ \frac{2 <n_1>}{1+e^{\beta\delta}} + \frac{8 <n_1>}{1+e^{\beta\Delta}} + \frac{8 <n_2>}{1+e^{\beta(\Delta-\delta)}} \Big], \qquad (4)$$

and the third-order perturbation term is finally given by

$$\frac{\sigma^{(3)}}{(\sigma^{(2)})^2} = 8\, J^3\, n(E_F) \Bigg\{ \Gamma^1(0,0) \left[\frac{\langle n_1\rangle}{1+e^{\beta\delta}} + \frac{10\langle n_1\rangle}{1+e^{\beta\Delta}} + \frac{10\langle n_2\rangle}{1+e^{\beta(\Delta-\delta)}} \right]$$

$$+ \Gamma^1(\delta,0) \left[\langle n_1\rangle\langle n_2\rangle + \langle n_3\rangle(\langle n_1\rangle + \langle n_2\rangle) + \frac{\langle n_1\rangle}{1+e^{\beta\delta}} \right.$$

$$\left. + \frac{2\langle n_1\rangle}{1+e^{\beta\Delta}} + \frac{2\langle n_2\rangle}{1+e^{\beta(\Delta-\delta)}} \right] + \Gamma^1(\Delta,0) \left[1 + 4\langle n_3\rangle \right.$$

$$- (\langle n_1\rangle - \langle n_3\rangle)^2 - (\langle n_2\rangle - \langle n_3\rangle)^2 + \frac{4\langle n_1\rangle}{1+e^{\beta\delta}}$$

$$\left. + \frac{12\langle n_1\rangle}{1+e^{\beta\Delta}} + \frac{12\langle n_2\rangle}{1+e^{\beta(\Delta-\delta)}} \right] \Bigg\} . \tag{5}$$

Thus, the equations (1) - (5) allow us to determine the magnetic resistivity ρ in the ferromagnetic or paramagnetic region and with or without an applied magnetic field. The mean value $\langle S\rangle$ of the spin S is determined self-consistently by equation (1) with $T_N = 3.85$ K and $S = 1/2$; the results are almost independent of the S value. The resistivity is then computed numerically with the parameters used in Ref. 1 for $CeAl_2$: $J = -0.089$ eV, $n(E_F) = 2.2$ st./eV-at. [$n(E_F)$ is the density of states for one spin direction], $m = 3$ a.u., $v_o = 234$ a.u., $c = 1/3$, $2j+1 = 6$. Figures 1 and 2 show the plots of the magnetic resistivity versus T/T_N for several fields ranging from 0 to 100 kG ; we have plotted only the resistivity given by (3), where we have substracted the first term in $\mathcal{U}^2$ of $(\sigma^{(2)})^{-1}$.

The plots are obtained for the values of the parameters used in Ref. 1 for $CeAl_2$ and with $\Delta = 100$ K in figure 1 and $\Delta = 50$ K in figure 2. For $H = 0$, the resistivity increases rapidly up to T_N where there is a very narrow maximum and then decreases up to a broad minimum at a temperature of order 3 to 5 T_N. Above this minimum, the resistivity increase is due to the crystalline field effect as explained in Ref. 1. When the magnetic field increases, the resistivity decreases, thus yielding a negative magnetoresistivity, and the resistivity curve becomes smoother. For the Δ value of 100 K used in Ref. 1 for $CeAl_2$, there is still a minimum

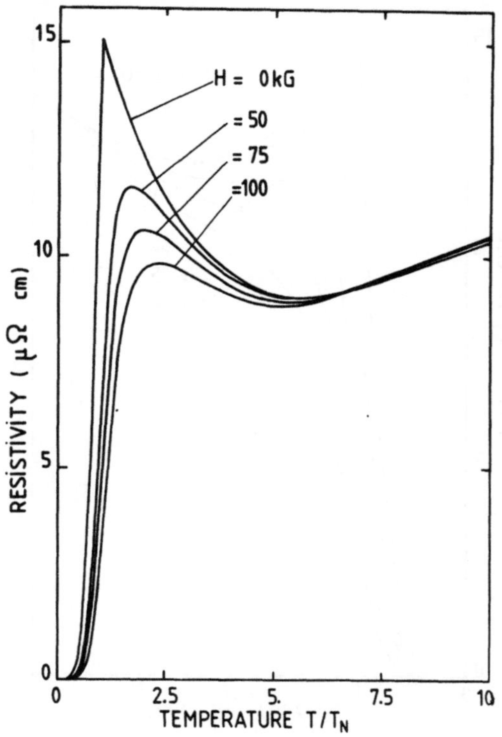

Fig. 1 : Magnetic resistivity plotted versus T/T_N for Δ = 100 K and several magnetic fields, using the parameters given in text.

for H = 100 kG, while for Δ equal to 50 K, the resistivity minimum disappears above 50 kG.

Let us now compare the theoretical curves of figures 1 and 2 to experiment. The theoretical curves are in good agreement with the experimental curves obtained for $CeAl_2$ with an applied magnetic field; however the agreement is much better with the curves of figure 2 where the resistivity minimum disappears above 50 kG. This would correspond to Δ = 50 K, while the Δ = 100 K value had been used in Ref. 1 to fit the high temperature part of the resistivity. However, we can argue that the present model is certainly crude, especially with the assumption of a ferromagnetic ordering but we can conclude that there is a qualitative agreement with experiment in the case of $CeAl_2$.

The case of CeB_6 is probably more complex since Δ is smaller, probably of order 15–20 K, and cannot be regarded as much larger

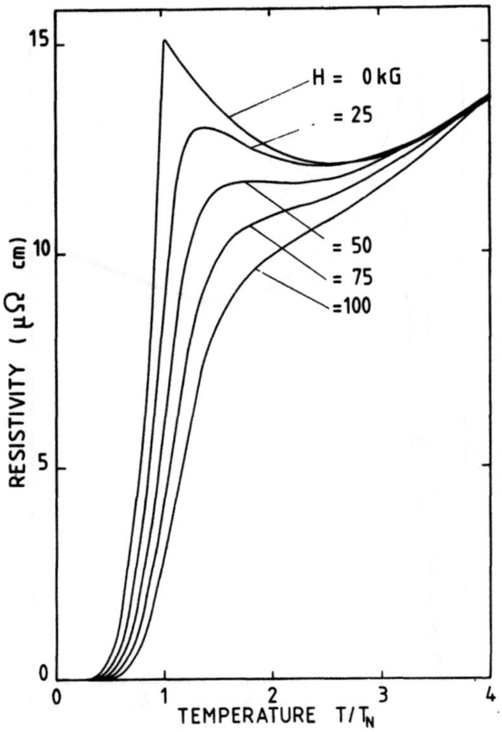

Fig. 2 : Magnetic resistivity plotted versus T/T_N for Δ = 50 K and several magnetic fields, using the parameters given in text.

than δ and T_N. However, theoretical curves, with a Δ value much smaller than in $CeAl_2$, can yield a good estimate of the resistivity curves of CeB_6. Work is in progress in that direction.

Finally, a similar behaviour can be obtained for the relaxation rate T_1. The experimental value of $1/T_1T$ decreases below the Neel temperature in $YbBe_{13}$ [9] and this behaviour can be also accounted for by the present calculation in the framework of the model describing both crystalline field and Kondo effects [10] .

Thus, the present calculation extends the previous one to low temperatures in the magnetically ordered regime and moreover yields a negative magnetoresistivity, in good agreement with experiment in $CeAl_2$.

REFERENCES

1. B. Cornut and B. Coqblin, Phys. Rev. B5, 4541 (1972).
2. F. Lapierre, P. Haen, B. Coqblin, M. Ribault and F. Holtzberg, Proceedings of LT16 Conference. Los Angeles (1981).
3. K. Andres, J.E. Graebner and H.R. Ott, Phys. Rev. Lett. 35, 1779 (1975).
4. I. Zoric, J. Markovics, L. Kupferberg, M. Croft and R.D. Parks, Proceedings of the Conference on "Valence Instabilities and Related Narrow-Band Phenomena", ed. by R.D. Parks, Plenum Press, Rochester, p.479 (1977).
5. T. Kasuya, K. Takegahara, Y. Aoki, K. Hanzawa, M. Kasaya, S. Kunii, T. Fujita, N. Sato, H. Kimura, T. Komatsubara, T. Furuno and J. Rossat-Mignod, Proceedings of the Int. Conf. on "Valence Fluctuations in Solids", ed. by L.M. Falicov, W. Hanke and M.B. Maple, North-Holland Publishing Co., p.215, (1981); A. Takase, K. Kajima, T. Komatsubara and T. Kasuya, Sol. State Comm. 36, 461 (1980).
6. B. Barbara, J.X. Boucherle, J.L. Buevoz, M.F. Rossignol and J. Schweizer, J. Physique 40, C5-321 (1979).
7. J. Rossat-Mignod, P. Burlet, T. Kasuya, S. Kunii and T. Komatsubara, to be published in Solid State Commun.
8. M.T. Beal-Monod and R.A. Weiner, Phys. Rev. 170, 552 (1968) and Phys. Rev. B3, 145 (1971).
9. P. Bonville, Thesis, University of Orsay (1981).
10. F. Gonzalez-Jimenez, B. Cornut and B. Coqblin, Phys. Rev. B11, 4674 (1975).

COMMENTS

KASUYA: It is very difficult to evaluate the p-f and d-f mixing matrix correctly; in particular, from the first principle standpoint. The best way, I think, is to evaluate them from the rigorous band calculations, including the f bands.

COQBLIN: Yes, indeed it would be certainly a very good way. But obviously, it is not easy to have a very good and reliable band calculation, especially in the case of f bands.

LOW TEMPERATURE MAGNETIC PHASE TRANSITIONS OF CeBi and CeSb STUDIED BY MAGNETOELASTICS

Tetsuo Nakajima*, Takashi Suzuki†, Masafumi Sera† and Tadao Kasuya†

*RIISOM, Tohoku University, Sendai 980, Japan
†Department of Physics, Tohoku University, Sendai 980 Japan

INTRODUCTION

Among the cerium monopnictides, all of which show different magnectic character[1], CeBi and CeSb exhibit extremely unusual magnetic properties[2]. Tetragonal distortions take place in both materials at the Néel temperature[3], below which the alignments of all sublattice magnetic moments hold fast to the [001] direction through the many successive magnetically ordered phases induced by external magnetic fields. In both compounds, the effective crystalline field is unusually small compared with other cerium monopnictides. Nevertheless, CeBi and CeSb show strong magnetic anisotropies favoring the [001] direction[4,5]. Recently we proposed a model based on the anisotropic p-f mixing mechanisms and obtained a better understanding of the common salient features observed in both compounds[6].

In order to understand the microscopic character of the above properties in more detail, we present the results of magnetostriction and thermal dilatation measurement in magnetic fields at low temperatures. We pay particular attention to aspects of the anisotropic p-f mixing mechanism.

EXPERIMENTAL DETAILS

Single crystals of CeBi and CeSb were prepared with the same methods employed before[7]. The crystals used for magnetostriction and thermal expansion measurements were in the form of rectangular parallelepipeds of approximately 1.5 x 1.5 x 1 mm.[3] Using a highly sensitive capacitance dilatometer (Gen Rad model 1615-A) the longitudinal and transverse magnetostrictions and the thermal expansions

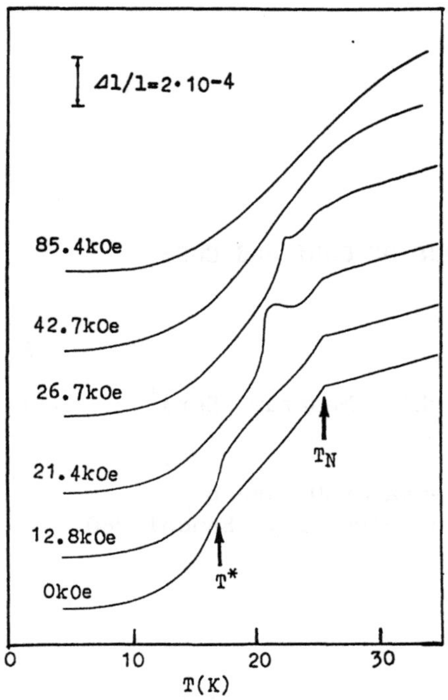

Fig. 1: Temperature variations of $(\Delta l/l)_c$ for CeBi measured in the c-direction in various magnetic fields parallel to the c-axis.

were measured between 4.2 and 70 K and in fields up to about 85 kOe.

RESULTS AND DISCUSSION

CeBi

Figure 1 shows the thermal expansions along the c-direction, $(\Delta l/l)_c$, in CeBi for various magnetic fields. To obtain the single domain state with moments aligned in the c-direction, a magnetic field of about 50 kOe is applied in the measuring direction [001] before beginning the measurements. A large change at $T^*(H=0)=16.7$ K corresponds to the first magnetic transition of its kind observed.

In Figure 2, the thermal expansion in the a-direction, $(\Delta l/l)_a$, is shown in fields parallel to the c-axis after a similar treatment to get a single domain specimen. It is remarkable that in phase II shown in Fig. 3 the a-axis lattice constant is independent of H and is equal to that of the c-axis for H=0. This is different to the result reported before but seems to be more consistent as will be shown. The first order phase transition between phases II and IV in Fig. 3 is sharp but that between II and III is not so sharp. It is also remarkable that except in phase II the c-axis crystal shrinks relative to the a-axis. This means that the crystal field effect dominates the anisotropic p-f mixing effect. In this

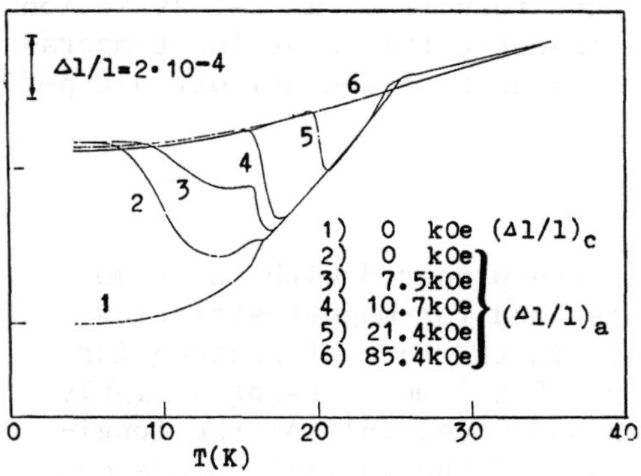

Fig. 2: Temperature variations of $(\Delta l/l)_a$ for CeBi in various magnetic fields parallel to the c-axis. For comparison, the results of $(\Delta l/l)_c$ for H=0 are shown by curve 1.

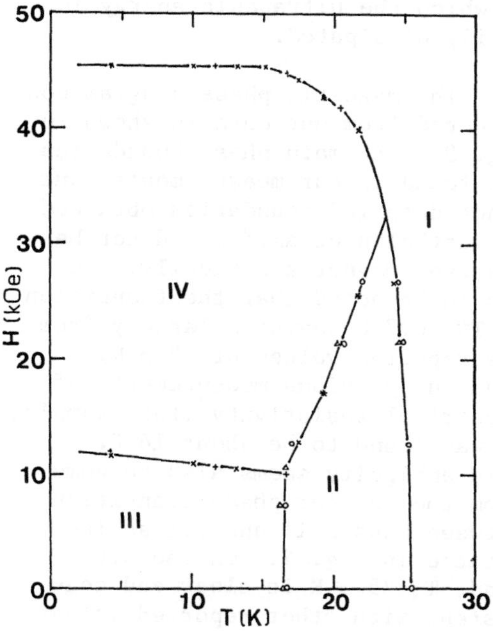

Fig. 3: H-T phase diagram of CeBi derived from the magnetostriction and the thermal expansion. x: longitudinal magnetostriction of the c-axis; +: transverse magnetostriction of the a-axis; o=$(\Delta l/l)_c$, Δ=$(\Delta l/l)_a$.

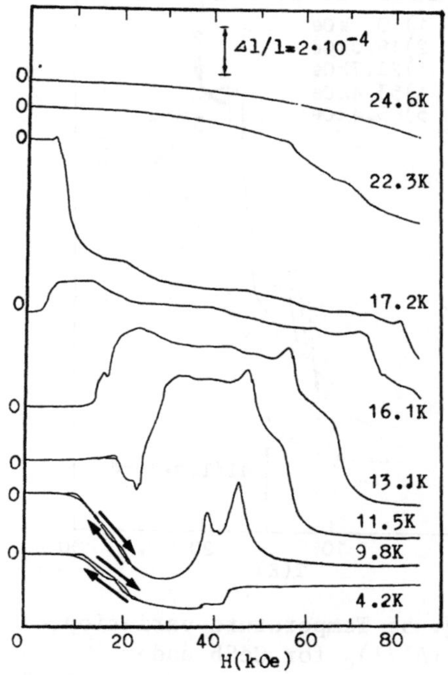

Fig. 4: Field dependences of the magnetostriction of the c-axis at various temperatures

respect the character of phase II is different from the others and requires more detailed investigation.

Figure 4 shows the field dependences of the longitudinal magnetostriction along the c-axis at various temperatures. The magnetostriction below T^* displays irreversibilities only for new traces starting from zero applied field. These irreversible behaviors should correspond to the rotation at a certain H of domains with the c-axis perpendicular to the field direction. Hence, the reverse curve expresses the magnetostriction of a single domain and, after that this curve is reproducible. However, such irreversible magnetostrictions could not be observed between T^* and T_N. This seems to be due to the fact that, as is shown in Fig. 2, the crystal remains cubic in this region and the domain rotation induces nearly no strain. This fact also seems to explain the experimental result demonstrated by Lüthi et al.[8], in which ultrasonic echoes for the C_{11}-C_{12} mode coupling to $(2\varepsilon_{zz}-\varepsilon_{xx}-\varepsilon_{yy})/\sqrt{6}$ and $(\varepsilon_{xx}-\varepsilon_{yy})/\sqrt{2}$ vanish between T^* and T_N. This is because the ultrasonic waves easily cause domain rotation

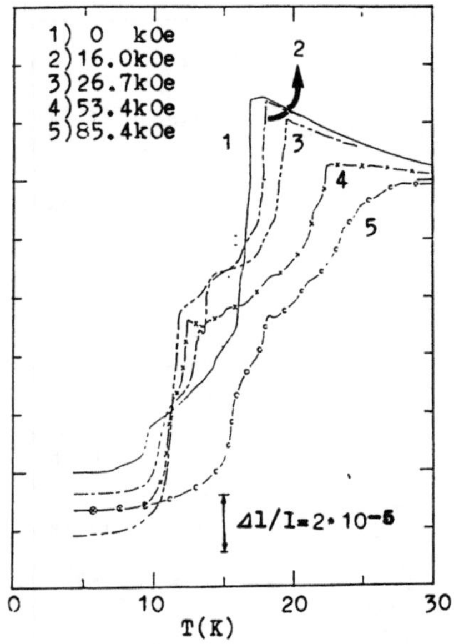

Fig. 5: Temperature variations of $(\Delta l/l)_c$ for CeSb under various magnetic fields parallel to the c-axis.

in which the ultrasonic energy is easily dissipated.

The magnetic phase diagram constructed from our data is shown in Fig. 3. The main phase boundaries are found in our measurements, but other detailed boundaries obtained by Bartholin et al.[9] could not be observed magnetoelastically. It should be noted that the transition at T*=16.7 K deviates largely from the reported values of 12.5 K.[4] From our previous measurements of electrical resistivity[7], for example, T* was found to be about 14 K. This ambiguity seems also to come from the unclear phase transition between phases II and III as indicated in Fig. 2. On the other hand, T_N=25.5 K is clear and consistent with other reported values.

CeSb

In Fig. 5, $(\Delta l/l)_c$ for CeSb is shown for magnetic fields parallel to the c-axis. The same treatment as for CeBi was done to prepare a single domain specimen. The curves $(\Delta l/l)_c$ in Fig. 5 show many kinks corresponding to the successive magnetic phase transitions. A small jump of $(\Delta l/l)_c$ is observed at 7.0 K and a larger one exists at 9.0 K. For $(\Delta l/l)_c$ over an expanded temperature scale, see Fig. 6. These phase boundaries seem

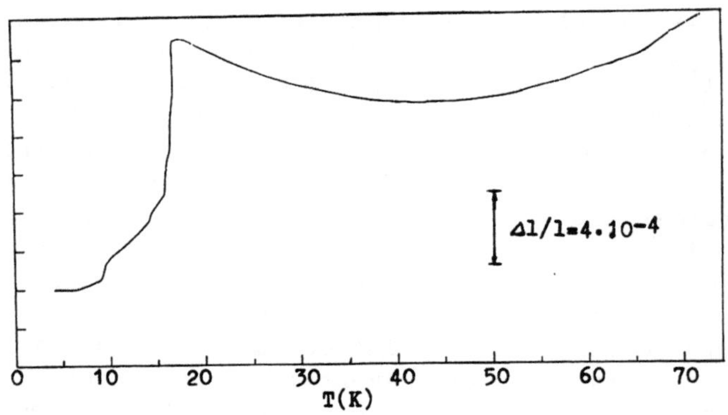

Fig. 6: Negative thermal expansion $(\Delta l/l)_c$ of CeSb at H=0 reproduced up to 70 K from Fig. 5.

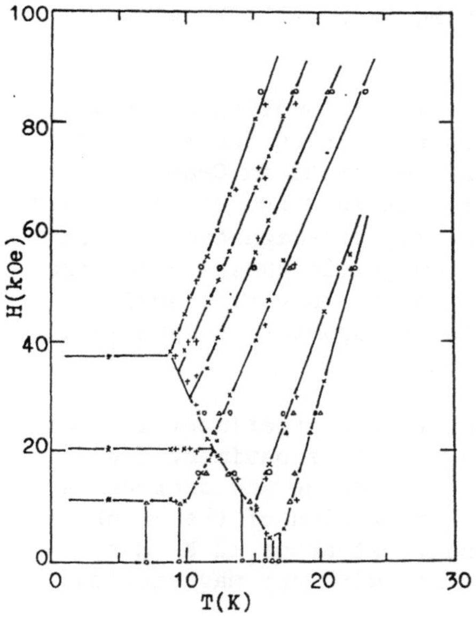

Fig. 7: H-T phase diagram of CeSb derived from the magnetostriction and the thermal expansion. Symbols are the same as in Fig. 4.

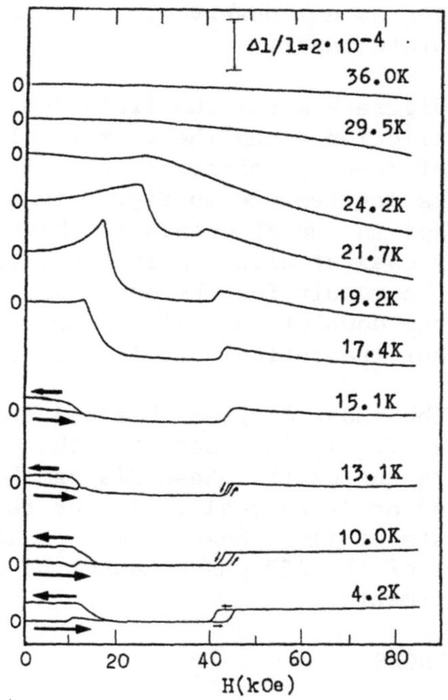

Fig. 8: Field dependences of longitudinal magnetostriction of the c-axis at various temperatures.

to correspond to a large change of lattice constant at 7 K already reported[3]. The phase boundaries obtained from the present results are shown in Fig. 7.

It is remarkable that, unlike CeBi, $(\Delta l/l)_c$ of CeSb has a peak at T_N for low applied fields. The results up to 70 K are shown in Fig. 6, where a minimum is clearly seen at about 45 K. The most natural explanation of this negative thermal expansion is due to the combined effects of the crystal field and the anisotropic p-f mixing. In CeSb, the Γ_7 doublet is expected to be about 37 K below the Γ_8 quartet. Therefore, the population of Γ_7 begins to increase below about 50 K. It has been shown that there is strong p-f mixing with Γ_8 but very little with Γ_7.[10] Therefore, with increasing population of Γ_7, the p-f mixing due to Γ_8 decreases, causing a decrease of distance between Ce and Sb. This is also consistent with the fact that this phenomenon is suppressed in an applied magnetic field[10]. In CeBi, because of smaller crystal field splittings, this property seen for CeSb is not observed. More detailed analyses will be given in a separate paper.

Figure 8 shows the field dependence of the longitudinal magnetostriction along the c-axis. Irreversible curves due to alignment of domains, observed in CeBi, are not obtained in CeSb. An interesting feature in Fig. 8 is a remarkable increase of the lattice constant in the FP phases in which the nearly non-magnetic doublet phases coexist with nearly saturated ferromagnetic phases. Together with the result for the paramagnetic region, it is very plausible that the doublet is built up mostly by the Γ_7 doublet in agreement with our theoretical prediction[10].

The phase diagram obtained from the present experiment is shown in Fig. 7. It is remarkable that in the present experiment the differences of the phase diagram between increasing and decreasing fields[4] are very small. In our results, the region of the FP phases consists of five phases. The phase boundary at about 12 K in the region of the AFP phase cannot be determined with our magnetoelastic measurements.

REFERENCES

1. F.Hulliger, in:"Handbook on the Physics and Chemistry of Rare Earths," K.A.Gschneidner,Jr. and L.Eyring, eds., North-Holland, Amsterdam (1979)
2. J.Rossat-Mignod, P.Burlet, S.Quezel and O.Vogt, Physica 102B, 237 (1980).
3. F.Hulliger, M.Landolt, H.R.Ott and R.Schnelczer, J.Low Temp. Phys. 20,269 (1975).
4. J.Rossat-Mignod, P.Burlet, J.Villain, H.Bartholin, W.Tcheng-Si and D.Florence, Phys.Rev. B16,440 (1977).
5. R.J.Birgeneau, E.Bucher, J.P.Maita, L.Passell, K.C.Turberfield, Phys.Rev. B8,5345 (1973).
6. K.Takegahara, A.Yanase and T.Kasuya, J.de Phys. 41,C5-327 (1980), Erratum ibid 41,1231 (1980). K.Takegahara, H.Takahashi, A.Yanase and T.Kasuya, J.Phys.C: Solid State Phys. 14,737 (1981). T.Kasuya, K.Takegahara, M.Kasaya, Y.Ishikawa, H.Takahashi, T.Sakakibara and M.Date, in:"Physics in High Magnetic Fields," S.Chikazumi and N.Miura, eds., Springer, Berlin (1981). K.Takegahara, H.Takahashi, A.Yanase and T.Kasuya, Solid State Commun. 39,857 (1981).
7. T.Suzuki, M.Sera, H.Shida, K.Takegahara, H.Takahashi, A.Yanase and T.Kasuya, in:"Valence Fluctuations in Solids," L.M.Falicov, W.Hanke and M.B.Maple, eds., North-Holland, Amsterdam (1981).
8. B.Lüthi, P.S.Wang and E.Bucher, in:"Conference on Crystalline Electric Field Effects in Metals and Alloys," Université de Montreal, 1974.

9. H.Bartholin, P.Burlet, S.Quezel, J.Rossat-Mignod and O.Vogt, J.de Phys. 40,C5-130 (1979).
10. K.Takegahara, H.Takahashi, A.Yanase and T.Kasuya, to be published in this proceedings.

COMMENTS

COOPER: (An answer to Dr. Rossat-Mignod's question about the nature of the Ce^{3+} states on the non-magnetic phases just below T_N in CeSb). With an anisotropic interaction of the CS-type plus CEF splitting, there are three doublets: $|\pm 5/2\rangle$. $|\pm 3/2\rangle$, and $|\pm 1/2\rangle$. However, it is essential to realize that an anisotropic exchange does not act on these states like an effective magnetic field, but rather acts like an additional crystal field, and therefore does not split this degeneracy.

ANOMALOUS BEHAVIOUR OF CERIUM IN RMg_3 AND $RInAg_2$ COMPOUNDS

R.M. Galera*, A.P. Murani† and J. Pierre*

*Laboratoire Louis Néel, C.N.R.S., 166X,
38042 Grenoble-cédex, France
†Institut Laue-Langevin, 156X, 38042 Grenoble-cédex

INTRODUCTION

We have studied light rare earth intermetallic compounds RMg_3 (DO_3 structure) and $RInAg_2$ (Heussler-type $L2_1$ structure), in order to compare the behaviour of cerium to that of "normal" rare earths in these compounds. Some anomalous properties are observed, which will be explained by the occurrence of a Kondo effect in $CeMg_3$ and $CeInAg_2$.

Lattice parameters and high temperature susceptibility data (1,2) show trivalency for all rare earth ions in the compounds studied. At low temperatures, the susceptibilities deviate from a Curie-Weiss behaviour, due to crystal field effects. Neutron inelastic scattering experiments have been performed at the high flux reactor of the Institut Laue-Langevin using the IN4 spectrometer. The crystal field parameters could be obtained without ambiguity (Table 1) and give a good account for the temperature dependence of the susceptibilities. A noticeable enhancement of the fourth order term $A_4\langle r^4\rangle$ is found for both cerium compounds, which cannot be explained by the larger spatial extent of the 4f shell only. This fact has been observed previously in other Ce compounds (3). The ground state is the non-magnetic orbital doublet Γ_3 for $PrMg_3$ and $PrInAg_2$. These Van Vleck paramagnets do not show any magnetic order or Jahn-Teller effect down to 1.5 K. Ce and Nd compounds order antiferromagnetically between 2 and 6.2 K (Table 1). The interactions between 4f spins may be deduced from the paramagnetic and antiferromagnetic molecular field constants (Table 1), obtained from the fit of susceptibilities and from the Néel temperatures, using a molecular field Hamiltonian (2). Interactions between spins are found much larger for Ce compounds than in the corresponding Pr or Nd compounds.

Table 1. Lattice parameters, Néel temperatures, crystal field (CEF) parameters (Kelvin), molecular field interactions between spins (Oe^2-mol-erg^{-1}).

Compound	a (Å)	T_N (K)	Ground state	$A_4\langle r^4\rangle$	$A_6\langle r^6\rangle$	n'_p	n'_{AF}
$CeMg_3$	7.444	4	Γ_7	85	-	-550	710
$PrMg_3$	7.415	-	Γ_3	52	-17	-184	-
$NdMg_3$	7.391	6.2	Γ_6	36	-19	-178	45
$CeInAg_2$	7.108	2.2	Γ_7	7.6	-	-280	100
$PrInAg_2$	7.075	-	Γ_3	- 0.3	-15	-112	-
$NdInAg_2$	7.044	2	Γ_6	+ 0.4	-12.5	- 62	13

RESULTS

Electrical resistivities were measured for all compounds between 1.5 and 300 K. The resistivities of La compounds are not exactly representative of the phonon contribution for all magnetic compounds. However the magnetic contributions have been evaluated for $CeMg_3$ and $CeInAg_2$ (1,2); they have an anomalous temperature dependence, namely a decrease with increasing temperatures over some ranges of temperature. Assuming a Kondo-type behaviour and using the theoretical analysis of Cornut and Coqblin (4), we deduce the negative coupling constant between the 4f shell and the conduction band multiplied by the density of states: $\Gamma n(E_F) \simeq -0.04$ and -0.08 for $CeMg_3$ and $CeInAg_2$ respectively.

We have investigated the magnetic structure of Ce and Nd compounds, using the D1B diffractometer at the Institut Laue-Langevin at temperatures between 1.8 and 20 K (5). For $CeMg_3$, faint magnetic lines appear at 2, 2.5 and 3 K, which may be indexed in the primitive cell; the magnetic structure of the f.c.c. cerium array is a first-type antiferromagnetism (Fig. 1a). The value of the moment reaches $0.59 \pm 0.03\ \mu_B$ at 2 K, which is less than the theoretical value ($0.77\ \mu_B$) calculated from a crystal field plus Heisenberg exchange Hamiltonian. The magnetic structure for $CeInAg_2$ is the same; the higher value of the moment ($0.95 \pm 0.05\ \mu_B$ at 2 K) is due to the stronger mixing of Γ_7 and Γ_8 wavefunctions by the exchange field, due to the small CEF splitting (Δ = 18 K). For $NdMg_3$, magnetic reflections of the type (h/2, k/2, 1/2) appear at low temperatures; this implies a doubling of the cell along three directions. The magnetic structure is a second-type antiferromagnetism (as in MnO) and is similar to that found in $TbMg_3$ by Buschow et al (6) (Fig. 1b). The experimental moment reaches $1.3 \pm 0.04\ \mu_B$ at 2 K, which is slightly less than the theoretical value ($1.45\ \mu_B$); the too small experimental value may be attributed

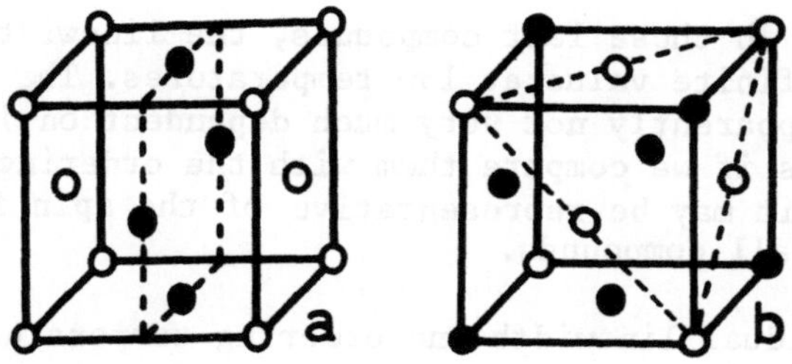

Fig. 1. Magnetic structures : a) first-type AF ($CeMg_3$, $CeInAg_2$) ; b) second-type AF ($NdMg_3$).

to a negative polarization of the conduction band. In the case of $CeMg_3$, it possibly arises from the Kondo effect. We thus observe a change in the ordered structure between $CeMg_3$ and the other two compounds $NdMg_3$ and $PrMg_3$. Both the magnitude and the range of indirect magnetic interactions are modified, due probably to the hybridization of 4f and conduction band states in $CeMg_3$.

The linewidth of crystal field excitations has been studied in several cerium compounds, in order to get information on the spin dynamics of these systems (1,7,8). We have extended previous measurements using the IN4 and IN5 spectrometers at I.L.L. For $CeMg_3$, the quasielastic linewidth γ_{qe} has been measured from T_N up to 780 K (Note that the CEF splitting is $\Delta = 195$ K). It is proportional to the temperature (Korringa law) above about 100 K (Fig. 2), with a slope of 0.082 meV/K which corresponds to $|\Gamma n(E_F)| \sim 0.045$ according to the theoretical predictions of Becker et al (10). This is in contrast to the behaviour of other dense Kondo systems such as $CeAl_3$ (8), $CeAl_2$ and $CeCu_2Si_2$ (7), where γ_{qe} has been found to be proportional to $T^{1/2}$ above certain low temperatures.

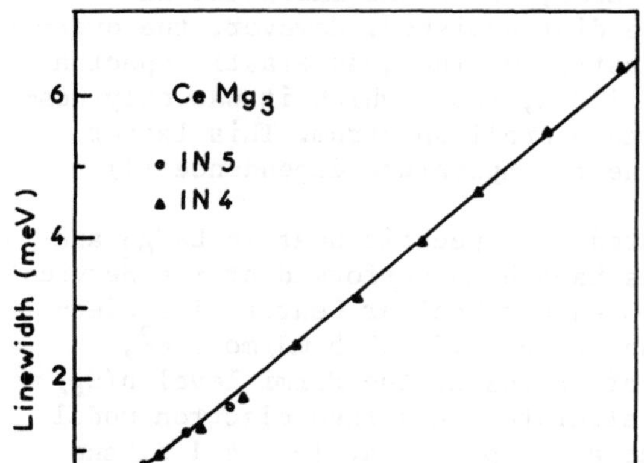

Fig. 2. Quasielastic linewidths for $CeMg_3$. IN4 data were taken with incident energy $E_i = 12.5$ meV, IN5 data with $E_i = 3.64$ meV.

However, as in those last compounds, the linewidth for $CeMg_3$ tends towards a finite value at low temperatures. The residual linewidths are apparently not very much dependent on indirect spin-spin interactions if we compare them with the ordering temperatures (Table 2). Thus it may be representative of the spin fluctuation temperature for all compounds.

Table 2. Residual linewidth and ordering temperature of some cerium compounds

Compound	γ_{qe} (meV)	T_N(K)	Reference
CeMg	0.55	20	9
$CeMg_3$	0.5	4	
$CeAl_2$	0.5	3.8	7
$CeAl_3$	0.5	-	8
$CeCu_2Si_2$	0.85	-	7

We note that in systems such as CeMg and $CeMg_3$, where the linewidth increases linearly with temperature, the CEF splitting Δ between the Γ_7 and Γ_8 states is large (190 and 195 K respectively), and the linewidths of the quasielastic and inelastic lines are smaller than Δ even at the highest temperatures investigated. Hence the overlap between the inelastic and quasielastic spectra is small. In other systems, such as $CeAl_2$, $CeAl_3$, $CeCu_2Si_2$, where γ_{qe} increases as $T^{1/2}$, we note that the overlap is significant due to the smaller magnitude of the crystal field splitting compared to the linewidths. In this respect, we recall that our earlier measurements (1) on CeZn (CEF splitting $\Delta \sim 65$ K) showed a downwards curvature of γ_{qe} at low temperatures, where the inelastic and quasielastic components could be distinguished. However, the overlap is very strong and the inelastic and the quasielastic spectra merge into a single line around 100 K, above which it was only possible to obtain the width for the overall spectrum. This latter width shows an approximately linear temperature dependence (1).

Finally, we have investigated the specific heat in $LaMg_3$ and $CeMg_3$ (Fig. 3); the experiments have been performed at the Service des Basses Températures of the Grenoble Nuclear Center. The electronic term C/T in $LaMg_3$ is found to be 3.7 ± 0.5 mJ/mole-K^2, which corresponds to a density of states at the Fermi level $n(E_F) = 1.57$ e$^-$/eV, smaller than that calculated in a free electron model. For $CeMg_3$, the specific heat has a sharp peak at $T_N = 4.1$ K, and decreases with increasing temperature up to 9 K. The magnetic entropy under T_N is much smaller than Rln2 as expected for the Γ_7 ground state; it reaches this value only at 25 K. This may be interpreted in terms of a splitting δ of the Γ_7 level above T_N,

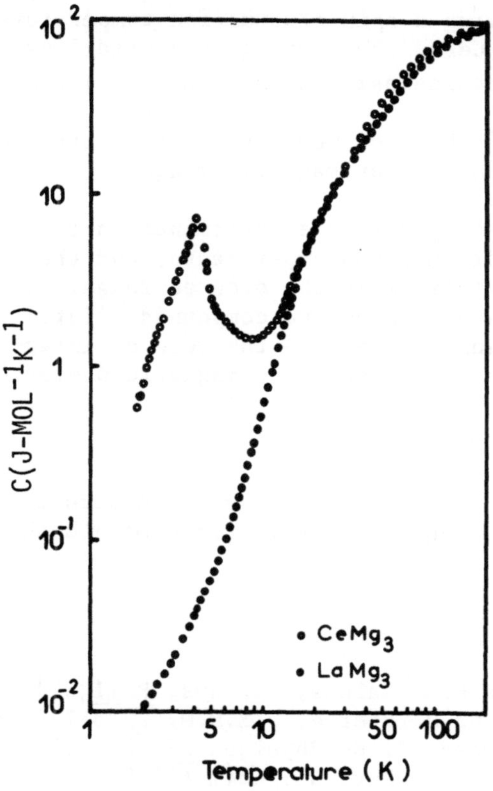

Fig. 3. Specific heats of $LaMg_3$ and $CeMg_3$ (logarithmic scale). Only a small fraction of experimental points is shown

which may be due either to short range order or, and more probably, to the Kondo effect. Fitting the magnetic specific heat as a Schottky anomaly, we obtain $\delta \simeq 6$ K, which is in reasonable agreement with the residual linewidth of 0.5 meV. In this respect, the system looks very similar to $CeAl_2$ (11).

CONCLUSION

We have tried to characterize the anomalous behaviour of cerium in these compounds. Several features may be related to the hybridization of the 4f and conduction band wavefunctions, leading to a Kondo effect:

1 - The magnetic resistivity decreases over some ranges of temperature for the Ce compounds.
2 - The crystal field parameter $A_4\langle r^4\rangle$ is anomalously high.
3 - Indirect magnetic interactions are strongly enhanced and their variation with distance modified.
4 - The magnetic moment is reduced in the ordered range.
5 - The quasielastic linewidth tends at low temperatures towards a

finite value for $CeMg_3$, giving a spin fluctuation temperature $T_{sf} \simeq 6$ K. For $CeInAg_2$ and CeZn, the damping of excitations leads to the collapse of the inelastic structure at T larger than the CEF splitting.

6 - The low temperature specific heat of $CeMg_3$ reveals a possible splitting of the Γ_7 level in the paramagnetic range.

This study will be followed up with the determination of the 4f level binding energy via photoemission experiments, and the study of the dispersion of excitations in the ordered range. We also hope to carry out measurements on cerium compounds diluted with La or Y, in order to distinguish between the role of interaction effects and Kondo effect in the observed magnetic properties.

ACKNOWLEDGMENTS

We thank J. Pannetier for his help during magnetic structure determinations. Specific heat measurements were performed thanks to R. Lagnier and P. Bertrand.

REFERENCES

1. J. Pierre, A.P. Murani and R.M. Galera, J. Phys. F 11,679(1981).
2. R.M. Galera, A.P. Murani and J. Pierre, J.M.M.M. 23,317 (1981).
3. P. Lethuillier and J. Chaussy, J. de Physique 37,123 (1976).
4. B. Cornut and B. Coqblin, Phys. Rev. B 5, 4541 (1972).
5. R.M. Galera, J. Pierre and J. Pannetier, submitted to J. Phys. F.
6. K.H.J. Buschow, G. Will and M.O. Bargouth, J. Phys. C 11,2405 (1978).
7. M. Loewenhaupt and E. Holland-Moritz, J. Appl. Phys. 50,7456 (1979) and J.M.M.M. 14,227 (1979).
8. A.P. Murani, K. Knorr, K.H.J. Buschow, A. Benoit and J. Flouquet, Sol. Stat. Commun. 36,523 (1980).
9. J. Pierre and A.P. Murani in "Crystalline Electric Field and Structural Effects in f-Electron Systems", J.E. Crow, R.P. Guertin and T.W. Mihalisin eds, Plenum Press (1980).
10. K.W. Becker, P. Fulde and J. Keller, Z. Physik B28,9 (1977).
11. J. Aarts, F.R. de Boer, S. Horn, F. Steglich and D. Meschede, P. 301 in "Valence Fluctuations in Solids", L.M. Falicov et al, eds, North Holland Publishing Co (1981).

COMMENTS

COQBLIN: The difference between the resistivities of $NdMg_3$ and $LaMg_3$ seems to decrease with increasing temperature at high temperatures. I do not believe too strongly that $NdMg_3$ could be a "Kondo system". But have you an estimate of the exchange integral Γ in order to see if Γ could be larger than for normal rare earths?

PIERRE: The difference between the slopes of the resistivities for $NdMg_3$ and $LaMg_3$ in the temperature range between 200 and 300 K probably comes from different Debye temperatures (contrary to the case of $CeMg_3$ and $LaMg_3$). I think that the exchange integral for Nd compounds may be slightly larger than for Gd, but it is far from the anomalously high values found for cerium.

MAGNETISM AND CRYSTAL FIELDS IN TERNARY SUPERCONDUCTORS*

G. K. Shenoy, G. W. Crabtree, D. Niarchos, F. Behroozi,**
B. D. Dunlap, D. Hinks and D. R. Noakes

Argonne National Laboratory
Solid State Science Division
9700 So. Cass Avenue
Argonne, Illinois 60439 USA

INTRODUCTION

Ternary compounds exhibiting superconductivity have been under intensive investigation in the last few years, and some of the related properties were reviewed in the last proceedings of this conference series by several authors.[1-3] Among the more stimulating phenomena is the observation that many of these materials exhibit superconductivity even when one of the constituents is strongly magnetic, such as a rare earth (RE) ion. This is in sharp contrast to the behavior usually found in binary compounds, where very small concentrations of magnetic ions are adequate to totally suppress the superconducting transition. In the ternary materials, the presence of large concentrations of magnetic ions has also provoked renewed interest in questions concerning the ways in which superconductivity and magnetism affect one another, and whether the two types of order can simultaneously occur. Much of the work carried out in the past on the coexistence question utilized pseudobinary systems, and it is now generally believed that those experiments showed the coexistence of superconductivity with spin-glass type ordering, rather than with true, long-range magnetism.[4] The ternary compounds, being structurally homogeneous, will not show that behavior, and so are much more ideally suited to stimulate and test theoretical predictions on the interaction between magnetism and superconductivity.

---

* Work supported by the U.S. Department of Energy.

**Permanent address: University of Wisconsin - Parkside
Kenosha, Wisconsin 53141

In the past few years, ternary and pseudo-ternary compounds showing both magnetism and superconductivity in various combinations have been studied utilizing techniques such as a.c. susceptibility, resistivity, magnetization, specific heat, magnetostriction, ultrasonic attenuation, tunneling, neutron scattering, EPR, NMR and the Mössbauer effect.[5] Not surprisingly, those materials containing RE atoms frequently show significant effects on physical properties due to the presence of crystalline electric field (CEF) interactions. The most extensively studied materials have been the Chevrel phase compounds, generically given by $RE_xMo_6Ch_8$ (Ch = S, Se; x = 1.0, 1.2), and the $RERh_4B_4$ compounds. In this paper we will emphasize measurements performed on $RERh_4B_4$ compounds which highlight the influence of the CEF, and discuss the difficulties involved in their evaluation. More briefly, discussion will also be given to the Chevrel phase materials.

CEF IN $RERh_4B_4$

Before evaluating the experimental information concerning the CEF in these compounds, it is necessary to consider the form of the crystal field Hamiltonian. The compounds have a tetragonal structure belonging to the space group $P4_2/nmc$.[6] The point symmetry of the RE site is $\bar{4}2m$, resulting in the Hamiltonian

$$H_{CF} = B_2^0 O_2^0 + B_4^0 O_4^0 + B_4^4 O_4^4 + B_6^0 O_6^0 + B_6^4 O_6^4 \qquad (1)$$

where the O_n^m are Stevens operators and the B_n^m are the CEF parameters. Point-charge calculations based on nearest neighbor Rh interactions with the RE indicate[7] that the term involving $B_2^0 O_2^0$ makes the primary contribution, with that due to $B_4^0 O_4^0$ being roughly one order of magnitude smaller. Thus, one may anticipate that a model which truncates the Hamiltonian to only the B_2^0 term may show many of the features of the experimental data, although it will not be correct in detail. In the following we will first describe a determination of the value of the first term in Eq. (1) from a Mössbauer effect experiment in $GdRh_4B_4$, apply the result to a discussion of the properties of other $RERh_4B_4$ compounds, and indicate the shortcomings in such an approach.

EXPERIMENTAL DETERMINATION OF B_2^0

The most direct way of observing the CEF levels is by inelastic neutron scattering. When such measurements are possible, the energy and intensity of various transitions can then be used to evaluate CEF level splittings. The large neutron absorption in $RERh_4B_4$ by both B and Rh have made such experiments difficult. An alternate method is to use hyperfine measurements to provide some of the information.

Charges distributed throughout the lattice not only produce CEF interactions at the RE site, but also set up an electric field gradient (EFG) which interacts with the nuclear quadrupole moment, and this quadrupole interaction can be measured using the Mössbauer effect. Since the nuclear quadrupole moment is usually known from other results, it is then possible to obtain a value for the EFG. In S-state ions such as Gd^{3+}, the EFG is generated completely by those charges external to the atom being measured. This "lattice EFG" has an operator form Y_2^0 and is also proportional to the CEF term $B_2^0O_2^0$.[8] The Hamiltonian describing the quadrupole interaction for a site in $\overline{4}2m$ symmetry is[9]

$$H_Q = \frac{e^2QV_{zz}}{4I(2I-1)} [3\hat{I}_z^2 - \hat{I}^2] \tag{2}$$

where Q is the nuclear quadrupole moment, $\hat{I}_z$ and $\hat{I}$ are nuclear spin operators, and V_{zz} is the principal component of the EFG tensor at the nucleus. For S-state ions this is related to B_2^0 through

$$\begin{aligned} V_{zz} &= -4A_2^0\,(1-\gamma_\infty) \\ B_2^0 &= \alpha_J A_2^0 \langle r^2\rangle (1-\sigma_2) \quad . \end{aligned} \tag{3}$$

Here γ_∞ and σ_2 are electronic shielding parameters, α_J is the Stevens factor and $\langle r^2\rangle$ is the second moment of the 4f electron distribution for the RE ion. Using calculated values of $1-\gamma_\infty=60$ and $1-\sigma_2=0.4$ for all the heavy rare earths[10-11] and tabulated values[12] for $\langle r^2\rangle$, the value of A_2^0 appropriate to Gd can be obtained. Assuming that A_2^0 is the same for all RE ions in an isostructural series, the values of B_2^0 for the other compounds can then be calculated.

Experiments[13] carried out on $GdRh_4B_4$ using the Mössbauer effect in ^{155}Gd above the magnetic transition temperature gave a value of $e^2QV_{zz} = -8.8 \times 10^{17}\ V/cm^2$, resulting in a value of $A_2^0 = +1165$ K/a.u. In Table 1, B_2^0 values calculated from this for other rare earth ions are tabulated. The uncertainty in these values are approximately 20%, primarily due to uncertainties in the values of the shielding parameters. Table 1 also provides a comparison of the maximum splitting of the CEF levels, calculated with only the B_2^0 term in Eq. (1), with that determined from specific heat data.

Table 1. Value of B_2^0 in $RERh_4B_4$ Deduced from the EFG Measurement in $GdRh_4B_4$ and Values of Maximum CEF Splittings

RE	B_2^0 (K)	Maximum CEF splitting (K)	
		B_2^0 Model	Specific Heat data[a]
Tb	-3.6	389	>100
Dy	-2.1	353	>100
Ho	-0.8	154	173
Er	+0.8	134	>35
Tm	+3.3	356	>80

a. Ref. 14-17

INFLUENCE OF CEF ON PHYSICAL PROPERTIES OF $RERh_4B_4$

Specific heat measurements for the $RERh_4B_4$ compounds display the CEF levels in the most direct way.[14-16] After subtracting the lattice and electronic contributions, the resulting specific heat data show a sharp low temperature peak in those compounds which undergo magnetic ordering, and a broad Schottky anomaly at higher temperatures due to the CEF interactions. The entropy under the magnetic peak gives information on the degeneracy of the low lying levels, and the temperature dependence of the Schottky specific heat contains information on the positions of all the CEF states. Such data are illustrated in Fig. 1, showing the magnetic and Schottky peaks for $ErRh_4B_4$.[14] In this case the CEF Hamiltonian of Eq. (1) will lift the degeneracy of the Hund's rule $4I_{15/2}$ state for Er^{3+} to produce eight Kramers doublets. Analysis of the magnetic peak shows that the associated entropy is close to Rln4, indicating that the ground state consists of two nearly degenerate Kramers doublets. Analysis of the Schottky peak indicates that seven Kramers doublets are distributed in energy up to about 35K, with the remaining doublet at roughly 100K. Such data, however can only provide information on the overall disposition of the CEF levels, and gives no information on the electronic or magnetic character of those states.

The makeup of the ground state can be more completely specified

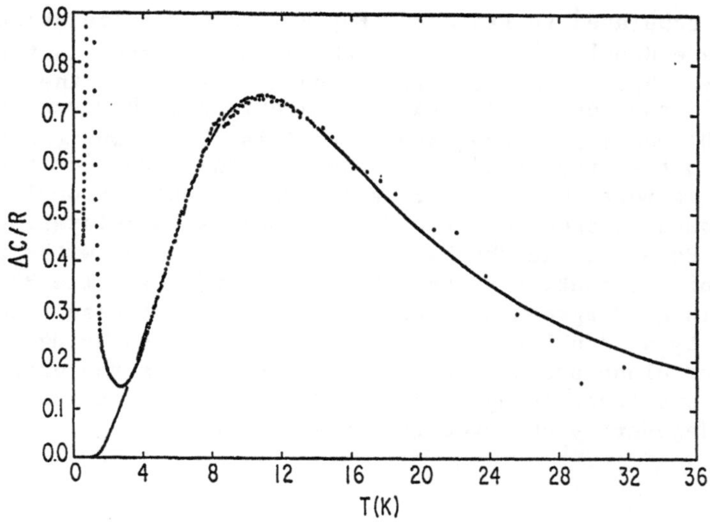

Fig. 1. Magnetic heat capacity as a function of temperature for $ErRh_4B_4$ showing the magnetic ordering at about 1K and the Schottky anomaly due to crystal field levels, (Ref. 14).

by measurement of the magnetic moment. Magnetization data in these cases may be difficult if only powder samples are available, due to the large single-ion anisotropies frequently present,[17] and it is often important to have data obtained in other ways. In the case of $ErRh_4B_4$, data on magnetic properties are now available from neutron diffraction, Mössbauer spectroscopy and single crystal magnetization work. This is summarized in Fig. 2. From neutron diffraction measurements both on powder[18] and single-crystal[19] samples, the material is found to be ferromagnetic with a saturation ordered magnetic moment of 5.6 μ_B, lying in the basal plane of the crystal. This is in contrast to a result obtained by Mössbauer spectroscopy,[20] which gave a moment of 8.3 μ_B. Single crystal magnetization measurements taken at 0.5K have shown[21] that the spontaneous moment along the a direction is about 6 μ_B. One may obtain a possible resolution of these various results by observing that the Mössbauer experiment provides a single ion property only, in this case being the total magnetic moment on an individual Er atom. However, the neutron diffraction and magnetization both measure the ordered component of the moment, i.e., that portion entering into the long-range ordering. This may imply that the true moment on the Er atom is 8.3 μ_B but that only the component of 5.6 μ_B is ferromagnetically ordered, with the remaining component being polarized by the application of small magnetic fields.

If, as above, we assume that B_2^0 is the only non-zero CEF parameter, then we may use the values of Table 1 to calculate the magnetic properties of the various RE ions in $RERh_4B_4$ compounds.

These are tabulated in Table 2. For Er, such a calculation gives a ground state doublet $|J_z = \pm 1/2\rangle$, which has a moment in the basal plane of 4.8 μ_B. Such a result may be indicative of the moment observed in the neutron diffraction experiments, however it is clear that the B_2^0 model is inadequate to explain the known results. In addition to the unresolved discrepancy in the magnetic moment measurements, we note that a $|\pm 1/2\rangle$ state will couple strongly to the electron bath to produce a rapid electronic spin relaxation of the Er ions. However, the Mössbauer spectra clearly show the presence of long spin relaxation times between 1.0 and 10K. The Mössbauer data are in good agreement with the specific heat result showing two close lying doublets with the remaining states being above 10K. The B_2^0 term alone produces an isolated doublet as the ground state with the next level lying at 22K, and in no case can provide the fourfold degeneracy observed in the experiments.

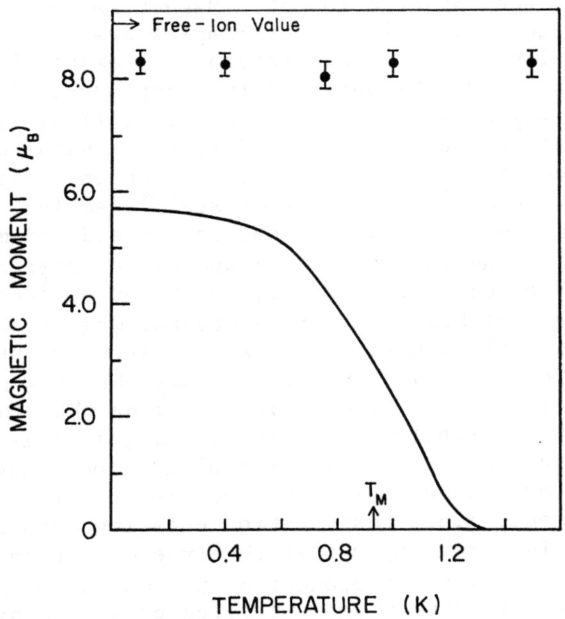

Fig. 2. Temperature dependence of Er moment (solid circles) as determined by Mössbauer spectroscopy and coherent moment (solid line) as determined by neutron diffraction on the Er atom in $ErRh_4B_4$.

Table 2. CEF Properties of $RERh_4B_4$ from Experiment and from B_2^0 Term in Eq. (1)

RE	CEF Ground State		Ground State Moment (μ_B)		Ordering and Moment Direction	
	B_2^0 Model	Experiment	B_2^0 Model	Experiment	B_2^0 Model	Experiment
Tb	±6	Doublet	9.0	--	c-axis	F
Dy	±15/2	Two doublets[a,b]	10.0	9.2 ± 0.2[b]	c-axis	F
Ho	±8	Doublet	10.0	8.7 ± 0.3[d]	c-axis	c-axis (F)
Er	±1/2	doublet [a,b]	4.8[c]	8.3 ± 0.2[b,e] 5.6 ± 0.1[f]	basal plane	basal plane (F)
Tm	0	Four levels[a,b]	0	6.4 ± 0.2[b]	no ordering	AF

a. specific heat (Ref. 14-16)
b. Mössbauer (Ref. 22)
c. calculated along a-axis
d. neutron diffraction (Ref. 23)
e. single-crystal magnetization (Ref. 21)
f. neutron diffraction (Ref. 18-19)
F. ferromagnetic but complex (Ref. 20,24)
AF. antiferromagnetic

In other $RERh_4B_4$ compounds, the specific heat and Mössbauer spectra can provide an estimate of the first excited state through the observation of paramagnetic relaxation effects.[22] Table 2 summarizes some of these results, and compares them with the predictions of the B_2^0 model. While some features of the data can be understood in this context, it is clear that more detailed information on the total CEF Hamiltonian is required in order to completely understand the data.

In Fig. 3, the magnetic and superconducting transition temperatures have been given for the $RERh_4B_4$ compounds. The most striking feature of this figure is that the values of T_M do not scale with the deGennes factor $(g_J-1)^2J(J+1)$, as is commonly observed in RE systems having CEF interactions which are weak compared with magnetic interactions. Most notably, these data show the largest T_M for $DyRh_4B_4$ rather than for $GdRh_4B_4$. Within the framework of a molecular field model[13], one can show that strong deviations from the deGennes scaling is to be expected, however one again needs detailed information on the CEF parameters to provide a quantitative evaluation of those effects. It is also possible that some of these

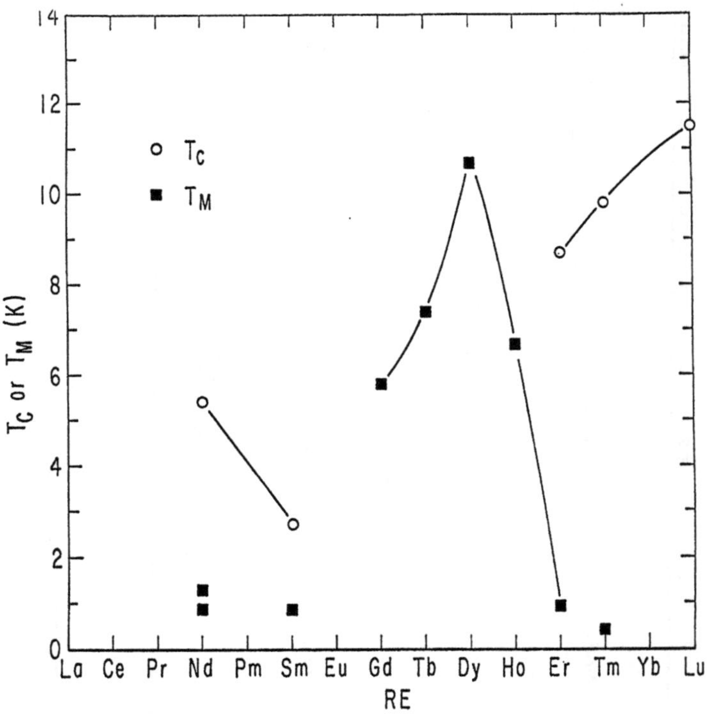

Fig. 3. Superconducting (T_c) and magnetic (T_M) transition temperatures of $RERh_4B_4$ compounds, (Ref. 5).

effects arise from an anisotropic exchange coupling. Table 1 shows the orientation of the magnetic moments in the ordered state, as obtained either from experiment or from the B_2^0 model. One notices that the value of T_M is large when the moments are parallel to the c axis, and small when in the basal plane. This may indicate a weaker exchange interaction in the basal plane, which would also favor the occurrence of superconductivity in those cases.

Within the framework of the isotropic RKKY model, the change from ferromagnetic to antiferromagnetic ordering when going from Er to Tm would imply a change in the exchange interaction, and this is very unlikely.[24] However, one may also note that in these cases, the values of T_M are on the order of those expected from spin-dipolar interactions. It is therefore likely that $TmRh_4B_4$ shows antiferromagnetism due to a direct classical dipolar coupling which is of comparable magnitude to the weak exchange coupling. It is also very possible that these two competing couplings give rise to the complex magnetism observed in $ErRh_4B_4$.[20] Magnetization measurements on powdered $HoRh_4B_4$ also indicate magnetic ordering which is more complex than a simple ferromagnetism along a single direction.[17]

CEF IN CHEVREL PHASE COMPOUNDS

There has been considerable effort using neutron scattering to determine the CEF in the Chevrel phase ternary superconductors. This was reviewed by Lynn.[2] Because of a small rhomohedral distortion of the otherwise cubic structure, the analysis of CEF is done using a Hamiltonian with cubic symmetry. The interpretation of data on $REMo_6Se_8$ compounds has met with considerable success, while in $REMo_6S_8$ no CEF transitions have been observed. For this reason it would be ideal to use other techniques to advance our understanding.

One such attempt was made using the Mössbauer effect in $YbMo_6S_8$.[25] In this compound Yb resides with both di- and tri-valent charges. The quadrupole interaction and magnetic interaction data indicate a large rhombohedral CEF at the Yb^{2+} site while a near-cubic CEF at the Yb^{3+} site, which is unusual. Further work in this area is now in progress in our laboratory.

CONCLUSIONS

In this paper, the present state of knowledge on CEF in two important classes of ternary superconductors has been described. It is clear that in understanding the superconducting and magnetic behavior of $RERh_4B_4$, the CEF plays a very important role. However, further experimental data are needed for better evaluation of the CEF parameters.

Considerable importance has been given to the specific heat Schottky anomalies in deducing the position and degeneracy of various CEF levels. Interpretation of these data is made difficult because of complicated subtraction of lattice, electronic and superconducting specific heats. Furthermore, the purity of the sample is important in such studies. It is known that a few percent of $Rh-B_x$, $RERh_3B_2$ and $RERh_6B_4$ are commonly present in $RERh_4B_4$, while $Mo-Ch_x$, $RE-Ch_x$ and RE_2O_2Ch phases occur in Chevrel phase compounds. Only single-crystal samples will lead to dependable specific heat data.

REFERENCES

1. M. B. Maple, H. C. Hamaker, L. D. Woolf, H. B. MacKay, Z. Fisk, W. Odoni and H. R. Ott in Crystalline-Electric Field and Structural Effects in f-Electron Systems, Eds. J. E. Crow, R. P. Guertin and T. W. Mihalisin, Plenum Press, New York (1980), p. 533
2. J. W. Lynn in Ref. 1, p. 547
3. S. Maekawa and C. Y. Huang in Ref. 1, p. 561
4. S.Roth, Appl. Phys. 15, 1 (1978)
5. G. K. Shenoy, B. D. Dunlap and F. Y. Fradin, Eds. Ternary Superconductors, North-Holland, New York, 1981
6. J. M. Vandenberg and B. T. Matthias, Proc. Natl. Acad. Sci. 74, 1336 (1977)
7. L. D. Woolf, Ph.D. Thesis, University of California-LaJolla (1980) Unpublished
8. S. Ofer, I. Nowik and S. G. Cohen in Chemical Applications of Mössbauer Spectroscopy, Eds. V. I. Goldanskii and R. H. Herber, Acadamic Press, New York, 1968, p. 428
9. See for example, G. K. Wertheim, Mössbauer Effect, Academic Press, N.Y., 1964
10. R. P. Gupta and S. K. Sen, Phys. Rev. A7, 850 (1973)
11. J. Blok and D. A. Shirley, Phys. Rev. 143, 278 (1966)
12. A. J. Freeman and R. E. Watson, Phys. Rev. 127, 2058 (1962)
13. G. K. Shenoy, D. R. Noakes, D. Hinks, D. Niarchos and B. D. Dunlap, to be published
14. L. D. Woolf, D. C. Johnston, H. B. MacKay, R. W. McCullum and M. B. Maple, J. Low Temp. Phys. 35, 651 (1979)
15. H. B. MacKay, L. D. Woolf, M. B. Maple and D. C. Johnston, J. Low Temp. Phys. 41, 639 (1980)
16. K. Kumagai, Y. Inoue and K. Asayama, J. Phys. Soc. Japan 47, 1363 (1979)
17. F. Acker and H. C. Ku, J. Low Temp. Phys. 42, 449 (1981)
18. D. E. Moncton, D. B. WcWhan, J. Eckert, G. Shirane and W. Thomlinson, Phys. Rev. Lett. 39, 1164 (1977)
19. S. K. Sinha, G. W. Crabtree, D. G. Hinks and H. Mook, to be published

20. G. K. Shenoy, B. D. Dunlap, F. Y. Fradin, S. K. Sinha, C. W. Kimball, W. Potzel, F. Pröbst and G. M. Kalvius, Phys. Rev. B9, 3886 (1980)
21. G. W. Crabtree, F. Behroozi and S. A. Campbell, (to be published)
22. G. K. Shenoy, P. J. Viccaro, D. Niarchos, J. D. Cashion, B. D. Dunlap and F. Y. Fradin in Ref. 5, p. 163
23. G. H. Lander, S. K. Sinha and F. Y. Fradin, J. Appl. Phys. 50, 1990 (1979)
24. H. C. Hamaker, H. B. MacKay, L. D. Woolf, M. B. Maple, W. Odoni and H. R. Ott, Phys. Lett. 81A, 91 (1981)
25. P. Bonville, J. A. Hodges, P. Imbert, G. Jehanno, R. Chevrel and M. Sergent, Revue Phys. Appl. 15, 1139 (1980)

COMMENTS

BABERSCHKE: I would like to comment on the comparison between pseudobinary and ternary superconductors if we count the concentration of magnetic ions per lattice site (and not per formula unit). We get almost the same numbers. For example for $La_{1-x}Gd_xOs_2$, x=0.155, x/lattice = 5.2%; for $Gd_{1.2}Mo_6S_8$, x/lattice = 7.9%; for $ErRh_4B_4$, x/lattice = 11%; and for $La_{1-x}Er_xOs_2$, x=0.30, x/lattice = 10%. The physics seems to be the same, namely low density of states at the Fermi energy and weak exchange.

SHENOY: I agree with the above facts. But in ternaries the magnetic atom has a unique crystallographic site with the same number of magnetic neighbors, and in pseudo-binaries this is not true. That is why the pseudo-binaries never show antiferromagnetism, for example. Hence the physics is not the same.

FULDE: Have attempts been made to measure the CEF by point contact techniques? They may be interesting in view of the fact that tunneling experiments are generally difficult.

SHENOY: Yes, attempts have been made by Dr. Pobell (Julich), by Dr. Alan Goldman (Minnesota) and by Dr. R. C. Dynes (Bell Labs) on thin films and bulk samples of ternary superconductors. On $ErRh_4B_4$ there is no published work.

ALEKSKEVSKII: The Chevrel phases can have the lattice distortions when temperature is lowered. How can such distortions influence the data you demonstrated?

SHENOY: I can not answer this question now. But I can only say that the most reliable data are obtained on the borides.

A NOVEL KIND OF METAL-RICH LANTHANIDE COMPOUND

Arndt Simon

Max-Planck-Institut für Festkörperforschung
Heisenbergstrasse 1, D - 7000 Stuttgart 80, FRG

INTRODUCTION

It is a well known fact that the lanthanides are trivalent with a few exceptions of di- and tetravalency depending on electronic configuration.

As an introduction to the topic one might ask two questions:

(a) Will it be possible to prepare compounds (preferably binaries) of monovalent lanthanides, e.g. monohalides LnX ?

(b) If yes, could such monohalides possibly have a rock salt structure ?

Question (b) is easily answered. A thermodynamic estimate shows that a monohalide with purely ionic structure must have such a high energy, that it decomposes into metal and trihalide.

Question (a) is much more difficult to answer. If the lanthanides are capable of forming homonuclear bonds (like carbon), such compounds might exist. Metal-metal bonds resulting in the formation of metal clusters is the "trick" which is used especially by transition metals to stabilize low oxidation states. This "trick" is illustrated with the halides of niobium in Figure 1.

Oxidation state	5.00	4.00	2.67	2.50	2.33	1.83
Cluster type	single atom	pair	tri- angle	rhombus	octa- hedra	octa- hedra
Examples	$NbCl_5$	$NbCl_4$	Nb_3Cl_8	$CsNb_4Cl_{11}$	Nb_6Cl_{14}	Nb_6I_{11}

Fig. 1. Oxidation states and metal cluster types in niobium halides

Niobium has five valence electrons; in $NbCl_5$ they are all involved in chemical bonding. In $NbCl_4$ one d-electron is left with each Nb atom and two neighbouring atoms use their electrons to form a single bond. The larger the number of d-electrons left with the metal atoms in low oxidation state, the more metal bonds are formed or (as one of the possible consequences) the larger the clusters become (Fig.1). Compounds like Nb_6Cl_{14} [1] and Nb_6I_{11} [2-4] contain octahedral metal clusters which are surrounded by 12 (cuboctahedron) Cl atoms in the Nb_6Cl_{12} unit and 8 (cube) I atoms in the Nb_6I_8 unit. The latter cluster type is well known to many physicists from the so-called Chevrel phases, which are molybdenum chalcogenides of the general formula $Mo_6X_8M_x$ (X = S, Se, Te; M = Cu, Pb, Ln etc.) [5].

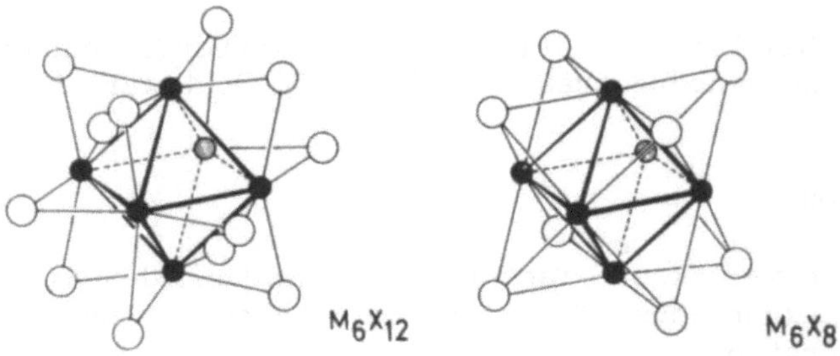

Fig. 2. Octahedral metal clusters with different non-metal atom enviroments.

Both cluster types shown in Figure 2 differ by the number of electronic states available for metal-metal bonding. It is well understood from an early molecular orbital treatment [6] that the M_6X_{12} cluster is most stable with 16 electrons and the M_6X_8 cluster with 24 electrons in metal-metal bonding states.

Despite the fact that the lanthanides have a small valence electron concentration they are capable of forming metal clusters like the more electron-rich transition metals, thus opening a new field of metal-rich compounds.

METAL-RICH LANTHANIDE HALIDES

The compound $GdCl_{1.5} = Gd_2Cl_3$ was the first to show a lanthanide element in an oxidation state below two [7-9]. In the meantime a large number of halides LnX_n ($1 \leq n \leq 2$) have been found, which contain metal clusters [10]. The present knowledge of the existing compounds is summarized in Table I; similar compounds exist with the 3d element scandium [18-21] and the 4d element yttrium [22].

The structural principle of the cluster compounds Ln_7I_{12} is illustrated in Figure 3.

Table I Metal-rich lanthanide halides LnX_n ($n < 2$)

n	1.71	1.60	1.50	1.43	1.25	1.17	1.00
Cluster type	M_6X_{12}	M_6X_{12}	M_6X_8	M_6X_{12}	M_6X_{12}	M_6X_{12}	M_6X_8
Compounds	La_7I_{12}[11] Ce_7I_{12}[11] Pr_7I_{12}[11] Gd_7I_{12}[11] Tb_7I_{12}[11] Er_7I_{12}[11] Lu_7I_{12}[11]	Gd_5Br_8[12] Tb_5Br_8[12]	La_2Cl_3[13] Gd_2Cl_3[7-9] Gd_2Br_3 [9] Tb_2Cl_3 [9] Tb_2Br_3[10] Er_2Cl_3[10] Lu_2Cl_3[10] Tm_2Cl_3[10]	Er_7I_{10}[14]	Ce_4Br_5[15] Er_4I_5 [14]	Tb_6Br_7[16] Er_6I_7 [16]	LaCl[10,13] LaBr[17] CeCl[10] CeBr[10] PrCl[10] PrBr[17] NdBr[10] GdCl[17] GdBr[17] TbCl[17] TbBr[17] DyBr[10] HoCl[10] HoBr[17] ErCl[10] ErBr[17] LuBr[10]
structural formula	$Ln_6X_{12}\cdot Ln$	$Ln_4X_8\cdot Ln$	Ln_2X_3	$Ln_6X_{10}\cdot Ln$	Ln_4X_5	Ln_6X_7	LnX
structural unit	isolated cluster	single cluster chain	single cluster chain	double cluster chain	single cluster chain	double cluster chain	cluster layer
No. valence electrons (per Ln) for Ln-Ln bonding	1.50	1.50	1.50	1.83	1.75	1.83	2.00

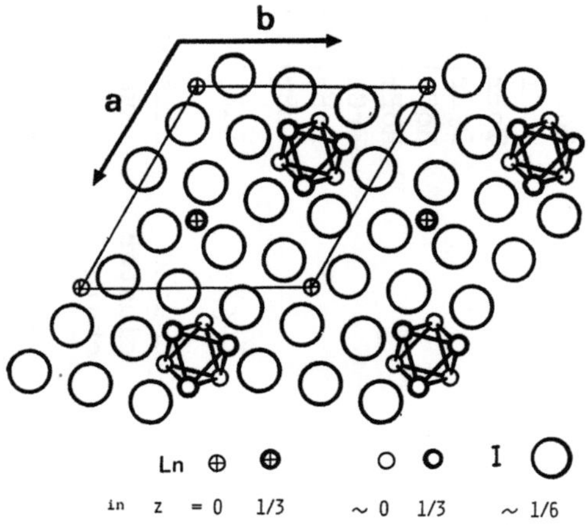

Fig. 3. Projection of a layer of the Ln_7I_{12} structure, which contains isolated octahedral Ln_6 clusters and additional Ln atoms.

The structure consists of fcc packed iodine atoms; a part of them is replaced by Ln_6 clusters. An alternative description leads to fcc packed Ln_6X_{12} clusters. Additionally single Ln atoms occupy octahedral voids (between iodine atoms!).

A qualitative view of the chemical bonding in $Ln_7I_{12} = Ln_6I_{12}\cdot Ln$ results from a comparison with similar compounds of d-metals. The M_6X_{12} type cluster needs 16 electrons for M-M-bonding as in

Nb_6Cl_{14}

$6x5-14 = 16.$

Obviously the cluster is still stable with less electrons as in

Zr_6I_{12} [18]

$6x4-12 = 12.$

But the electron deficiency is too large for a cluster

Ln_6I_{12}

$6x3-12 = 6.$

The additional "intercalated" Ln atoms in Ln_7I_{12} add more electrons to the cluster according to

$Ln_6I_{12}\cdot Ln^{3+}$

$6 + 3 = 9.$

These conditions are similar to what is found in the (electron deficient) Chevrel phases

$Mo_6S_8Ln_x^{3+}$

$6x6-16+3x = 20 + 3x$ (instead of 24 in the case of the M_6X_8 cluster).

The used electron counting procedure (on the basis of a quasi-ionic limit) might seem rather naive, yet it is fully justified by recent band structure calculations for transition metal cluster compounds [23,24].

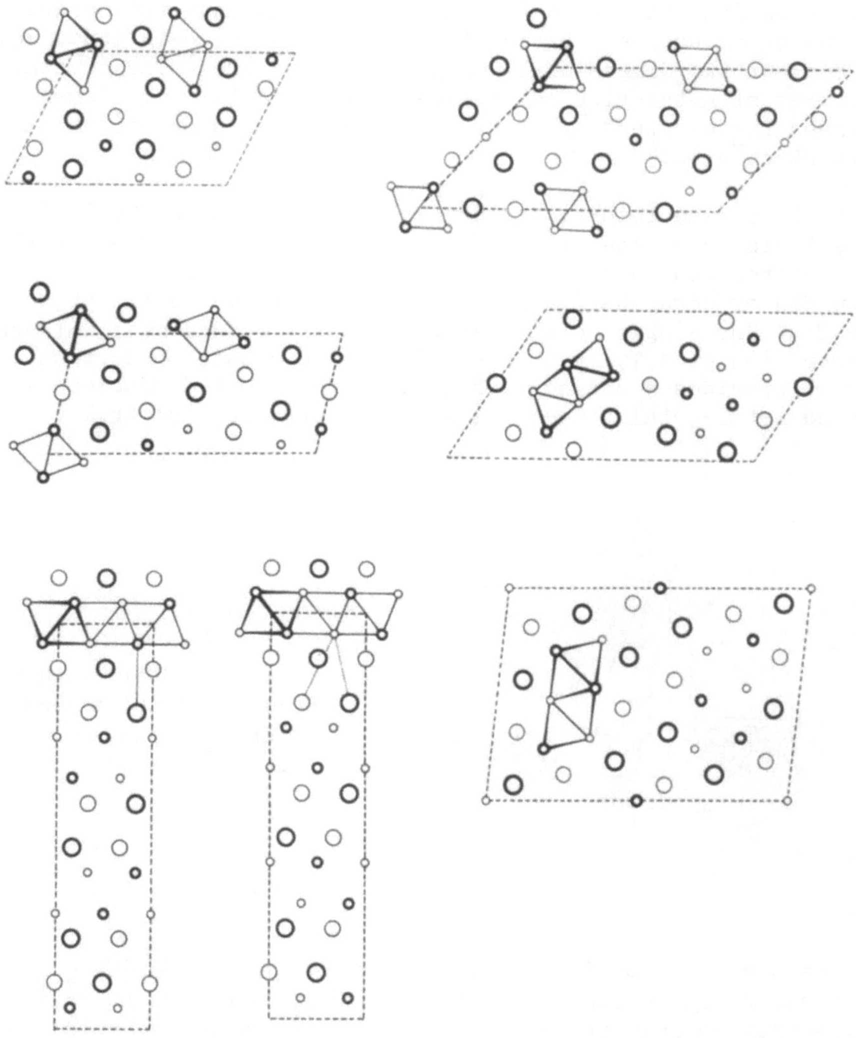

Fig. 4. Projections of the crystal structures of LnX_n ($n < 2$; X = Cl, Br or I). The unit cells are drawn by dashed lines, small circles represent Ln atoms and large circles X atoms. The height z = 0 or 1 and z = 1/2 are indicated by different line strengths. (Taken from [10]). a: Gd_2Cl_3, b: Tb_5Br_8, c: Er_4I_5, d: Er_6I_7, e: Er_7I_{10}, f, g: stacking variants of TbCl.

Defining the role of the single Ln atoms in the structures of metal-rich lanthanide halides as electron donors one arrives at the number of valence electrons per Ln atom which can be used for Ln-Ln bonds. These numbers (last line in Table I) serve as a guideline to order and understand the structures of these compounds. An increase in electron concentration in the cluster at the expense of the halide content leads to a further increase of the cluster size. It is important to note that in all compounds the octahedral Ln_6 cluster occurs as a basic structural unit which is condensed into one- or two-dimensionally extended structures. The crystal structures of metal-rich lanthanide halides containing elements of condensed Ln_6 clusters are shown in Figure 4 as projections along the shortest axes.

As discussed before, the number of electrons in the Ln-Ln bonds of the isolated cluster in Ln_7I_{12} is 9 (1.5 per Ln atom). At the same electron concentration but decreased halide content the infinite chain of edge-sharing Ln_6 octahedra is found in the structure of Gd_2Cl_3. The single octahedral chain still occurs at an electron concentration of 1.75 (with a modified inter-chain connection). Figure 5 provides a stereographic view of the one-dimensionally extended metal-metal bonded units in the structure of Er_4I_5.

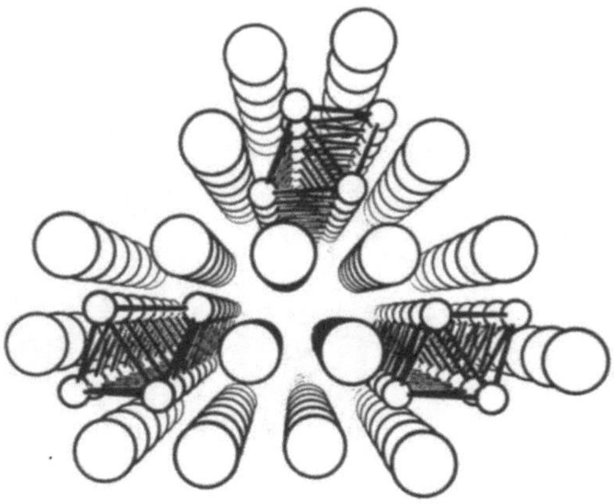

Fig. 5. Stereographic drawing of the structure of Er_4I_5. Lines connect the metal atoms.

The structure of Ln_5Br_8 also contains the single chain of Ln_6 octahedra, which is expected both according to the Br:Ln ratio and the electron concentration of 1.75 calculated with the same assumption as was made for Ln_7I_{12}. When the electron concentration is slightly increased to 1.83, a critical value seems to be passed. Now, two chains of octahedra condense to show the first step into the direction of a two-dimensionally extended array of Ln_6 units. Such fascinating double chains occur in the structures of Er_6I_7 and Er_7I_{10} at rather different X:Ln ratios but at the same electron concentration for Ln-Ln bonding.

A more detailed analysis of the crystal structures derived from condensed Ln_6X_{12} clusters show that the compounds Ln_4X_5, Ln_6X_7 and Ln_5X_8, Ln_7X_{10} belong to different series of a general formula type $Ln_{2a+2}X_{2a+3}$ and $Ln_{2a+3}X_{2a+6}$, where "a" denotes the number of condensed chains. The structures of further members of these series can be predicted in detail, but have not been found so far. Only the modified structures of the final members representing the formula LnX and formed by condensation of an infinite number of chains into layers, could be prepared and characterized.

In the structure of lanthanide monohalides, close-packed double layers of Ln atoms alternate with double layers of close-packed X atoms (Figs. 4 f, g; 6). As the structure suggests, the lanthanide monohalides behave mechanically like graphite and are two-dimensional metals. The electrical conductivity in the plane is larger by a factor of 10^4 as compared to the one in the vertical direction.

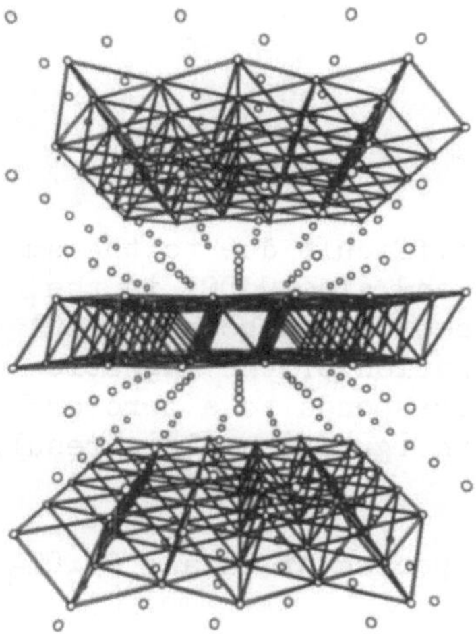

Fig. 6. Stereographic view of the crystal structure of lanthanide monohalides LnX. Lines connect the metal atoms.

CHEMICAL BONDING

The crystal structures discussed in Chapter II are a novelty for lanthanide compounds, yet they are closely related to structures of metal-rich compounds of early (3d), 4d and 5d metals [10]. The structural aspect suggests that we deal with the lanthanides as d-metals in this new field. Some rather preliminary measurements of the physical properties, especially of Gd compounds, verify this point of view, as they clearly indicate a core with a Gd $4f^7$ configuration as in the Gd^{3+} ion.

(a) Magnetic properties: The experimental value of 7.89 μ_B per Gd atom in Gd_2Cl_3 [25] is nearly identical with the theoretical value of 7.94 μ_B for Gd^{3+}. The additional 1.5 electrons per Gd atom do not contribute significantly to the magnetic moment. The large negative Weiss constant (-166 K) indicates a tendency to antiferromagnetic ordering which occurs near 60 K [26]. The reduced magnetic moment of 5.63 μ_B between 25 and 60 K has to be discussed in terms of a spin pairing of 50% of the Gd atoms. Most probably those pairs of Gd atoms are involved which represent the interconnecting edges of the Gd_6 octahedra and have the shortest distances d_{Gd-Gd} = 337 pm in the structure. At 25 K a second process of magnetic ordering is observed, the nature of which is not yet clear.

(b) Mössbauer spectra: ^{155}Gd spectra of $GdCl_3$, Gd_2Cl_3 and GdCl at 4.2 K yield the data summarized in Table II [27].

Table II

Compound	isom.shift [mm·sec^{-1}]	magn.hyperf.field [kØe]
$GdCl_3$	0.632(2)	-
Gd_2Cl_3	0.534(5)	258(5)
GdCl	0.42(1)	211(5)

The accurate evaluation of the spectra is difficult due to the small but not exactly known amounts of Gd, $GdCl_3$ and also GdOCl in the reduced samples. The occurrence of magnetic ordering both in GdCl and Gd_2Cl_3 nevertheless allows the determination of the magnetic hyperfine field which in both compounds corresponds to a zero orbital magnetic moment according to a $4f^7$ configuration. This result emerges also clearly from:

(c) Photoelectron spectroscopy: The He II spectra of Gd and Gd_2Cl_3 are plotted in Figure 7 [28].

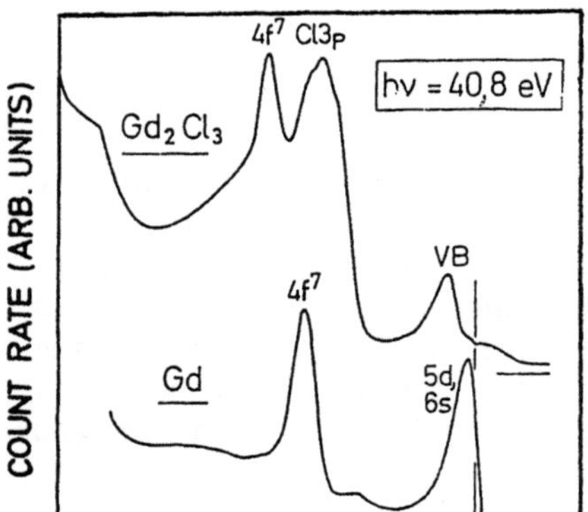

Fig. 7. He II spectra of Gd and Gd_2Cl_3

The spectra of the metal-rich halides contain two features of interest. First, the photoemission from the f-band shows up as a sharp single structure in both Gd and Gd_2Cl_3. This result is expected if the electrons are excited from a core with a (half-filled) Gd $4f^7$ configuration. Otherwise, as in the case of Tb and Tb_2Cl_3, the f-band exhibits a pronounced multiplet structure. Second, no electron density is found at the Fermi level for Gd_2Cl_3. This result is surprising, as the compound contains chains of Gd_6 octahedra with closely spaced metal atoms (337 to 390 pm) and a non-integral number of valence electrons involved in metal-metal bonding. One would therefore expect that Gd_2Cl_3 is metallic (with rather one-dimensional character), similar to GdCl which is a (two-dimensional) metal. But the UPS result has been verified by conductivity measurements which show that Gd_2Cl_3 is a semiconductor with a band gap of 1.1 eV [29]. Although a band structure calculation for Gd_2Cl_3 [30] is in agreement with this result and shows that essentially 5d electrons are involved in metal-metal bonding, further work is necessary to understand the chemical bonding in more detail.

PROSPECTS

Two new results will demonstrate that the metal-rich lanthanide halides offer interesting aspects both in (a) physics and (b) chemistry.

(a) The compounds provide a system of two-, one- and zero-dimensional metal-metal bonded units. The lanthanide ions are strongly coupled via directional bonds which are often shorter than in the pure metals themselves. This leads to interesting magnetic exchange interactions; Figure 8 gives just one example [26].

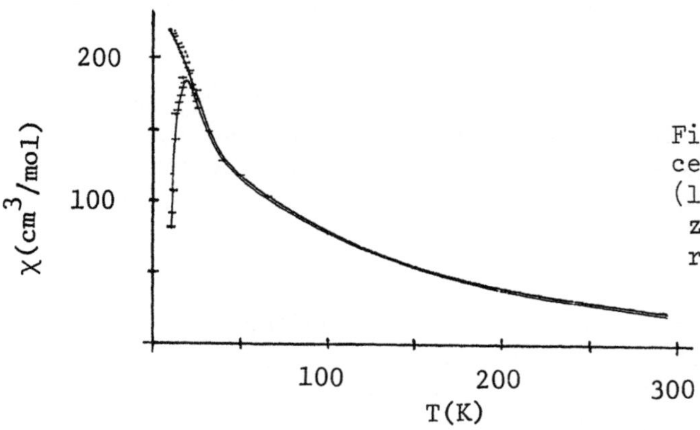

Fig. 8. Magnetic susceptibility of Tb_7I_{12} (lower curve represents zero-field cooled results).

Tb_7I_{12} shows a behaviour which is similar to that of a spin glass. With an applied field of 1 kG the susceptibility increases steadily with decreasing temperature, but a sharp maximum in $\chi = f(T)$ occurs for zero-field cooling. Below 20 K the magnetisation is time dependent. As Tb_6I_{12} contains isolated clusters, intra- and/or inter-cluster exchange might provide an explanation for the magnetic behaviour.

(b) The compounds contain octahedral Ln_6 clusters, in an isolated or condensed state. The new phase $Gd_5Cl_9C_2$ [31, 32] plays the role of a missing link, since it is composed of clusters of two condensed Ln_6X_{12} units. The resulting $Gd_{10}Cl_{18}$ group lies in between the isolated cluster in Ln_7I_{12} and the infinite chain as it occurs e.g. in Er_4I_5 (Fig.5). Furthermore, each cluster center is occupied by two carbon atoms which are linked by (approximately) a single bond.

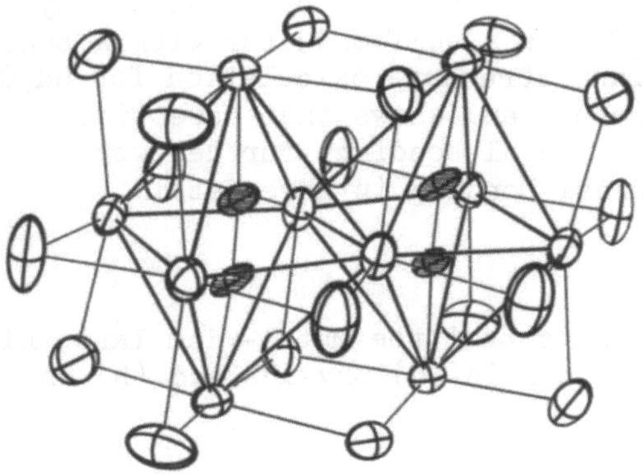

Fig. 9. $Gd_{10}Cl_{18}(C_2)_2$ cluster in the structure of $Gd_5Cl_9C_2$. The metal atoms of the two Ln_6 octahedra are connected by strong lines; the atoms of the interstitial C_2 units are hatched (taken from [32]).

The idea is that the C_2 units add electrons to the cluster system through covalent bonds. $Gd_5Cl_9C_2$ is the first example of lanthanide cluster compounds with interstitial atoms, but (in the meantime) not the only one we have found. In conclusion, we note that metal-rich lanthanide halides seem to be an interesting subject for chemists and physicists.

My thanks are due to W. Bauhofer, K. Berroth, G. Czjzek, G. Ebbinghaus, R. Eger, N. Holzer, R. Masse, Hj. Mattausch, E. Warkentin and W. J. Westerhaus for their valuable contributions to these investigations.

REFERENCES

[1] A. Simon, H. G. Schnering, H. Wöhrle, H. Schäfer Z. anorg. allg. Chem. 339, 155 (1965)
[2] L. R. Bateman, J. F. Blount, L. F. Dahl, J. Amer. Chem. Soc. 88, 1082 (1966)
[3] A. Simon, H. G. Schnering, H. Schäfer, Z. anorg. allg. Chem. 355, 295 (1967)
[4] H. Imoto, A. Simon, Inorg. Chem. [in press]
[5] K. Yvon, Curr. Top. Mater. Sci. 3, 55 (1979)

[6] F. A. Cotton, T. E. Haas, Inorg. Chem. 3, 10 (1964)
[7] J. E. Mee, J. D. Corbett, Inorg. Chem. 4, 88 (1965)
[8] D. A. Lokken, J. D. Corbett, Inorg. Chem. 12, 556 (1973)
[9] A. Simon, N. Holzer, Hj. Mattausch, Z. anorg. allg. Chem. 456, 207 (1979)
[10] A. Simon, Angew. Chem. 93, 23 (1981) [Intern. Ed. 20, 1 (1981)]
[11] K. Berroth, Thesis Stuttgart (1980)
[12] Hj. Mattausch, A. Simon, R. Eger, Rev. Chim. Minér. 17, 516 (1980)
[13] R. E. Aranjo, J. D. Corbett, Inorg. Chem. 20, 3082 (1981)
[14] K. Berroth, A. Simon, J. Less-Common Met. 76, 41 (1980)
[15] Hj. Mattausch, R. Eger, A. Simon, unpublished (1981)
[16] K. Berroth, Hj. Mattausch, A. Simon, Z. Naturforsch. B 35, 626 (1980)
[17] Hj. Mattausch, A. Simon, N. Holzer, R. Eger, Z. anorg. allg. Chem. 466, 7 (1980)
[18] J. D. Corbett, R. L. Daake, K. R. Poeppelmeier, D. H. Guthrie, J. Amer. Chem. Soc. 100, 652 (1978)
[19] K. R. Poeppelmeier, J. D. Corbett, Inorg. Chem. 16, 1107 (1977)
[20] K. R. Poeppelmeier, J. D. Corbett, J. Amer. Chem. Soc. 100, 5039 (1978)
[21] K. R. Poeppelmeier, J. D. Corbett, Inorg. Chem. 16, 294 (1977)
[22] Hj. Mattausch, J. B. Hendricks, R. Eger, J. D. Corbett, A. Simon, Inorg. Chem. 19, 2128 (1980)
[23] O. K. Andersen, W. Klose, H. Nohl, Phys. Rev. B 17, 1209 (1978)
[24] H. Nohl, O. K. Andersen, Inst. Phys. Conf. Sec. 55, 61 (1980)
[25] J. D. Greiner, J. F. Smith, J. D. Corbett, F. J. Jelinek, J. Inorg. Nucl. Chem. 28, 971 (1966)
[26] W. J. Westerhaus, A. Simon, to be published
[27] G. Czjzek, N. Holzer, A. Simon, to be published
[28] G. Ebbinghaus, A. Simon, to be published
[29] W. Bauhofer, A. Simon, to be published
[30] D. W. Bullet, Inorg. Chem. 19, 1780 (1980)
[31] R. Masse, A. Simon, Mat. Res. Bull. 16, 1007 (1981)
[32] A. Simon, E. Warkentin, R. Masse, Angew. Chem. [in press]

COMMENTS

BABERSCHKE: How large is the external field that would change the magnetization curve so dramatically?

SIMON: Approximately 3 kOe.

MAGNETIC PROPERTIES AND QUADRUPOLAR INTERACTIONS IN PrAg

P. Morin and D. Schmitt

Laboratoire Louis Néel, C.N.R.S., 166X
38042 Grenoble-cédex, France

INTRODUCTION

The rare earth intermetallic compound PrAg has the cubic CsCl-type structure (a = 3.718 Å). Bulk magnetization measurements on polycrystalline samples have been interpreted as indicating a ferromagnetic behaviour (1) (T_c = 14 K) whilst neutron diffraction experiments (2) have clearly shown the existence of an antiferromagnetic ordering below T_N = 14 K. On the other hand, the field dependence of the magnetization on a polycrystalline sample above T_N has revealed a behaviour which has been explained by the presence of biquadratic exchange coupling (3).

We have therefore investigated more thoroughly the bulk magnetic properties of annealed poly- and single crystals of PrAg. From the whole set of experimental results (neutron diffraction, magnetization, susceptibility) it has been found that a weak ferromagnetic component ($\mu_F \sim 0.3\ \mu_B$) coexists with a large antiferromagnetic component ($\mu_{AF} = 2.4\ \mu_B$), removing the apparent inconsistency of the previous papers. In addition the detailed analysis of the magnetization curves along the [100] and [111] directions above T_N has been made within the third-order magnetic susceptibility formalism (4). Strong negative quadrupolar interactions have been revealed.

EXPERIMENTS ON A POLYCRYSTALLINE SAMPLE

Three types of experiments have been carried out in order to state the magnetic properties and the crystal field precisely: resistivity, neutron spectroscopy and neutron diffraction.

Resistivity measurements

Resistivity measurements were performed between 4.2 and 300 K on a staggered grid in order to obtain a good accuracy. The low temperature part indicates clearly a second order transition at T_N = 10.4 K. No anomaly was detected between 4.2 K and T_N.

Neutron spectroscopy

Neutron inelastic scattering experiments were performed on the IN4 time-of-flight spectrometer of the Institut Laue-Langevin (Grenoble). Spectra were taken at 20 and 40 K with incident energies E_o = 12.6 and 50.4 meV. The analysis of the observed transitions (energy transfer, relative intensity and temperature dependence) confirms without ambiguity the previous crystal field scheme (2,3). With the usual notation, W = -5.4 K and x = -0.79.

Neutron diffraction

Neutron diffraction experiments were performed on the multidetector of the SILOE neutron reactor (Grenoble). Numerous spectra were taken between 2 and 20 K, with an incident neutron wavelength of 1.323 Å. The main magnetic structure is confirmed with regard to a previous determination (2): it is antiferromagnetic, with a propagation vector $\vec{q}$ = (1/2 1/2 0) 2π/a; the magnetic moments reach μ_{AF} = 2.4 ± 0.2 μ_B at 2 K and lie in the (001) plane if the structure is collinear. However, other multiaxis (non-collinear) configurations may exist (5), giving the same neutron diffraction diagrams. In particular, solutions exist with magnetic moments lying along the [100], or the [110] or the <111> directions. The temperature dependence of the (1/2 1/2 0) reflection confirms the Néel temperature T_N = 10.4 K. However a detailed analysis of the nuclear reflections shows a very weak increase of their intensity at 2 K relative to the spectra taken at 5.8 K and above. This corresponds to a ferromagnetic moment of 0.3 ± 0.2 μ_B.

EXPERIMENTS ON A SINGLE CRYSTAL

These experiments have consisted of first- and third-order magnetic susceptibility and magnetization measurements along the main symmetry directions.

First-order magnetic susceptibility

Magnetic susceptibility measurements have been carried out for the three main cubic symmetry directions in 1045 Oe. In addition the [100] direction has been investigated in very low fields (53 and 103 Oe).

In the paramagnetic range the susceptibility follows a Curie-Weiss law and the curve is fitted with the crystal field parameters given above and a bilinear exchange parameter $\Theta^* = 11 \pm 0.5$ K. It is surprising to observe such a strongly positive value while the structure is essentially antiferromagnetic.

Below T_N the susceptibility is anisotropic (Fig. 1a). Along the [100] direction there is no susceptibility maximum at T_N in 1045 Oe. However this maximum occurs in very low fields (Fig. 1b). In addition in the range 7 K - 10.4 K the susceptibility increases with the field indicating a metamagnetic behaviour; but below 7 K the susceptibility is the largest in low fields, as observed for a ferromagnet.

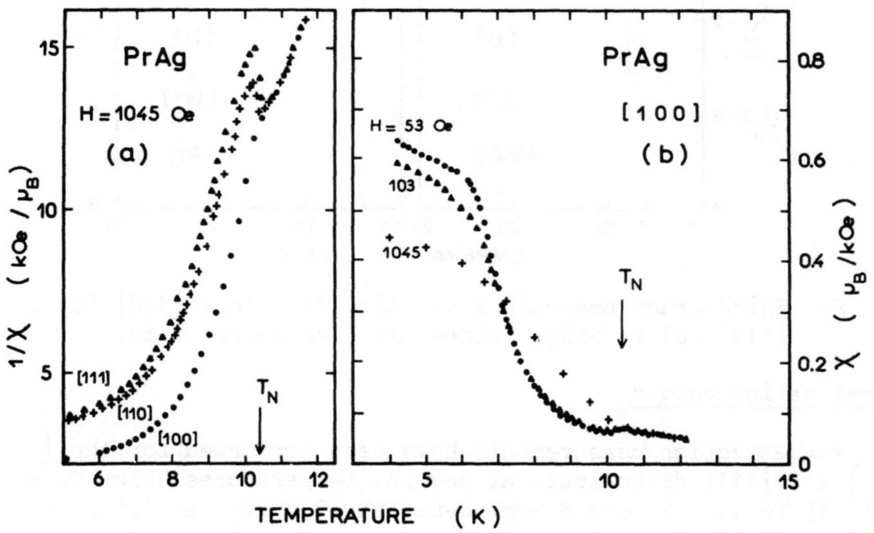

Fig. 1. (a) Reciprocal magnetic susceptibility along the three symmetry directions in PrAg ; (b) susceptibility along [100] for different fields.

Third-order magnetic susceptibility

The third-order magnetic susceptibility has been extracted from a detailed analysis of M/H versus H^2 curves in the temperature range 11 K - 25 K, for the [100] and [111] directions; it has been analysed within an analytical formalism (4).

For the [100] direction (Fig. 2a), the fit is not very selective relative to the tetragonal quadrupolar parameter G_1 (6) which should be between 0 and -1 K. However, values between 0 and -0.1 K are more likely, because larger values would lead to instabilities of the magnetization. For the [111] direction (Fig. 2b) fits are more selective and we can evaluate the trigonal quadrupolar parameter (6) $G_2 = -0.6 \pm 0.2$ K. We note that the third-order magnetic susceptibility is very anisotropic, the experimental values being roughly twice larger along [100] than along the [111] direction in the range 11-20 K.

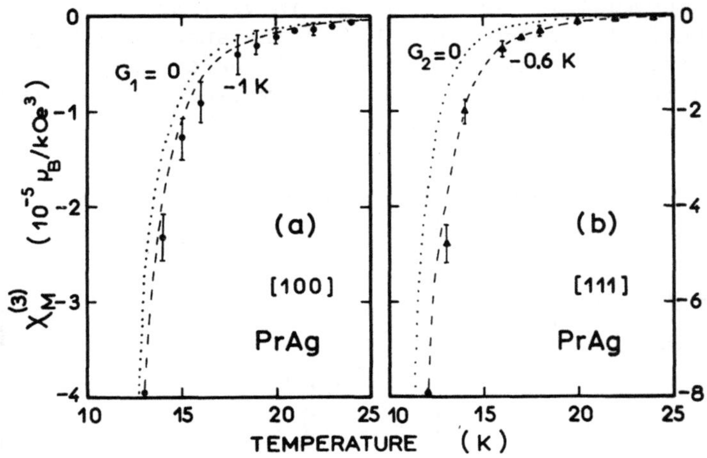

Fig. 2. Third-order magnetic susceptibility along [100] (a) and [111] (b) in PrAg; lines are theoretical fits.

Magnetization curves

Magnetization measurements have been performed for the [100], [110] and [111] directions, at several temperatures below T_N and up to 15 Teslas. Figure 3 represents the 3 curves at 1.6 K, showing critical fields the values of which range from 4.5 to 12 Teslas according to the field direction. In addition, low field measurements have shown a critical field of the order of 0.1 T at 5 K along [100] (see below). The temperature dependence of these critical fields is reported on the inset of figure 3. Note that in Ref. 3, only one critical field was observed at 0.45 T with a polycrystalline sample. These critical fields may be associated with spin-flop transitions between the zero-field magnetic structure and either a ferromagnetic state or other intermediate (canted) structures.

The low field curves (below 0.5 T) present an anomalous behaviour (Fig. 4). At 9 K the curvature is positive for the [100] direction and negative for the [110] and [111] directions, but there is no evidence of spontaneous magnetization in zero field. At 5 K the behaviour is quite ferromagnetic for the 3 directions, a transition occurring at 0.1 T for the [100] curve. The sponta-

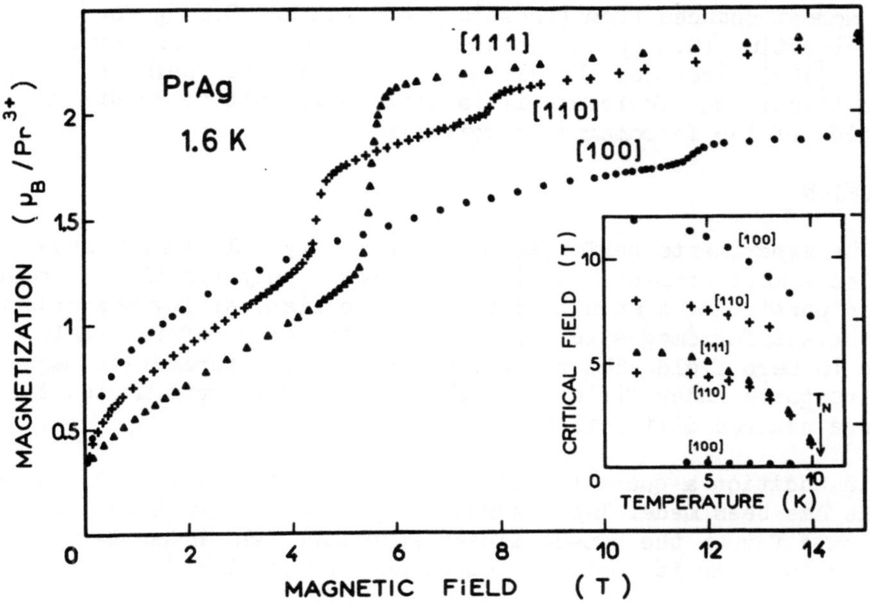

Fig. 3. Magnetization curves along the three symmetry directions at 1.6 K, in PrAg. Inset : temperature variation of the critical fields (no value is available for the low critical field along [100] below 4 K).

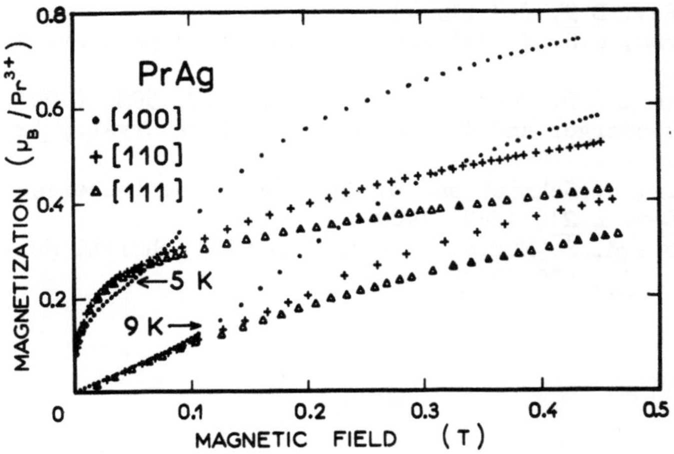

Fig. 4. Magnetization curves of PrAg along the three symmetry directions in low fields, at 5 and 9 K.

neous moment deduced from Arrott's plots reaches 0.2 μ_B for the [111] direction (0.3 μ_B at 1.6 K), this moment vanishing around 7 K. For the [100] direction the situation is less clear perhaps because of the transition. Therefore it is still difficult to secure the direction of the ferromagnetic component.

CONCLUSION

The experiments performed on a single crystal of PrAg have revealed a more complex behaviour than for a polycrystalline sample (3). In particular a spontaneous moment coexists at low temperature with the antiferromagnetic structure. PrAg is therefore weakly canted in zero field. The determination of the intermediate magnetic structures under field require complementary experiments, for instance neutron diffraction under field.

In addition a quantitative evaluation of the quadrupolar interactions has been made. The quadrupolar parameters are negative. This could favour the non-collinear structure, which has been observed in other isomorphous compounds such as TbMg (7).

ACKNOWLEDGMENTS

We would like to thank Dr. R. Aléonard and Dr. P. Lethuillier for their assistance with the resistivity, magnetization and susceptibility measurements.

REFERENCES

1. R.E. Walline and W.E. Wallace, J. Chem. Phys., 41, 3285 (1964).
2. T.O. Brun, G.H. Lander, D.L. Price, G.P. Felcher and J.F. Reddy, Phys. Rev. B 9, 248 (1974).
3. T.O. Brun, J.S. Kouvel and G.H. Lander, Phys. Rev. B 13, 5007 (1976).
4. P. Morin and D. Schmitt, Phys. Rev. B 23, 5936 (1981).
5. M. Wintenberger and R. Chamard-Bois, Acta Cryst., A28, 341 (1972).
6. P. Morin, D. Schmitt and E. du Trémolet de Lacheisserie, Phys. Rev. B 21, 1742 (1980).
7. R. Aléonard, P. Morin, J. Pierre and D. Schmitt, Sol. Stat. Commun., 17, 599 (1975).

EXPERIMENTAL DETERMINATION OF THE ELECTROSTATIC CONTRIBUTION TO THE CRYSTALLINE ELECTRIC FIELDS IN NON-CUBIC METALS

R.A.B. Devine*† and Y. Berthier§

*Physics Department, University of Miami, Coral Gables, Florida 33124 USA
§Laboratoire de Spectrométrie Physique, CNRS, BP 53 X, 38041 Grenoble Cedex, France

It was demonstrated very quickly that the point charge model for the crystal field seen by 4f electrons in metals failed[1]. Progressing naturally, the effects of conduction electrons were next included taking into account both Coulomb and exchange terms[2]. However, comparison between experimental and theoretical results is still poor even when using sophisticated theoretical approaches[3] (for example full band structure calculations). From a detailed study of S state ions in metals and insulators[4] it has been concluded that there may be yet another important contribution to the crystal field, omitted to date, namely covalency. Experimentally it is very difficult to devise experiments from which we can hope to deduce the individual contributions to the crystal field in a direct manner. The purpose of this paper is to demonstrate that it may in fact be possible to gain some insight into this problem using nuclear quadrupole hyperfine interaction measurements together with neutron diffraction or other crystal field data.

The nuclear quadrupolar Hamiltonian for a rare earth nucleus of spin, I, in axial symmetry is given by

$$\mathcal{H}_Q = P_{4f}(3I_z^2 - I^2) + P_L(3I_z^2 - I^2) \qquad (1)$$

where P_{4f} arises from the electric field gradient[5] (EFG) at the nucleus due to the localised 4f electron. In the simplest (insulator like) model, P_L arises from field gradients at the nucleus due to charges in the surrounding lattice. P_L can be written[6] (under these assumptions)

---

†Present address: CNET, B.P. 42, 38240 Meylan, France.

$$P_L = \frac{e^2Q}{4I(2I-1)} \frac{(1-\gamma_\infty)}{(1-\sigma)} \left(\frac{-4A_2^\circ}{e^2} \right) . \quad (2)$$

A_2° is the second order crystalline electric field parameter. (We shall consider that the effects of conduction electrons non-local to the rare earth nucleus can be engulfed in a point charge like term). Note that the quadrupolar splitting manifests directly the presence of a crystalline electric field effect both in the P_{4f} and P_L terms. Deduction of P_L, present in symmetry lower than cubic, should access directly the second order crystal field parameter.

Following Bleaney[5] P_{4f} is written (for $P_{/\!/}$ see Ref. 5):

$$P_{4f} = \frac{P_{/\!/}}{3} \frac{\langle 3J_z^2 - J(J+1) \rangle}{J(2J-1)} . \quad (3)$$

In order to deduce P_L, and hence A_2°, we must separate the two terms in equation 1. P_{4f} depends upon temperature (through variation of $\langle 3J_z^2 - J(J+1) \rangle$) and P_L does not, so that in principle, measurements as a function of temperature extrapolated to $T = \infty$ or temperature dependence fits, yield P_L. In reality, it is sufficient to go to temperatures larger than the overall crystal field splitting to obtain P_L (relaxation time permitting). Various authors[6,7,8] have used this and similar techniques to deduce crystal field parameters in metallic systems. Whether we consider a metallic or insulating system has no consequence because we can always separate our quadrupolar energy into one part coming from the 4f electrons and one from the "rest" ie. lattice charges plus conduction electrons if present.

The question now arises as to whether or not there is a direct relationship between the second order crystal field parameter, A_2°, determined by this method and by other methods. The most direct approach is to compare results obtained by, for example, neutron diffraction and hyperfine parameter measurements. Unfortunately there exist few hyperfine results for dilute, paramagnetic 4f systems having axial symmetry. Mössbauer measurements using the source method, have been made on ^{166}Er diluted in Y by Stohr and Wagner[9]. Detailed susceptibility[10] and neutron diffraction[11] measurements have also been made on this system so we can directly compare values for A_2°.

Because of relaxation effects the Mössbauer spectrum of ^{166}Er could not be measured to high temperatures and the $T = \infty$ extrapolation made. We must therefore deduce the 4f contribution to the measured quadrupolar spectrum by other means.

Using crystal field parameters from references 10 and 11 we have calculated $<3J_z^2 - J(J+1)>$ as a function of temperature from which, using equation (3), the 4f part of the quadrupolar energy has been deduced. We have deduced ${}^{166}P_{\parallel}$ from Bleaney's value for ${}^{167}P_{\parallel}$ by scaling with appropriate nuclear spins and moments. We deduce P_{4f} to be 2.02 MHz at 1.4 K. It should be noted that this value differs significantly from that deduced by Stohr and Wagner[9] because the crystal field parameters available at that time[12] were later found to be incorrect[10]. The measured value[9] is P_{EXP}=14.5±2MHz so that by subtraction we deduce

$$P_L = P_{EXP} - P_{4f} = 12.5 \pm 2 \text{ MHz} \tag{4}$$

where we assume P_L contains lattice and conduction electron effects. Since all charges contributing to the EFG in P_L may be safely considered to lie well away from the nucleus, we can deduce A_2° from the experimental value of P_L using equation (2). (Note that this condition is not in general satisfied when considering the electric field seen by 4f electrons which are much less localised.) Following Sternheimer[13] we take $\gamma_\infty = -80$ and $\sigma = 0.63$ from Gupta et al[14]. Crystal field parameters deduced from other experiments are usually obtained in the product form $A_n^\circ <r^n>$. Since they pertain to the crystal field seen by a particular ion, therefore we deduce $A_2^\circ <r^2>$ for Er by multiplying our value of A_2° by[15] 0.666 a.u.. We obtain $A_2^\circ <r^2> = 384 \pm 61$ K.

The value of $A_2^\circ <r^2>$ obtained from susceptibility[10] and neutron diffraction[11] results is $A_2^\circ <r^2> = -111 \pm 13$ K and we note immediately that there is both a size and sign disagreement between this value and that we have obtained. At first glance, there would appear to be a gross inconsistency in that we used the value of $A_2^\circ <r^2> = -111$ K to deduce the 4f wavefunction which subsequently enabled deduction of P_{4f}. We believe that this is not in fact inconsistent for reasons upon which we shall elaborate. We must first ask the question, are all of the component parts making up the crystal field seen by a 4f electron equally seen by the nucleus? This leads to the question, what is seen by the 4f electron? It has been assumed that there are three major contributions to the crystal field felt by a 4f electron. Limiting ourselves to second order terms we have

$$A_2^\circ \text{ (total)} = A_2^\circ \text{ (P.C.)} + A_2^\circ \text{ (C.E.C.)} + A_2^\circ \text{ (C.E.E.)}. \tag{5}$$

P.C. = Point Charge
C.E.C. = Conduction electron Coulomb
C.E.E. = Conduction electron exchange

The last term is an effective crystal field parameter due to the exchange interaction between the 4f and conduction electrons[16]. It is clear that since the nucleus senses only charge effects, this latter contribution must be absent in any value of A_2° we deduce from nuclear measurement. (It is contained however, in as much as it influences $<3J_z^2 - J(J+1)>$ which means it comes in through the P_{4f} term). If we now subtract the value of A_2° obtained from the Mössbauer results from the magnetisation /neutron results we should obtain an estimate of the conduction electron/ 4f exchange contribution. We obtain

$$A_2^\circ <r^2> \text{(C.E.E.)} = -495 \pm 74 \text{ K} \tag{6}$$

(bearing in mind that the Mössbauer result was obtained using a theoretical value of $<r^2>$).

In order to go further we must obtain a value for $A_2^\circ <r^2>$ (Point Charge). This can be calculated and yields $A_2^\circ<r^2>=-40.6\ Z'K$ where Z' is the effective neighbouring charge. A priori we have no idea of the sign or magnitude of Z'. However, we can obtain some insight into the point charge term by examining nuclear hyperfine results for non-4f ions. For a non-4f ion such as Sc not only are the 4f electrons absent (A_2°(C.E.E.) = 0) but also the important 5d electrons which give rise to significant[2] contributions to A_2°(C.E.C.) as seen by the 4f electron. A non-4f ion then sees only A_2°(P.C.) where we do however assume that conduction electron contributions non-local to the Sc ion are present and simply form part of A_2°(P.C.).

Nuclear magnetic resonance measurements have been made[17] on ^{45}Sc in Y. From the measured quadrupolar splitting the EFG due to the Y environment can be deduced and we find A_2°(P.C.)$=-3.55\times10^{19}$ K/cm^2. The measured splitting of 5.2 MHz is undetermined as to sign so that we must admit that $A_2^\circ = \mp 3.55 \times 10^{19}$K/cm^2. Assuming nothing happens to this term when we replace Sc by Er

$$A_2^\circ \text{ (P.C.)} <r^2> = \mp 66\text{K} \tag{7}$$

(note we again use $<r^2>$ as per reference[15])
This value should correspond to what we have termed in equation 5, the point charge value. Bearing in mind imprecisions which may be due to our lack of exact knowledge of $<r^2>$ we obtain by subtraction.

$$A_2^\circ\text{(C.E.C.)} = \begin{cases} 450 \pm 87 \text{ K } (A_2^\circ\text{(P.C)} <0) \\ 218 \pm 87 \text{ K } (A_2^\circ\text{(P.C.)}>0) \end{cases} \tag{8}$$

There is clearly a significant range of values for $A_2^0 <r^2>$ (C.E.C.). Some of the uncertainty could easily be removed if we could determine the sign of the quadrupolar splitting in Sc. We might bear in mind the point charge calculations which gave $A_2^0 \langle r^2 \rangle$ (P.C.) = -40.6 Z'K. Since it seems reasonable to assume the Y ions are positively charged we sould expect $A_2^0 \langle r^2 \rangle < 0$. In fact the value of -66 K is not unreasonable since it predicts Z' $\simeq$ 1.5 positive charge which is consistent with trivalency allowing for some conduction electron screening of neighbouring ions. Most probably then we should deduce

$$A_2^0 \langle r^2 \rangle \text{ (C.E.C.)} \simeq + 450 \pm 87 \text{ K}. \qquad (9)$$

According to atomic calculations[2], the ratio

$$\frac{A_2^0 \langle r^2 \rangle \text{ (C.E.E.)}}{A_2^0 \langle r^2 \rangle \text{ (C.E.C.)}} \simeq -0.85\,. \qquad (10)$$

Our crude values give - 1.1 which is of correct sign but too large. This may be because we have made too many simplifying assumptions or that some other term is present. At the present time one cannot state with certainty that some other term is present but if one admits the possibility, then it seems most likely that the covalency recently mentioned[4] may well indeed be contributing.

In conclusion we find that nuclear hyperfine measurements can yield different information than the susceptibility or neutron data on the crystal field parameters. It is not inconsistent that they be different, indead there are physical justifications and these differences may be exploited to yield detailed information on the origin of the crystal field. More measurements are needed to explore these points.

REFERENCES

1. G. Williams and L.L. Hirst, Phys. Rev. 185 407 (1969)
2. H.C. Chow, Phys. Rev. B7 3404 (1973)
3. D. Schmitt, Thèsis, Université Scientifique et Médicale de Grenoble (unpublished) (1979)
4. S.E. Barnes, K. Baberschke and M. Hardiman, Phys. Rev. B18 2409 (1978)
5. B. Bleaney in "Physics of Rare-Earth Metals" ed. R.J. Elliott, Chap. 8 (1972)
6. R.G. Barnes, R.L. Mössbauer, E. Kankeleit and J.M. Poindexter, Phys. Rev. 136 A 175 (1964)
7. D.L. Uhrich and R.G. Barnes, Phys. Rev. 164 428 (1967)
8. L.H. Hirst and E.R. Seidel, J. Phys. Chem Sol. 31 857 (1970)
9. J. Stohr and W. Wagner, J. Phys. F5 812 (1975)

10. P. Touborg, Phys. Rev. B16 1201 (1977)
11. O. Rathmann and P. Touborg, Phys. Rev. B16 1212 (1977)
12. J. Høg and P. Touborg, Phys. Rev. B9 2920 (1974)
13. R.M. Sternheimer, Phys. Rev. 146 140 (1966)
14. R.P. Gupta, B.K. Rao and S.K. Sen, Phys. Rev. A3 545 (1971)
15. This value of $<r^2>$ for Er^{3+} is open to some discussion but is the value usually quoted from the theoretical calculations A.J. Freeman and R.E. Watson, Phys. Rev. 127 2058 (1962)
16. N.L. Huang Liu, K.J. Ling and R. Orbach, Phys. Rev. B14 4087 (1976)
17. R.G. Barnes, F. Borsa, S.L. Segel and D.R. Torgeson, Phys. Rev. 137 A 1828 (1965)

COMMENTS

WOHLLEBEN: What other experimental methods can you think of to tackle this very important problem of identifying the various contributions to the CEF parameters?

BERTHIER: It seems that the simplest way to separate the different contributions is to use two different probes which can sense differently the electrostatic and the exchange part of the conduction electron contribution. Furthermore, these probes must be sensitive enough in order to be used for dilute alloys. For these reasons, very few experimental data yielding useful information to this problem can be found.

CRYSTAL FIELD INFLUENCE ON THE SPECIFIC HEAT AND SCHOTTKY EFFECT IN RARE EARTH MONOSULFIDES

L.N. Vasil'ev, A.V. Golubkov, A.G. Gorobetz, V.S. Oscotsky, I.A. Smirnov and V.V. Tikhonov

A.F. Ioffe Physico-Technical Institute, Academy of Sciences of the USSR, Leningrad, USSR

INTRODUCTION

The rare earth monochalcogenides have a wide homogeneity range[1-3]. In the monosulfides (LnS_x) it ranges from x=1 to x=0.90 (LaS_x) to x=0.84 (PrS_x) and to x=0.78 (NdS_x). We have investigated the influence of the deviation from stoichiometry on the specific heat and the Schottky effect in RE monosulfides.

EXPERIMENTAL DETAILS

LnS_x (Ln=La, Pr and Nd) were prepared by synthesis from elements as described in Ref. 4. RE metals of 99.9% purity and sulfur with an impurity content of less than 100 ppm were used. Chemical and x-ray phase analysis, density measurements and investigations by a metallographic microscope to control the sample quality were performed. All samples were well formed and had an NaCl-type crystal structure. The density of the stoichiometric samples determined from x-ray analysis coincides with the experimental ones. The lattice constant values for all samples were in good agreement with those from Ref. 5.

The heat capacity was measured in a vacuum adiabatic calorimeter over a temperature range from 2 to 300 K[6].

EXPERIMENTAL RESULTS AND DISCUSSION

The compositions $LaS_{1.0}$, $LaS_{0.90}$, $PrS_{1.0}$, $PrS_{0.94}$, $PrS_{0.89}$, $NdS_{1.0}$, $NdS_{0.89}$ and $NdS_{0.79}$ were all investigated. The Schottky contribution to the specific heat was obtained, as usual, by subtraction of the LaS_x specific heat from the specific heat of

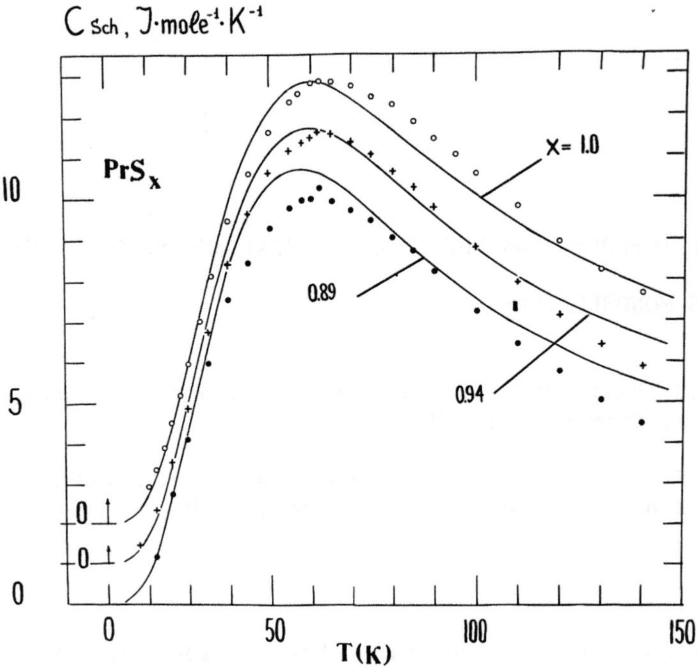

Fig. 1: The Schottky heat capacity of PrS_x. Solid lines give theoretical calculations. The C_{Sch} scales for $PrS_{0.94}$ and $PrS_{1.00}$ are shifted by 1 and 2 J/mole-K respectively.

PrS_x or NdS_x. The Schottky specific heat (C_{Sch}) was used for the determination of the paramagnetic levels of the RE ions, split in a cubic crystal field. The Hamiltonian for the 4f - electrons has the form[7]

$$H=W[x \frac{O_4}{F(4)} + (I-|x|) \frac{O_6}{F(6)}]. \qquad (1)$$

Here, W and x are parameters characterizing a splitting of the energy levels in a cubic crystal field, O_4 and O_6 are the equivalent operators of the fourth and sixth order[8], and F(4) and F(6) are constants. W and x were determined numerically by least square fitting through a comparison of experimental and calculated values of C_{Sch} at different temperatures. Results are given in Figs. 1-3 and Table 1.

The splitting scheme of the ground state depends on its term

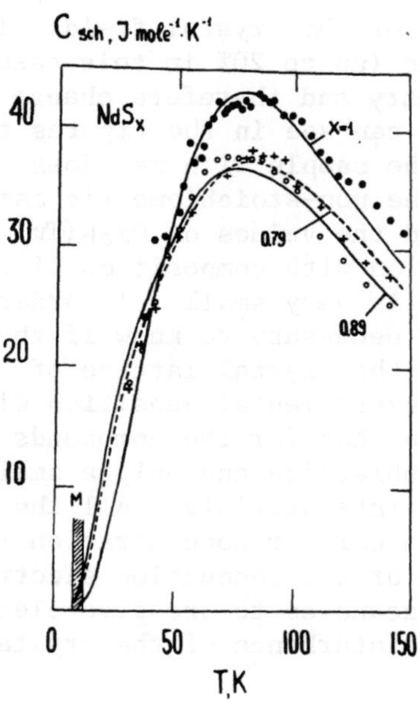

Fig. 2: The Schottky heat capacity of NdS_x. Solid and dashed lines are theoretical calculations; M is the range of the magnetic phase transitions.

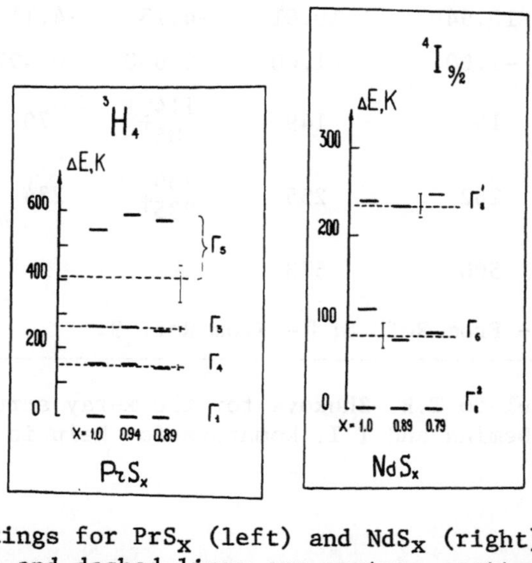

Fig. 3: Splittings for PrS_x (left) and NdS_x (right). Heavy lines are our data and dashed lines are neutron scattering results for stoichiometric samples (Ref. 8).

and the local symmetry of the crystal field. The presence of a large number of defects (up to 20% in this case) must significantly change the local symmetry and therefore change the magnitude of a splitting. But as one can see in the figures the temperature dependence of C_{Sch} for the sample with x=1 does not differ significantly from C_{Sch} for the non-stoichiometric samples, the position of the maximum (T_m) and the values of $C_{Sch}(T_m)$ being practically the same. Thus the variation with composition of the position of the RE paramagnetic levels is very small. In order to explain this important result it is necessary to know if there are Frenkel or Schottky defects in the crystal lattice of samples with x<1. The comparison of the experimental densities with the x-ray ones leads to the conclusion that for the compounds LnS_x (x<1), vacancies are created in the S-sublattice and only a small number (5-12%) of the excess RE ions are interstitials. All the compounds investigated are metals with a carrier concentration (1-2) x $10^{22}cm^{-3}$. The space distribution of the conduction electrons has a maximum at positively charged S-vacancies to preserve electroneutrality. This leads to only a small disturbance of the crystal field near a vacancy.

Table 1: Crystal Field Parameters and Splittings for the 4f electron levels of PrS_x and NdS_x

	PrS	$PrS_{0.94}$	$PrS_{0.89}$	NdS	$NdS_{0.89}$	$NdS_{0.79}$
W (K)	10.047	10.94	10.61	-4.13	-4.11	-4.37
x	-0.979	-1.00	-1.00	0.630	0.692	0.687
E_1(K)	153 153*	153	149	114 85†	79	87
E_2(K)	262 262*	262	255	239 235†	234	249
E_3(K)	544 408*	590	573	-	-	-

* - From Ref. 8; † - From Ref. 9.

We are grateful to T.B. Zhukova for the x-ray structure analysis and M.A. Demina and T.I. Komarova for help in preparing the samples.

REFERENCES

1. G.F. Trammel, Phys. Rev. 131, 932 (1963).
2. K.W.H. Stevens, E. Pytte, Sol. State Commun. 13, 101 (1973).
3. M. Guittard, Compt. Rend. Acad. Sci. 261, 2109 (1965).
4. A.V. Golubkov, V.M. Sergeeva, T.B. Zhukova, Neorg. Mater. 2, 77 (1966).
5. A. Bruzzone, G.L. Olcese, Editiones CNRS Colloque 157, 387 (1965).
6. R.G. Mitarov, V.V. Tikhonov, L.N. Vasil'ev, A.V. Golubkov, I.A. Smirnov, Phys. Stat. Sol. (a) 30, 457 (1975).
7. K.R. Lea, M.J.M. Leask, W.P. Wolf, J. Phys. Chem. Sol. 23, 1381 (1962).
8. K.C. Turberfield, L. Passell, R.L. Birgeneau, E. Bucher, J. Appl. Phys. 42, 1746 (1971).
9. A. Furrer, E. Warming, J. Phys. C: Solid State Phys. 7, 3365 (1974).

INTERPRETATION OF THE ζ-HOLMIUM SESQUISELENIDE MAGNETIC SUSCEPTIBILITY

L. Pawlak, M. Duczmal and S. Pokrzywnicki

Institute of Inorganic Chemistry and Metallurgy
of Rare Elements, Technical University
50-370 Wrocław, Poland

The magnetic properties of the heavier lanthanide sesquiselenides were studied as early as 1937 by Klemm and Koczy [1] and later in the 1970's by a group of Soviet investigators[2,3]. The measurements were carried out down to liquid nitrogen temperature, and they gave linear plots of the temperature dependence of the reciprocal magnetic susceptibility. The widening of the measurement range down to helium temperatures makes a more detailed interpretation of magnetic properties possible, relating magnetic susceptibility to crystal field effects. Such an analysis becomes simplest in the case of compounds with a high symmetry of the magnetic ion environment. This is the reason for our interest in sesquiselenides with the Sc_2S_3 crystal structure, since they are the ones which demonstrate an octahedral environment of metal ions. The orthorhombic Sc_2S_3 structure arises from the NaCl structure by removing 1/3 of the cations in an ordered manner. The first coordination sphere of each metal ion contains six selenide ions forming a slightly distorted octahedron. The deviations of the lattice constant ratios a/b and c/b from their ideal values, being equal to $\sqrt{2}$ and 3 respectively, can be taken as a measure of that distortion.

The Sc_2S_3 structure is characteristic of the lanthanide sesquiselenides of the elements from holmium to lutetium as well as Sc_2Se_3 and Y_2Se_3[4]. Magnetic susceptibility studies of Yb_2Se_3, Tm_2Se_3 and Er_2Se_3 in the liquid helium to room temperature range were done earlier[5-7]. In this paper we present the results obtained for the last isostructural compound in the series, i.e. holmium sesquiselenide.

EXPERIMENTAL DETAILS

Ho_2Se_3 was obtained by a direct reaction of holmium metal (99.9%, Koch Light Labs.) with elemental selenium (5 N, POCH) in sealed quartz ampules. The syntheses were carried out at 1200 K and then the samples were homogenized at 1050 K for one week. The polycrystalline Ho_2Se_3 obtained was dark brown. Lattice constants measured by the Guinier method were: a=1.141 nm, b=0.813 nm and c=2.427 nm, which is in good agreement with the literature data[4,8]. A slight deviation of the lattice constants ratios, a/b=1.404 and c/b=2.985, from their theoretical values ($\sqrt{2}$ and 3, respectively) indicate an insignificant distortion of the octahedral symmetry for the holmium ions. This enabled us to approximate the crystal field in Ho_2Se_3 with a cubic field for subsequent analysis.

Magnetic susceptibility was measured by the Faraday method, using $HgCo(SCN)_4$ as a standard. The measurement within the temperature range 4.2-300 K was carried out in a 50 mT magnetic field. The magnetic susceptibility of the sample showed no dependence on magnetic field at room temperature. The results of the measurements are shown in Fig. 1. No magnetic ordering was found to occur down to the lowest temperatures.

The magnetic moment and Weiss constant were determined using the linear part of the $\chi^{-1}(T)$ plot, and they are equal to 10.67 μ_B and 0 K, respectively. The theoretical value of the free ion magnetic moment is 10.61 μ_B, whereas those given in the literature were 10.78[2] and 11.00 μ_B[3].

DISCUSSION

The ground term of the Ho^{3+} ion, $^5I_8$, splits in a regular crystal field into seven sublevels, viz. one singlet, two doublets, and four triplets. The energies of those sublevels can be calculated as eigenvalues of the cubic crystal field Hamiltonian:

$$\hat{H}_{cf}=B_4(\hat{O}_4^0+5\hat{O}_4^4)+B_6(\hat{O}_6^0-21\hat{O}_6^4), \tag{1}$$

where $\hat{O}_n^m$ stands for equivalent operators, and B_4 and B_6 are the crystal field parameters. The latter are connected with the frequently used W and x parameters of Lea, Leask and Wolf, by the following relations[9]:

$$B_4=A_4\langle r^4\rangle\beta=\frac{Wx}{F(4)};\quad B_6=A_6\langle r^6\rangle\gamma=\frac{W(1-|x|)}{F(6)}, \tag{2}$$

where F(4), F(6), β and γ are numerical coefficients. The crystal field parameters $A_n\langle r^n\rangle$ can be calculated employing the point charge model. For an octahedral environment of the central ion, one obtains

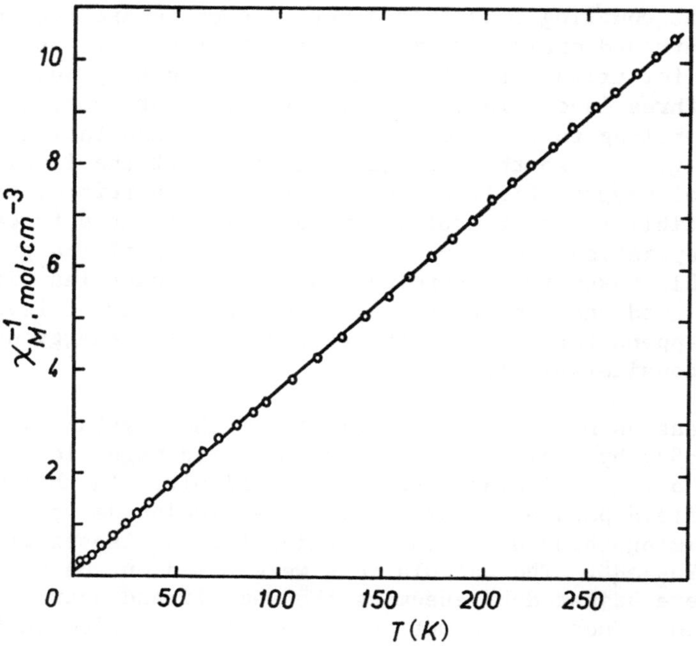

Fig. 1: Experimental (circles) and calculated (solid line) reciprocal molar susceptibility of ζ-Ho_2Se_3.

$$A_4\langle r^4\rangle = \frac{7}{16}\frac{Ze^2}{R^5}\langle r^4\rangle, \qquad A_6\langle r^6\rangle = \frac{3}{64}\frac{Ze^2}{R^7}\langle r^6\rangle, \qquad (3)$$

where R is the central ion-ligand distance, Ze is the ligand charge and $\langle r^n\rangle$ is the expectation value of r^n.

When the crystal field parameters are known then the sequence of the lowest energy levels can be determined. Then making use of the Van Vleck formula the magnetic susceptibility can be calculated. The reverse procedure, i.e. the determination of crystal field parameters from the known magnetic susceptibility vs. temperature meets a number of difficulties however. The problem is least complicated when the number of effectively populated energy levels is small over the temperature range of interest. In the case of the heavier lanthanides this condition is satisfied by the ions with a small value of the J quantum number for the ground state, since large values of the spin-orbit coupling constants assure a good separation between the ground term and excited terms. From this there arises a peculiar significance for trivalent ytterbium compounds whose ground term splits into three levels in the regular field. Moreover, all of these levels belong to different irreducible representations of O_h and T_d point groups, so that the wave functions of the levels are independent of crystal field parameter values. Therefore, the magnetic susceptibility of ytterbium compounds is solely a function of the two separations between the three sublevels of the ground term. All this makes the determination of their energies with sufficiently good accuracy possible. When the number of levels is larger, as happens for holmium, the accuracy of the energy determination is considerably less.

These reasons mean that the estimate of the crystal field parameters in Ho_2Se_3 by a direct fit of the measured magnetic susceptibility plot to that calculated could be difficult. In this work, the crystal field parameters of holmium sesquiselenide were calculated utilizing those determined earlier for the isostructural compound, ζ-Yb_2Se_3[5]. The calculations were based on the relations (3), which were adjusted by assuming the same ligand charges in both selenides. Such an assumption may not be satisfied in the isostructural series of rare earth compounds. Large differences are expected especially between lighter and heavier lanthanides. For instance, Birgeneau et al.[10], who analyzed the crystal field in rare earth monopnictides, concluded that the effective ligand charge for the compounds from the first half of the series is approximately -1.2, whereas for the remainder it approaches -2. Such a large difference in lanthanide compounds is difficult to explain. Nonetheless, there are certain reasons suggesting that in the heavier rare earth ions the radial wave function is more extended than results from theoretical calculations[11]. Therefore, the differences between the ions of lighter and heavier rare earths can be explained more in terms of the $\langle r^n \rangle$ radial integral changes than ligand charge alterations.

The calculation performed in this work is based on the "effective" $\langle r^n \rangle$ value determined by Devine and Berthier from analysis of the crystal field in a wide variety of isostructural series of lanthanide compounds[12]. The values obtained by them for Ho^{3+} and Yb^{3+} ions are 2.7 and 2.0 a.u., respectively. These values are approximately two times larger than those calculated for the free ions, but their ratio differs from the theoretical one by less than 10%. The expectation value of $\langle r^6 \rangle$ used in the calculations was taken from the paper of Freeman and Desclaux[13] since the experimental data are considerably scattered[12].

Substituting the known values of the crystal field parameters for Yb_2Se_3, $A_4\langle r^4 \rangle = 103.9\ cm^{-1}$ and $A_6\langle r^6 \rangle = 3.07\ cm^{-1}$, one obtains the following values for Ho_2Se_3: $A_4\langle r^4 \rangle = 131.7\ cm^{-1}$ and $A_6\langle r^6 \rangle = 3.70\ cm^{-1}$ (i.e. $x = 0.799$, $W = -0.329\ cm^{-1}$). These parameters make possible the calculation of the splitting pattern of the $^5I_8$ term in the crystal field: $\Gamma_3(2)$-0, $\Gamma_4(2)$-1.9, Γ_1-5.8, $\Gamma_5(2)$-140, $\Gamma_4(1)$-141, $\Gamma_3(1)$-184 and $\Gamma_5(1)$-188 (all values are in cm^{-1}). The simultaneously calculated eigenvectors of the crystal field operator (1) enabled us to apply the Van Vleck formula to the description of the temperature dependence of the magnetic susceptibility (Fig. 1).

With regard to the small separations between the three lowest sublevels and their mutual intersection at approximately x=0.83, an unequivocal determination of the ground level is impossible. This problem becomes even more complicated when we realize that both the possible ground levels, Γ_1 and $\Gamma_3(2)$, are non-magnetic ones.

The data in the literature concerning crystal field in the holmium compounds are rather modest. From this point of view the monopnictides[14] are the most extensively studied compounds. It follows from our studies that the crystal field in the rare earth sesquiselenides (measured by a total splitting of the ground term) is stronger than that found in phosphides, but weaker than that in the nitrides. Holmium sesquiselenide also satisfies this rule.

Good agreement of the calculated magnetic susceptibility with the experimental data reveals that the applied method of crystal field parameter determination makes possible a quite satisfactory interpretation of the magnetic properties of Ho_2Se_3. A more detailed analysis could be possible with the use of methods permitting a direct determination of the lowest-lying electronic energy levels.

REFERENCES

1. W. Klemm and A. Koczy, Z. Annorg. Allg. Chemie 233, 84 (1937).
2. I.P. Skripka, V.D. Levandovskij, V.P. Linnik and V.A. Obolonchik, Dopov. AN URSR, A11, 1041 (1976).
3. G.V. Lashkarev, V.P. Fedorchenko, V.A. Obolonchik and I.P. Skripka, Poroshk. Metall. 5, 79 (1976).
4. J.P. Dismukes and J.G. White, Inorg. Chem. 4, 970 (1965).
5. L. Pawlak, M. Duczmal, S. Pokrzywnicki and A. Czopnik, Sol. Stat. Commun. 34, 195 (1980).
6. S. Pokrzywnicki, L. Pawlak and A. Czopnik, Physica 86-88B, 1141 (1977).
7. L. Pawlak and M. Duczmal, to be published.
8. J. Flahaut and P. Laruelle, in: Progress in the Science and Technology of the Rare Earths", vol. 3, LeRoy Eyring, ed., Pergamon Press (1968).
9. K.R. Lea, M.J.M. Leask and W.P. Wolf, J. Phys. Chem. Solids 23, 1381 (1962).
10. R.J. Birgeneau, E. Bucher, J.P. Maita, L. Passel and K.C. Turberfield, Phys. Rev. B8, 5345 (1973).
11. T.O. Brun and G.H. Lander, Phys. Rev. Lett. 29, 1172 (1972).

12. R.A.B. Devine and Y. Berthier, Sol. Stat. Commun. 26, 315 (1978).
13. A.J. Freeman and J.P. Desclaux, J. Magn. Magn. Mat. 12, 11 (1979).
14. F. Hulliger, J. Magn. Magn. Mat. 8, 183 (1978).

ELECTRONIC STRUCTURE AND CRYSTAL FIELD IN Sm_3Se_4 AND Sm_3Te_4

M. Sugita*, S. Kunii*, K. Takegahara*, N. Sato*, T. Sakakibara†, P.J. Markowski*, M. Fujioka§, M. Date† and T. Kasuya*

*Department of Physics, Tohoku University, Sendai, Japan
†Department of Physics, Osaka University, Osaka, Japan
§Cyclotron Radioscope Center, Sendai, Japan

INTRODUCTION

The Th_3P_4 cubic crystal R_3X_4 (X denotes S, Se or Te) with Eu and Sm for R are interesting materials because the Fermi level is just at the 4f level which is situated in the gap between the conduction and valence bands. The transport of electrons occurs by thermally induced hopping between the equivalent R^{++} and R^{+++} sites which coexist in the ratio 1:2, respectively. In the case of Eu_3S_4 this mechanism is supported by the results of a Mössbauer experiment[1] in which the shape of the spectra between 83 K and 300 K could be interpreted in terms of valence fluctuations in the time range of 10^{-7}-10^{-11} s. The charge ordering of Eu^{++} and Eu^{+++} occurs at a temperature of about 160 K.[2] For Sm_3X_4 compounds, however, the problem is more complex because the charge ordering has not yet been observed even at the lowest temperature down to 1.5 K.[3-6] This may mean that some kind of valence fluctuation occurs at the lowest temperature. Furthermore, the data analyses of the magnetic susceptibility presented so far are highly controversial,[3,4,7] again related to the possible existence of valence fluctuations. We have made fairly big single crystals of Sm_3Se_4 and Sm_3Te_4, and we examined the above problems more carefully, both experimentally and theoretically.

RESULTS AND DISCUSSION

The lattice constants of Sm_3Se_4 and Sm_3Te_4 were measured to be 8.880 Å and 9.494 Å, respectively. The molar susceptibility of Sm_3Se_4 at room temperature is just between those of Sm_3S_4 and

Sm_3Se_4 of the IBM group[4], and our Sm_3Te_4 is just the same as their Sm_3Se_4. These results indicate that our samples are good in the stoichiometry. The susceptibility of Sm_3Se_4 and Sm_3Te_4 were measured from 1.73 K to 300 K. Hereafter, we treat the value corrected by the diamagnetic part due to the divalent anions, and these data are shown in Fig. 1. It is seen that all three kinds of crystals have nearly the same magnetic susceptibility.

Magnetization measurements were performed in pulsed fields up to 380 kOe at 1.3 K for powdered Sm_3Se_4 and bulk Sm_3Te_4, and up to 95 kOe in static fields at 0.6 K, 1.3 K and 4.2 K for Sm_3Se_4 and at 1.6 K and 4.2 K for Sm_3Te_4. For the static field the induction method using a superconducting magnet was used. The error of the data obtained does not exceed 2%. Its calibration was carried out by doing the same experiments on a pure Ni standard. The facility at Osaka University was used for the pulsed field measurements. The error of the high field data themselves does not exceed 10% and it was reduced by the calibration by the low field one. Figure 2 shows the magnetization for various temperatures for Sm_3Se_4 and Sm_3Te_4. The magnetization first increases gradually up to 100 kOe and then increases rather linearly up to 380 kOe. Note that the magnetization at 1.3 K is nearly the same as that at 0.6 K and 1.6 K, indicating saturation below 1.6 K and suggesting a kind of antiferromagnetic ordering.

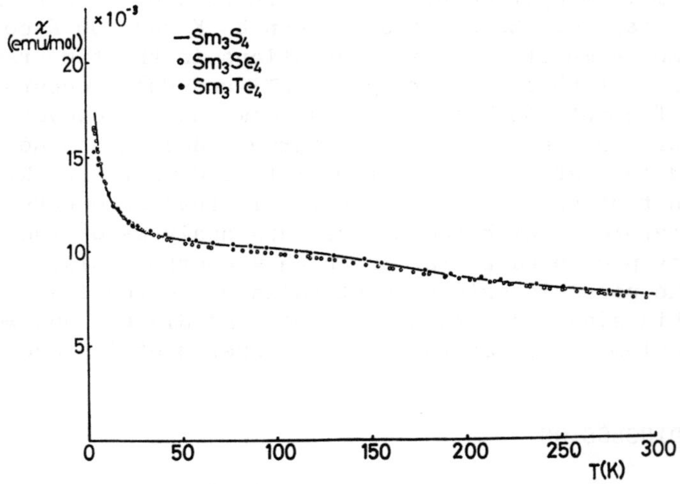

Fig. 1: Molar magnetic susceptibility of Sm_3Se_4 (o) and Sm_3Te_4 (•); the solid lines after Ref. 4.

At first, we tried to fit the susceptibility data. Because the susceptibilities of these samples are very similar, we should assume that the effect of the exchange interaction is negligibly small because the exchange constants should depend on anions fairly strongly as was observed in Sm and Eu-chalcogenides[8]. At first, the high temperature data are considered, where it is clear that the data should be the sum of Sm^{++} and Sm^{+++} contributions in the ratio of 1:2. In Sm^{++}, the S=3, L=3 configuration makes a J=0 singlet ground state and the J=1 first excited state is situated about 420 K above the ground state level for the Sm atom. We note that it was reported that even in the valence fluctuating SmS the above discussed splitting is unchanged[9,10]. Therefore we use here also the above value for the splitting. In Sm^{+++}, the S=5/2, L=5 configuration makes J=5/2 the ground state and J=7/2 the first excited state, which is situated about 1400 K above the ground state in the atomic state. Therefore, the susceptibility of Sm^{+++} is given mostly by two terms, the Curie part of J=5/2 and the Van Vleck term due to J=7/2 mixing. However, here it is impossible to fit the data with the atomic value of the splitting. We should use much smaller values. For the best fits, λ=1100 K for Sm_3Se_4 and λ=1000 K for Sm_3Te_4 were obtained. The origin of this discrepancy is not clear.

It is more convenient to plot χT as a function of T at low temperature when the exchange interaction is small. The result for Sm^{+++} is shown in Fig. 3, in which the dashed slope gives the Van Vleck term and the extrapolated segment on the χT axis gives the Curie constant of the ground state. In both cases, the experimental results give small values of the Curie constant and they are fitted only by the lowest possible value for the crystal field split level, that is, the cubic Γ_7 doublet. The local symmetry of each Sm atom

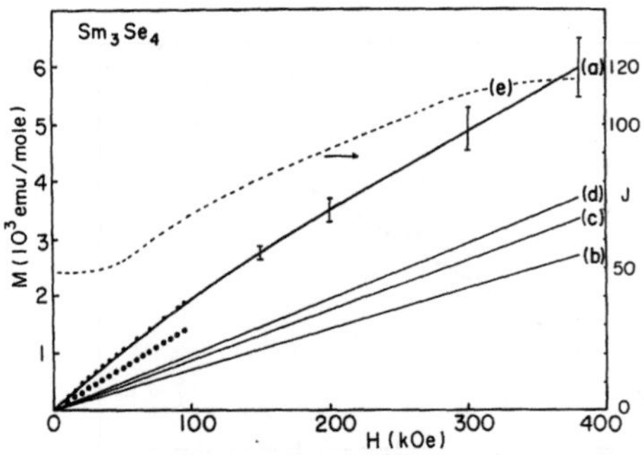

Fig. 2-1: High field magnetization at 1.3 K (a), at 0.6 K (•), and at 4.2 K(o).

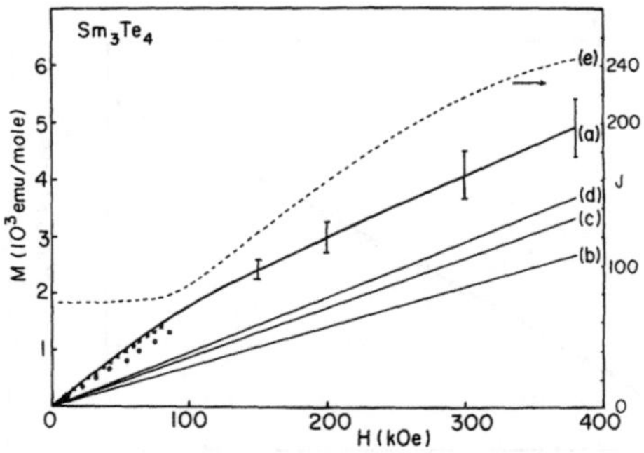

Fig. 2-2: High field magnetization at 1.3 K (a), at 1.6 K (x), and at 4.2 K (o).

(b): Magnetic moment of Sm^{++}. (c) minus (b): van Vleck paramagnetic moment of Sm^{+++} due to J=7/2. (d) minus (c): due to quartet of J=5/2. (e): exchange constant J (arb. units).

is S_4 and the point charge model calculation shows that the uniaxial term dominates, making $J_z=\pm1/2$, which makes the quartet in the cubic symmetry, lowest. Therefore, to make the cubic Γ_7 level lowest, a large amount of opposite mechanisms such as p-f mixing should cancel the uniaxial term making the cubic term dominant. Anyway, the experimental result seems to indicate that in all three S, Se and Te compounds the main effective crystal field term is cubic with Γ_7 lower. Furthermore, to explain the behavior in the lowest temperature region below 10 K, some amount of antiferromagnetic exchange interaction should be introduced into the Curie part. These values are determined by the best fit procedure. The fitting results are shown by the solid lines in Fig. 3, in which the following parameters were used: for sulphide the crystal field splitting between Γ_7 and Γ_8, $\delta=280$ K, and the exchange constant J=30 emu-mole and for the telluride $\delta=240$ K and J=30 emu-mole. Note that the exchange field is defined here by -JM in which M is the induced magnetic moment corresponding to the Cruie part. Note also that the corresponding Curie temperature is -1.5 K. The present model is different from that of the IBM group in which the Γ_8 quartet was assumed to be lower and split slightly into two doublets due to the uniaxial field[4]. The main reason for this picture is the maximum of specific heat at about 1 K, common in all kinds of Sm compounds studied to date. To check this point, a more detailed study of specific heat is in progress by the Sasaki group. A preliminary result shows no other peak down to 0.3 K except the one agreeing with the present model. Then the peak at 1.2 K should be interpreted as due to the common dipole-dipole interaction.

The result of the high field magnetization was analyzed by the above model and is also shown in Fig. 2. When all the van Vleck terms are subtracted, we have the moment corresponding to the Curie

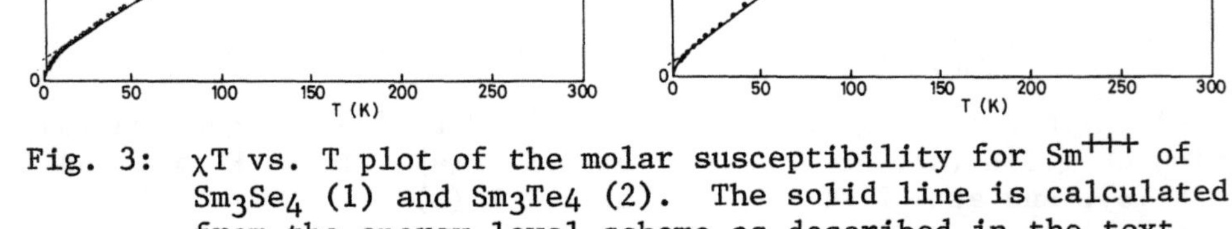

Fig. 3: χT vs. T plot of the molar susceptibility for Sm^{+++} of Sm_3Se_4 (1) and Sm_3Te_4 (2). The solid line is calculated from the energy level scheme as described in the text. See the text for the dashed line.

term as shown in Fig. 2 by (a)-(d). Surprisingly, however, the moment value never shows saturation and is substantially smaller than the saturation value even at 380 kOe: 85% for the sulphide and 52% for the telluride. To explain this result in the present model, we should assume that the exchange constant is field, or induced moment, dependent. The values of the exchange constant to fit the data are also shown in Fig. 2. Such an unusual behavior is hard to understand using usual mechanisms and seems to indicate strongly the existence of some kind of partial quantum tunneling valence fluctuation at lowest temperatures.

The Mössbauer absorption spectra of Sm_3Se_4 were also obtained in a standard transmission geometry, using a conventional constant-acceleration spectrometer. Measurements were peformed using the 22.5 keV γ-transition in ^{149}Eu. The γ-rays were detected by a pure germanium detector. The source, consisting of ^{149}Eu, was produced in a (p,2n) reaction using the cyclotron accelerator at Tohoku University. The proper thin-thickness of an absorber and a source were chosen to eliminate additional line broadenings using Debye temperature (200 K) and experimental trails. To eliminate the base line concave- or convex-form in the Mössbauer data, we used the velocity in triangular wave form so that we could nearly eliminate problems by adding right- and left-half spectra. The sample velocity was calibrated by detecting the laser light reflected from a mirror attached to the rear end of the source driver, and often by a standard ^{57}Fe sample. The temperature of the absorber was changed from 4.2 K to room temperature, and at twelve different temperatures the Mössbauer data were taken.

The behavior above 77 K is understood in the way similar to Eu_3S_4[1]. However, because of the small difference between the isomer shifts of Sm^{++} and Sm^{+++} in comparison with the experimental line width, the observed spectrum is a single broad peak rather than two well-resolved peaks. Below 14 K, however, the linewidth of Sm^{+++} shows broadening compared with higher temperatures, indicating

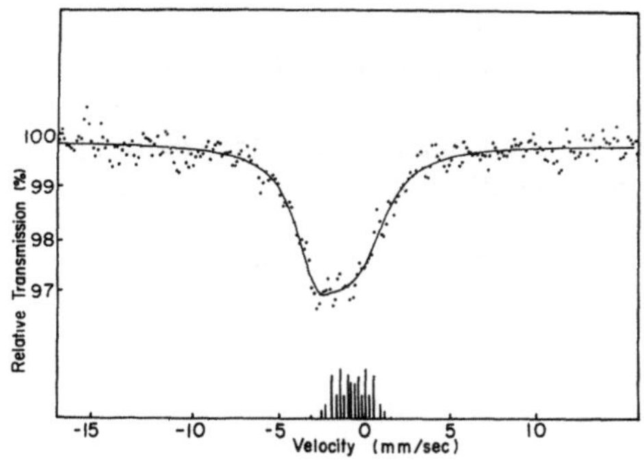

Fig. 4: Mössbauer spectra of Sm_3Se_4 at 4.2 K. The dots are the experimental results, and the solid lines are the calculated spectra by least-squares fit. It is assumed that the intensity ratio of Sm^{++} and Sm^{+++} is 1:2, and Sm^{+++} gives hyperfine splittings as shown in the lower part of the figure.

the beginning of short range spin ordering. This is contradictory to the model of the IBM group but consistent with the present model in which the ground Γ_7 doublet undergoes some kind of ordering. Mössbauer spectra at 4.2 K are plotted in Fig. 4. The data are not fitted by a singlet Lorentzian line because of their antisymmetrical line shape. Around the peak position, moreover, the plateau region has been seen, in contrast to higher temperature spectra in which we could fit the data adequately by a single Lorentzian line[11]. However, fitting by two Lorentzian curves is much worse because fitting results in an intensity ratio and a separation between Sm^{++} and Sm^{+++} which are unreasonable. Thus this should be attributed to the nuclear level splittings of Sm^{+++} ions by an internal field. We can fit these data by a hyperfine field of 54.6 kOe as shown in Fig. 4. It was reported that the hyperfine field for the fully spin polarized Sm^{+++} is 4700 kOe[12]. Therefore the above value corresponds to a 1.2% polarization of the Γ_7 moment, which is not inconsistent with the large tail of the specific heat peaked at 1.2 K. Much short range order seems to some extent partly due to the randomness of Sm^{+++} and Sm^{++} ions, but also partly due to the intrinsic character of the dipole-dipole interaction.

REFERENCES

1. O. Berkooz, M. Malamud and S. Shtrikman, Sol. Stat. Commun. 6, 185 (1968).
2. H.H. Davis, I. Bransky and N.M. Tallan, J. Less-Comm. Met., 22, 193 (1970).
3. P. Wachter, Phys. Letters 58A, 484 (1976).
4. J.M.D. Coey, B. Cornut, F. Holtzberg and S. von Molnar, J. Appl. Phys. 50, 1923 (1979).
5. F. Holtzberg, Phil. Mag. B42, 491 (1980).
6. P. Wachter, Phil. Mag. B42, 497 (1980).
7. T.K. Bolkonskaya, S.A. Kizhaev, and I.A. Smirnov, Sov. Phys. Sol. State 21, 726 (1979).
8. E. Bucher, V. Narayanamurti and A. Jayaraman, J. Appl. Phys. 42, 1741 (1971).
9. M.I. Nathan, F. Holtzberg, J.E. Smith Jr., J.B. Torrance and J.C. Tsang, Phys. Rev. Letters 34, 467 (1975).
10. S.M. Shapiro, R.J. Birgeneau and E. Bucher, Phys. Rev. Letters 34, 470 (1975).
11. M. Sugita, S. Kunii, M. Kawakami, M. Fujioka, P.J. Markowski, H. Horita, T. Suzuki and T. Kasuya, CYRIC Ann. Rep., p.131, (1980).
12. M.E. Caspari, S. Grankel and G.T. Wood, J. Appl. Phys. 33, 1089 (1962).

STRONG CRYSTAL FIELD EFFECTS IN $TmNi_5$

D. Gignoux*, B. Hennion† and A. Naït Saada*

*Laboratoire Louis Néel, C.N.R.S., 166X 38042 Grenoble-cédex, France
†Laboratoire Léon Brillouin, B.P.2, Gif-sur-Yvette, France

INTRODUCTION

This work on $TmNi_5$ is part of a wider study devoted to the quantitative determination of the rare earth 4f and nickel 3d magnetism in the hexagonal compounds RNi_5 (R = rare earth) (1-3). Because of the filling of the 3d band due to transfer of 4d electrons contributed by the rare earth the Stoner criterion of the onset of ferromagnetism is not satisfied for the Ni atoms, which thus exhibit only an exchange enhanced Pauli paramagnetism (4). Thus the main contribution to the magnetism of the RNi_5 compounds originates from the rare earth on which the crystal field effect (CEF) plays a preponderant role (if $L \neq 0$). In $TmNi_5$ the sign and the value expected for the second-order term of the CEF should lead to a huge uniaxial anisotropy. In that case it is worth studying the effect of such anisotropy on the Curie temperature.

In this paper, we present the magnetization and susceptibility measurements performed on a $TmNi_5$ single crystal along the three main symmetry axes (a, b and c) of the orthohexagonal cell as well as neutron spectroscopy and specific heat measurements performed on a polycrystalline sample.

The RNi_5 compounds crystallize in the hexagonal $CaCu_5$-type structure (space group P6/mmm) (5). Rare earth atoms lie in the 1a site (0, 0, 0) while Ni atoms lie in the 2c and 3g sites. The $TmNi_5$ single crystal was prepared by the Czochralski method and the sample was spark cut into a 3.5 mm diameter sphere. Magnetization measurements were performed at the Service National des Champs Intenses of Grenoble in fields up to 150 kOe. Neutron spectroscopy measurements were performed on the statistical chopper time-of-flight spectrometer IN7 at the I.L.L., Grenoble. Finally the specific heat measurements were performed at the Service des Basses Temperatures of the C.E.N. Grenoble.

EXPERIMENTAL RESULTS

The thermal variations of the reciprocal susceptibilities measured along and perpendicular to $\vec{c}$ are reported in figure 1. These variations are linear only above 200 K. Along $\vec{c}$ the susceptibility is always much higher than in a perpendicular direction and diverges at 4.5 K which is the Curie temperature of $TmNi_5$.

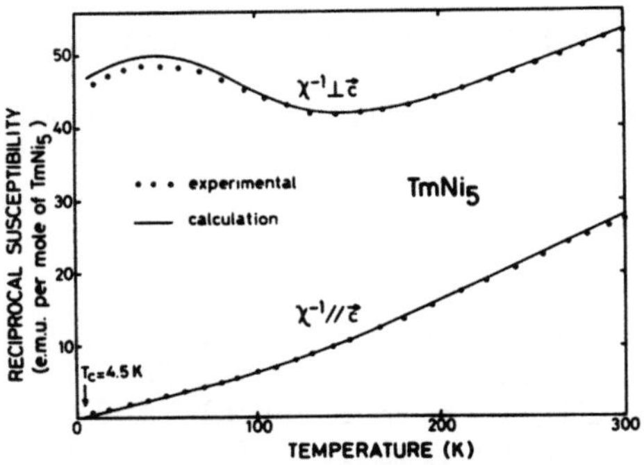

Fig. 1. Thermal variations of the reciprocal susceptibilities parallel and perpendicular to the six fold axis.

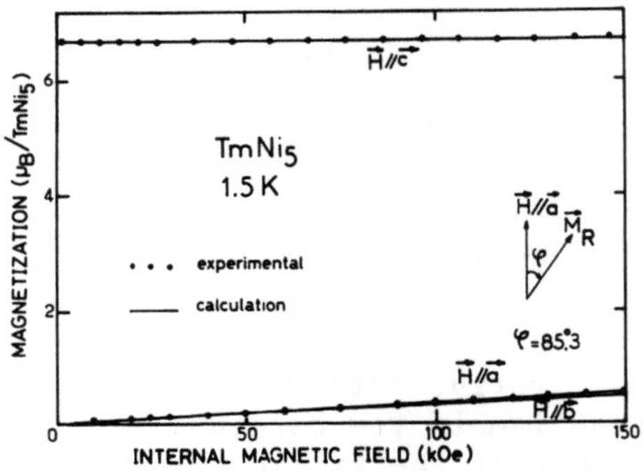

Fig. 2. Magnetization versus the internal field at 1.5 K along the three symmetry axes $\vec{a}$, $\vec{b}$ and $\vec{c}$ of the orthohexagonal cell.

The magnetization versus internal field measured at 1.5 K along the three symmetry axes ($\vec{a}$, $\vec{b}$ and $\vec{c}$ of the orthohexagonal cell) is reported in figure 2. $\vec{c}$ is the easy magnetization axis; the spontaneous magnetization is 6.76 $\mu_B/TmNi_5$. When the field is increased, the magnetization varies slowly and linearly and reaches 6.82 $\mu_B/TmNi_5$ under 150 kOe. The superimposed susceptibility is 20×10^{-4} e.m.u./mole which must be compared to that of $LaNi_5$ (20×10^{-4} e.m.u./mole) and YNi_5 (23×10^{-4} e.m.u./mole) at 4.2 K (4). A huge uniaxial anisotropy is observed: along the $\vec{a}$ and $\vec{b}$ axes, the magnetizations are almost the same and vary slowly and linearly with the field; however $\vec{b}$ is the most difficult axis. In a 150 kOe field the magnetizations are 0.55 and 0.52 $\mu_B/TmNi_5$ along $\vec{a}$ and $\vec{b}$ respectively. The extrapolation of the variation along $\vec{a}$ should lead to an anisotropy field of 2×10^3 kOe. The inelastic neutron scattering spectra obtained at 20 and 150 K are given in Fig. 3. The energy of the incident beam leading to the elastic peak is 45.84 meV. The continuous lines correspond to an averaged expe-

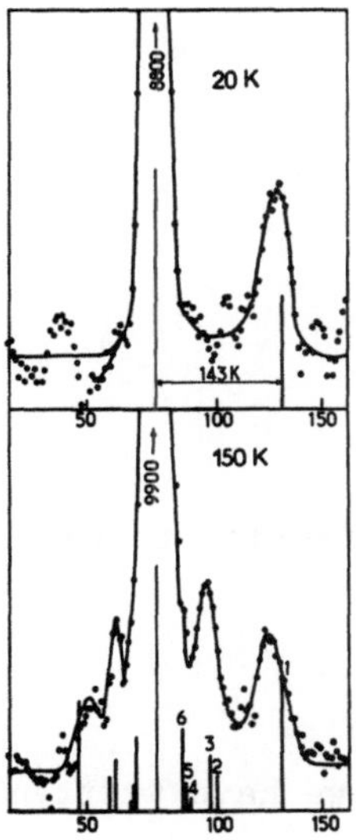

Fig. 3. Neutron energy spectra of $TmNi_5$ at 20 and 150 K.

rimental variation. At 20 K one excitation peak is observed at 12.3 meV (143 K) on the right of the elastic peak. At 150 K, excited states are occupied and apart from the previous peak a new excitation peak at 5.9 meV (68 K) and two de-excitation peaks at the same energies are observed. The increase of the width of the elastic peak must correspond to small energy transitions.

The thermal variation of the specific heat of $TmNi_5$ between 1.5 and 300 K is reported in Fig. 4. As evidenced in the insert, a λ anomaly is observed at T which is the Curie temperature. In dashed line we have represented the lattice and electronic contributions deduced from the specific heat of $LaNi_5$. A large magnetic contribution to the specific heat is evi-enced denced above the Curie temperature. As shown in Fig. 5 this contribution corresponds to a strong Schottky anomaly.

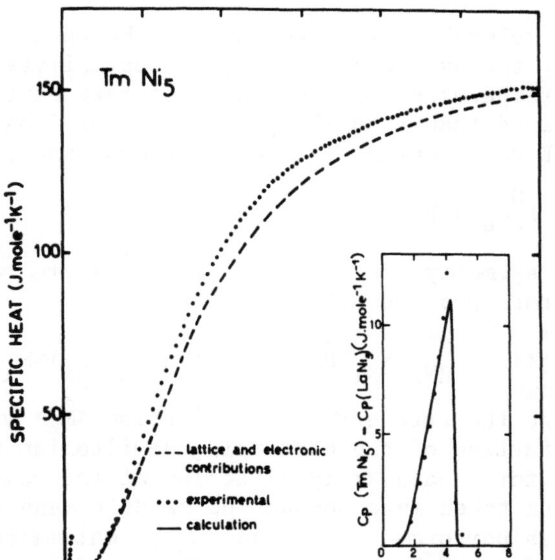

Fig. 4: Thermal variation of the specific heat C_p of $TmNi_5$.

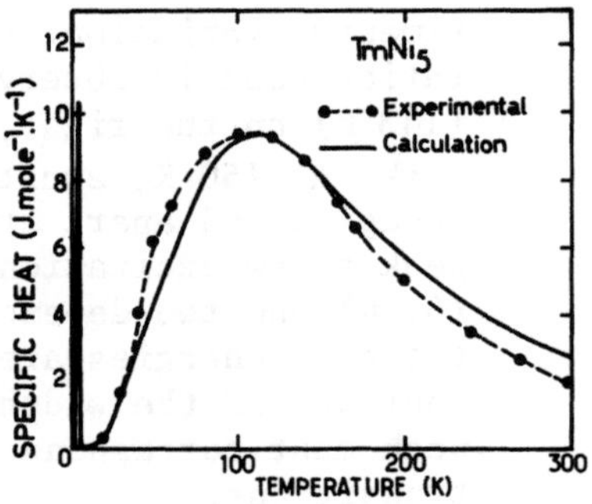

Fig. 5: Thermal variation of the magnetic contribution to the specific heat of $TmNi_5$.

ANALYSIS OF THE RESULTS

In order to analyse the magnetic measurements we have written the total magnetization $\langle\vec{M}_{RNi}\rangle$ as the sum

$$\langle\vec{M}_{RNi}\rangle = \langle\vec{M}_R\rangle + \langle\vec{M}_{Ni}\rangle \quad (1)$$

The 3d magnetization $\langle\vec{M}_{Ni}\rangle$ described in the itinerant model may be written

$$\langle\vec{M}_{Ni}\rangle = \chi_{Ni}(\vec{H} + n_{RNi}\langle\vec{M}_R\rangle) \quad (2)$$

where n_{RNi} is the molecular field coefficient between rare earth and nickel and χ_{Ni} the exchange enhanced susceptibility of Ni which is assumed to be temperature independent. The rare earth magnetization $\langle\vec{M}_R\rangle$ is obtained thanks to the perturbing Hamiltonian which is the sum of the CEF Hamiltonian $\mathcal{H}_c$ and the Zeeman Hamiltonian $\mathcal{H}_H$:

$$\mathcal{H}_c = B_2^0 O_2^0 + B_4^0 O_4^0 + B_6^0 O_6^0 + B_6^6 O_6^6 \quad (3)$$

for the hexagonal symmetry. The O_l^m's and the B_l^m's are, respectively, the Stevens operators and the CEF parameters.

$$\mathcal{H}_H = g_J \mu_B \vec{J}\left[(1 + n_{RNi}\,\chi_{Ni})\vec{H} + (n_{RR} + n_{RNi}^2\,\chi_{Ni})\langle\vec{M}_R\rangle\right]. \quad (4)$$

At each temperature the rare earth magnetization $\langle\vec{M}_R\rangle$ was obtained from the diagonalization of the perturbing Hamiltonian in a three dimensional space for an arbitrary direction of the magnetization. Thermal effects are taken into account using Boltzmann statistics. The calculation was performed in a self-consistent manner until the calculated value of $g_J\mu_B\langle\vec{J}\rangle$ was equal to $\langle\vec{M}_R\rangle$ used in the expression of the molecular field.

The total magnetization as well as the total susceptibility

depend then on seven parameters: the B_l^{m}'s and χ_{Ni}, n_{RNi} and n_{RR} We have determined these parameters by simultaneous fits of the magnetization curves, the thermal variation of the reciprocal susceptibility and the neutron spectroscopy spectra. In the case of $TmNi_5$ it is possible to directly determine the two parameters χ_{Ni} and n_{RNi} because in zero applied field the Tm magnetization reaches its maximum value. Indeed at 1.5 K the superimposed susceptibility is the same as in YNi_5 or $GdNi_5$. Moreover if we use the B_l^m parameters deduced with a scaling law from those of $TbNi_5$, the ground state is the $\pm|6\rangle$ doublet which is much below the first excited state $(\pm|5\rangle)$. Then the superimposed susceptibility at 1.5 K along $\vec{c}$ is a direct measurement of χ_{Ni}. In zero field we have a direct measurement of $\langle M_{Ni}\rangle$ (0.24 $\mu_B/TmNi_5$) and hence from (1): n_{RNi} = -17.1 (e.m.u./mole)$^{-1}$.

The determination of the other parameters has led to: B_2^0 = -3.80 K, $B_4^0 = -1.26 \times 10^{-2}$ K, $B_6^0 = 1.83 \times 10^{-4}$ K, $B_6^6 = -7.6 \times 10^{-4}$K and $n_{RR} = -0.33$ (e.m.u./mole)$^{-1}$. The variations of M(H) and χ(T) then calculated with these parameters are drawn in full lines on figures 1 and 2. The agreement is excellent. On Fig. 6, we have reported the overall splitting of the $^3H_6$ ground state multiplet

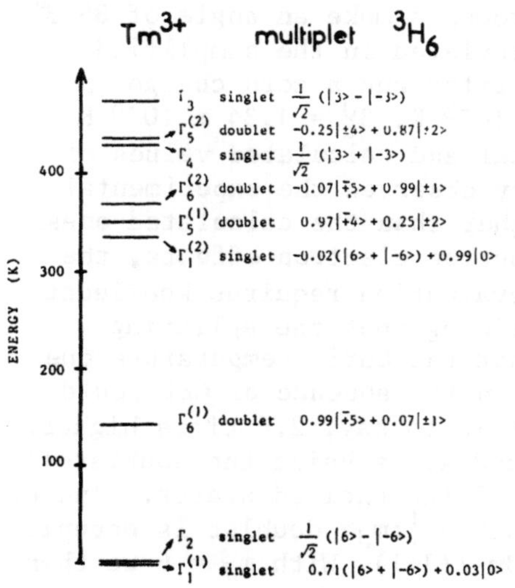

Fig. 6. $TmNi_5$: splitting of the $^3H_6$ multiplet of the Tm^{3+} ion by the crystalline electric field CEF.

N°	Transitions	Energy	Intensity (excitation) arbitrary unit
1	$\Gamma_1^{(1)} \rightarrow \Gamma_6^{(1)}$ $\Gamma_2 \rightarrow \Gamma_6^{(1)}$	143	5.0
2	$\Gamma_5^{(1)} \rightarrow \Gamma_4$	79.1	1.3
3	$\Gamma_6^{(2)} \rightarrow \Gamma_5^{(2)}$	68.2	2.0
4	$\Gamma_1^{(2)} \rightarrow \Gamma_6^{(2)}$	32.0	0.4
5	$\Gamma_5^{(1)} \rightarrow \Gamma_6^{(2)}$	17.7	1.0
6	$\Gamma_4 \rightarrow \Gamma_5^{(2)}$	6.8	3.0

Table 1. Calculated transitions between the energy level of the Tm^{3+}.

of the Tm^{3+} ion by the crystal field. As expected the ground state is the doublet-like $|\pm 6>$. The first excited level at 143 K is an almost pure state $|\pm 5>$. The transition observed at 20 K by inelastic scattering corresponds to a transition between these levels. On Fig. 3 we have schematized by vertical bars the calculated transition. Their heights correspond to the intensity of the calculated transitions. These transitions are summarized in the Table. The agreement between the observed and calculated spectra is quite good. The calculated thermal variation of the specific heat at low temperature is in perfect agreement with the observed one (insert of Fig. 4). Especially at T_C the calculated magnetic entropy reaches $R\ell n2$ as observed. Indeed at T_C only the doublet like $|\pm 6>$ is occupied. In the paramagnetic range as shown in Fig.5 the calculated magnetic contribution to the specific heat gives very good account for the strong observed Schottky anomaly.

DISCUSSION

In $TmNi_5$ the magnetic properties are dominated by the strong CEF effect which is responsible of the huge uniaxial anisotropy and the strong Schottky anomaly observed in the specific heat measurements. These properties are very well accounted for by the model used. In particular, the calculation allows us to understand the variation of magnetization as a function of the field below the Curie temperature. Along $\vec{a}$ and $\vec{b}$ the variation of the observed magnetization corresponds to a very slow rotation of the Tm moments. Whereas in zero field the Tm moment is along the $\vec{c}$ axis, in an internal field of 150 kOe applied along $\vec{a}$ the moments make an angle of 85.3° with the field. The CEF parameters calculated in the simplified point charge model, with +3e on rare earths and a zero charge on nickel, are $B_2^0 = -2.53$ K, $B_4^0 = -1.95 \times 10^{-3}$ K, $B_6^0 = 1.34 \times 10^{-6}$ K and $B_6^6 = 9.25 \times 10^{-6}$ K. The experimental and calculated values of B_2^0 are not very different. As generally observed the experimental values of the other terms are much higher than the calculated ones. These differences arise from the conduction electron effects, the contribution of which is high. Their evaluation requires knowledge of the band structure. It is worth noticing that the splitting of the multiplet by the CEF is such that the Curie temperature due to the same exchange interactions but in the absence of CEF would be $\Theta_p = 1.75$ K while T_C is 4.5 K, that is to say, 2.6 times higher. This is characteristic of the CEF ground state being the doublet-like state $|\pm 6>$, which lies well below any of the excited states. Indeed considering that at low temperature only a $|\pm m_J>$ doublet is occupied, we can easily establish that $T_C/\Theta_p = 3m_J/(J+1)$. With $m_J = 6$ we then obtain $T_C/\Theta_p=2.6$.

At 1.5 K and in zero applied field the Ni contribution to the magnetization is $-0.050\ \mu_B/Ni$. As generally observed in the R-Co and R-Ni rare earth compounds (6), its sign corresponds to a negative interaction between the spins of the rare earths and those of the nickel atoms. The nickel susceptibility being very close to that in $GdNi_5$, YNi_5 and $LaNi_5$, the band structure must be very similar in all these compounds. The value of n_{RR} is very small and negative suggesting weak negative interactions between Tm ions. The positive coupling between Tm ions would occur through Ni-3d electrons. However, this result must be considered with caution because the error bar on n_{RR} is of the order of its absolute value.

REFERENCES

1. P. Escudier, D. Gignoux, D. Givord, R. Lemaire and A.P. Murani, Physica 86-88B,197 (1977).
2. D. Gignoux, A. Naït Saada and R. Perrier de la Bâthie, J. Physique C5,188 (1979).
3. G. Aubert, D. Gignoux, B. Hennion, B. Michelutti and A. Naït Saada, Sol. Stat. Commun. 37,741 (1981).
4. D. Gignoux, D. Givord,and A. Del Moral, Sol. Stat. Commun. 19, 891 (1976).
5. J.H. Wernick and S. Geller, Acta Cryst. 12,662 (1959).
6. B. Barbara, D. Gignoux, D. Givord, F. Givord and R. Lemaire, Int. J. Magn. 4,77 (1973).

COMMENTS

GREIDANUS: Have I understood you correctly that Ni carries a moment below the phase transition?

GIGNOUX: The magnetization localized on the Ni atom is induced by the rare earth. In $TmNi_5$ this magnetization reaches 0.05 μ_B/Ni at 1.5 K. With non-magnetic rare earths like Y and La, Ni does not carry a magnetic moment: YNi_5 and $LaNi_5$ are exchange enhanced Pauli paramagnets.

ROSSAT-MIGNOD: I think that you do not observe the enhancement of the ordering temperature, but it results only from a crossover between a high temperature Heisenberg behavior and a low temperature Ising behavior.

GIGNOUX: If you prefer we can say that the CEF leads to an increase of the ordering temperature due to the existence of a $|\pm J\rangle$ ground state far below the first excited state. This situation is opposite to that observed in some Pr based compounds where the ground state due to the CEF is a non-magnetic ground state.

COQBLIN: Generally in rare earth-nickel compounds, nickel does not carry any magnetic moment. Are there, however, some such compounds where Ni is magnetic?

GIGNOUX: In R_2Ni_{17} and RNi_3 Ni carries a magnetic moment. Y_2Ni_{17} is ferromagnetic. YNi_3 is a very weak itinerant ferromagnet.

McEWEN: Have you measured any dispersion in the crystal field excitations, either at 20 K or below T_C?

GIGNOUX: Not yet, but we plan to make these measurements.

MAGNETIC PROPERTIES OF SOME SOLID AMORPHOUS RARE EARTH ALLOYS

A. Apostolov, H. Hristov, T. Mydlarz,
M. Mihov and V. Skumriev

Faculty of Physics, University of Sofia
1126 Sofia, Bulgaria

INTRODUCTION

The amorphous state is a metastable state, its structure characterized by a topological short-range order and an absence of long-range order. Nevertheless, in the majority of amorphous substances containing magnetic atoms or ions a strong long-range order exists under certain conditions. The magnetic properties of these substances show a prevailing difference of the temperatures of magnetic transitions from those of the respective crystal analogues, and the values of the spontaneous magnetizations in the amorphous state are smaller than those in a crystalline state.

To explain these effects Handrich[1] suggests a random distribution for the values of the exchange interactions. However, the variations in the values of the exchange integrals for the amorphous alloys of rare earths (R) with iron-group metals (T) are too small and insufficient to explain their anomalous magnetic behavior. The magnetic properties of these alloys can be explained starting with the random character of the electrostatic field at different points in the sample, a consequence of the distorted symmetry. The model of random magnetic anisotropy presented by Harris and other authors[2] assumes that an axial electrostatic (crystal) field acts at the position of each rare earth magnetic moment, its orientation changing from site to site. This randomness in the magnetic field orientation, a consequence of topological disorder of quasi-close packed groups of atoms, leads to a random local magnetic anisotropy. The Hamiltonian is

$$\hat{H} = -\frac{1}{2} J \sum_{ij} \vec{I}(i)\vec{I}(j) - D\sum_{i}[I_{z_i}(i)]^2 \qquad (1)$$

where J is the exchange integral between nearest neighbors, $\vec{I}(i)$ is the atomic angular momentum vector at point j, $I_{z_i}(i)$ is the projection of the angular momentum vector at point i on the local axis of magnetic anisotropy z_i, and C is the magnetic anisotropy parameter, assumed to be a constant.

The further evolution of the random magnetic anisotropy theory[3-5] reveals three possible magnetic states for the amorphous R-T alloys with ferromagnetic interactions: the paramagnetic, the ferromagnetic and the spin-glass-like state. The latter is due to the random local magnetic anisotropy and is characterized by the randomly oriented elementary magnetic moments, lower spontaneous magnetization and a large coercive force. By lowering the temperature a successive transition can be achieved from the paramagnetic into the ferromagnetic and then to the spin-glass-like state. The existence and position of the last transition are essentially dependent on the parameter $\alpha = D/vJ$, where v is the number of the nearest neighbors.

More recent theoretical investigations reveal that the RMA model is a good approximation for the rare earths with high I. With relatively low I-values ($I \leq 4$) and primarily for non-Kramers ions, the non-axial components of the atomic angular momentum, I_x and I_y, should also be considered when the magnetic anisotropy is included.

The above mentioned models give only a qualitative explanation for some of the observed magnetic properties of the amorphous R-T alloys. The present paper deals with the magnetic properties of the series of amorphous alloys R_4Co_3 (R=Y,Tb,Dy,Er,Gd,Sm). Tb is a non-Kramers ion with I=6, Dy and Er are Kramers ions with I=15/2, and Gd with I=7/2 is in an S-state. Sm is an ion with I=5/2, while Y is non-magnetic. The majority of these alloys possesses crystalline analogues.

EXPERIMENT AND EXPERIMENTAL RESULTS

The solid amorphous alloys R_4Co_3 were obtained by rapid cooling, an argon jet discharging the melt on a rapidly rotating copper cylinder[7]. The x-ray amorphousness was determined with CrK_α-radiation and by means of electron diffraction. The magnetic susceptibility above 77K was studied by a translation magnetic balance in a field of 7.8 kOe. The magnetization for $4.2 \leq T \leq 300K$ was done in fields to 60 kOe by means of a vibration magnetometer and up to 140 kOe with a ballistic magnetometer. (Some of the measurements were taken at MLSMPNT-Wroclaw, Poland.)

At high temperatures the alloys are in a paramagnetic state. The magnetic susceptibility of Sm_4Co_3, Tb_4Co_3, and Er_4Co_3 follows the Curie-Weiss law, with the Curie constant C and asymptotic Curie temperature θ_p, as indicated in Table 1 below. The magnetic susceptibility of Dy_4Co_3 and Gd_4Co_3 does not conform to the Curie Weiss law

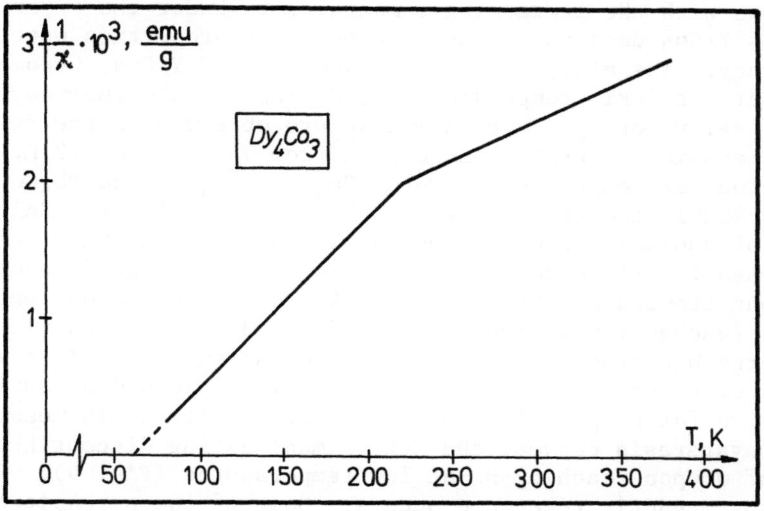

Fig. 1: Temperature dependence of the magnetic susceptibility for Dy_4Co_3 above the Curie temperature.

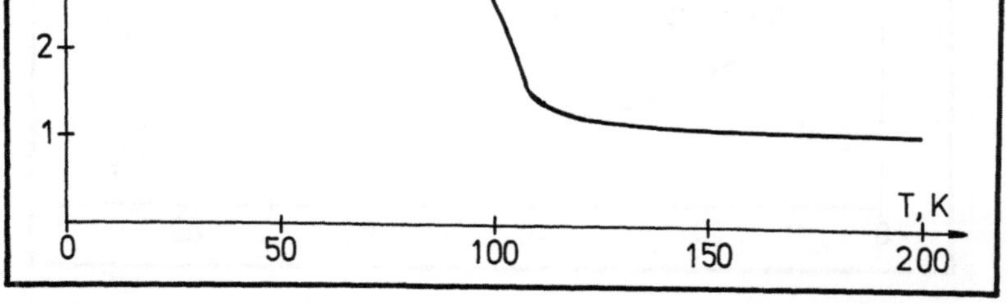

Fig. 2: Temperature dependence of the magnetization of Sm_4Co_3 in zero magnetic field.

(Fig. 1). Even in the lowest temperature range the Curie constant is higher than the maximum possible one for those compositions and still grows with the increasing temperature. Under the experimental conditions Y_4Co_3 does not reveal paramagnetic properties within our accuracy. The alloys, with the exception of Y_4Co_3, become ferromagnetic at the Curie temperature θ_C, defined by the thermodynamic method of Belov-Gorjaga. No anomalies are observed in the ferromagnetic behavior of Er_4Co_3 and Gd_4Co_3 from θ_C down to 4.2 K. However, the ferromagnetic nature of Sm_4Co_3, Dy_4Co_3 and Tb_4Co_3 occurs within a certain temperature range only - from θ_C to T_f. Below T_f a number of anomalies can be seen: at T_f the magnetization measured in zero magnetic field reaches its maximum value (Fig. 2) accompanied by a strong temperature hysteresis. At T_f the spontaneous magnetization σ_s reaches its maximum as well (Fig. 3). The magnetic anisotropy energy W_{an} steepens below this temperature. (W_{an} is measured with the curves of initial magnetization by the method of surfaces and coercive force H_c). The coercive force H_c itself is measured with the hysteresis curves, the latter manifesting discontinuous changes of σ upon reaching H_c at low temperatures (Fig. 4). Below T_f a strong magnetic remanence occurs. Some of the parameters obtained as a result of the magnetic measurements are summarized in Table 1.

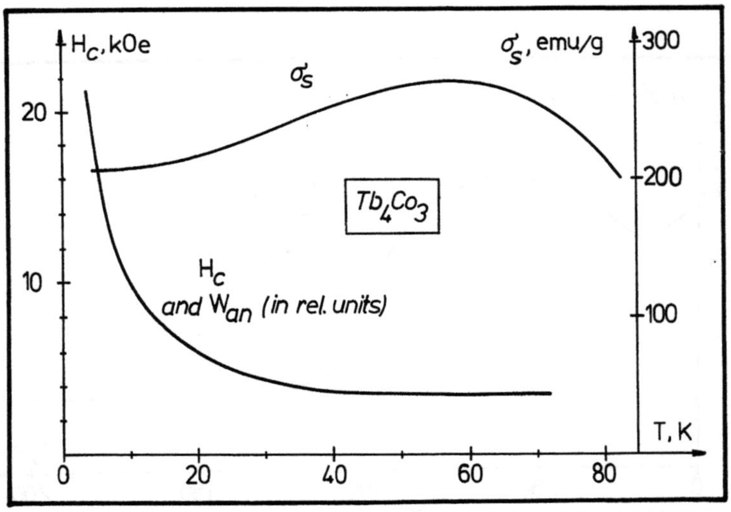

Fig. 3: Temperature dependence of the spontaneous magnetization σ_s, the coercive field H_c and the magnetic anisotropy energy, W_{an}, in arbitrary units for Tb_4Co_3.

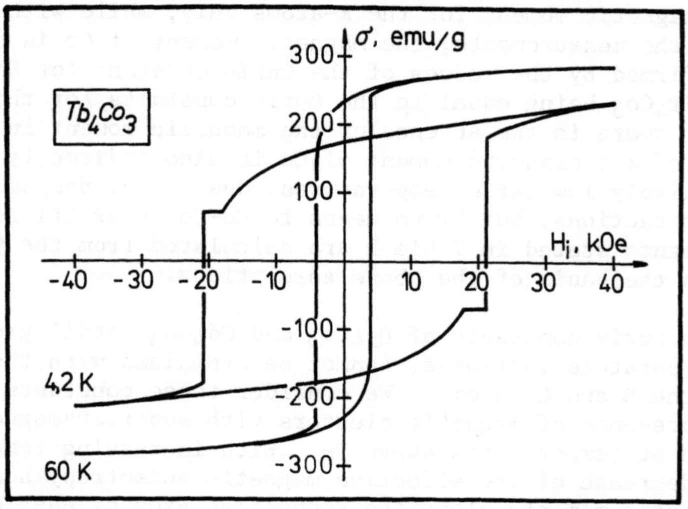

Fig. 4: Hysteresis curves for Tb_4Co_3.

Table 1: Magnetic Properties of some Amorphous R_4Co_3 Systems

Alloy	θ_p [K]	θ_c [K]	C_{mol}	M_{eff} μ_B	T_f [K]	H_c [kOe]	W_{an} [erg/cm^3]
Sm_4Co_3	110	114	1	1.4	55	–	–
Tb_4Co_3	135	137.0	48.8	9.9	60	21	2.9×10^7
Dy_4Co_3	61*	69.5	66.8*	11.5*	40	7	1.6×10^7
Er_4Co_3	22	23	42.8	9.3	–	2	1.1×10^7
Gd_4Co_3	212*	230	49.7*	10.0*	–	–	$\sim10^5$

*These alloys do not conform to the Curie-Weiss law. The values are obtained from the lowest temperature part of the curve above the Curie temperature.

DISCUSSION

The research carried out with the R-T compounds shows the presence of a magnetic moment for the R-atoms only, while within the accuracy of the measurements, the magnetic moment of Co is zero. This is confirmed by the values of the Curie constant for Sm_4Co_3, Tb_4Co_3 and Er_4Co_3 being equal to the Curie constants for the R-atoms. In addition, there is the absence of the magnetic moment in Y_4Co_3. The absence of the magnetic moment of Co is also indirectly confirmed by the relatively low Curie temperatures, due to the comparatively weak R-R interactions, but by no means to Co-Co interactions. The magnetic moments stated in Table 1 are calculated from the Curie constants on the basis of the above assumptions.

The high Curie constants of Dy_4Co_3 and Gd_4Co_3, still growing when the temperature increases, cannot be explained with the magnetic moments of the R and Co atoms. We consider these constants as being due to the presence of magnetic clusters with superparamagnetic behavior, even at temperatures above T_C. With increasing temperature there is a decrease of the effective magnetic anisotropy holding the cluster magnetic moments along the respective axes of easy magnetization. The moments are released while their contribution to the paramagnetism is increased. The fact that the above effect is observed not only in Dy_4Co_3 but also in Gd_4Co_3, where Gd is in an S-state (defining the low magnetic anisotropy below T_C) indicates that the effective magnetic anisotropy in the superparamagnetic state is not a consequence of the anisotropic properties of the magnetic atoms, but rather a peculiarity of the whole superparamagnetic formation. This is probably either a shape or an exchange anisotropy.

The positive Curie temperatures and the nature of the magnetization curves below T_C, reveal prevailing ferromagnetic interactions between the elementary magnetic moments of the R-atoms. The special features in the magnetic behavior of Sm_4Co_3, Tb_4Co_3 and Dy_4Co_3 below T_f suggest a transition from a ferromagnetic to a spin-glass-like state at this temperature. This is a mixed state, energetically described by two competing components - a ferromagnetic and a spin-glass-like one[4]. With this state the ferromagnetically interacting elementary magnetic moments are related to the chaotically disoriented local axes of random magnetic anisotropy. As a result the magnetization decreases below T_f, which is typical for a pure spin-glass-like state. On the other hand, the ferromagnetic component does not tend to zero at $T \to 0$, but remains fairly large.

The spin-glass-like state is a rather stable one; even 140 kOe stationary magnetic fields do not alter the character of the σ dependence, (Figs. 2 and 3). The relevant magnetic anisotropy starts to increase exponentially at T_f (Fig. 3). This shows that the magnetic anisotropy parameter in the Hamiltonian cannot be considered to be a constant, as is accepted in most theories.

The hard magnetization directions conditioned by the random magnetic anisotropy act as potential barriers against the changing of the direction of the elementary magnetic moments during the remagnetization in an external magnetic field. This accounts for the form of the observed hysteresis curves and the jump of the magnetization when reaching the coercive field H_c, in agreement with Refs. 4 and 8. The most typical feature of the hysteresis curves is the increase of the coercive force at T_f, reaching exceptionally high values for some alloys. The temperature dependence of H_c which can be approximated as an exponent is quite similar (in a certain temperature range) to that of the magnetic anisotropy. This illustrates again the connection between the magnetic properties of the amorphous rare earth alloys.

The absence of a transition to a spin-glass-like state in Gd_4Co_3 is understandable, Gd being in an S-state, having thus rather weak magnetic anisotropy. The latter is strong with Er_4Co_3, a transition to a spin-glass-like behavior can arise at temperatures below 4.2 K.

It is evident from the research work that the unusual magnetic properties of the solid amorphous R-T alloys are qualitatively explained by the thoery of random magnetic anisotropy (RMA).

REFERENCES

1. K. Handrich, Phys. St. Sol. B53, K17 (1952).
2. R. Harris, M. Plischke and M. Zuckermann, Phys. Rev. Lett. 31, 160 (1973).
3. R. Harris and D. Zobin, J. Phys. F: Metal Phys. 7, 2, 337 (1977).
4. R. Harris, S.H. Sung and M.J. Zuckermann, IEEE Trans. Magn. MAG-14, 5, 725 (1978).
5. R. Ferrer, R. Harris, S.H. Sung, M.J. Zuckermann, J. de Phys. 40, C5-221 (1979).
6. A. Fert and I.A. Campbell, J. Phys. F:Metal Phys. 8, 3, L57 (1978).
7. A. Apostolov, R. Damjanov, M. Mihov, V. Skumriev, N. Stanev, and P. Fournadziev, C.r.Ac.Sci.Bulg., (1981), in press.
8. E. Callen, Y.J. Lin and J.R. Cullen, Phys. Rev. B16, 263 (1977).

MAGNETIC PHASE DIAGRAMS OF SOME URANIUM MONOPNICTIDES AND MONOCHALCOGENIDES

J. Rossat-Mignod*, P. Burlet*, S. Quézel*, O. Vogt† and H. Bartholin§

*DRF/DN, Centre d'Etudes Nucléaires, 85 X, 38041 Grenoble Cedex, France
†E.T.H. - CH 8093 Zürich, Switzerland
§Université de Toulon, 83130 La Garde, France and SNCI 166 X, 38042 Grenoble Cedex, France

INTRODUCTION

The uranium monopnictides, with the NaCl crystal structure, exhibit highly unusual magnetic behaviors which have many similarities with those of cerium monopnictides[1]. The characteristic magnetic properties originate from their extended nature of the cation f-electron wave functions and their proximity to the Fermi level[2,3] which gives rise to a p-f hybridization with the electrons of the anion p-band. Among this series UAs is certainly the compound which presents the most complex behavior[1,4,5]. At about 1/2 T_N UAs undergoes a first order transition from a +-+- (Type I) to a ++-- (Type IA) stacking of ferromagnetic (001) planes[6,7]. This is accompanied by an increase of the magnetic moment of the uranium ion[7]. Moreover, no significant tetragonal distortion was detected at T_N or at the Type I-Type IA transition[8]. Measurements in high magnetic fields have evidenced a multiphase behavior[9,10], in particular a "ferrimagnetic" phase appears to exist close to T_N[4]. In trying to clarify the situation and shed some light on the puzzling magnetic properties of this compound, an extensive invetigation of the magnetic phase diagram (H,T) of UAs has been undertaken, including neutron diffraction and magnetization experiments. The main results for a magnetic field applied along the <001> and <110> directions are reported in this paper.

Solid solutions between uranium antiferromagnetic monopnictides and ferromagnetic uranium monochalcogenides likewise present complicated magnetic properties. The availability of single crystals has allowed us to investigate in detail the USb system either by diluting antimony by tellurium[11,12,13], or by diluting uranium by nonmagnetic ions such as thorium or yttrium. The temperature-concentration phase diagrams are presented here. The unusual magnetization processes for the antiferromagnetic structures as evidenced by the neutron and magnetization experiments will be emphasized.

EXPERIMENTAL DETAILS

All the investigated samples were single crystals which were grown at E.T.H. Zürich by a recrystallization process in which the material is kept ≃50°C below the melting temperature. Magnetization measurements have been performed at E.T.H. Zürich up to 90 kOe and at the Service National des Champs Intenses at Grenoble up to 200 kOe. Neutron experiments have been carried out at Grenoble Siloe reactor using a two-axis spectrometer equipped with a moving up counter arm. A vertical magnetic field was provided by a superconducting magnet. The cryostat allows a temperature variation in between 1.5 K and 300 K. The highest available field value is 100 kOe, which can be obtained with the superfluid helium bath at atmospheric pressure[14].

We emphasize[1] that in a neutron experiment only the amplitude and the direction of the Fourier component $\vec{m}_k$ are determined and not the moment distribution $\vec{m}(\vec{R}_n)$. The latter may be deduced from the relation

$$\vec{m}(\vec{R}_n)=\sum_k \vec{m}_k e^{2\pi i\vec{k}\cdot\vec{R}_n} .$$

In uranium monopnictides only longitudinal waves are observed ($\vec{m}_k // \vec{k}$), thus $\vec{m}_k$ can be written as:

$$\vec{m}_k=\frac{1}{2}A_k e^{i\phi_k}\,\hat{u}_k,$$

where A_k is the amplitude, $\hat{u}_k$ the polarization, and ϕ_k the phase of the wave.

(H,T) MAGNETIC PHASE DIAGRAM OF UAs

In Figure 1 the magnetic (H,T) phase diagram of UAs is reported for a magnetic field applied along a [001] axis. The transition lines have been determined by scans at constant T in the region below T=62 K and by scans at constant H above this temperature. Besides the antiferromagnetic Type I and IA phases, ferrimagnetic phases are seen to be induced by a magnetic field at both high and low temperatures.

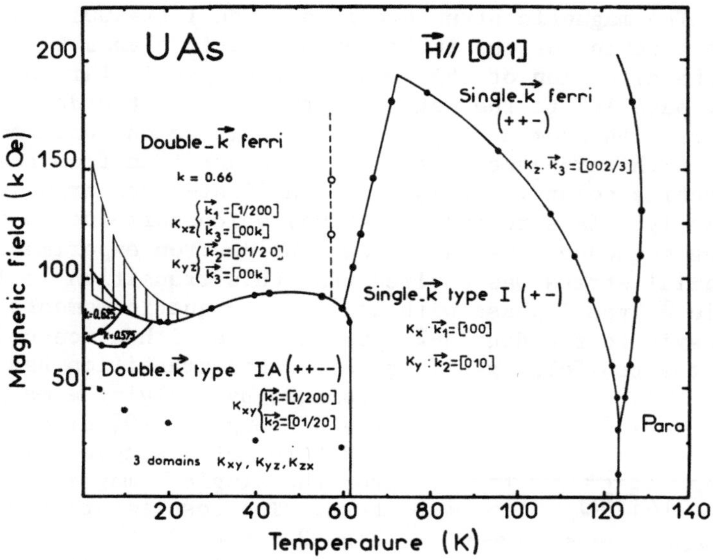

Fig. 1: (H,T) magnetic phase diagram of UAs for a magnetic field applied along a [001] axis.

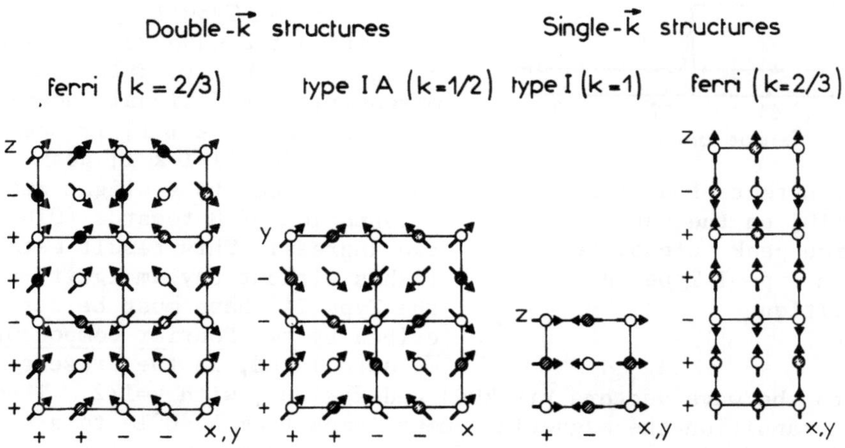

Fig. 2: Magnetic structures of the main phases of the (H,T) phase diagram of UAs for $\vec{H}$// [001]. Shaded symbols refer to projection of atoms in the adjacent plane.

Antiferromagnetic Phases:

For H=0 UAs orders at T_N=124.5±1 K with a pronounced first order transition. The magnetic structure is of Type I ($\vec{k}$=<001>) which amounts to a stacking of the (001) ferromagnetic planes in a +-+- sequence. The direction of the moments is perpendicular to the sheets, i.e. parallel to the wave vector. The first-order transition most likely results from a competition between second and fourth-order terms in the exchange interaction, rather than from the vicinity of the Lifschitz point[5]. At T_0=62±1 K a first-order transition occurs to the Type IA structure (k=<00½>) which corresponds to a ++-- stacking sequence. We have shown[4] by neutron experiments in which a uniaxial stress was applied that this transition is indeed from a single-$\vec{k}$ Type I phase with collinear magnetic moments along a four-fold axis to the double-$\vec{k}$ Type IA phase with a non-collinear order along the two fold axes (Fig. 2). This transition has been re-examined by applying a magnetic field along a [1$\bar{1}$0] direction in order to get a single domain state for the single-$\vec{k}$ magnetic structure. The results are given in Fig. 3 for H=80 kOe and decreasing T. In the Type I phase only the superlattice peak [1 1 1-k] with k=1 has been observed, indicating that the antiferromagnetic direction lies along the [001] axis perpendicular to the field direction. At the Type I-Type IA transition the wave vector $\vec{k}$=[00k] changes from k=1 to k=1/2 but, moreover, an additional superlattice peak [1-k 1 1] is seen to appear. The [1 1-k 1] peak did not appear due to a slight misorientation of H towards [010] of two degrees. This result establishes without any ambiguity that the Type IA phase must be described by <u>two Fourier components</u> $\vec{m}_k$ associated, in the present case, to the wave vectors $\vec{k}_1$=[k00] and $\vec{k}_3$=[00k] with k=1/2. Then at the transition the magnetic moment jumps from a <001> to a <110> direction inducing an increase of the magnetic moment value. A similar behavior has recently been found in UP; the results will be published elsewhere. In UP, however, the single-$\vec{k}$ double-$\vec{k}$ transition at T_0=22.5 K does not entail a change of wave vector value. Thus the low temperature transition in UAs and UP appear to be of the same nature; the Type IA phase, observed in UAs, may be the

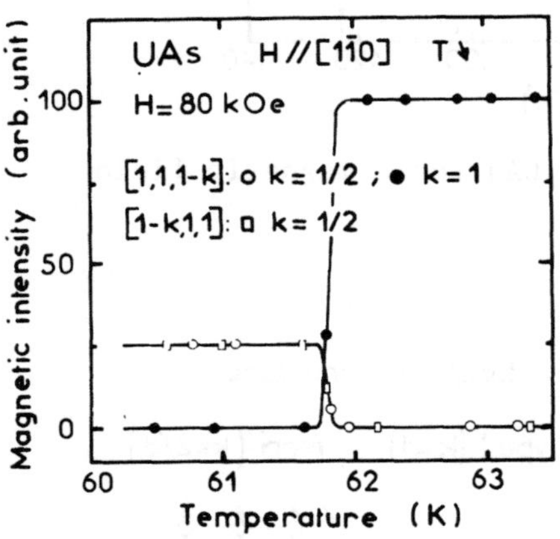

Fig. 3: Effect of a field H ∥[1$\bar{1}$0] on the superlattice peak intensities at the Type I-Type IA transition

result from a modification of the exchange integrals induced by the onset of the double-$\vec{k}$ structure.

For H applied along [001] the domain distribution is different. For T>62 K and H>10 kOe there exist two domains noted by K_x and K_y in Fig. 1, whereas for T<62 K a single-domain K_{xy} associated with a double-$\vec{k}$ Type IA structure is obtained for fields in the range of 25-50 kOe (dotted line in Fig. 1). Indeed a double-$\vec{k}$ order also gives rise to the three domains, denoted by K_{xy}, K_{yz}, K_{zx}, which are found in the lower field region for T<62 K. In the K_{xy} domain (Fig. 2) the moment distribution can be written as:

$$\vec{m}(R_n)=m_o[\cos(2\pi\vec{k}_1\cdot\vec{R}_n+\phi_1)\hat{x}+\cos(2\pi\vec{k}_2\cdot\vec{R}_n+\phi_2)\hat{y}]$$

Since the phases ϕ_1 and ϕ_2 can both take the values $\pm(\pi/4)$ or $\pm(3\pi/4)$ a large number of phase domains may exist.

High Temperature Ferrimagnetic Phase

The high temperature ferrimagnetic phase extends above the triple point (T_t=124±1 K, H_t=30 kOe) both for H applied along [001] and [1$\bar{1}$0]. For $\vec{H}$ // [001] the neutron experiments evidence superlattice peaks associated to a single wave vector $\vec{k}_3=[00\frac{2}{3}]$. Thus the ordering corresponds to a modulated collinear structure which is

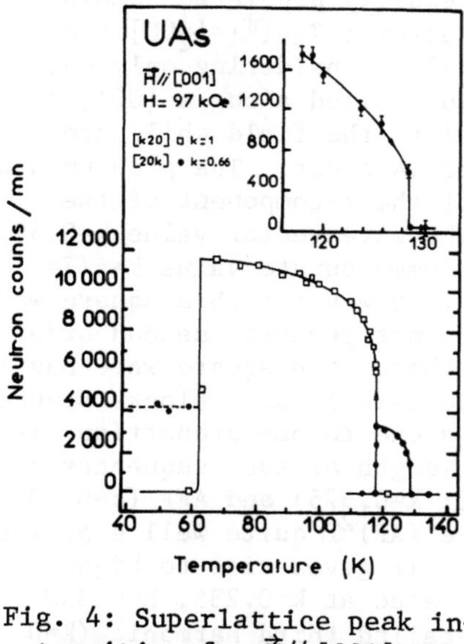

Fig. 4: Superlattice peak intensities for $\vec{H}$// [001] at 97 kOe as a function of temperature

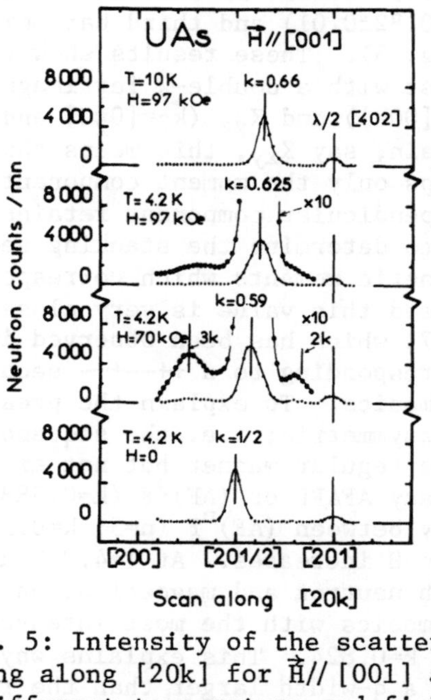

Fig. 5: Intensity of the scattering along [20k] for $\vec{H}$// [001] at different temperature and field values

close to a ++- stacking sequence of the (001) ferromagnetic planes (Fig. 2). Indeed the modulation is not a complete square wave, for example at T=120 K and H=97 kOe a modulation amplitude A_k=1.7±0.2 μ_B is obtained whereas the ferromagnetic component is only 0.3 μ_B (a ratio of four is needed for a complete square wave). As shown in Fig. 4, the transitions from the ferrimagnetic phase to the Type I phase and to the paramagnetic phase are both first order. In the latter case the slope of the transition line becomes vertical for H≃130 kOe, so that for H>130 kOe a decrease of the magnetization at the ordering temperature is found, in full agreement with the Clausius-Clapeyron equation.

Low Temperature Ferrimagnetic Phases

For T<60 K a ferrimagnetic state is induced by a field H≃70-90 kOe with a net magnetization of ≃0.44 μ_B at 150 kOe[9]. However, for T<15 K a step-like behavior in the magnetization curves occurs indicating more complicated processes on which we will concentrate first.

At T=4.2 K the double-$\vec{k}$ Type IA structure is destabilized by a field H_{C1}=70 kOe. Above H_{C1} the superlattice peaks associated with the wave vectors $\vec{k}_1=[\frac{1}{2}00]$ and $\vec{k}_2=[0\frac{1}{2}0]$ remain but are a factor of two less intense. Furthermore scans along [20k] (Fig. 5) indicate new superlattice peaks associated with a wave vector $\vec{k}_3$=[00k] with k=0.590±0.005; in addition to this fundamental component, second (k=0.82±0.01) and third harmonics (k=0.27±0.01) are also detected (Fig. 5). These results show that two equally populated domains exist with a double-$\vec{k}$ ferrimagnetic structure: K_{xy}($\vec{k}_1=[\frac{1}{2}00]$ and $\vec{k}_3$=[00k]) and K_{yz} ($\vec{k}_2=[0\frac{1}{2}0]$ and $\vec{k}_3$=[00k]). Considering only one domain, say K_{xy}, this means that a field applied along a [001] axis flips only the moment component parallel to the field while the perpendicular component retains the Type IA order. The problem now is to determine the stacking sequence of the z-component of the magnetic moments which corresponds to the wave vector value k=0.59. Indeed this value is very close to the commensurate value k=4/7= 0.571 which has been observed in CeSb[15]. However such a square wave, corresponding to a ++--++- sequence does not generate second order harmonics. To explain the presence of these, the square wave has to be asymmetric; i.e. the sequence must contain A and F blocks stacked in a regular manner but not exactly in a one to one proportion: let us say AFAFF or $(AF)^2F$ (k=0.588). The length of such sequences can vary between $(AF)^nF$ (n=5: k=0.579; n=10: k=0.575) and AFF (k=0.60) when H increases. At T=4.2 K a sequence $(AF)^2F$ quite well explains both neutron and magnetization results. It gives rise to higher harmonics with the most intense ones located at k=0.235, k=0.353 and k=0.824. This explains why the so-called third harmonic (k=0.27) has a q-width larger than the resolution because it is indeed the sum of two harmonics of equal intensity at k=0.235 and k=0.353. Moreover, such a sequence gives rise to a net magnetization

$\mu_F=0.176\ m_U/\sqrt{2}=0.28\mu_B$ ($m_U=2.25\ \mu_B$[16]) in good agreement with the experimental value 0.27 μ_B at H=70 kOe. A sequence ABAB without defects would give $\mu_F=0.23\ \mu_B$. By further increasing the field the wave vector is found to shift to a vlue of k=0.625 at H_{c2}=75 kOe (Fig. 5) and remains constant up to 97 kOe. Such a value is explained by an AF^4 sequence. So the magnetization processes would result from an increase in the number of F-blocks at the cost of the A-blocks in the stacking.

At T=10 K and $H=H_{c1}$ the wave vector has a slightly smaller value (k=0.574±0.005) indicating that the actual ordering is closer to the ideal stacking AF (++−−++−). This phase persists up to 90 kOe where the wave vector jumps directly to the value k=0.66=2/3 (Fig. 5). Indeed at H=100 kOe, with increasing T, this value remains the same until the Type I phase is reached. The main low temperature phase is therefore a double-$\vec{k}$ ferrimagnetic phase described, for the K_{xz} domain, by the wave vector $\vec{k}_1$=[1/2 0 0] and $\vec{k}_3$=[0 0 2/3] (Fig. 2). However, the magnetic structure entails a net magnetization of $m_U/3\sqrt{2}=0.53\mu_B$, while the experimental value is only $0.44\mu_B$ at 150 kOe. To reconcile the neutron and the magnetization results we may assume the (++−) sequence to contain defects (or solitons), for example (+−−) blocks which will reduce the ferromagnetic component without changing the wave vector value. Such defects are reminiscent of (++−−) sequences (++− +|−− ++|− ++−), and their density would depend on field and temperature. Additional neutron and magnetization experiments are needed to confirm this unusual behavior and to understand the weak magnetization anomaly observed at T≃55 K and H>100 kOe (dashed line in Fig. 1). Finally the transition line between the double-$\vec{k}$ ferrimagnetic and the Type I phases has a constant slope of 8 kOe/K.

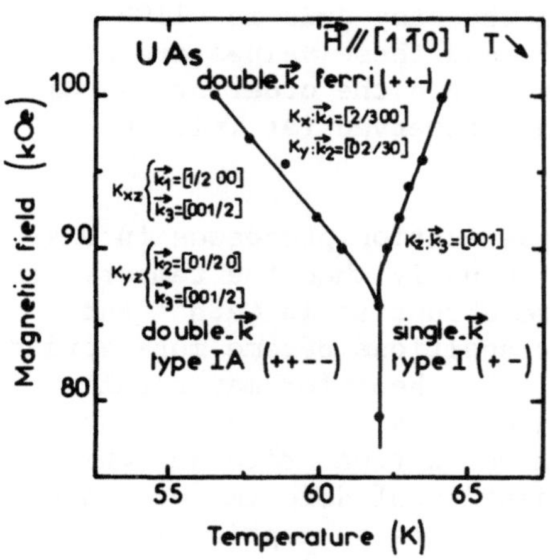

Fig. 6: (H,T) phase diagram of UAs for $\vec{H}$// [1$\bar{1}$0] near the Type I-Type IA transition

(H,T) Phase Diagram for $\vec{H}$// [1$\bar{1}$0]

For a magnetic field applied along [1$\bar{1}$0] the phase diagram has the same feature as that obtained for $\vec{H}$// [001], except for two points: the slope between the L.T. ferrimagnetic and the Type I phase is smaller (6 kOe/K) and the critical fields observed for T<60 K are all about 20 kOe larger. Therefore, the neutron experiments could only map two small regions of the phase diagram corresponding to T<20 K and T∿60 K.

Below T=20 K the double-$\vec{k}$ Type IA structure (domains K_{xz} and K_{yz}) is destabilized for

$H \sim 80$ kOe to give a double-$\vec{k}$ ferrimagnetic structure (domain K_{yz}) similar to that obtained for $\vec{H}$// [001], i.e. characterized by the wave vectors $\vec{k}_2$=[0 k 0] with k=0.575±0.005 and $\vec{k}_3$=[0 0 1/2] and with a net magnetization $\mu_F \sim 1/7\ 1/\sqrt{2}\ m/\sqrt{2}=0.16\ \mu_B$ in agreement with the experimental value (0.16 μ_B). Like $\vec{H}$// [001], the moment components perpendicular to H remain antiferromagnetically coupled.

The phase diagram obtained with decreasing T in the vicinity of T=62 K is reported in Fig. 6. The ferrimagnetic ordering has a different nature because no AF component remains. Indeed we have observed both the wave vectors $\vec{k}_1$=[k00] and $\vec{k}_2$=[0k0] with k=2/3, indicating a ++- sequence along both [100] and [010]. So there are two possible structures: i) a collinear ferrimagnetic structure with moments along a fourfold axis and two domains, ii) a double-$\vec{k}$ ferrimagnetic structure with a non-collinear order but with μ_F// [1$\bar{1}$0]. By comparing the ferromagnetic component to the experimental value ($\sim$0.5 μ_B at T=60 K) the single-$\vec{k}$ structure (μ_F=0.46 μ_B) is favored compared to the double-$\vec{k}$ structure (μ_F=0.75 μ_B). But at low temperatures μ_F becomes larger ($\sim$0.6 μ_B at T=10 K and H=150 kOe) and, similarly to $\vec{H}$// [001], defects may be invoked to reduce the ferromagnetic component. Thus, the double-$\vec{k}$ ferrimagnetic structure seems to be the most probable one.

Magnetization Processes

The knowledge of the structures of the L.T. ferrimagnetic phase for $\vec{H}$// [001] and [1$\bar{1}$0] enables us to understand why μ_F is $\sqrt{2}$ times larger for $\vec{H}$// [110] than [001]. If $\vec{H}$// [001] the magnetic field flips only one component of the double-$\vec{k}$ structure ($\vec{\mu}_F$=[0 0 P/$\sqrt{2}$]) whereas for $\vec{H}$// [1$\bar{1}$0] both components are flipped ($\vec{\mu}_F$=[P/$\sqrt{2}$ P/$\sqrt{2}$ 0]). Therefore the critical field along [001] is lower than along [110], as observed experimentally, even though the easy axis is <110> (at T=40 K $H_{c[001]}$=92 kOe, $H_{c[110]}$=108 kOe), and these values are not in a simple $\sqrt{2}$ ratio. On the other hand, when the ordering is collinear along [001] the $\sqrt{2}$ ratio is almost observed (at T=115 K $H_{c[001]}$=100 kOe, $H_{c[110]}$=130 kOe).

Another unusual feature of the magnetization processes in UAs is the behavior observed below $T \sim 20$ K. Usually when the temperature is decreased the transitions become more abrupt as in CeSb[11] and CeBi[17,18]. Contrastingly, in UAs the transitions became more spread as shown by the shaded area in Fig. 1. This behavior may result from the double-$\vec{k}$ structure which contains a large number of phase domains. The latter could have complex relaxation processes which vary drastically below 20 K. This relaxation of defects (or phase domains) may also be invoked to explain why the susceptibility of the double-$\vec{k}$ ferrimagnetic phase is much larger at low temperature, contrary to expectation ($\sim$ a factor of two between T=4.2 K and T=20 K).

MAGNETIC PROPERTIES OF $USb_{1-x}Te_x$, $U_{1-x}Th_xSb$ and $U_{1-x}Y_xSb$

USb is an antiferromagnet of Type I (T_N=213±2 K[13]) with an ordered moment at 4.2 K of 2.82 μ_B[19] whereas UTe is a ferromagnet with moments aligned along a <111> axis[20] (m_U=2.25μ_B, T_C=104 K).

$USb_{1-x}Te_x$ [11,12,13]

For USb-UTe solid solutions a detailed investigation of the temperature-concentration phase diagram has been undertaken. For tellurium concentration x>0.27 the solid solutions order ferromagnetically along a <111> axis whereas for x<0.06 the Type I structure of USb remains. For 0.06<x<0.27 an incommensurate order is established at T_N∿210 K and a first order transition towards a commensurate state occurs at T_{IC}∿160 K. The latter corresponds to a Type IA order (++−−; k=1/2), except for x=0.10 where a ferrimagnetic order (++−; k=2/3) has been found to appear. Indeed this ferrimagnetic structure and a neutron experiment under uniaxial stress on USb has lead us to conclude that USb as well as the $USb_{1-x}Te_x$ antiferromagnetic solid solutions order with a triple-$\vec{k}$ structure. This implies that for any tellurium concentration, the uranium moments always have a <111> easy axis. The main effect of the tellurium substitution is then to destroy drastically the antiferromagnetic coupling between nearest neighbors.

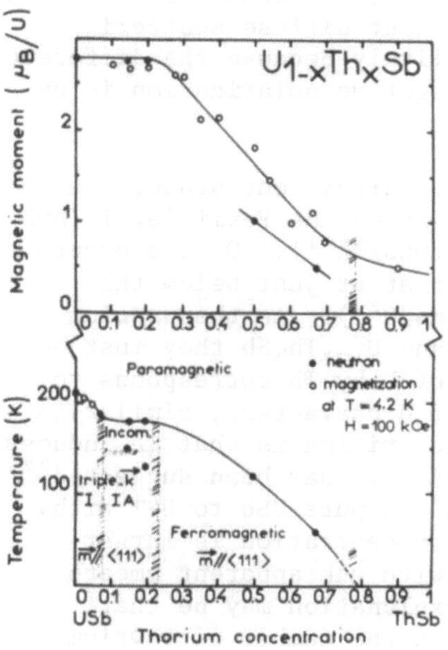

Fig. 7: Magnetic phase diagram of $U_{1-x}Th_xSb$ and dependence of the U moment on Th.

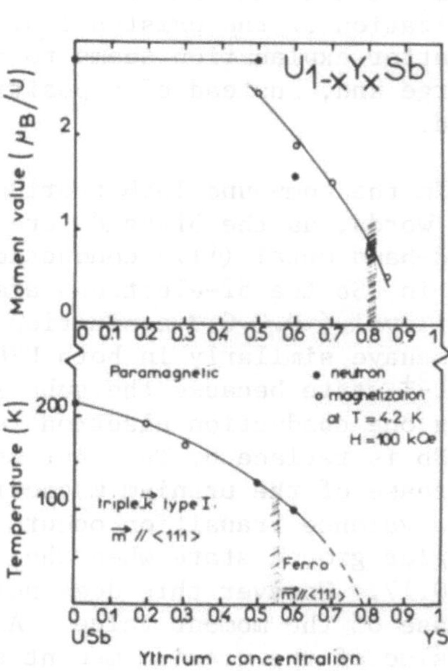

Fig. 8: Magnetic phase diagram of $U_{1-x}Y_xSb$ and dependence of the U moment on Y.

$\underline{U_{1-x}Th_xSb}$

In Fig. 7 the magnetic phase diagram of $U_{1-x}Th_xSb$ solid solutions is reported as determined by neutron and magnetization measurements. The dilution of uranium by thorium gives exactly the same result as the substitution of antimony by tellurium. The Type I phase disappears for x>0.07 and the ferromagnetic state is found to occur for x>0.25. In between an incommensurate order develops at T_N which locks into a commensurate triple-$\vec{k}$ Type IA phase at lower temperature. For small impurity concentration even a one-to-one correspondence appears to exist between Te and Th dopings: at T_N the wave vector value is k=0.515 for $U_{0.85}Th_{0.15}Sb$ and k=0.505 for $USb_{0.85}Te_{0.15}$. However, since thorium is non-magnetic, it depresses the ordering temperature ($T_C \sim 60$ K for x=0.66) and long-range order does no longer exist above x≃0.80 (which corresponds indeed to the percolation limit for an f.c.c. lattice).

The most surprising result is the dependence on the thorium concentration of the uranium magnetic moment given in Fig. 7. In the antiferromagnetic structures (x<0.25) it remains nearly constant ($m_U=2.75\pm0.1\mu_B$ for x=0.15), whereas in the ferromagnetic state it decreases almost linearly with the thorium concentration. This follows from both magnetization (at T=4.2 K and H=100 kOe) and neutron experiments. However the latter give a moment value lower by ≃40% ($m_U=1.0\pm0.1$ μ_B for x=0.5 and $m_U=0.5\pm0.2$ μ_B for x=0.66). This difference may be accounted for by either a conduction electron polarization of the existence of an important diffuse scattering. The latter explanation seems to be most likely because the difference is large and, instead of a positive, a negative polarization is expected.

In the compound ThSb thorium is in a tetravalent state. In other words, as the 5f-bands are located above the Fermi level ThSb is a d-band metal (∿1.5 conduction electrons/Th[21]). On the other hand, in USb the 5f-electrons are located at or just below the Fermi level (∿0.2-0.3 conduction electrons/U[21]). As the uranium ions behave similarly in both $USb_{1-x}Te_x$ and $U_{1-x}Th_xSb$ they must be in a U^{3+} state because the substitution of U by Th corresponds to adding one conduction electron (mainly of d-character), similar to when Sb is replace by Te. But what is surprising is that Th induces a decrease of the uranium magnetic moment. It has been suggested[22] that a valence transition occurs from U^{3+} in pure USb to U^{4+} with a singlet ground state when the thorium concentration is larger than 0.17. However this does not agree with the apparent smooth decrease of the moment value. Another explanation may be that the value of the uranium moment depends on the number of thorium atoms surrounding a uranium atom.

$\underline{U_{1-x}Y_xSb}$

In order to check whether there is any effect of the valence state of the nonmagnetic cation[23], the phase diagram of $U_{1-x}Y_xSb$ solid solutions have also been investigated. As shown in Fig. 8 the behavior is different from $U_{1-x}Th_xSb$, the magnetic order remaining of Type I with a triple-$\vec{k}$ structure up to x=0.55, above which it changes directly to ferrimagnetism. The ordering temperature decreases smoothly with x and disappears for x∿0.8 similar to Th doping. The high field magnetization data (T=4.2 K, H=100 kOe) indicate the uranium moment decreases almost linearly with the yttrium concentration (∿2.4 μ_B for x=0.5 and ∿0.5 μ_B for x=0.85). On the other hand, neutron experiments performed on single crystals with a composition near x=0.55 indicate that the moment value drops at this critical concentration: 2.8±0.1 μ_B for the Type I order at x=0.5 and 1.57±0.1 μ_B in the ferromagnetic state at x=0.6. However, neutron experiments on single crystals are needed to clarify the behavior for larger yttrium concentration.

In the case of dilution by yttrium it is more difficult to understand the behavior of the magnetic moment because yttrium is certainly in a Y^{3+} state so that its introduction does not entail additional conduction electrons as in the case of Th or Te. This result also strongly supports a U^{3+} ground state. Lattice contraction effects appear to be unlikely because USb and YSb have almost the same lattice parameters; experiments with La and Lu dilution would be interesting. Indeed, this result is very difficult to understand because in both the antiferro- and the ferromagnetic states, the moments align along the <111> axes, whereas the moment value seems to change only in the ferromagnetic state. One possibility is to invoke the virtual valence fluctuation mechanism proposed by T. Kasuya[24] to explain the unusual and very analogous behavior of the cerium monopnictides. This model is based on a strong interaction between f-electrons located just below the Fermi level and the anion p-band, which must touch the Fermi level. This mechanism gives rise to an extra contribution to the crystal field and may account for the large anisotropic exchange interactions present in these compounds. This strong p-f hybridization has two effects: i) to push up the valence band thereby increasing the number of conduction electrons (∿0.2-0.3 e/U[21]); ii) to move down selectively the crystal field levels, in the present case a Γ_8^1 quartet, which explains the strong single-ion anisotropy yielding a <111> easy axis. In the uranium compounds the p-f hybridization is certainly more important than in cerium compounds. Therefore the doping of USb with Th or Te will correspond to the addition of conduction electrons, i.e. by shifting up the Fermi level the p-band is filled. Consequently the p-f hybridization decreases and the exchange integrals are modified. To put Y in USb would have the same effect as pushing down the valence band because there are no f-electrons. Such an effect has been observed in diluted CeSb compounds[25]. So a decrease of the p-f mixing would move up the Γ_8^1 ground state closer to the Fermi level and uranium would enter an intermediate valence regime. Such a behavior does not occur in CeSb or CeBi in which the 4f electrons are further away from the Fermi level (∿2 eV). Of course quantitative calculations and more experimental work are needed to confirm the above explanation.

Magnetization Processes in Triple-$\vec{k}$ Structures

Although the $USb_{0.85}Te_{0.15}$ and $U_{0.85}Th_{0.15}Sb$ solid solutions order in the compensated antiferromagnetic Type IA structure, the low temperature magnetization curves are more resemblant to those of a ferro- or a ferrimagnet than that of an antiferromagnet, as illustrated by the hysteresis loop presented in Fig. 9. Even for $U_{0.5}Y_{0.5}Sb$, which orders in the Type I structure, such a behavior was observed but in that case without the multistep behavior which appears to be characteristic of the Type IA phase. Indeed this metastable behavior is found in each of the solid solutions at low enough temperatures and for arbitrary field direction. With increasing temperature the value of the remanent moment decreases to zero, however hysteresis effects remain present. Another peculiarity of the triple-$\vec{k}$ structures is that although the easy axis is always found to be a <111> axis, the critical field has a lower value along <100>. In order to understand these special features, neutron experiments with a magnetic field along <111> have been performed on both $USb_{0.85}Te_{0.15}$ and $U_{0.85}Th_{0.15}Sb$.

For $USb_{0.85}Te_{0.15}$ magnetization and neutron results at T=100 K are summarized in Fig. 10. For H=0 the three Fourier components associated with $\vec{k}_1$=[1/2 0 0], $\vec{k}_2$=[0 1/2 0] and $\vec{k}_3$=[0 0 1/2] are equivalent. At H=20 kOe the superlattice peak [1/2 1 1] disappears and $\vec{k}_1$ shifts to $\vec{k}_1$=0. Neutron experiments indicate that the ferromagnetic component is indeed along [100] with a value of 1.55±0.1 μ_B, giving along [111] a value μ_F=0.90 μ_B in good agreement with magnetization data (0.92 μ_B). Thus for H>20 kOe the ordering corresponds to a triple-$\vec{k}$ structure in which one moment component is ferromagnetically ordered ($\vec{k}_1$=0) whereas the other two components retain the antiferromagnetic ++-- sequence ($\vec{k}_2$=[0 1/2 0] and $\vec{k}_3$=[0 0 1/2]). This state remains unchanged until H>42 kOe; then the [1 1/2 1] and [1 1 1/2] peaks successively disappear and above H=50 kOe a collinear ferromagnetic state is reached. The latter two transitions are too close to be discerned in the present magnetization measurements. We remark that in the intermediate phase the ferromagnetic component is along [110].

Thus our neutron experiments prove without any ambiguity that the antiferromagnetic USb-UTe solid solutions order in a triple-$\vec{k}$ structure. In the investigated case, a magnetic field applied along

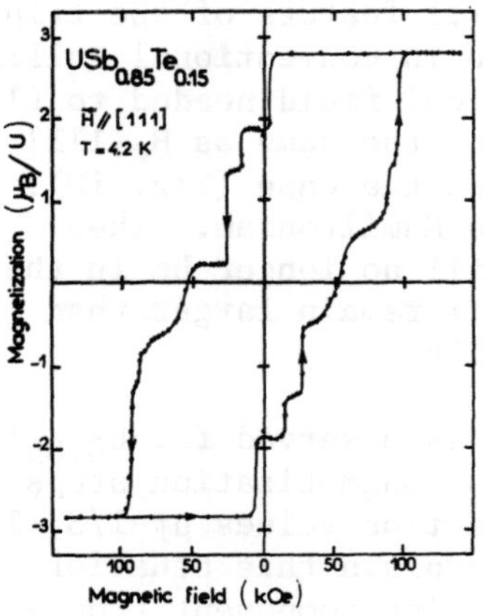

Fig. 9: Hysteresis loop at T=4.2 K of $USb_{0.85}Te_{0.15}$

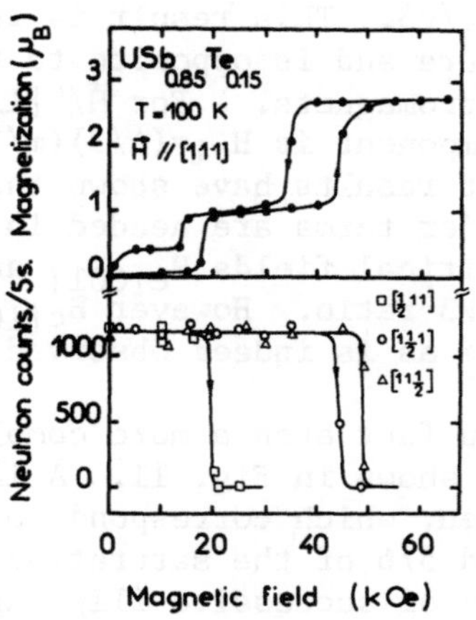

Fig. 10: Neutron and magnetic data showing the magnetization processes in the triple-$\vec{k}$ structure of $USb_{0.85}Te_{0.15}$

[111] flips successively one, two and three Fourier components to give a net magnetization of respectively $[m\ \sqrt{3},0,0]$, $[m/\sqrt{3},m/\sqrt{3},0]$ and $[m/\sqrt{3},m/\sqrt{3},m/\sqrt{3}]$. This peculiar behavior explains the complex hysteresis loop presented in Fig. 9. On the other hand, a field applied along [001] leaves the antiferromagnetically ordered moment components perpendicular to the field unchanged, whereas the parallel component yields a net magnetization $\mu_F=m/\sqrt{3}$. Therefore in the triple-$\vec{k}$ structures the critical fields will be lower for $\vec{H}//$ [100] than for $\vec{H}//$ [111], since only part of the antiferromagnetic interactions are broken. To understand this result, we will consider an exchange Hamiltonian of second order which can be written in Fourier space as

$$\sum_k T^{\alpha\beta}(\vec{k})m_k^{\alpha}m_{-k}^{\beta}(\alpha,\beta=x,y,z).$$

Since there is no coupling between the three Fourier components, $\vec{m}_{k1}$, $\vec{m}_{k2}$ and $\vec{m}_{k3}$, a field $\vec{H}//$ [001], which flips only one Fourier component, has to overcome only one third of the A.F. exchange interaction $\Delta=E_{IA}-E_F$ needed to reach the collinear ferromagnetic state. The moments being fixed to <111>, the ferromagnetic component // [001] is only $\mu_F=m/\sqrt{3}$ instead of $\mu_F=m$ for $\vec{H}//$ [111]. Therefore the critical field needed to flip all three components along <111>, $H_{c[111]}=\Delta/m$ will be $\sqrt{3}$ larger than $H_{c[001]}=(\Delta/3)(m/\sqrt{3})^{-1}$ =

$H_{c[111]}/\sqrt{3}$. This result is indeed a general feature of the triple-$\vec{k}$ structure and is opposite to the situation in conventional collinear antiferromagnets. For $\vec{H}//$ [111] the critical field needed to flip one component is $H_{c1}=(\Delta/3)(m/3)^{-1}=\Delta/m$, i.e. the same as H_c[111]. But our results have shown that this is not the case (Fig. 10), so 4th order terms are needed in the exchange Hamiltonian. Then the critical fields $H_{c[001]}$ and $H_{c[111]}$ will no longer be in the ideal $\sqrt{3}$ ratio. However $H_{c[100]}$ will still remain larger than $H_{c[111]}$ as is indeed observed experimentally.

In fact even a more complex behavior is observed for $U_{0.85}Th_{0.15}$ Sb, as shown in Fig. 11. A large number of magnetization steps are seen, which correspond to net magnetization values μ_F=1/3, 1/2, 2/3 and 5/6 of the saturation value. To explain this behavior instead of successive flipping of each Fourier component from a (++--) directly to a (++++) state, an +++- intermediate sequence must occur. For example, the phase with μ_F=m/2 is accounted for by a triple-$\vec{k}$ ferrimagnetic structure, in which the x-component of the moment propagates with a ++++ sequence, the y-component with a +++- sequence and the z-component with a ++-- sequence. The above-mentioned series of values for the magnetization correspond indeed to triple-$\vec{k}$ ferrimagnetic phases with successive ferromagnetic components $[m/\sqrt{3},0,0][m/\sqrt{3},m/2\sqrt{3},0],[m\sqrt{3},m\sqrt{3},0]$ and $[m/\sqrt{3},m\sqrt{3},m/2\sqrt{3}]$.

CONCLUSION

In this paper we have described a series of magnetization and neutron experiments which show that UAs and USb solid solutions exhibit highly unusual magnetic properties.

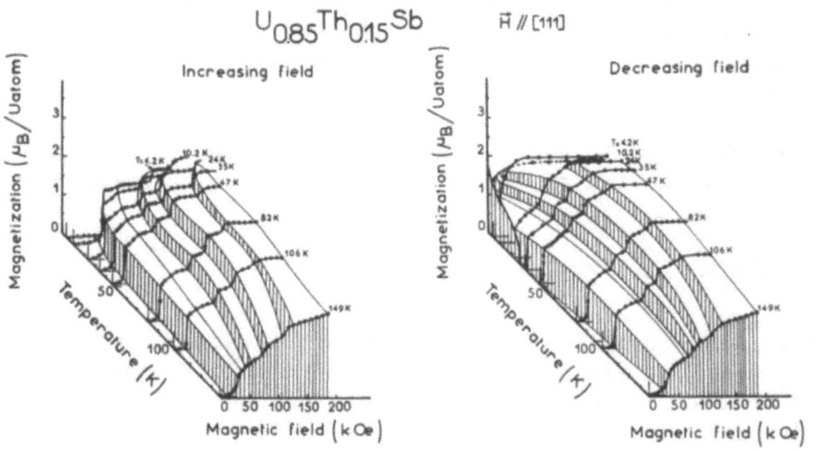

Fig. 11: Magnetic phase diagrams of $U_{0.85}Th_{0.15}Sb$ in increasing T and magnetic field

In USb and $USb_{1-x}Te_x$, $U_{1-x}Th_xSb$ and $U_{1-x}Y_xSb$ triple-$\vec{k}$ magnetic structures have been found to exist up to T_N. In these non-collinear structures the moments are along <111> directions in agreement with polarized neutron data[19], which indicates that the magnetization density associated with the quartet Γ_8^1 is oblate. Moreover, for USb, the triple-$\vec{k}$ Type I structure that we propose allows understanding[26] of the unusual features of the inelastic neutron scattering experiments[27]. The large anisotropy which favors <111> is likely a single-ion crystal field contribution. Then the triple-$\vec{k}$ structure results from a compromise between this single-ion anisotropy along <111> and anisotropic exchange interactions which favor a <001> easy axis. So the behavior of USb seems to be well described by a localized picture with f-electrons located below the Fermi level in a Γ_8^1 quartet state. In the cerium monopnictides, the strong exchange anisotropy is thought[24] to arise from p-f mixing, and a similar argument might apply to the uranium monopnictides where this mixing is even stronger. However an alternative explanation of the exchange anisotropy has been given[28] by using a Coqblin-Schreiffer interaction. Moreover in order to explain the critical field values, this anisotropic exchange interaction must contain 4th order terms, possibly of the quadrupolar type. Higher-order anisotropic terms must also be considered in explaining the magnetic behavior of UAs. To account for the double-$\vec{k}$ structure observed at low temperatures, the single-ion anisotropy must favor a <110> easy axis. So the ground state of UAs is certainly different from that of USb, although the magnetization density is oblate in both cases[5,19]. However, the double-$\vec{k}$ Type IA structure does not remain up to T_N, since at $T_N/2$ a double-$\vec{k}$ single-k transition is observed. The existence of this transition implies that the 4th order terms in the exchange Hamiltonian must be even large enough to overcome the <110> single-ion anisotropy, so that a collinear structure with moment along <001> results. The magnetic phase diagram of UAs is complicated, with a multiphase behavior, and is similar in many respects to that of CeBi[17], except for the nature of the main ferrimagnetic phase (++- instead of +++- in CeBi).

Finally, we want to emphasize the anomalous behavior of the uranium magnetic moment in the USb solid solutions when diluting U by either Th or Y. An intermediate velence state or a modification of the ground state level may be invoked to explain this puzzling result.

ACKNOWLEDGEMENT

We want to thank Dr. J. de Jongh for a critical reading of the manuscript.

REFERENCES

1. J. Rossat-Mignod, P. Burlet, S. Quézel and O. Vogt, Physica 102 B:237 (1980)
2. J. Schoenes, Physica 102 B:45 (1980)
3. Y. Baer, Physica 102 B:104 (1980)
4. J. Rossat-Mignod, P. Burlet, H. Bartholin, R. Tchapoutian, O. Vogt, C. Vettier and R. Lagnier, Physica 102 B:177 (1980)
5. S.K. Sinha, G.H. Lander, S.M. Shapiro and O. Vogt, Physica 102 B:174 (1980) and Phys. Rev. 23:4556 (1981)
6. J. Leciejewicz, A. Murasik and R. Troć, Phys. Status Solidi 30 :157 (1968)
7. G.H. Lander, M. H. Mueller and J. F. Reddy, Phys. Rev. B6:1880 (1972)
8. H.W. Knott, G.H. Lander, M.H. Mueller and O. Vogt, Phys. Rev. B 21:4159 (1980)
9. G. Busch, O. Vogt and H. Bartholin, J. de Phys. C4-40:64 (1979)
10. O. Vogt, Physica 102 B+C:206 (1980)
11. J. Rossat-Mignod, P. Burlet, O. Vogt and G.H. Lander, J. Phys. C12:1101 (1979) and J. de Phys. C4-40:70 (1979)
12. P. Burlet, S. Quézel, J. Rossat-Mignod, G. H. Lander and O. Vogt J. Magn. Magn. Mater. 14:309 (1979)
13. P. Burlet, S. Quézel, J. Rossat-Mignod, O. Vogt and G.H. Lander Physica 102 B:271 (1980)
14. G. Claudet, B. Gravil, J. Rossat-Mignod and S. Burgess, submitted to Cryogenics
15. J. Rossat-Mignod, P. Burlet, J. Villain, H. Bartholin, Wang Tcheng-Si, D. Florence and O. Vogt, Physica 86-88 B:129 (1977) ; Phys. Rev. B 16:440 (1977)
16. The moment value 2.25 μ_B obtained from powder experiments[7] has been used because our single crystal data gives a larger value ($\sim$2.4 to 2.5 μ_B) indicating the presence of extinction effects.
17. H. Bartholin, P. Burlet, S. Quézel, J. Rossat-Mignod and O. Vogt, J. de Phys. C5-40:130 (1979)
18. H. Bartholin, D. Florence, Wang Tchen-Si and O. Vogt, Phys. Status Solidi (a) 24:631 (1974)
19. G.H. Lander, M. H. Mueller, D. M. Sparlin and O. Vogt, Phys. Rev. B 14:5035 (1976)
20. A. Delapalme, G. Busch, O. Vogt and G.H. Lander, J. de Phys. C4-40:74 (1979)
21. J. Schoenes, private communication
22. B.R. Cooper, O. Vogt and R. Siemann, J. Magn. Magn. Mat. 15-18: 1249 (1980)
23. B. R. Cooper, O. Vogt and R. Siemann, Physica 102 B:41 (1980)
24. K. Takegahara, H. Takahashi, A. Yanase and T. Kasuya, J. Phys. C 14:737 (1981)

25. H. Bartholin, J.M. Effantin, P. Burlet, J. Rossat-Mignod and O. Vogt, submitted to this conference
26. J. Jensen, P. Bak, preprint
27. G. H. Lander and W.G. Stirling, Phys. Rev. B 21:436 (1980)
28. R. Siemann and B.R. Cooper, Phys. Rev. Lett. 44:1015 (1980)

COMMENTS

KASUYA: Because of a very large mixing effect the real valency seems to be nonintegral. How did you obtain the conclusion that in USb the uranium is trivalent? In addition, it is dangerous to guess the valency from the point charge model because of large anisotropic mixing effects which may be the main anisotropy mechanism.

ROSSAT-MIGNOD: There are several evidences for U to be trivalent in USb: band calculations, photoemission and neutron measurements. The phase diagram of $USb_{1-x}Te_x$ is similar to that of $U_{1-x}Th_xSb$ and not to that of $U_{1-x}Y_xSb$. For such a valency the point charge model gives a <100> easy axis. Indeed, a [111] single-ion anisotropy has been demonstrated which may originate from the $\Gamma_8^{(1)}$ quartet. This can be favored, as in CeSb, by the p-f mixing which must in fact be larger. Consequently I argue that the uranium valency will not be an integral number, but smaller than three.

SUSKI: I think that the idea of explaining the moment jump by the change of the electronic structure cannot be completely rejected because both moment jump and change of magnetic structure could be driven by a change in the electronic structure.

ROSSAT-MIGNOD: I do not think that the single $\vec{k}$ - double $\vec{k}$ transition comes from an electronic structure change. The most likely explanation is competition between a single-ion anisotropy (CEF) and the fourth order anisotropic exchange term.

TRZEBIATOWSKI: In pseudobinary systems formed by ferro- and antiferromagnetics, the area of ferromagnetic coupling is always much larger than the antiferromagnetic. It seems therefore that in the uranium systems this kind of coupling is stronger. Rather small percentages of the ferromagnetic component introduced in antiferromagnetic solid solutions are sufficient to reverse this kind of coupling.

COOPER: I think the $U_xY_{1-x}Sb$ experiments showing the decrease of moment with dilution are important and interesting. For some reason the changeover from Kondo-like to mixed valence-like behavior seems to be very close in the uranium pnictides. These two regimes of behavior in theory correspond to an increasing degree of f-electron band electron (d and p) hybridization. But the question is, why do these two regimes of behavior fall so close together? With regard to Dr. Suski's remark on UP we should remember the work on electric resistivity done in Poland some years ago showing normal magnetic ordering anomalies for a magnetic metal.

ROSSAT-MIGNOD: No special comments; I agree with the above remarks.

WOHLLEBEN: When the moment is changing, as a function of concentration, are you then looking at the inelastic magnetic scattering to see "where the missing moment is going"?

ROSSAT-MIGNOD: Up to now we have not performed an energy analysis of the magnetic scattering. Experiments have been done on a 2-axis spectrometer.

LIGAND FIELD OF URANIUM (4+) ANTIPRISMATIC CLUSTER IN LCAO MO APPROACH

J. Mulak and Z. Gajek

Institute for Low Temperature and Structure Research
Polish Academy of Sciences, 50-950 Wrocław, Poland

INTRODUCTION

The physical origin of the crystalline field in ionic compounds has been the subject of intensive theoretical investigations. It has been proved[1-2] that the widely used electrostatic description is applicable only in those few cases when the contributions derived from other sources of crystalline field are mutually cancelled. These are, among others, the effects connected with the exchange interaction between central ion and ligand electrons, overlap and covalent mixing of their orbitals.

The present paper concerns the case of the antiprismatic cluster $(UO_8)^{12-}$. The calculations have been done for the compound $U(SO_4)_2\,4H_2O$ and the temperature dependence of paramagnetic susceptibilities[3] constitutes their experimental verification.

METHOD

The method applied is the perturbation approach of LCAO MO model. Its detailed description may be found elsewhere[4,5]. Therefore we present only its basic ideas and definitions which may be useful for further considerations.

First consider an ionic cluster in which a central ion is surrounded by ligands, the whole having point symmetry G. Let ϕ^{Γ} and χ^{Γ} denote free-ion symmetry orbitals transforming according to the irreducible representation Γ of the G group. We take into account only outer electrons of the central ion and ligands respectively. According to LCAO MO formalism, their appropriate combinations form antibonding $\psi^{\Gamma A}$ and bonding $\psi^{\Gamma B}$ molecular orbitals:

$$\begin{aligned} \psi^{\Gamma A} &= N^{\Gamma A} \ (\phi^{\Gamma} - \lambda^{\Gamma}\chi^{\Gamma}) \\ \psi^{\Gamma B} &= N^{\Gamma B} \ (\chi^{\Gamma} + \gamma^{\Gamma}\phi^{\Gamma}) \end{aligned} \tag{1}$$

where $N^{\Gamma A}$ and $N^{\Gamma B}$ are the normalization factors. Ionic character of the complex considered permits treatment of the mixing coefficients λ^{Γ} and γ^{Γ} as small parameters. Such an assumption considerably simplifies further procedures. As a result the one-electron energies of the antibonding states up to the second order have the following form (for the sake of simplicity indices Γ have been omitted):

$$\begin{aligned} \varepsilon = \langle\psi^{A}|h|\psi^{A}\rangle &= \langle\phi|h|\phi\rangle + S^2(\langle\phi|h|\phi\rangle + \langle\chi|h|\chi\rangle) - \\ &- 2S\ \langle\phi|h|\chi\rangle + \gamma^2(\langle\phi|h|\phi\rangle - \langle\chi|h|\chi\rangle) \end{aligned} \tag{2}$$

where

$$S = \langle\phi|\chi\rangle$$

$$\gamma = \frac{S\langle\chi|h|\chi\rangle - \langle\phi|h|\chi\rangle}{\langle\phi|h|\phi\rangle - \langle\chi|h|\chi\rangle}$$

and h is the H - F Hamiltonian of the system in the zeroth order approximation. The first term of ε is simply the diagonal energy. It includes the classic point charge potential, Kleiner's correction[6] and ligand exchange terms. The next two terms are contributions to the overlap energy and the last one comes from the covalent mixing effects.

The $(UO_8)^{12-}$ Cluster

We applied the simple method described above to the compound $U(SO_4)_2 4H_2O$. The local point symmetry at the uranium ion (4+) in this compound is very nearly D_{4d} (antiprism) and we neglect the slight deviation from it[3]. The average U-O and the smallest O-O distances in the antiprism are equal to 2.40 Å and 2.76 Å respectively. The ninefold degenerate $^3H_4$ ground term of the U^{4+} ion is split into the four doublets: $|\pm4\rangle$, $|\pm3\rangle$, $|\pm2\rangle$, $|\pm1\rangle$ and one singlet $|0\rangle$.

The Matrix Elements

In order to evaluate the overlap and energy integrals, we have used the central ion's 5f and ligand's 2s and 2p free-ion radial functions obtained from the Dirac-Slater procedure adopted by Koelling[7].

All the two-centre integrals were computed using the Sharma formula[8] of the expansion of the function about a displaced centre and a numerical integration scheme. The non-negligible three-centre integrals were approximated by the two-centre ones . The four-centre integrals were neglected. The calculations of the matrix elements were simplified by collecting together those terms which correspond to the energies of the electrons of the free metal or ligand ions, and substituting the known free-ion Dirac-Slater energies.

RESULTS

We relate the effectively one-electron energies obtained to the usual CFP by means of known formulas. The values of these parameters and the forming of their contributions are listed in Table 1. It is seen that the relatively small values of the B_2^0 and B_6^0 parameters result from the mutual compensation of particular contributions. In the case of the B_4^0 parameter, only the electrostatic terms compensate, whereas the others determine its high value and therefore decide in an important way about the total splitting. Finally, we get the following splitting scheme for the $^3H_4$ term (in cm^{-1}):

$E_{|\pm3>} = -970$, $E_{|\pm2>} = -500$, $E_{|\pm1>} = 430$, $E_{|\pm4>} = 610$, $E_{|0>} = 860$.

This leads to the following dependence of the paramagnetic susceptibility expressed as a function of temperature for low T (in emu/mole):

$$\chi_M(T) = 0.72/T + 0.0014 \ . \tag{3}$$

For comparison, this dependence using the point-charge model is expressed by

$$\chi_M(T) = 0.72/T + 0.0009. \tag{4}$$

Table 1. Various Contributions to the Crystal Field Parameters: B_2^0 , B_4^0 , B_6^0 in cm^{-1} for $U(SO_4)_2 4H_2O$

Origin	B_2^0	B_4^0	B_6^0
Point-charge	-5.56	0.559	0.0041
Kleiner's correction	4.26	-0.576	-0.0069
Exchange	-0.02	0.419	-0.0028
Overlap	-0.80	0.277	0.0045
Covalency	0.86	0.087	-0.0008
Total	-1.26	0.766	-0.0002

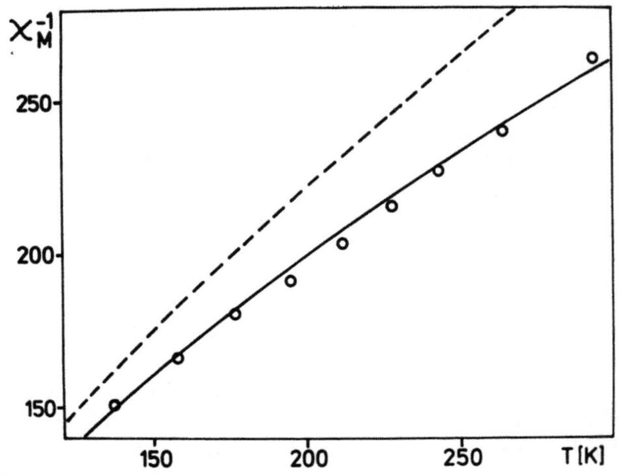

Figure 1: Reciprocal molar susceptibility of $U(SO_4)4H_2O$ vs. temperature. o - experimental points; solid line - the theoretical line based on these calculations; dashed line - the theoretical line based on the point-charge model (mole/emu).

Both the above calculated numbers as well as the experimental results are presented in Fig. 1.

DISCUSSION

The resultant values of the crystal field parameters are in good agreement with experiment. They originate mainly from exchange and overlap contributions. The covalent mixing effects produce a contribution of as much as 1/3 of the splitting resulting from the overlap. The electrostatic terms (point-charge model and the Kleiner's correction) cancel each other giving nearly zero splitting.

The agreement obtained with the experiment itself is not sufficient proof of adequacy of the model employed. Many effects[1], even as important as shielding and overlap effects associated with the uranium 6s and 6p electrons or the influence of the cluster's external environment, have not been taken into account. As far as shielding is concerned, it can be supposed that its influence is negligible because the B_2^0 parameter, the most sensitive to it, has a relatively low value. The remaining effects are to be considered in detail in the near future.

ACKNOWLEDGEMENTS

The authors would like to thank Mrs. Z. Pawlowska and Mr. S. Daniuk for their facility with the Dirac-Slater procedure and their aid in numerical calculations. We are indebted to Mr. Z. Zolnierek for helpful discussion.

REFERENCES

1. D. J. Newman, Adv. Phys. 20,197 (1971)
2. S. Sugano, R. G. Shulman, Phys. Rev. 130,517 (1963)
3. J. Mulak, A. Czopnik, Bull. Acad. Polon. Sci., Ser. Sci. Chim. 20,209 (1972)
4. C. J. Ballhausen, "Introduction to Ligand Field Theory", McGraw-Hill, New York (1962)
5. R. E. Watson, A. J. Freeman, Phys. Rev. 134A,1526 (1964)
6. W. H. Kleiner, J. Chem. Phys. 20,1784 (1952)
7. D. Koelling, private communication.
8. R. P. Sharma, Phys. Rev. 13A,517 (1976).

MAGNETIC STRUCTURE AND LATTICE DEFORMATION IN UO_2

V.L. Aksenov, Th. Frauenheim and V. Sikora

Joint Institute for Nuclear Research, Dubna, USSR
H.P.O. Box 79, Moscow, USSR

INTRODUCTION

One of the fundamental problems in the theory of magnetism in actinide systems is to decide whether the electrons are localized or delocalized, and to determine the mechanism which is responsible for the formation of the magnetic ordered state[1]. From this point of view uranium is one of the most interesting actinide elements, with a large overlap of 5-f orbitals in the pure metal and a decreasing overlap upon the formation of compounds.

During the last two decades UO_2 has been intenstively studied both experimentally and theoretically. From the optical data the localized character of the 5f - electrons in UO_2 is well established[1]. In recent work by Solt and Erdös[2] it was shown that one had to include a direct quadrupole-quadrupole coupling of U^{4+} ions in order to explain the lattice deformation mechanism in UO_2. This, however, indicates the possibility of direct overlap of the 5f-wave functions in this compound. Performing the symmetry analysis of the phase transition in UO_2[3] we derive the magnetic structure and the lattice deformation in the low temperature phase in comparison with the neutron diffraction data[4,5]. From this we give an explanation for the occurrence of a homogeneous and inhomogeneous deformation of the magnetic phase transition and show that the Solt and Erdös result[2] is uncertain due to an incomplete lattice deformation model for the oxygen sublattice[5].

MODELS FOR THE LATTICE DEFORMATION IN UO_2

The detailed neutron diffraction study of UO_2[4] showed that in the magnetic ordered state below T_N=30.8 K a noncollinear 4-sublattice antiferromagnetic (AFM) structure is realized. (The magnetic moment of the U ions is aligned along the [110] direction.) Theoretical calculations of the magnetic cross sections using a number of possible states (including J-mixing effects) were unable to reproduce the experimental data at large momentum transfer[5]. Subtracting the calculated magnetic cross section from the observed scattering cross section ($d\sigma_{obs}-d\sigma_{cal}$) Faber and Lander[5] determined that additional scattering is present only for reflections with h,k even and l odd. To explain this behavior they concluded that the magnetic transition is strongly connected with a displacive distortion of the oxygen sublattice, for which a model of collinear displacements of the oxygen ions along one of the cubic axes was proposed (Faber-Lander model). The calculated extra contribution to the nuclear cross section due to this deformation is

$$d\sigma_D = \left| b_o e^{-W_o} \sum_{n=1}^{8} e^{i\vec{\tau}_{hkl}\cdot\vec{\Delta}_n} \right|^2 \qquad (1)$$

(b_o=scattering length for oxygen, $\exp(-W_o)$-Debye-Waller factor, $\vec{\tau}_{hkl}=2\pi/a(hkl)$, $\vec{\Delta}_n$-distortion of the nth oxygen ion.) This leads to a relative displacement of $\Delta/a=(2.6\ 0.2)\times10^{-3}$ within a least square fit of $d\sigma_D$ and ($d\sigma_{obs}-d\sigma_{cal}$) and a value of $\chi^2=1$.

This model, which has been widely used up to now, has served as a basis for the explanation of the lattice deformation mechanism in UO_2. In the paper by Siemann and Cooper[6] this distortion, which corresponds to a frozen-in optical phonon mode at the Brillouin-zone (BZ) boundary (M5-mode), was analyzed in connection with the Allen-A-mode[7], which describes a combination of a frozen-in mode in the BZ-center with an acoustic phonon. After the calculation of the electrostatic energy gain and the elastic energy loss for the two cases, they concluded by using the lattice dynamic parameters deduced by Dolling et al.[8], that the M_5-distortion (Faber-Lander distortion) should occur as a result of the Jahn-Teller mechanism. This explanation is uncertain for two reasons as discussed by Solt and Erdös[2]. First, from the measured phonon data up to now no soft M_5-mode could be observed. Second, the A-mode is softer than that calculated in Ref. 6 due to a neglected dynamical coupling between the $\vec{q}=0$ optical phonon and the homogeneous strain. Solt and Erdös reanalyzed the problem of competing deformations in the low temperature phase of UO_2, favoring the M_5-mode becoming soft as a result of the direct quadrupole-quadrupole coupling, which will be increased due to the distorted oxygen sublattice. This means, however, that the 5f-electronic wave functions must overlap, which is in contradiction to the experimental data and again the question arises of localized versus itinerant 5f-behavior.

The main aspect for the inconsistency of the models predicted for the phase transition in UO_2 by Siemann and Cooper[6] and Solt and Erdos[2] is that, by believing in the Faber-Lander-distortion, they are in conflict with the fact that there is an experimentally established cubic symmetry of UO_2 below T_N.[9] This distortion corresponds to the transition along one of the star arms $\{k_{10}\}$ (space group O_h^5) which leads to the tetragonal symmetry[10] and favors the two sublattice AFM-structure[11].

A clue to solving this problem can be found in Ref. 3, where it was shown by Izyumov that cubic symmetry remains only at the transition point if the oxygen ions are displaced along the cubic space diagonals. These types of displacive distortions are strongly connected with the model suggested in Ref. 5 of the noncollinear 4-sublattice AFM structure and the orientation of magnetic moments of the U ions along the diagonals of the cube. From the symmetry analysis it can be shown that the magnetic structure in the low temperature phase of UO_2 can be described by the one- and two-dimensional representations τ_3 and τ_9 with the following orientations of magnetic moments of the U ions in the elementary cell:

U Positions	$S_1(0,0,0)$	$S_2(0,1/2,1/2)$	$S_3(1/2,0,1/2)$	$S_4(1/2,1/2,0)$
τ_3	(+++)	(+−−)	(−+−)	(−−+)
τ_3^1	(+++)	(−−+)	(+−−)	(−+−)
τ_3^2	(+++)	(−+−)	(−−+)	(+−−)

Corresponding to the magnetic structure we also have three possible models for the oxygen displacements along the cubic space diagonals which are presented in the figure below:

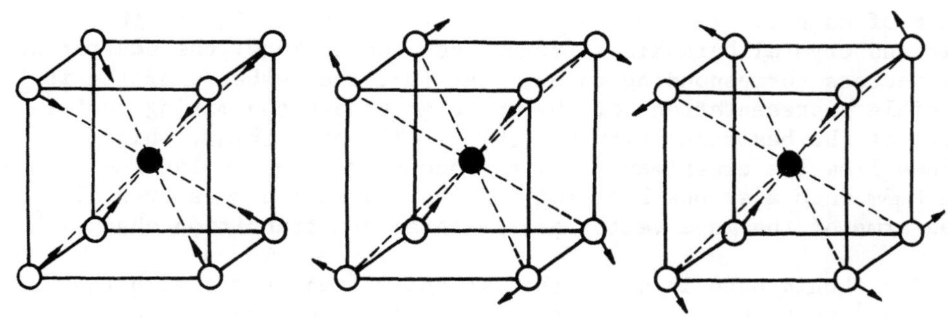

τ_3 τ_9^1 τ_9^2

All these representations correspond to the space group $Fm3m(O_h^5)$ and describe the three arm character of the magnetic and crystalline structure, where the phase transition takes place over all the three arms of the star $\{k_{10}\}$.

To decide what is the real crystalline and magnetic symmetry of UO_2 below T_N we have calculated the extra contribution for the nuclear scattering cross section $d\sigma_D$ due to the deformation of the oxygen sublattice for the different triple-k-structures[3], comparing with experiment[4,5].

In the representation τ_3 the deformation of the oxygen sublattice gives rise to an addition scattering $d\sigma_D \sim (\Delta/a)^2$ for all ℓ-odd reflections. However, the calculated contributions are large for reflections with small ($d\sigma_{obs}-d\sigma_{cal}$) and small for reflections with large ($d\sigma_{obs}-d\sigma_{cal}$), so that the fit to the experimental data is quite poor. In the representation τ_9^2 we obtain additional scattering contributions $d\sigma_D \sim (\Delta/a)^2$ for one half of the ℓ-odd reflections which are unable to give a satisfactory explanation for the experiment. The structure τ_9^1 also leads to an extra scattering $d\sigma_D \sim (\Delta/a)^2$ for all ℓ-odd reflections, except (001). In this case the agreement with the experimental result is good. The best fit, however, is poorer than for the Faber-Lander model because the ℓ-odd reflections with $h,k \neq 0$ (even) are underestimated in the representation τ_9^1.

Summarizing, none of the considered triple-k structures satisfies the experimental data or gives an excellent fit as in the Faber-Lander case, which however contradicts the cubic crystal symmetry in the magnetic ordered state. For this reason we have done a more detailed symmetry analysis of the phase transition in UO_2.

SYMMETRY ANALYSIS

We follow the method[3] used for the determination of the transition channel and the corresponding magnetic structure from a system of magnetic reflections. For a full explanation of the magnetic and crystallographic structure one has to know: the scattering wave vectors corresponding to one star {k}, the number ν of the irreducible representations of the space group and the mixing coefficients of the basis function set $\{C_\lambda^{L\nu}\}$. The star $\{k_{10}\}$, which follows from the experiments[4,5], may occur from an fcc lattice structure Fm3m into a sc one Pm3m (O_h^1) if the transition goes over all three arms of the wave vector star, a three arm transition channel[10].

One should have in mind that in phase transitions with the star $\{k_{10}\}$ we have the translations of the symmetrical phase (t1, t2,t3), which are forbidden in the new phase-lost translations. Therefore, one needs to take into account the prephase subgroup (of the initial phase) in which the generators are the elements connected with the lost translations. In other words, one should consider not only the zero unit generators, (the space group elements of type (h/0), but also the generators (h,t_1).

As a result of the representation τ_9, for example, the displacements of the eight oxygen ions in the cubic cell are obtained in the form:

$\vec{U}(1)$	$\vec{U}(3)$	$\vec{U}(5)$	$\vec{U}(7)$		
(111)	(111)	(111)	(111)	Pa3	τ_9^1
(111)	(111)	(111)	(111)	Pa3	τ_9^2

$\vec{U}(2)=-\vec{U}(1)$, $\vec{U}(4)=-\vec{U}(3)$, $\vec{U}(6)=-\vec{U}(5)$, $\vec{U}(8)=-\vec{U}(7)$.

The displacements above coincide with those reported in Refs. 3 and 11. In the phase described by the group Pa3 the U ions have the positions (4a) and the oxygen ions (8c). But the analysis, carried out in this work led to the conclusion that the structure should be described by the combination of the two lines[4] and not only by one of them as was proposed in Ref. 3. In other words, the system is divided into translational domains. Note that in the case of a three-arm transition channel, the structure reveals the mono-domain state in the commonly accepted sense of rotational domains. Within the other representations the displacements of oxygen ions are also coincident with those obtained in Refs. 3 and 11, and the corresponding groups have the following form: for τ_{10} - two types of translational domains with the symmetry group $P2_13(T^4)$, and for τ_1 and τ_4, single domain states with the groups Pm3m and P43m (T_d^1) respectively.

The magnetic symmetry groups can be found in the same way. As a result, in the τ_9 representation the structure consists of two translational domains with symmetry group Pa3, and for τ_3 a single domain state with the Pm3m group.

In this way the transition into the low symmetry phase may lead to the following groups: 1) a magnetic single domain structure with Pm3m and a single domain crystal structure with 1a) Pm3m and 1b) P43m; 2) a two domain magnetic structure Pa3 and two domain crystallic ones 2a) Pa3 and 2b) $P2_13$. The calculated scattering cross sections were compared with the spectrometric data[5] and showed that the experiment corresponds to case 2) with equal probability of the translational domains and with the same accuracy as in Ref. 5. (The least square fit for $d\sigma_D$ and $(d\sigma_{obs}-d\sigma_{cal})$ gives $\chi^2=1$.) In this case the structures 2a) and 2b) are not distinguishable. However, the appearance of highly symmetric structures seems more probable for the conformity of magnetic and crystallic symmetries, i.e. Pa3-Pa3. The value of the displacement obtained from the comparison of experimental and calculated data is $\Delta/a=(3.34\pm0.3)\times10^{-3}$ (a is the lattice constant), i.e. it is 1.28 times as large as that reported in Ref. 5.

CONCLUSIONS

In conclusion let us estimate the magnetostriction energy. The free energy expansion in an order parameter series looks the same for both cases of oxygen distortions, 2a) and 2b). According to (2) the magnetostriction energy is as follows:

$$F_{ms}=a(\eta_1^2+\eta_1^2)+A\eta_3^2+B(\eta_1 S_1^2+\eta_2 S_2^2)+B\eta_3 S_1 S_2 \qquad (2)$$

where $\eta_{1,2}$, $S_{1,2}$ are the deformations and magnetic order parameters in the translational domains. $\eta_{1,2}$ are inhomogeneous deformations (corresponding to a frozen optical phonon at the boundary of the Brillouin zone); η_3 is the homogeneous internal deformation (corresponding to a frozen optical phonon at the center of the Brillouin zone, the A-mode). The homogeneous internal deformation occurs due to the appearance of translational domains. Its existence is in agreement with the energy estimations[6,9] which state that the Jahn-Teller mechanism should lead to the A-mode.

The results reported here allow us to give a new interpretation of the neutron diffraction data[5]. In the antiferromagnetic phase there is realized a three-arm noncollinear magnetic structure with two types of translational domains. More precise neutron diffraction measurements of the magnetic transition under deformation as well as the domain structure study with polarized neutrons should be done in the future. Diffraction methods from high power sources might serve an additional experimental test.

REFERENCES

1. J. Schoenes, Phys. Rep. 63, 301 (1980).
2. G. Solt, P. Erdös, Phys. Rev. B22, 4718 (1980).
3. Yu.A. Izyumov, Usp. Fiz. Nauk. 131, 887 (1980).
4. J. Faber Jr., G.H. Lander and B.R. Cooper, Phys. Rev. Lett. 35, 1770 (1975).
5. J. Faber Jr., and G.H. Lander, Phys. Rev. B14, 1151 (1976).
6. R. Siemann and B.R. Cooper, Phys. Rev. B20, 2869 (1979).
7. S.J. Allen, Phys. Rev. 166, 530 (1968).
8. G. Dolling and R.A. Cowley, Phys. Rev. Lett., 683, (1966).
9. G.H. Lander and M.H. Mueller, Phys. Rev. B10, 1994 (1974) and J. Faber Jr and M.H. Mueller, (unpublished).
10. Y.E. Naish, Y.N. Syromiatnikov, Kristallographiy 21, 1085 (1976) in Russian.
11. I.E. Dzyaloshinsky, Commun. on Phys. 2, 69 (1977).

COMMENTS

ROSSAT-MIGNOD: My feeling is that in UO_2 the phase transition is not towards a triple $\vec{k}$ magnetic and triple $\vec{k}$ crystallographic structure as you have suggested. I think that the most likely explanation corresponds to a transition with a double $\vec{k}$ magnetic structure coupled with a single $\vec{k}$ crystallographic structure. This explanation agrees with neutron results of G.H. Lander et al. and with the theory proposed by B.R. Cooper et al.. Of course a neutron experiment with an applied stress or a magnetic field is needed to clarify the situation.

COOPER: The behavior of UO_2 can be understood rather easily on the basis of an internal shear distortion picture where the uranium sublattice retains an undisturbed cubic lattice and the oxygen sublattices have a shear distortion. The theory of Siemann and myself treated such a situation, and this model is consistent with observed behavior in the neutron experiments and makes sense from the symmetry point of view.

FRAUENHEIM: In the symmetry analysis (Ref. 8), it was shown that the Faber-Lander distortion, which corresponds to a transition along one of the arms of the wave vector star $\{k_{10}\}$, could lead in the case of a coupled magnetic-structural transition to a tetragonal symmetry, favoring a colinear two-sublattice antiferromagnetic structure. This, however, is not predicted by the experiment (Ref. 4).

CRYSTAL FIELD AND p-f MIXING EFFECTS IN URANIUM PNICTIDES

K. Takegahara, A. Yanase and T. Kasuya

Department of Physics
Tohoku University
Sendai 980, Japan

INTRODUCTION

The triuranium tetrapnictides, U_3X_4, crystallize in a cubic body centered crystal structure of Th_3P_4 type.[1] From the neutron diffraction experiments,[2] the uranium in these compounds is in the tetravalent configuration with $5f^2$, S=1, L=5, J=4 and g=0.8 in the Russell-Saunders coupling. Main features of the magnetic properties are as follows: 1) The value of the ferromagnetic moment is about a half of that for J=4.[3] 2) The magnetic anisotropy in the ordered state is very large with the easy axis in the <111> direction.[4,5] 3) The ordered structure is multiaxial and the direction of each sublattice moment tilts about 20° from the <111> direction to each uniaxial direction corresponding to one of the <001> axes.[1,6] 4) Nevertheless, the increase of the moment under strong external magnetic fields is very small. For example, in U_3As_4, the increase is less than 1% under 20T at 4.2K.[7] 5) In U_3As_4, the spin orientational transition is observed at the external magnetic field of 20T in the <001> direction.[5,7] This is very sensitive to a small tilt of the field direction from the <001> direction.[7] The theoretical investigation was done by Troč et al.[1] according to the point charge model of the crystal field. However, the point charge model can not explain the points 3) and 5).

To explain the features mentioned above, we introduce the p-f mixing effect which was investigated by us to explain the unusual magnetic properties of the Ce-monopnictides.[8] The essential point of the p-f mixing effect in the Ce-monopnictides is the presence of a small amount of valence band holes. It had been shown already that the non-magnetic Th_3X_4 are all narrow gap semiconductors.[9] When some Th is substituted by U, the systems become semimetals in which the number of conduction electrons is much smaller than that of substituted U atoms.[10] This indicates strongly that the situation is similer to that in the Ce-monopnictides and, because of the p-f mixing effect, the top of the valence bands is pushed up and overlaps with the conduction bands with an increase of U concentration. It is not so clear exactly how many holes exist in U_3X_4 but the initial gap of several tenths of eV in Th_3X_4[9] and the larger p-5f mixing energy, about twice of the p-4f mixing in the Ce-compounds, suggest that it will be about 10% per pnictogen atom, which is also consistent with one recent de Haas-van Alphen experiment.[11] Therefore, a more enhanced effect than the Ce-monopnictides is expected. The situation is more complicated here because U^{4+} has two 5f electrons corresponding to Pr^{3+}. The present study should also give important suggestions on Pr-compounds.

PARAMAGNETIC REGION

In the U_3X_4 crystal, the U ions locate at the six symmetry-equivalent positions with eight nearest anions forming the local point symmetry S_4 along the x, y and z axes. By the simple symmetry argument, the crystal field potential in the point charge model is given in the following form for the S_4 symmetry of the z axial site.

$$V_{PC}=e^2\langle r^2\rangle A_{20}C_{20}/a^3+e^2\langle r^4\rangle[A_{40}C_{40}+A_{44}C_{44}+A_{44}^{*}C_{4-4}]/a^5$$
$$+e^2\langle r^6\rangle[A_{60}C_{60}+A_{64}C_{64}+A_{64}^{*}C_{6-4}]/a^7, \quad (1)$$

where $C_{lm}=[4\pi/(2l+1)]^{1/2}Y_{lm}$.[12] In the calculation of the coefficients A_{lm}, the contributions from all U^{4+} and X^{3-} ions are considered. The values of $B_{lm}=e^2\langle r^l\rangle A_{lm}/a^{l+1}$ for U_3As_4 (a=8.521A) are also calculated by using the calculated results of mean radial powers of 5f electrons $\langle r^l\rangle$.[13] These results are listed in Table 1.

The p-f mixing effect is treated by simple second order perturbation. Due to the lack of detailed information about the valence bands, the valence holes are assumed to be concentrated at the Γ point and have the same phase. The energy shift of the $5f^2$ states is written as

$$V_{pf}=\Delta E[V_{\sigma\sigma}+V_{\sigma\pi}(pf\pi)/(pf\sigma)+V_{\pi\pi}(pf\pi)^2/(pf\sigma)^2],$$
$$\Delta E=x(pf\sigma)^2/6(E_{5f}-E_p), \quad (2)$$

where x means the number of valence holes per pnictogen, E_{5f} the energy level of $5f^2$ states, E_p the energy of the valence hole at the Γ point and $(pf\sigma)$ and $(pf\pi)$ are Slater-Koster's integrals.[14] The coefficients V_{ij} are written in the same form as V_{PC}:

Table 1. Coefficients for effective crystal field

l m		A_{lm}	$B_{lm}(K)$	$D^{\sigma\sigma}_{lm}$	$D^{\sigma\pi}_{lm}$	$D^{\pi\pi}_{lm}$
0 0				0.717	-1.719	1.082
2 0		14.94	2308.2	-4.266	10.711	-6.463
4 0		643.8	1433.7	3.267	-7.436	4.437
4 4	Re	756.4	1684.5	0.394	-1.605	1.375
	Im	-1232.6	-2744.8	-0.675	2.751	-2.357
6 0		-11841.2	-636.6	0.346	0.990	-1.732
6 4	Re	-4571.0	-245.7	-0.570	2.324	-1.991
	Im	1984.0	106.7	0.978	-3.984	3.413

$$V_{ij}=\sum_{lm} D^{ij}_{lm} C_{lm}\,. \tag{3}$$

The non-zero values of D^{ij}_{lm} are listed in Table 1. The value of ΔE is estimated to be about -200K by using the following values; $E_{5f}-E_p$=-1eV, (pfσ)=1eV and x=0.1.

The effective crystal field acting on the $5f^2$ states is the sum of V_{PC} and V_{pf}. Because of the strong covalency of these compounds, the crystal field of the point charge model should be reduced by a factor Z, which may be around 0.2~0.3 from the results of the rare earth monopnictides.[15] It should be noted that in the point charge model C_{20} is the most dominant term as was already shown,[1] which, cannot however, explain self-consistently the experimental facts mentioned above. This picture can not explain also the observed specific heat in the paramagnetic region. Then, we consider ΔE, (pfπ)/(pfσ) and Z as fitting parameters and analyze the specific heat[16,17] and the susceptibility[1] experiments of U_3As_4. The best fit results of the temperature dependence of the specific heat, the entropy, and the susceptibility over the paramagnetic region, from 200K to 1000K, are as follows; ΔE=-100K, (pfπ)/(pfσ)=-0.75 and Z=0.2. These are very reasonable values. The details will be given in a forthcoming paper.

FERROMAGNETIC REGION

The ferromagnetic ground state is investigated with the following simple model. At first, at each U site, there is the $5f^2$ state of the crystal field in the point charge model. The state is denoted by the z component, M_{J_2}, of the total angular momentum $J_2=4$, $|f^2 M_{J_2};i\rangle$ at the i-th site. The quantization axis is chosen so as to be the easy axis of the crystal field at each U site. Second, we consider the hole states. The valence hole is the p state at the pnictogen site and denoted as $|pm_j;n\rangle$ where m_j means the z component of the total angular momentum j at the n-th site. The p-f mixing in j=3/2 is larger than that in j=1/2 and thus we consider only the j=3/2 state in which the quantization axis is in the <111> direction. The 5f hole state is written as $|f^1 M_{J_1};i\rangle$ where M_{J_1} means the z component of the total angular momentum J_1 of the $5f^1$ state. Again, the p-f mixing in $J_1=5/2$ is larger than that in $J_1=7/2$ and thus we consider only the $J_1=5/2$ state where the quantization axis is the same as that of the $5f^2$ state. In various interactions among them, we consider the mixing term between the p state and 5f state and the crystal field of the point charge model acting on the $5f^1$ and $5f^2$ states. For simplicity, we consider only the C_{20} term. The level of the p hole is chosen as the energy origin and the $5f^1$ state is at $E_f\sim-1$eV. Then the variational function of the ferromagnetic ground state is written as

$$|g.s.\rangle=\sum_{iM_{J_1}} c(M_{J_1};i)|f^1 M_{J_1};i\rangle \Pi_{i'\neq i}\sum_{M_{J_2}} a(M_{J_2};i')|f^2 M_{J_2};i'\rangle$$
$$+\Pi_i \sum_{M_{J_2}} a(M_{J_2};i)|f^2 M_{J_2};i\rangle \sum_{nm_j} b(m_j;n)|pm_j;n\rangle \qquad (4)$$

where $a(M_{J_2};i)$, $c(M_{J_1};i)$ and $b(m_j;n)$ are variational parameters and normalized as follows

$$\sum_{M_{J_2}} |a(M_{J_2};i)|^2=1$$
$$\sum_{iM_{J_1}} |c(M_{J_1};i)|^2+\sum_{nm_j} |b(m_j;n)|^2=1. \qquad (5)$$

The actual calculation is performed on six U ions and their nearest eight pnictogens because we consider the nearest neighbour p-f mixing. In the normalization conditions, the summation i is performed on 6/8x uranium sites and on 1/x pnictogen sites. The variational parameters are determined so as to minimize the total energy. It should be noted that in the present calculation all the p bands at the Γ point are assumed to be at the same energy because of the lack of any detailed band calculation.

When there is no crystal field of the point charge model, the moment direction of the $5f^2$ state tilts too much, that is, between the <111> and <110> directions. Note that the value of the moment on each site is nearly the full moment of J=4. Next, the crystal field of the point charge model moves the direction of the moment to the <001> direction in good agreement with the experiments at a reasonable value of the parameters. In this case too, however, the magnetic moment does not shrink.

The most important shortcoming in the present calculation is clearly the neglect of the p band splitting at the Γ point. Detailed band calculations are now in progress and the inclusion of this effect is expected to remove the above disagreement. Preliminary calculations show that the present model is also hopeful in explaining the experimental facts of 4) and 5) mentioned above. The detailed results will appear in the forthcoming paper.

REFERENCES

1. R. Troć, J. Mulak, and W. Suski, Phys. Stat. Sol., (b)43:147 (1971).
2. P. Burlet, J. Rossat-Mignod, R. Troć, and Z. Henkie, Solid State Commun., 39:745 (1981).
3. W. Trzebiatowski, in:"Ferromagnetic Materials," Vol. 1, E. P. Wohlfarth, ed., North-Holland, Amsterdam (1980).
4. C. F. Buhrer, J. Phys. Chem. Solids, 30:1273 (1969).
5. V. E. Bril, R. Z. Levitin, R. E. Osipova, V. L. Yakovenko, and M. Zelený, Phys. Stat. Sol., (a)57:393 (1980).
6. L. Dobrzynski and J. Przystawa, Phys. Stat. Sol., 42:K15 (1970).
7. R. Troc, J. Sznajd, P. Novotný, and T. Mydlarz, J. Mag. Mag. Mat., (in press).
8. K. Takegahara, A. Yanase, and T. Kasuya, J. de Phys., 41:C5-327 (1980), Erratum J. de Phys., 41:1231 (1980).
K. Takegahara, H. Takahashi, A. Yanase, and T. Kasuya, J. Phys. C:Solid State Phys., 14:737 (1981).
T. Kasuya, K. Takegahara, M. Kasaya, Y. Isikawa, H. Takahashi, T. Sakakibara, and M. Date, in:"Physics in High Magnetic Fields," S. Chikazumi and N. Miura, eds., Springer-Verlag, Berlin, (1981).
T. Kasuya, in:"Electron Correlation and Magnetism in Narrow-Band Systems," T. Moriya, ed., Springer-Verlag, Berlin, (1981).
K. Takegahara, H. Takahashi, A. Yanase, and T. Kasuya, Solid State Commun., 39:857 (1981).
9. Z. Henkie and P. J. Markowski, J. Phys. Chem. Solids, 39:39 (1978).
10. P. J. Markowski, Z. Henkie, and A. Wojakowski, Solid State Commun., 32:1119 (1979).
11. Z. Henkie, W. R. Johanson, G. W. Crabtree, D. H. Dye, A. J. Arko, and C. Bazan, in:"Actinides 81," (in press).
12. B. G. Wybourne, "Spectroscopic Properties of Rare Earths," Interscience Pub., New York, (1965).
A. J. Kassman, J. Chem. Phys., 53:4118 (1970).
13. J. P. Desclaux and A. J. Freeman, J. Mag. Mag. Mat., 8:119 (1978).
14. K. Takegahara, Y. Aoki, and A. Yanase, J. Phys. C:Solid State Phys., 13:583 (1980).
15. R. J. Birgeneau, E. Bucher, J. P. Maita, L. Passell, and K. C. Turberfield, Phys. Rev., B8:5345 (1973).
16. A. Alles, B. G. Falk, E. F. Westrum, Jr., F. Grønvold, and M. R. Zaki, J. Inorg. Nucl. Chem., 39:1993 (1977).
17. A. Blaise, R. Lagnier, R. Troć, Z. Henkie, P. J. Markowski, and M. J. Mortimer, J. Low Temp. Phys., 39:315 (1980).

COMMENTS

COQBLIN: The saturation magnetic moments of uranium pnictides are different from the ionic values corresponding to definite $4f^n$ configurations; this point indicates probably a strong k-f "hybridization" (or "covalent mixing"). Thus, I have the feeling that the Yang-Cooper mechanism is certainly important in uranium pnictides and perhaps larger than the p-f mixing mechanism that you consider here. What do you think of this?

TAKEGAHARA: The situation of U_3P_4 is very similar to that of CeSb but we have a more complex magnetic structure due to the more complicated crystal structure in U_3P_4. Because the 5f-d mixing should be more important than 4f-d mixing, there is a possibility that the 5f-d mixing interaction is also an important effect. So far, we have been treating the 5f-p mixing effect in a similar way as CeSb. Therefore, I cannot answer your question right now. As concerns the large reduction of the moment in the U site, it is possible to explain it within the 5f-p mixing mechanism.

COOPER: Could you please say more about how you introduce the p-f mixing starting with a band structure? My understanding is that you start with a band structure that does not include f-bands. Then, on a phenomenological basis you introduce an ^2f state and, especially, the p-holes in the band structure. Is this correct?

TAKEGAHARA: First, the band of Th_3P_4 is calculated rigorously including 5f electrons which are situated above the Fermi energy. The p-f mixing matrix is obtained from this band structure. In U_3P_4 we assume U^{4+} and thus two 5f electrons are below the Fermi energy, separated from the main part. We cannot treat this situation by the usual band calculation, and thus we artificially put the $5f^2$ level 1 eV below the Fermi level. Then the p-f anisotropic mixing is treated.

MAGNETIC PROPERTIES OF THE URANIUM TRICHALCOGENIDES

B. Janus,[†] W. Suski[†] and A. Blaise*

[†]Institute for Low Temperature and Structure Research
Polish Academy of Sciences, 50-950 Wroclaw, Poland
*DRF/PhS, Centre d'Etudes Nucléaires, Grenoble, France

INTRODUCTION

The magnetic investigations of the monoclinic uranium trichalcogenides until now have been carried out on powder samples only (Ref. 1 and references herein). They have revealed that the temperature dependence of the reciprocal susceptibility does not follow the Curie - Weiss law. UTe_3 and USe_3 exhibit maxima at 19.5 and 50 K, respectively, in the temperature dependence of the magnetic susceptibility, while US_3 shows temperature independent paramagnetism below about 15 K. It is widely accepted that the low symmetry in which this group of compounds crystallizes does not support the creation of a long range magnetic order. Moreover, the temperature independent paramagnetism observed in the sulfide suggested that any magnetic ordering existing in the trichalcogenides should be of the induced type. Thus, the observed maxima have been discussed in terms of crystal field theory and the Fermi liquid model[1]. However, these preliminary experiments here show that this family of uranium compounds is extremely sensitive to oxidation, etc. Therefore, one can suspect that a few of the observed anomalies are parasitic effects. On the other hand, the examination of Pr has proven that the sample form has substantial importance for the magnetic properties (see Ref. 2 for other singlet-ground state materials). For this reason we have started the present investigation based on single crystal material. We have previously published the results of the electrical resistivity measurements obtained on the pseudosingle crystal of UTe_3[3]. From this examination we have concluded that UTe_3 exhibits semiconducting character without any anomalies at low temperatures. In this paper we present the magnetic properties of uranium tritelluride and trisulphide examined in the temperature range 4.2 - 300 K and in magnetic fields up to 50 kOe. We have obtained also single crystals of USe_3. However, because of the high reactivity of this compound we have not been able until now to provide safe conditions for measuring it.

EXPERIMENTAL

All the compounds were obtained by the chemical transport method using iodine as a transporting agent and applying a temperature gradient of about 50 C in the temperature range 600 - 800 C according to Ref. 4. The crystals are in the form of thin plates corresponding to the (001) plane. The lattice constants listed in Table I are in fair agreement with the previous results. The magnetic measurements have been performed perpendicular and parallel to the (001) plane, the equipment being now standard in our laboratory.

Table I. Lattice constants and magnetic data of the uranium trichalcogenides

Compound	Lattice constants in [Å]				T_N (K)	Θ (K)	μ_{eff} (μ_B)
	a	b	c				
US_3	5.48	4.00	9.065	98	54 46	-154 -131	3.42 3.41
USe_3	5.63	4.06	9.55	98			
UTe_3	6.090	4.226	10.302	98	11 --	-62 -71	3.61 2.94

RESULTS

The results of magnetic measurements are collected in Table I and in Figs 1 - 3. The reciprocal magnetic susceptibility of both compounds follows the Curie - Weiss law, which seems to contradict the conclusion of Ref 1. In Ref. 5, however, the temperature dependence of the reciprocal susceptibility is also linear below 300 K. At low temperatures these results are different from those observed previously. As mentioned above the maximum in the temperature dependence of the susceptibility of UTe_3 has been observed at 19.5 K[1]. At present, the maximum in the temperature dependence of magnetization is observed at 11 K when the (001) plane is oriented perpendicularly to the magnetic field direction. The maximum does not appear when the measurement is carried out with this plane parallel to the field direction. From Table I it follows

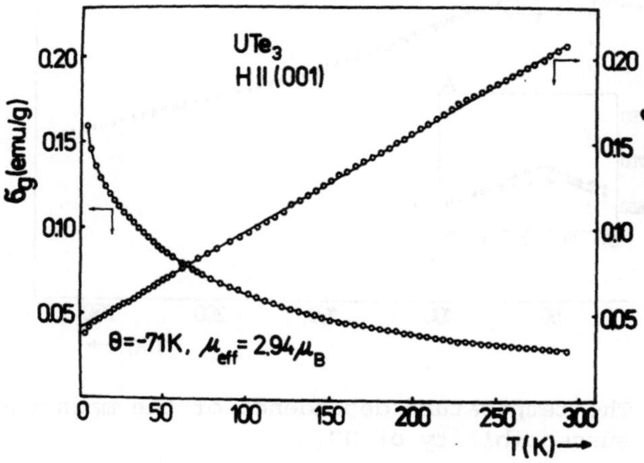

Fig. 1. The temperature dependence of the magnetization and the reciprocal susceptibility of UTe_3: (001) plane || to the magnetic field direction.

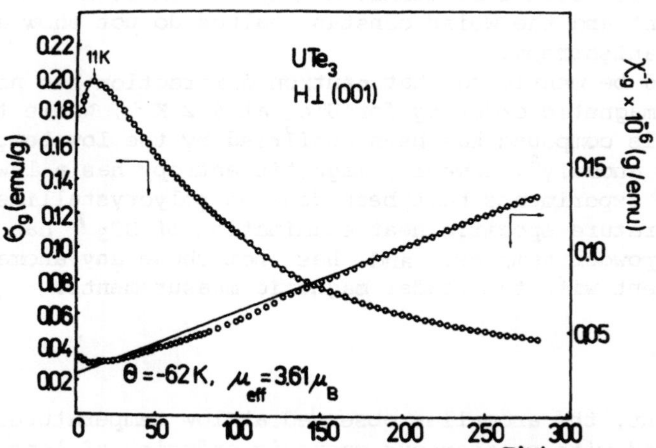

Fig. 2. The temperature dependence of the magnetization and the reciprocal susceptibility of UTe_3: plane ⊥ to the magnetic field direction.

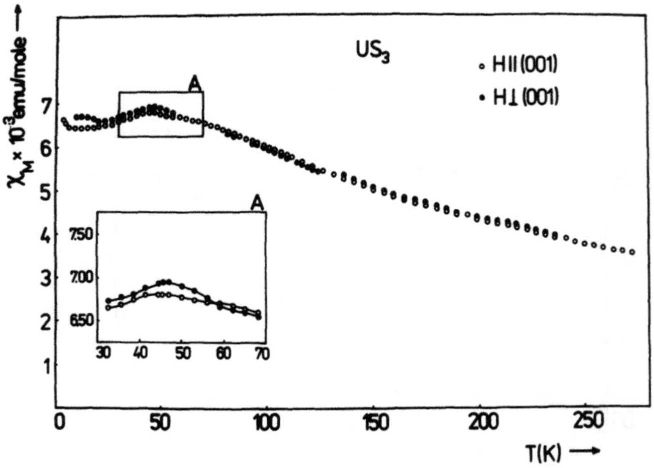

Fig. 3. The temperature dependence of the magnetic susceptibility of US_3 .

that there is some anisotropy in the magnetic moment and the Weiss constant values. Concerning US_3 the discrepancy with the previous results[1] is even more pronounced. In the powder sample temperature independent paramagnetism was observed[1] , whereas at present there is a diffuse maximum, in principle at the same temperature - of about 50 K for both directions. As can be seen from Table I the magnetic moment and the Weiss constant values do not show any considerable anisotropy.

It should be mentioned that neutron diffraction has not revealed any magnetic ordering for UTe_3 at 4.2 K[5] , while the maximum at 11 K in this compound has been confirmed by the low temperature specific heat anomaly[6]. However, magnetic entropy has a low value. Both of these experiments have been done on polycrystalline samples. The low temperature specific heat examination of US_3 [7] has been also performed on powder samples, and has not shown any anomalies, being consistent with the powder magnetic measurements.

DISCUSSION

At present, the anomalies observed at low temperatures seem to be unconnected with spontaneous magnetic ordering of long range. We cannot explain why in UTe_3 there is a difference in the appearance of a maximum depending on direction. One can assume that the puzzling properties observed in the uranium trichalcogenides may result from two reasons.

1. The monoclinic trichalcogenides crystal structure ($ZrSe_3$-type), having chains of uranium atoms with the uranium - uranium dis-

tance close to 4 Å and the interchain distance approaching 5 Å, favors the existence of strong interactions along the chains while the interchain interactions are much weaker. Thus, for the uranium trichalcogenides one can expect an onset of a low dimensional short range magnetic ordering. Usually in such a case the temperature dependence of low temperature specific heat exhibits two maxima: one diffuse at the higher temperature, due to the low dimensional order, and the other at the low temperature corresponding to the onset of a long-range order. The results for UTe_3 [6] show only one rather small anomaly which at present is difficult to explain.

2. In the uranium trichalcogenides among the polychalcogenides the U^{4+} ion seems to be firmly established. Then the formula unit can be written as $U^{4+}X^{2-}X_2^{2-}$. This idea is confirmed by the paramagnetism of these compounds while their isostructural analogs of the IV group are diamagnetic (see e.g. Ref. 8). Then the U^{4+} ion in a coordination polyhedron of low symmetry could have a singlet as a crystal field ground state. Thus, the magnetic anomalies, lacking magnetic ordering as demonstrated by the neutron diffraction experiment performed without magnetic field[5] and having a low value of magnetic entropy[6], could result from this special scheme of crystal field levels with an induced magnetic ordering. It has been shown[9,10] that for a considerable ratio of the crystal field interaction to the exchange interaction the value of magnetic entropy might be extremely low.

Unfortunately, we are not able to produce any exact calculation on the basis of the available data. Any fitting procedure in the light of the shortcomings of known models (see e.g. Ref. 11), especially for low symmetry materials, is not reasonable. Both of these explanations, however, may be rejected. Therefore, further experiments are needed to decide which of these models - or possibly some other - is valid.

REFERENCES

1. W. Suski, Bull. Acad. Polon. Sci., Sér. Sci. Chim., 24,75 (1976).
2. J. W. Cable, R. M. Moon, W. C. Koehler and E. O. Wollan, Phys. Rev. Letters, 12,553 (1976); R. J. Birgeneau, J. Als-Nielsen and E. Bucher, Phys. Rev., B6,2724 (1972); J. G. Houmann, M. Chapelier, A. R. Macintosh, P. Pak, O. D. McMasters and K. A. Gschneidner Jr., Phys. Rev. Letters, 34,587 (1975).
3. A. Blaise, B. Janus and W. Suski, Solid State Commun., 37,417 (1981).
4. V. K. Slovyanskikh, G. V. Ellert, E. I. Yarembash and M. D. Korsakova, Izv. Ak. Nauk SSSR, Neorg. Mat., 2,973 (1966).
5. J. Leciejewicz, unpublished data.
6. A. Blaise, R. Lagnier, M. J. Mortimer, W. Suski and B. Janus, to be published.

7. F. Grønvold and E. F. Westrum Jr., J. Inorg. Nucl. Chem., 30, 2127 (1968).
8. H. Haraldsen, A. Kjekshus, E. Røst and A. Steffensen, Acta Chem. Scand., 17, 1283 (1963).
9. J. Grunzweig-Genossar, Solid State Commun., 8, 1673 (1970).
10. W. Suski, A. Czopnik and T. Mydlarz, Phys. Stat. Sol. (a) 12, 1673 (1972).
11. D. J. Newman, Adv. in Physics, 20, 197 (1971).

COMMENTS

WOHLLEBEN: Did you determine the magnetic entropy by subtracting the specific heat of a reference compound, for example, using Th instead of U?

SUSKI: Not yet. The magnetic entropy was approximated by subtracting the entropy from the total entropy using smooth lines. The experiment on ThS_3 is now in progress.

GREIDANUS: Do all these compounds show a phase transition?

SUSKI: Yes. However, it is difficult to determine the character of these transitions.

MAGNETIZATION OF U_3P_4 IN MAGNETIC FIELDS UP TO 500 kOe

K.G. Gurtovoj*, A.S. Lagutin*, R.Z. Levitin†, V.I. Ozhogin*

*Kurchatov Institute of Atomic Energy, 123182 Moscow USSR

†Moscow State University, 117234 Moscow, USSR

INTRODUCTION

Uranium pnictides with the cubic Th_3P_4-type structure (the space group $I\bar{4}3d$) attract much attention because of a possible non-collinear magnetic structure. It is supposed[1] that the magnetic structure of these crystals consists of three sublattices of U^{4+} ions with magnetic moments symmetrically tilted from the [111] direction at some angle. The magnetization jump has been observed[2] during measurements of the magnetization along the hard direction [100] on U_3As_4 crystals (in a field of about 200 kOe at the temperature 4.2 K). Extensive information on its magnetic anisotropy and phase transitions induced by a magnetic field have been obtained as a result of further experimental and theoretical investigations[3,4] of this substance. In U_3P_4, which is very similar to U_3As_4, the magnetization jump was not observed, however, up to 300 kOe[5]. The purpose of the present work consists in the investigation of U_3P_4 in still higher magnetic fields.

EXPERIMENTAL TECHNIQUE

The magnetization was measured using a system of four induction coils balanced in the absence of a sample. The voltage generated by these coils was recorded with a memory oscilloscope. Magnetic field pulses with amplitude up to 500 kOe were produced by discharge of a high voltage capacitor bank into a many-turn solenoid. The solenoid was wound with a copper strip reinforced with fibres of NbTi alloy[6]. The pulse solenoid was cooled with liquid nitrogen, so that the superconducting component of the winding material was in the normal state, and only its increased (2-3 times compared to copper) strength was used. The half-sinusoidal field pulse had a duration of 10 msec. The accuracy of calibration of the sample magnetization was about 10%, and the calibration of the field was equal to 5%.

The single crystal of U_3P_4 has been prepared by chemical vapor transport. The sample was a cylinder of 1.5 mm height and 2.5 mm in diameter. The cylinder base coincided with the (110) plane of the crystal. The crystal axes were oriented along the magnetic field with an error $\leq 2°$.

EXPERIMENTAL RESULTS AND DISCUSSION

Magnetization curves of the U_3P_4 single crystal in the [100] and [111] directions at 4.2 and 80 K are shown in Fig. 1. We observed that the magnetization increases monotonically with the field in the hard direction [100]. In the highest field the magnetization reaches 0.9 of the saturation value measured along the easy axis [111]. Thus for U_3P_4 in contrast to U_3As_4 the magnetization jump in the hard direction either does not exist or does not exceed 0.1 of the saturation magnetization.

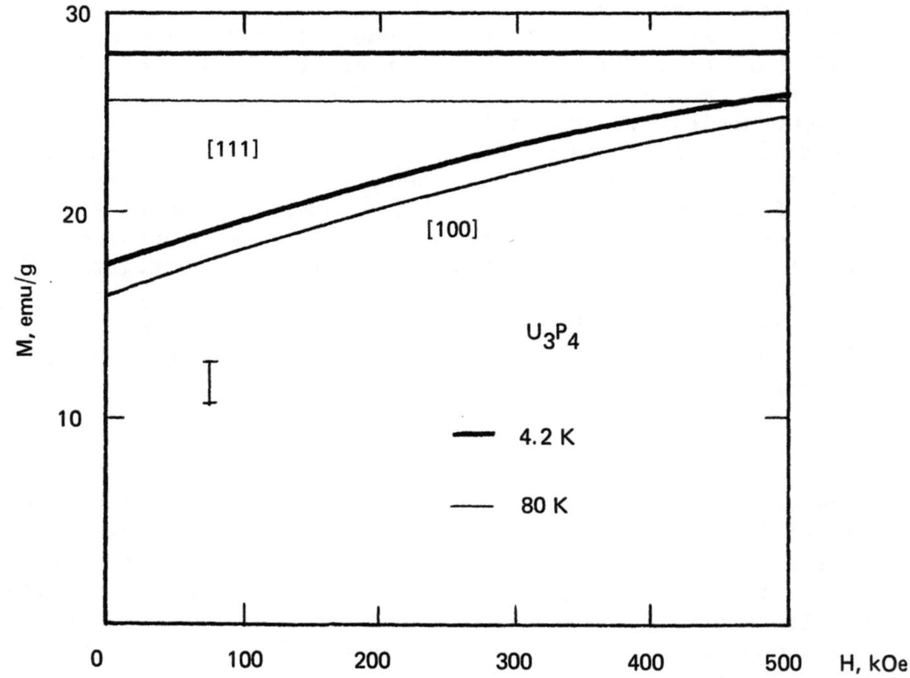

Fig. 1: Magnetization curves of the U_3P_4 single crystal in [100] and [111] directions at 4.2 and 80 K. The vertical bar represents the accuracy of calibration of the sample magnetization.

These experimental results can be explained on the basis of the theory[4] of phase transitions induced by a magnetic field in U_3X_4-type ferromagnets (X=P,As). In this theory the microscopic three-sublattice model is considered. The 12 U^{4+}-cations in the Th_3P_4 type cell are divided into three sublattices with local crystal field axes [100], [010] and [001], respectively. The terms of the second and fourth order are taken into account in the expansion of the crystal field energy in powers of an uranium ion U^{4+} effective spin S. The numerical calculations, made within the framework of the molecular field theory, show that the first order phase transition induced by the magnetic field directed along the hard axis [100] is possible provided $S \geq 2$. This phase transition into the saturated state is accompanied by the magnetization jump. It is known that the fourth order terms B_4 of the spin Hamiltonian are non-zero for $S \geq 2$. In the opposite case, the terms containing B_4 are identically zero, and only the second order phase transition occurs to the saturated state, where the magnetization of all three sublattices is parallel to the [100] direction.

The differences of the magnetization processes in U_3P_4 and U_3As_4 can be understood if one takes into account the differences of electronic structures of the U^{4+} ions in these crystals. According to the heat capacity measurements[7,8,9] only the ground pseudo-doublet[10], Γ_{t2}, Γ_{t3} is populated[7] in U_3P_4, when the temperature is not much greater than the Curie temperature T_C=140 K, and the uranium effective spin is 1/2. The magnetic entropy of U_3As_4 (T_C= 200 K) is much greater than that of U_3P_4 and reaches the value corresponding to the effective spin S=3/2 in accordance with estimates[8]. Thus it seems probable that in U_3As_4 the magnetic ordering process involves the doublet Γ_{t5} in addition to the ground pseudo-doublet. However, the value S=2 agrees better with the existence of the magnetization jump which accompanies the magnetization process in U_3As_4. This value of the effective spin may be obtained from the heat capacity data analysis using a more accurate estimate of the electronic coefficient γ. In any case, the uranium effective spin in U_3As_4 is 3-4 times greater than the effective spin in U_3P_4.

Thus the fourth order terms are absent in the spin Hamiltonian of U_3P_4 because of the small value of the uranium effective spin in accordance with the theory[4], and the magnetization process occurs continuously without a jump of magnetic moment. In U_3As_4 the fourth order terms contribute to the magnetic anisotropy, resulting in the discontinuous phase transition induced by a magnetic field.

REFERENCES

1. J. Przystawa, E. Praveczki, J. Phys. Chem. Solids 33, 1943 (1972).
2. P. Novotny, J. Sternberk, Phys. Stat. Sol.(a)33, K15(1976).

3. V.E. Bril, R.Z. Levitin, R.E. Osipova, V.L. Yakovenko, M. Zeleny, Phys. Stat. Sol.(a) 57, 393 (1980).
4. B. Kowalczyk, Phys. Stat. Sol. (b) 98, 37 (1980); J. Phys. C: Solid St. Phys. 13, 5749 (1980).
5. K.P. Belov, Z. Genke, A.S.Dmitrievskii, R.Z. Levitin, W. Trzebiatowski, Sov. Phys. JETP 37, 686 (1973).
6. A.S. Lagutin, V.I. Ozhogin, Pribory i tekhnika eksperimenta N3, 195 (1981).
7. B. Stalinski, Z. Bieganski, R. Troc, Phys. Stat. Sol. 17, 837 (1966); J.F. Counsell, R.M. Dell, A.R. Junkinson, J.F. Martin, Trans. Faraday Soc. 63, 72 (1967).
8. A. Blaise, R. Lagnier, R. Troc, Z. Henkie, P.J. Markowski, M.J. Mortimer, J. Low Temp. Phys. 39, 315 (1980).
9. A. Alles, B.G. Falk, E.F. Westrum, F. Gronvold, M.R. Zaki, J. Inorg. Nucl. Chem. 39, 1993 (1977).
10. R. Troc, J. Mulak, W. Suski, Phys. Stat. Sol. 43, 147 (1971).

TEMPERATURE DEPENDENCE OF MAGNETIZATION IN U_3P_4 AND U_3As_4 SINGLE CRYSTALS

P.J. Markowski*, S. Kunii†, K. Takegahara†, T. Suzuki†, Z. Henkie* and T. Kasuya†

*Institute for Low Temperature and Structure Research Polish Academy of Sciences, Wrocław, Poland
†Department of Physics, Tohoku University, Sendai, Japan

INTRODUCTION

The uranium pnictides which crystallize in the Th_3P_4 structure show various interesting properties and have attracted much attention[1,2,3]. Recently it was revealed by neutron diffraction that the ordered ferromagnetic structure is described by a canted structure in which the moment on each sublattice, with the uniaxial axis along one of the [001] axes, cants from the uniaxial direction by about 35 degrees to the [111] direction[4]. Furthermore, the net moment on each sublattice is substantially smaller than the saturation moment expected from the atomic tetravalent U atom[5]. This structure usually gives a fairly large induced additional moment with the magnetic field along [111]. However, the results reported so far usually give very small values, even though their values are scattered. Because we could prepare good single crystals of U_3P_4 and U_3As_4, we checked this point by careful measurements under high magnetic fields. Next, we measured the temperature dependence of the magnetization at low temperatures, expecting it would be very small because of the structure anticipated from the weak induced moment. The results obtained will be shown in the following, and possible models are discussed.

EXPERIMENTAL METHODS AND RESULTS

Technical problems arise in preparing single crystals of U_3P_4 and U_3As_4, which decompose before melting. It is not possible to use sublimation or growing from the melt. Therefore we used a chemical vapor transport reaction method to prepare single crystals[6]. Magnetization measurements with static magnetic fields up to 8.5T

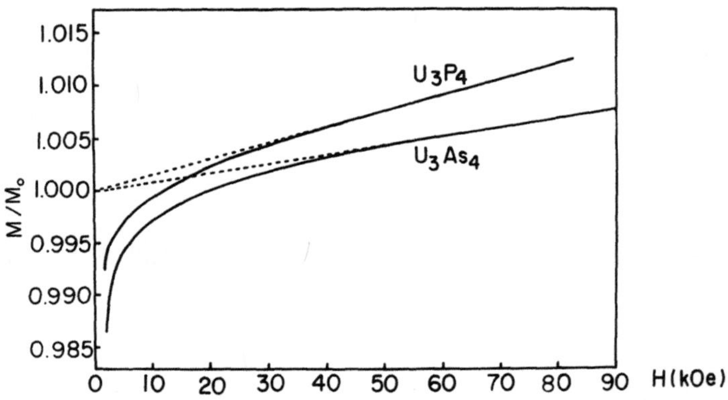

Fig. 1: Magnetic field dependence of the magnetization of U_3P_4 and U_3As_4 single crystals; magnetic field along the [111] axis at 4.2K. M_o is the saturation moment in zero external field at 4.2K.

were done by the extraction method. The boundaries for the integration of the signal induced in the pick-up coil were determined at the initial and final zero crossing points of the signal[7]. For high fields, pulsed fields up to 50T were applied by using the facility at Osaka University. To avoid eddy current problems during magnetic field changes, the sample was cut in thin plates with a width of about 0.1 mm.

The results with static fields at 4.2K up to 8.5T are given in Fig. 1. For fields above 4T the magnetization is nearly proportional to the applied field H. The following values for the relative change in slope dM/MdH were obtained: $1.4 \times 10^{-3} T^{-1}$ for U_3P_4 and $0.8 \times 10^{-3} T^{-1}$ for U_3As_4. Experimental results for U_3P_4 by pulsed fields up to 50T show[8] that the slope for applied fields along [111] is constant up to 50T and the value for dM/MdH at 77K is nearly equal to the value at 4.2K. Comparing to the slope for the [111] field, the slope for the field applied along [100] is nearly equal at 4.2K and at 77K, and a factor of two or three larger than that for [111]. However, we could not detect any sharp transition up to 50T in U_3P_4. In U_3As_4, a sharp increase of moment was observed in a 19T applied field along [100] at 4.2K, in agreement with the reported value[9].

The temperature dependence of the magnetization under a fixed magnetic field along [111] was measured. The results are shown in Fig. 2. It is clear that there are two steps for the temperature dependence. The initial step in the lower temperature region below 40 to 50K is very unusual. Firstly, we could not find any initial

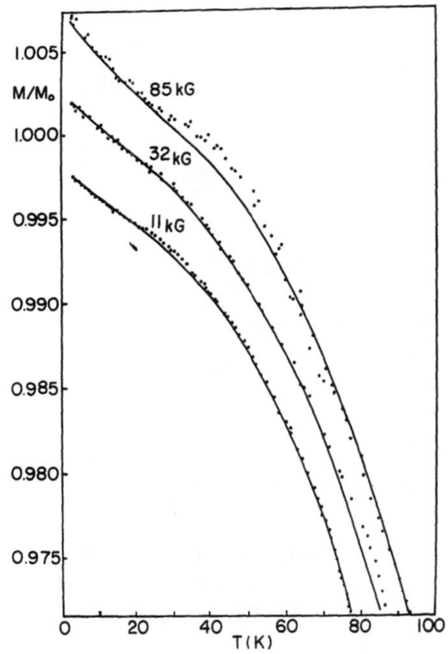

Fig. 2: Temperature dependences of magnetization under constant magnetic field along [111] axis for U_3As_4. Black points show observed value and solid line shows calculated value with case (a) and fitting parameter shown in Table 1.

exponential region even for 85 kOe of applied magnetic field, which should induce a large gap in the acoustic spin wave mode due to a large Zeeman energy. Secondly, the initial temperature dependence seems to be weaker than linear in T, even at 85 kOe of applied field. Previously, we found an unusual temperature dependence such as TlogT in MnP[7] but the behavior of weaker than linear in T is very unusual. Furthermore, such a strong decrease of magnetization in the lower temperature region apparently contradicts the previous experiment which showed definite structure and thus hard anisotropy affecting the magnetic moment on a given structure. Naturally we expected at least an exponential change in the initial stage with a fairly large gap.

An attempt was made to fit the initial parts of the data by various power laws of T and we found that $T^{3/4}$ seemed to be best for all the available data as shown in Fig. 3. The result for U_3P_4 is shown in Fig. 4. On the other hand, the behavior in the higher temperature region cannot be fit by the usual exponential term. Then the whole data were fitted by the following form,

$$M(T,H)/M(0,0)=A(H)T^{3/4}+B(H)T^2e^{-\Delta(H)/T}. \qquad (1)$$

The T^2 part in the second term has no physical basis, but the overall fitting is better by introducing this part. The best fitted values are listed in Table 1 in which two different cases are shown for the second term; the case (a) for a fixed value of Δ and the

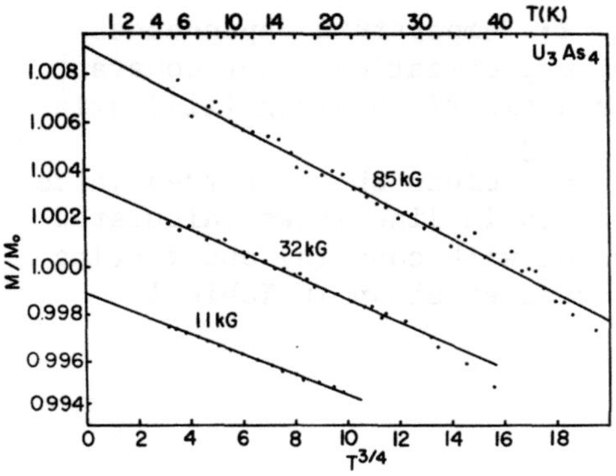

Fig. 3: The temperature dependence of the magnetization of U_3As_4. The data at constant field are well expressed by a $T^{3/4}$ dependence.

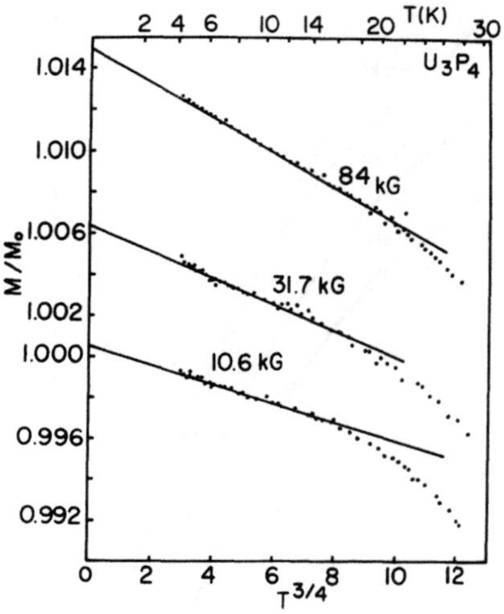

Fig. 4: Temperature dependence of magnetization under constant magnetic field of U_3P_4. The data are well expressed by a $T^{3/4}$ dependence.

case (b) for a fixed value of B. The calculated curves are shown in Fig. 2 by solid lines for case (a). The calculated curves in case (b) are not represented in Fig. 2 because the fitting is worse than that of case (a). For U_3P_4, because of a smaller temperature region, the second term was not accurately determined. In any case, the following facts are clear. The second term decreases with increasing value of field. This is natural because Δ is expected to increase with increasing field due to the Zeeman energy. On the other hand, the fact that A(H) increases with increasing field is again unusual and hard to understand.

DISCUSSION

First, let us consider the small induced moment observed. The simplest picture is the competition between the uniaxial field and the exchange field and it may be written as follows, for a sublattice of, say, a [001] uniaxial field:

$$U(\theta)= -V\cos^2\theta-\cos(\theta_{111}-\theta)[H_o+H_{ex}\cos(\theta_{111}-\theta)], \qquad (2)$$

in which the moment at each site is assumed to be constant and only rotates. The first term indicates the uniaxial crystal field energy. In the second term, H_o means the Zeeman energy due to the external field applied along [111] and H_{ex} the exchange field, assumed to be always in the [111] direction and proportional to the total moment. The angle θ_o for zero applied field is given by:

$$\theta_o=(1/2)\tan^{-1}2\sqrt{2}/(3V/H_{ex}-1),\text{ or}$$
$$(\theta_{111}-\theta_o)=(1/2)\tan 2\sqrt{2}/(3H_{ex}/V-1). \qquad (3)$$

From the neutron diffraction experiment, $\theta_{111}-\theta_o$ was found to be 25° and 13° for U_3As_4 and U_3P_4, respectively[4]. The H_o term is treated by perturbation and the change of θ by the applied field, $\delta\theta$, is easily estimated to be:

$$\delta\theta = \frac{H_o}{2H_{ex}}\,\frac{\sin(\theta_{111}-\theta_o)}{V/H_{ex}+\cos 2(\theta_{111}-\theta_o)}\,. \qquad (4)$$

From Eq. (3) H_{ex}/V is estimated to be 1.12 and 2.27 for U_3As_4 and U_3P_4, respectively, while H_{ex} is estimated from the Curie temperature, 200K and 140K for U_3As_4 and U_3P_4, respectively, to be 236K and 168K, respectively. Then for a field of 1T, $\delta\theta$ is 1.25×10^{-3} and 1.07×10^{-3} and $\Delta M/M$, given by $\delta\theta\tan(\theta_{111}-\theta_o)$, is 0.58×10^{-3} and 0.25×10^{-3} for U_3As_4 amd U_3P_4, respectively. Agreement with the experimental value is fairly good in U_3As_4 but not as good for U_3P_4. Note that in this model the constant reduction of moment was assumed. Physically, this reduction should be due to p-f mixing, which is, however, strongly anisotropic as shown by a previous paper[10]. Then there is a serious question as to whether the form of Eq. (2) is applicable or not.

The temperature dependence of the magnetic moment at low temperatures is difficult to understand. One possible model to explain the $T^{3/4}$ law seems to be the q^4 law for the magnon dispersion, $h\omega_q=cq^4+\Delta$. Then the change in the magnetic moment is described by:

$$\frac{\Delta M}{M} = \frac{a^3}{96\pi^2 J}\left(\frac{kT}{c}\right)^{3/4}\int_o^\infty \frac{dx}{x^{1/4}(e^{x+\Delta/kT}-1)}\,, \qquad (5)$$

and when the gap Δ in the magnon dispersion is set to zero the $T^{3/4}$ law seems to be realized. However, when Δ is zero, the integral diverges and thus Δ cannot be zero. When Δ is finite, the calculation shows that the T dependence becomes stronger than linear in T. The same situation exists for a higher power q dispersion. In the limits, we have the following dispersions: $h\omega_q=\Delta$ for $q\leq q_m$, and $h\omega_q=\infty$ for $q_m<q$. In this limiting case, we obtain a linear in T relation in the limit of $T\gg\Delta$ but a stronger dependence in other cases. Therefore, it is very difficult to obtain the result weaker than linear in T. Another possibility is to introduce a temperature dependence to Δ or c. However, then Δ and c should increase with in-

Table 1: The fitting parameter calculated by phenomenological formula Eq. (1) for the observed temperature dependence of the magnetization for U_3As_4 and U_3P_4.

U_3As_4	$A\times10^{-4}$	case (a)		case (b)		U_3P_4	$A\times10^{-4}$
		$B\times10^{-5}$	Δ(K)	$B\times10^{-5}$	Δ(K)		
85kOe	9.20	1.10	100	1.35	112	10.6kOe	6.44
32kOe	7.85	1.35	100	1.35	100	31.7kOe	4.84
11kOe	7.02	1.60	100	1.35	87	84 kOe	3.50

creasing temperature which seems unphysical. A temperature dependence of the canting angle may be another possibility but then we should rely on accidental cancellation. The p-f mixing model proposed in the previous paper[10] may provide the answer but further experimental and theoretical work is necessary.

REFERENCES

1. W. Trzebiatowski, R. Troc, Bull.Acad.Polon.Sci.ser.Sci.chim. 11, 661 (1963).
2. W. Trzebiatowski, A. Sepichowska and A. Zygmunt, Bull. Acad. Polon. Sci. Ser. Sci. Chim. 12, 687 (1964).
3. K.P. Belov, Z. Henkie, A.S. Dmitrievskii, R.Z. Levitin and W. Trzebiatowski, Zh. eksper. teor. Fiz. 61, 1522 (1971).
4. P. Burlet, J. Rossat-Mignod, R. Troc and Z. Henkie, Solid State Commun. 38, 745 (1981).
5. R. Troc, J. Sznajd, P. Novotny and T. Mydlarz, to be published.
6. Z. Henkie, Roczniki Chemii, 42, 363 (1968).
7. A. Takase and T. Kasuya, J. Phys. Soc. Japan 47, 491 (1979).
8. P.J. Markowskii, T. Sakakibara, S. Kunii, T. Suzuki, M. Date and T. Kasuya, to be published.
9. P. Novotny and J. Sternberk, Phys. Stat. Sol. (a) 33, K15 (1976).
10. K. Takegahara, A. Yanase and T. Kasuya, this conference.

COMMENTS

LOEWENHAUPT: You assume a sharp spin wave spectrum and cannot find a reasonable explanation of your data. Have you tried fitting your data assuming a strongly overdamped spectrum (broad dynamic magnetic responses) like, for example, in UAs?

MARKOWSKI: The main feature of our results consists of the $T^{3/4}$ dependence of M, and the range of these anomalies strongly depends on M. Such behavior may account for the spectra presented which were phenomenologically fitted since they were unknown. The over-damped spectra may perhaps be more realistic; however, we do not expect they could account for the strong field-dependent character of the M vs T curves.

GURTOVOJ: What was the duration of the magnetic field pulses in your experiments in Osaka? What was the error of the magnetic field orientation along the crystal axes in your high field measurements?

MARKOWSKI: The duration used was 10 ms. The orientation of the samples was performed within an accuracy of $\pm 2°$.

ROSSAT-MIGNOD: Neutron experiments have shown that in U_3P_4 the canting angle is smaller than for U_3As_4. Maybe this is the reason why there is no transition in U_3P_4 when the magnetic field is applied along <100>.

LIST OF SENIOR AUTHORS OF IV IC-CEFSE IN f-ELECTRON SYSTEMS

1. A. Andreeff, Sektion Physik, Technische Univeristät Dresden, 8027 Dresden, Mommsenstrasse 13, GDR

2. A. Apostolov, Faculty of Physics, Sofia University, Bul. A.Ivanov 5, 1126 Sofia, Bulgaria

3. K. Baberschke, Institut fur Atom- und Festkörperphysik, Freie Universität Berlin, D-1000 Berlin 33, FRG

4. H. Bartholin, Université de Toulon et du Vac, Château Saint-Michel, 83130 La Garde, France

5. Y. Berthier, Laboratoire de Spectrométric Physique, CNRS, BP 53X, 38041 Grenoble-Cedex, France

6. I.A. Campbell, Physique des Solides, Université Paris Sud, 91405 Orsay, France

7. B.R. Cooper, Department of Physics, West Virginia University, Morgantown, WV 26506

8. B. Coqblin, Laboratoire de Physique des Solides, Université Paris Sud, 91405 Orsay, France

9. R. Dagys, Lietuvos TSR Mokslu Akademi ja Puslaidiniukiu Fizikos Institutas, 52 K. Pozhelos, Vilnius 232600, USSR

10. H. Drulis, Instytut Niskich Temperatur i Badan Strukturalnych PAN, Pl. Katedralny 1, P.O. Box 937, 50-950 Wrocław, Poland

11. G. Feller, Sektion Physik der Karl Marx Universität, Leipzig, GDR

12. J.J.M. Franse, Natuurkundig Laboratorium der Universiteit van Amsterdam, Valckenierstraat 65, 1018 XE Amsterdam, Holland

13. T. Frauenheim, Joint Institute for Nuclear Research, Dubna, 101000 Moscow, P.O. Box 79, USSR

14. P. Fulde, Max-Planck Institut für Festkörperforschung, Heisenbergstrasse 1, 7000 Stuttgart 8, W. Germany

15. D. Gignoux, Laboratoire Louis-Néel CNRS, 166X, 38042 Grenoble-Cedex, France

16. E.A. Gorlich, Instytut Fizyki UJ, Ul. Reymonta 4, 30-059 Krakow, Poland

17. F.J.A.M. Greidanus, Kamerlingh Onnes Laboratorium der Rijksuniversiteit Leiden, Nieuwsteeg 18, The Netherlands

18. R.P. Guertin, Department of Physics, Tufts University, Medford, MA 02155

19. K.G. Gurtovoj, Kurchatov Institute of Atomic Energy, 123182 Moscow, USSR

20. A. Hasegawa, General Education Department, Niigata University, Niigata 950-21, Japan

21. B. Janus, Instytut Niskich Temperature i Badan Strukturalnych PAN, P. Katedralny 1, P.O. Box 937, 50-950, Wrocław, Poland

22. M. de Jong, Naturkundig Laboratorium der Universiteit van Amsterda Valckenierstraat 65, 1018 XE Amsterdam, Holland

23. T. Kasuya, Physics Department, Tohoku University, Sendai 980, Japan

24. J. Keller, Fakultät für Physik, Universität Regensburg, Postfach 397, 8400 Regensburg, W. Germany

25. J. Leciejewicz, Instytut Badan Jadrowych, 05-400 Swierk, Poland

26. L. Kowalewski, Zakład Teorii Ciała Stałego, Instytut Fizyki, Uniwersytet im. A. Mickiewicza, Poznan, Poland

27. M. Loewenhaupt, Institut für Festkörperforschung KFA-Jülich, GmbH, Postfach D-5170, Jülich 1913, W. Germany

28. A. Loidl, Institut fur Physik, Universitat Mainz, Jakob-Welder Weg, 11, D-6500 Mainz, W. Germany

29. B. Luthi, Physikalisches Inst. der J.W. Goethe University, 6000 Frankfurt am Main - 1, Robert Mayer Str. 2-4, FRG

30. P.J. Markowski, Inst. Niskich Temperature i Badan Strukturalnych PAN, pl. Katedralny 1, P.O. Box 937, 50-950 Wrocław, Poland

31. W. Matz, Zentralinstitut für Kernförschung der AdW der DDR, Rossendorf, 8051 Dresden, GDR

32. K.A. McEwen, Department of Physics, University of Salford, Salford, M5 4WT, England

33. J. Mulak, Inst. Niskich Temperatur i Badan Strukturalnych PAN, pl. Katedralny 1, P.O.Box 937, 50-950 Wrocław, Poland

34. L. Pawlak, Instytut Chemii Nieorganicznej i Metalurgii Pierwiastkow Rzadkich, Politechnika Wrocławska, Wybrzeze Wyspianskiego 27, 50-370 Wrocław, Poland

35. J. Pierre, Laboratoire Louis Néel, CNRS, 166X 38042 Grenoble Cedex, France

36. M. Piotrowski, Instytut Badan Jadrowych, 05-400 Swierk, Poland

37. P. Purreur, Physique des Solides, Université Paris Sud, 91405 Orsay, France

38. C. Rettori, Instituto de Fisica DEQ, Unicamp, Campinas 13100/SP/ Brazil

39. J. Rossat-Mignod, Centre d'Études Nucleaires, DRF/DN, 85X, 38041 Grenoble Cedex, France

40. I.P. Sadikov, Kurchatov Inst. of Atomic Energy, 123182 Moscow, USSR

41. W. Schlabitz, II Phys. Inst. der Univ. zu Köln, Zülpicher Str. 77, 5000 Köln 41, W. Germany

42. D. Schmitt, Laboratoire Louis Néel CNRS, 166X, 38042 Grenoble-Cedex, France

43. G.K. Shenoy, Argonne National Lab., Solid State Div., 9700 South Cass Avenue, Argonne, IL 60439

44. A. Simon, Max-Planck-Inst., Heisenbergstrasse 1, 7000 Stuttgart 8, W. Germany

45. I.A. Smirnov, A.F. Joffe Physico-Technical Inst., Academy of Sciences of the USSR, 194021 Leningrad, USSR

46. F. Steglich, Inst. für Festkörperphysik, Technische Hochschule, Hochschulstr. 2, D-6100 Darmstadt, W. Germany

47. G.A. Stewart, Inst. für Atom and Festkörperphysik, Freie Universitä Berlin, 1000 Berlin 33, FRG

48. T. Suzuki, Department of Physics, Tohoku University, Sendai 980, Japan

49. A. Szytuła, Inst. Fizyki UJ, ul. Reymonta 4, 30-059 Krakow, Poland

50. K. Takegahara, Department of Physics, Tohoku University, Sendai 98(Japan

51. U. Walter, II Phys. Inst. Univ. zu Köln, Zülpicher Str. 77, D-5000 Köln, FRG

52. W. Wegener, SCK-CEN, Mol, Belgium

53. H. Wehr, Inst. für Phys., Universität Mainz, Jakob Welder Weg 11, D-6500 Mainz, W. Germany

54. D. Wohlleben, II. Phys. Inst. Univ. zu Köln, Zülpicher Str. 77, D-5000 Köln, FRG

55. Z. Zołnierek, Inst. Niskich Temperatur i Badan Strukturalnych PAN, P. Katedralny 1, P.O.Box 937, 50-950 Wrocław, Poland

56. J. Zukrowski, Zakład Fizyki Ciała Stałego, AGH, al. Mickiewicza 30, 30-059 Krakow, Poland

57. I.A. Garifulin, Phys. Inst. of the Academy of Sciences USSR, Kaxan 420029, USSR

58. K. Knorr, Inst. für Phys., Universität Mainz, Jakob Welder Weg 11 D-6500 Mainz, W. Germany

59. E. Leyarowski, Inst. of Solid State Physics, Bulgarian Academy of Sciences, 1184 Sofia, Bulgaria

LIST OF PARTICIPANTS

BABERSCHKE, K.
Institut für Atom- und Festkörperphysik, Freie Universität Berlin, West Berlin

BARTHOLIN, H.
Dept. of Physics, University of Toulon, France

BAUMANN, J.
II. Phys. Inst. of Univ. zu Köln, D-5000 Köln, Zülpicher Str. 77, FRG

BERTHIER, Y.
Laboratoire de Spectrometrie Physique, B.P. 53 X, 38041 Grenoble Cedex, France

BIEGAŁA, L.
Institute for Low Temperature and Structure Research, Polish Academy of Sciences, Wrocław, Poland

CAMPBELL, I.A.
Laboratoire de Physique des Solides, Université Paris-Sud, 91405 Orsay, France

COOPER, B.R.
Department of Physics, West Virginia University, Morgantown, WV 26506, USA

COOBLIN, B.
Laboratoire de Physique des Solides, Université Paris-Sud, 91405 Orsay, France

CZOPNIK, A.
Institute for Low Temperature and Structure Research, Polish Academy of Sciences, Wrocław, Poland

DAGYS, R.S.
Institute of Semiconductor Physics, 232600 Vilnius K. Pozelos 52, USSR

DRULIS, H.
Institute for Low Temperature and Structure Research, Polish Academy of Sciences, Wrocław, Poland

DRULIS, M.
Institute for Low Temperature and Structure Research, Polish Academy of Sciences, Wrocław, Poland

DUCZMAL, M.
Institute of Inorganic Chemistry and Metallurgy of Rare Elements, Technical University, Wybrzeże Wyspiańskiego 27, 50-370 Wrocław, Poland

FEILE, R.
Institut für Physik, Univ. Mainz, D-6500 Mainz, FRG

FRANSE, J.J.M.
Natuurkundig Laboratorium der Universiteit van Amsterdam, Valckenierstraat 65, 1018 XE Amsterdam, The Netherlands

FRAUENHEIM, T.
Laboratory of Neutron Physics, Joint Institute for Nuclear Research Dubna, 101000 Moscow, USSR

FRICK, B.
Institute für Festkörperforschung, KFA-Jülich, FRG

FULDE, P.
Max Planck Institut für Festkörperforschung, Heisenberg Str. 1, D-1 Stuttgart 80, FRG

GIGNOUX, D.
Laboratoire Louis Néel, C.N.R.S., 166 X, 38042 Grenoble-Cedex, France

GÖRLICH, E.A.
Institute of Physics, Jagiellonian University, Cracow, Poland

GREIDANUS, F.J.A.M.
Kamerlingh Onnes Laboratorium, Leiden, The Netherlands

GUERTIN, R.P.
Physics Department, Tufts University, Medford, MA 02155 USA

GURTOVOJ, K.G.
Kurchatov Institute of Atomic Energy, 123182 Moscow, USSR

HASEGAWA, A.
General Education Department, Niigata University, Niigata 950-21, Japan

HENKIE, Z.
Institute for Low Temperature and Structure Research, Polish Academy of Sciences, Wrocław, Poland

HOFFMAN, K.
Institute for Low Temperature and Structure Research, Polish Academy of Sciences, Wrocław, Poland

ILIEW, N.
Institute for Low Temperature and Structure Research, Polish Academy of Sciences, Wrocław, Poland

JANUS, B.
Institute for Low Temperature and Structure Research, Polish Academy of Sciences, Wrocław, Poland

JAROSZ, J.
Institute of Physics, Silesia University, Katowice, Poland

deJONG, M.H.
Natuurkundig Laboratorium der Universiteit van Amsterdam, Valckenierstraat 65, 1018 XE Amsterdam, The Netherlands

KACZMARSKA, K.
Institute of Physics, Silesia University, Katowice, Poland

KASUYA, T.
Department of Physics, Tohoku University, Sendai 980, Japan

KELLER, J.
Fakultät für Physik, Universität Regensburg, Postfach 397, 8400 Regensburg, FRG

KHARAKHASHYAN, E.
Physico-Technical Institute, Academy of Sciences of USSR, Kazan, USSR

KLETOWSKI, Z.
Institute for Low Temperature and Structure Research, Polish Academy of Sciences, Wrocław, Poland

KNORR, K.
Insitut für Physik, Univ. Mainz, D-6500 Mainz, FRG

KOWALEWSKI, L.
Solid State Theory Division, Institute of Physics, A. Mickiewicz University, Poland

KRAWCZYK, L.
Institute for Low Temperature and Structure Research, Polish Academy of Sciences, Wrocław, Poland

KWAPULINSKA, E.
Institute of Physics, Silesia University, Katowice, Poland

LATKA, K.
Institute of Physics, Jagiellonian University, Cracow, Poland

LECIEJEWICZ, J.
Institute of Nuclear Research, Swierk Research Establishment, Warsaw, Poland

LEHMANN-SZWEYKOWSKA, A.
Solid State Theory Division, Institute of Physics, A. Mickiewicz University, Poznan, Poland

LEYAROVSKA, L.
Institute of Solid State Physics, Bulgarian Academy of Sciences, 1 184 Sofia, Bulgaria

LEYAROVSKI, E.
Institute of Solid State Physics, Bulgarian Academy of Sciences, 1 184 Sofia, Bulgaria

LOIDL, A.
Institut für Physik, Univ. of Mainz, D-6500 Mainz, FRG

LOEWENHAUPT, M.
Institut für Festkörperforschung, KFA-IFF, Postfach 1913, D-5170 Jülich, FRG

LÜTHI, B.
Physikalisches Institut der Universitat, 6000 Frankfurt, FRG

MARKOWSKI, P.J.
Institute for Low Temperature and Structure Research, Polish Academy of Sciences, Wrocław, Poland

MASLANKA, R.
Institute for Low Temperature and Structure Research, Polish Academy of Sciences, Wrocław, Poland

MATZ, W.
Zentralinstitut für Kernforschung Rossendorf der AdW der DDR, 8051 Dresden, DDR

McEWEN, K.A.
Department of Physics, University of Salford, Salford M5 4WT, England

MICNAS, K.
Institute of Molecular Physics, Polish Academy of Sciences, Poznan, Poland

MÜHLE, E.
Sektion Physik der Karl Marx-Universität Leipzig, DDR

NAÏT-SAADA, A.
Laboratoire Louis Néel, C.N.R.S., 166 X, 38042 Grenoble-Cedex, France

NEUMANN, G.
II. Phys. Inst. of Univ. zu Köln, D-5000 Köln, Zülpicher Str. 77, FRG

PAWLAK, L.
Institute of Inorganic Chemistry and Metallurgy of Rare Elements, Technical University, Wybrzeze Wyspianskiego 27, 50-370 Wrocław, Poland

PIERRE, J.
Institute Laue-Langevin, 156 X, 38042 Grenoble Cedex, France

PIOTROWSKI, M.
Institute for Nuclear Research Swierk, Poland

PUREUR, P.
Laboratoire de Physique des Solides, Université Paris-Sud, 91405 Orsay, France

REGGENTIN, K.
II. Phys. Inst. of Univ. zu Köln, D-5000 Köln, Zülpicher Str. 77, FRG

ROSSAT-MIGNOD, J.
Laboratoire de Diffraction Neutronique, Département de Réchèrche Fondamentale, Centre d'Étude Nucleaire, 85 X, 38041 Grenoble Cedex, France

SADIKOV, I.
Kurchatov Institute of Atomic Energy, 123182 Moscow, USSR

SCHLABITZ, W.
II. Phys. Inst. of Univ. zu Köln, D-5000 Köln, Zülpicher Str. 77, FRG

SCHMITT, D.
CNRS, Laboratoire Louis Néel, Associé à l'USMG, 166 X, 38042 Grenoble Cedex, France

SHENOY, G.K.
Physics Department, Argonne National Laboratories, Argonne, IL 60439, USA

SIMON, A.
Max-Planck-Institute für Festkörperforschung, Stuttgart 80, FRG

SMIRNOV, I.A.
A.F. Ioffe Physico-Technical Institute, Academy of Sciences of the USSR, 194021 Leningrad, USSR

STALINSKI, B.
Institute for Low Temperature and Structure Research, Polish Academy of Sciences, Wrocław, Poland

STEGLICH, F.
Institut für Festkörperphysik, Technische Hochschule, D-6100 Darmstadt, FRG

STEWART, G.A.
Institut für Atom- und Festkörperphysik, Freie Universität, Berlin, West Berlin

SUSKI, W.
Institute for Low Temperature and Structure Research, Polish Academy of Sciences, Wrocław, Poland

SUZUKI, T.
Department of Physics, Tohoku University, Sendai 980, Japan

SZNAJD, J.Z.
Institute for Low Temperature and Structure Research, Polish Academy of Sciences, Wrocław, Poland

SZYTUŁA, A.
Institute of Physics, Jągiellonian University, Cracow, Poland Poland

TAKEGAHARA, K.
Department of Physics, Tohoku University, Sendai 980, Japan

THOMAS, M.
Solid State Theory Division, Institute of Physics, A. Mickiewicz Univeristy, Poznan, Poland

TRZEBIATOWSKA, B.
Institute for Low Temperature and Structure Research, Polish Academy of Sciences, Wrocław, Poland

TRZEBIATOWSKI, W.
Institute for Low Temperature and Structure Research, Polish Academy of Sciences, Wrocław, Poland

ULNER, J.
Institute for Low Temperature and Structure Research, Polish Academy of Sciences, Wrocław, Poland

WALTER, U.
II. Phys. Inst. of Univ. zu Köln, D-5000 Köln, Zülpicher Str. 77, FRG

WEGENER, W.
SCK-CEN, Mol, Belgium

WEHR, H.
Institut für Physik, Universität Mainz, D-6500 Mainz, FRG

WOHLLEBEN, D.
II. Phys. Inst. of Univ. zu Köln, D-5000 Köln, Zülpicher Str. 77, FRG

WOJCIECHOWSKI, R.J.
Solid State Theory Division, Institute of Physics, A. Mickiewicz University, Poznan, Poland

ZIPPER, E.
Institute of Physics, Silesia University, Katowice, Poland

ZOŁNIEREK, Z.
Institute for Low Temperature and Structure Research, Polish Academy of Sciences, Wrocław, Poland

ZUKROWSKI, J.
Institut für Atom- und Festkörperphysik, Freie Universität Berlin, West Berlin

ZYGMUNT, A.
Institute for Low Temperature and Structure Research, Polish Academy of Sciences, Wrocław, Poland

SUBJECT INDEX

MATERIALS INDEX

MIX
Papier aus verantwortungsvollen Quellen
Paper from responsible sources
FSC® C105338

If you have any concerns about our products,
you can contact us on
ProductSafety@springernature.com

In case Publisher is established outside the EU,
the EU authorized representative is:
Springer Nature Customer Service Center GmbH
Europaplatz 3, 69115 Heidelberg, Germany

Printed by Libri Plureos GmbH
in Hamburg, Germany